HORTICULTURE

HORTICULTURE

R. Gordon Halfacre, Ph.D., M.L.A.

Associate Professor of Horticulture
Clemson University

John A. Barden, Ph.D.

Associate Professor of Horticulture
Virginia Polytechnic Institute
and State University

McGraw-Hill Book Company

New York   St. Louis   San Francisco   Auckland   Bogotá   Düsseldorf
Johannesburg   London   Madrid   Mexico   Montreal   New Delhi
Panama   Paris   São Paulo   Singapore   Sydney   Tokyo   Toronto

For Angela and Robert Halfacre
and Cindy Lou and Jay Barden

This book was set in Memphis Light by Black Dot, Inc.
The editors were C. Robert Zappa and Susan Gamer;
the designer was Joan E. O'Connor;
the production supervisor was Dominick Petrellese.
The part opening and cover woodcuts were executed by Carla Bauer;
the drawings were done by Allyn-Mason, Inc.
R. R. Donnelley & Sons Company was printer and binder.

HORTICULTURE

234567890DODO7832109

Library of Congress Cataloging in Publication Data

Halfacre, R. Gordon, date
    Horticulture.

    Includes bibliographies and index.
    1. Horticulture.   I. Barden, John A., joint
author. II. Title.
SB318.H28   635   78-15116
ISBN 0-07-025573-3

CONTENTS

PART TWO   Plant Environment

PART THREE Horticultural Practices

PART FOUR Branches of Horticulture

PREFACE

The goal of this book is to introduce students to horticulture and provide information on its many facets. Horticulture was once largely an art, but recently it has evolved into a highly developed science as well. A student who is considering horticulture as a profession should explore its various areas—which range from the arts of landscape design and floral arrangement to the science of plant breeding and include, between these extremes, such diverse possibilities as crop production, storage, and marketing as well as many service and support industries.

This book was written primarily to serve as a text for college courses such as "Principles of Horticulture" or the equivalent; however, we have tried to make it of sufficient breadth and depth to serve as a reference. The material is organized into chapters, with as little overlap as possible, so that each subject is covered rather thoroughly in one place and can be readily located.

Depending on the offering, *Horticulture* may serve as a text for one or two basic or core courses. If only one such course is offered, the instructor may wish to select appropriate chapters for emphasis. For example, one teacher might concentrate on Part One (which covers basic principles of plant science) and Part Two (on factors of plant environment); another might emphasize Part Three (on principal horticultural practices) and Part Four (on principal horticultural crops and disciplines).

The reader will find that certain areas have been given extra emphasis. For instance, Chapter 7, Temperature Relations, provides a basic background in temperature as it relates to horticultural crops; in this chapter, principles are taken from physics, meteorology, and climatology as well as from horticulture. Water relations are covered in a similar manner in Chapter 8. We fully recognize that this approach is new and in some ways anticipates changes in course design. However, such material can be used to provide subject matter for more advanced horticultural courses as well as to provide essential principles at the beginning.

The future of horticulture depends on innovative, in-depth study and thought by people in the profession. Students of horticulture should acquire both as much basic knowledge as possible and a "feel" for field practices. The objectives of *Horticulture* are to provide (1) a base of information that will ease the transition from basic studies in arts and sciences to horticulture, (2) a reference source on fundamental horticultural principles and practices, and (3) an overview of each of the commercial areas of horticulture. Horticultural practices depend on an understanding of the principles of plant classification, structure, and physiology and of the environmental factors that influence plant growth. When the basic principles have been grasped, the practices follow logically. To put it simply, horticulture is a logical discipline, and its logic can be compared to the logic of the basic sciences of biology, chemistry, and physics. In addition, however, an understanding of horticulture requires a sound appreciation of aesthetic values.

ACKNOWLEDGMENTS

The authors offer their thanks to their many colleagues who have given so willingly of their time and efforts, in so many ways, to help make this book a reality. Among these are R. P. Ashworth, L. W. Baxter, Jr., D. A. Bender, D. W. Bradshaw, G. B. Briggs, N. D. Camper, A. Chadwell, M. A. Cohen, F. W. Cooler, D. C. Coston, R. J. Downs, D. A. Eckard, W. M. Epps, J. E. Fairey, J. P. Fulmer, M. F. George, B. Graves, J. R. Haun, C. R. Johnson, E. V. Jones, W. S. Jordan, M. W. Jutras, A. J. Lewis, III, J. W. Love, C. L. McCombs, C. H. Miller, E. C. Montgomery, W. L. Ogle, B. J. Parliman, D. J. Parrish, A. J. Pertuit, P. J. Porpiglia, D. Porter, E. T. Sims, G. E. Stembridge, and D. F. Wagner.

We also offer our sincere thanks to Vickie Klaff, Margarette Ogle, and Carolyn Halfacre for their excellent typing of the manuscript.

Our thanks are also extended to the many individuals, companies, and organizations who provided photographs.

Finally, we convey to our families our deepest gratitude for their encouragement, inspiration, cooperation, and understanding: to Angela and Robert Halfacre and Lela and Harvey Halfacre; and to Irma, Cindy Lou, and Jay Barden.

TO THE READER

Since this is a first edition, it may contain some errors. We will be grateful to readers for pointing them out to us, and indeed would appreciate constructive criticism in general from teachers and students, especially suggestions for improvements.

R. Gordon Halfacre
John A. Barden

CHAPTER 1
INTRODUCTION

 Horticulture is the intensive cultivation of plants. Etymologically, horticulture means "garden cultivation" (from the Latin *hortus* and *cultura*). The first-known use of the term *horticultura* is found in Peter Laurenberg's writings in 1631. Phillips, in *The New World of Words*, (1678 edition, printed in London) first used the term *horticulture* in English to refer to the science and art of growing fruits, vegetables, flowers, and ornamental plants.

Horticulture encompasses all life—especially humanity—and bridges the gap between science, art, and human beings. A horticulturist cannot simply deal with plants as a science or as an art but must study plants as food for people and as an aesthetic and functional part of the environment. Human beings survive through horticultural foods and through the landscape. In a sense a horticulturist touches the beauty, perfection, and serenity of nature and all its boundless moods and temperaments. All this is accomplished with the final and total effect of helping mankind survive in a beautiful world.

ORIGINS AND EARLY HISTORY OF HORTICULTURE

The field of horticulture had its beginning when human beings first began to cultivate gardens. Before then, people obtained their food by hunting, fishing, and gathering wild, edible plants and plant parts. The forerunners of our present fruit and cereal crops may have been found as individual plants, or in small patches, growing next to a primitive poppy or a thistle or parsley; a little farther on, a bunch of cocksfoot; and then again another cereal or fruit plant. These early people collected plants for their medicinal, cosmetic, and aphrodisiac properties as well as for their food value. Early evidence indicates that as far back as 7000 B.C. women may have begun to cultivate a few of these wild plants which they had sampled and found edible. With the passage of time, more plants were collected, cultivated, and propagated, usually by seed. Since the growing season for any crop was limited and early peoples had limited means of preserving or storing food from season to season, it was necessary to discover edible crops which matured in a short growing season. Over thousands of years the selection and propagation of better food-producing crops resulted in plants far superior to their wild ancestors; this was the beginning of plant breeding.

Starvation was common for centuries, until people slowly discovered and perfected methods of preserving food. Preservation of food makes human beings more independent of nature's forces. When people realized that they could preserve food and not spend each day in the wilds wandering in search of a daily ration, they began to build and establish permanent homes. Walls were often constructed around groups of permanent homes to form small cities which offered protection from enemies. Early people soon realized how impractical it was to walk many miles to gather food, and so they scratched an area of land within and around the walled cities and planted seed there. Since space was limited within the walls, grains and other crops that required extensive acreage were planted outside. However, for convenience, practicality, and perhaps even beauty, those vegetables and fruits that required intensive cultivation and constant care were grown within the walls. Thus horticulture began.

Early civilizations were basically agrarian and added much to today's knowledge through their trial-and-error discoveries. Farming had its roots in the "fertile crescent"—the valleys of the Euphrates, Tigris, and Nile rivers—more than seven thousand years ago. Some horticultural crops cultivated by the Egyptians in 3000 B.C. were figs, dates, bananas, cucumbers, grapes, olives, melons, lettuce, and lemons. As early as 1500 B.C. the Egyptians developed landscape gardens in which flowering plants, shade trees, and ornamental shrubs were planted and cultivated for their beauty alone. The Greeks were also very observant when it came to plants. Theophrastus (370–285 B.C.) was known for his studies of plants. He observed and noted such facts as the absorption of nutrients through the roots for use by the plants, the difference between the cotyledons of germinating beans and corn (dicotyledon versus monocotyledon), the increase in flowering due to root pruning, and the process of cross-pollination. After the Roman Empire defeated the Greeks, scientific thought dwindled as a way of life. The Romans simply adopted Greek techniques and improved on them, using such cultural techniques as planting legumes to improve the soil, applying manure to improve the nutritive content of the soil, and employing cultivation to control weeds. The Romans also devised a way to store fruits for the winter by placing them in straw in a dry, cool place, such as a cave. Some actual scientific research was carried on during this time, however, by such men as Dioscorides, who wrote *De Materia Medica* (77 A.D.). In his research Dioscorides identified plants and listed their medicinal value. In writing the book, he described the roots, stems, leaves, and sometimes the flowers of plants; his research was widely used as a reference for the next 1500 years.

There were very few new breakthroughs from the time of the Roman Empire until Linnaeus (1707–1778), a Swedish physician and botanist with a great ability to systematize and describe, developed the binomial system of classification of plants, which is still used today. This system gave each plant a name consisting of two parts. The first part is the *genus*, or generic part, and the second part is the *species*, or specific part, of the name. The next major development in plant research came when Charles Darwin published his work. Charles Darwin, noted for his theory of survival of the fittest in his 1859 publication of *On the Origin of Species by Means of Natural Selection or the Preservation of Favored Races in the Struggle for Life*, also published *The Power of Movement in Plants*. In this book, Darwin discussed *geotropism*, the movement of plants in response to gravity, and *phototropism*, the movement of plants in response to light. Also active in the 1800s was Gregor Mendel (1822–1884), who did classic work breeding sweet peas. He observed seven contrasting pairs of traits in the peas, e.g., tall plants and short plants. He crossed plants with the different characteristics; e.g., crossing a tall pea with a short pea gave a first filial or F_1 generation of all tall plants. He then crossed the F_1 plants with other F_1 plants and got three tall plants to one short plant in the second or F_2 generation. Before Mendel, other people had made similar crosses and obtained similar seedling plants, but Mendel wanted to know *why*. He described the plant forms in each generation and determined the ratio of each form to the total population of the generation, but like others, he was ahead of his time and his analysis of inheritance was not accepted until many years after his death.

Liberty Hyde Bailey (1858–1954) was the modern-day Dioscorides. Bailey devoted his life to studying plants and pioneering many practices of horticulture in America. His books are considered the supreme authority on cultivated-plant nomenclature, taxonomy, and pruning. Some of his books are *Cyclopedia of American Horticulture*,

Manual of Cultivated Plants, Hortus, Hortus Second, Hortus Third, and *How Plants Get Their Names.*

During the eighteenth and nineteenth centuries dozens of less-well-known persons including the gardener and the farmer added greatly to our knowledge of crop production and of the preservation and storage of food and feed crops. These discoveries have made it possible for us to produce and store food more efficiently. The ultimate effect of these advances has been that people are not forced to work constantly simply to prevent starvation. This increased efficiency has meant that less energy is required for survival; the result is that industrialization and the arts have become more prominent. People have spent more time planting and cultivating flowering plants, shade trees, and ornamental shrubs for their beauty.

AGRICULTURE AND HORTICULTURE IN THE NEW WORLD

Agriculture always has played and always will play a great role in the history and development of any given country. Early civilizations referred to agriculture as the cultivation of field crops. *Agri* is Latin for "fields"; the word agriculture in its earliest connotations was used exclusively for the cultivation of a field. The present view of agriculture includes the production of both plants and animals. Horticulture is only one part of the broad field of agriculture.

When the Norsemen discovered North America, they found an abundance of grapes and named the new continent "Vineland," associating this unknown place with a familiar plant rather than with some other recognized feature. Early French, Spanish, and English explorers and settlers recorded information in their journals about the many kinds of fruits and vegetables found in the new land. When the first settlers arrived, the Indians were growing crops such as corn, potatoes, pumpkins, squash, blueberries, and tobacco on the continent.

Plant introductions from all over the world have played an important part in the development of the horticultural industry (Figure 1-1). Species of plum and apple are native to various sections of the North American continent, but their fruits are inferior to those of the introduced species. Many of our present species are hybrids between native cultivars and cultivars introduced onto this continent.

For thousands of years vegetable gardens and fruit trees have been planted and cultivated to provide the family with its horticultural needs. Commercial horticulture, on the other hand, is a more recent innovation. In America it had its birth in the nineteenth century when people began moving from the farm to the city as part of the industrial revolution. A demand for horticultural products developed since people living in the cities usually had neither the space nor the time to

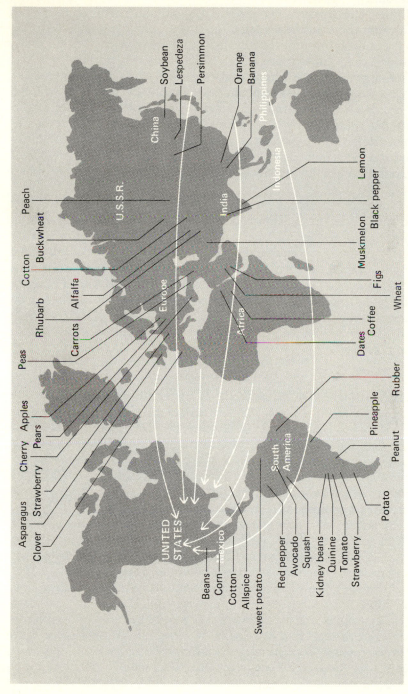

FIGURE 1-1 Some world contributions to the crop production of the United States. Many horticultural crops originated in other countries but have been developed extensively through research in the United States. (From the U.S. Department of Agriculture, Off. Foreign Agr. Relat. Adapted from T. K. Wolfe and M. S. Kipps, *Production of Field Crops*, 6th ed., McGraw-Hill, New York, 1959.)

devote to gardening. This demand grew slowly during the first half of the century.

In 1862 Congress created the U. S. Department of Agriculture to aid both the farmer and the consumer, and in the same year passed the Morrill Act, which established land-grant colleges that were primarily concerned with the teaching of agriculture and mechanical arts. The Hatch Act, passed in 1887, provided for the development of agricultural research programs in the various states. Finally the Smith-Lever Act of 1914 created the Cooperative Extension Service as a mechanism for taking the information developed by the research program to the people. These three functions characterize the Colleges of Agriculture of the land-grant institutions of the United States. Horticulture has been an integral part of these colleges and universities since their establishment.

Horticulture departments in all land-grant universities have provided programs in teaching, research, and extension services to supply the needed information for the growing horticultural industry. Through scientific research the horticultural industry has expanded to meet the ever-increasing demand for its products. The Cooperative Extension Service has disseminated the information obtained from this research and has bridged the gap between the researcher and the commercial horticulturist. The teaching programs have trained many in the use of up-to-date technology to produce the best and highest yields possible.

HORTICULTURE IN RELATION TO OTHER DISCIPLINES

Horticulture, as other sciences, consists of "a branch of knowledge which deals with systematically arranged facts which show the operation of natural laws." The addition of systematic facts and information comes through research and the scientific method. The scientific method consists of (1) recognizing the problem, (2) developing one or more hypotheses (a tentative assumption made in order to draw out and test its logical or empirical consequences) for the solution by using known facts, (3) testing the hypothesis, (4) identifying the best hypothesis based on the test results which then becomes a theory, and (5) continuing to test the theory under a wide range of conditions. If the theory proves to be valid in all dimensions it becomes the basis of a natural law.

Many solutions to problems in the field of horticulture require the interplay of both art and science to achieve the desired results. If this were more readily recognized, a closer working relationship between art and science and the basic and applied science areas probably could be established. An example of the interplay of art and science to achieve these desired results would be the solving of a drainage problem in a landscape design. It would take the science (engineering and soils) to solve the problem and the art to make whatever is done blend aesthetically into the landscape.

Horticulture is related to many other disciplines. Horticulture includes landscaping and the production, storage, and marketing of fruits, vegetables, floricultural products, and nursery crops. Agronomy encompasses the production of grains, fibers, and other field crops as well as soil science. Forestry includes forest management and wood production and utilization. All are applied sciences in the field of agriculture, and all are interrelated. Botany and its various branches, including plant physiology, are pure sciences dealing with the fundamental principles of plant life. Horticulturists, agronomists, and foresters generally apply these fundamental principles to the production and utilization of plants. They generate, assimilate, and disseminate information about solutions to the problems encountered in the industry. The field of horticulture relies on many other disciplines including chemistry, physics, botany, engineering, architecture, plant pathology, bacteriology, geology, meteorology, economics, entomology, genetics, and ecology. It is the incorporation of various aspects of each which makes horticulture such a diverse and exciting career. A complete understanding of horticultural problems involves research in several of the basic and applied sciences.

At one time agronomy, forestry, and horticulture were distinctly separated according to the purposes for which a particular crop was produced and the intensity of its cultivation, but as areas of the plant sciences have expanded, the lines of demarcation have blurred. The relationship between horticulture and the other disciplines in the agricultural plant sciences can be shown best through examples. When grown for its fruit or nuts, the pecan tree is associated with horticulture, but when grown for its wood, it is a forest crop. Grasses that are grown as turf are studied in both horticulture and agronomy. However, when grasses are used as pasture crops, they become strictly agronomic crops. Tobacco and rice are customarily classified as agronomic crops even though both are intensively cultivated and involve high growing costs with relatively large returns, which are characteristics of horticultural crops. Sweet corn and field corn do not differ greatly in their classification or method of production; however, sweet corn is considered a horticultural product because it is eaten as a vegetable, whereas field corn is an agronomic grain crop. Compared with agronomic and forest plants, horticultural crops are intensively cultivated and often have much higher economic and yield returned per hectare of production. However, because of expanded population and new production methods, the plantings of fruits and vegetables are increasing, and production is becoming more extensive, making these factors vague in defining a particular agricultural area. Some horticultural greenhouse crops produce gross annual returns up to $650,000 per hectare.

HORTICULTURE TODAY

The field of horticulture requires the static and demands the dynamic. On one hand, it is possible to relish the solitude of an herb garden of centuries ago and at the same time to enjoy the advantages gained from the fact that horticulture has far outstripped the serenity of the ancient walled gardens. It encompasses the intensive and extensive cultivation of quality food and plants for functional and aesthetic uses in our environment. The horticultural industry has the responsibility of providing much of the food for the masses in the most efficient manner while at the same time allowing individuals to grow plants for aesthetic enjoyment through a relationship with the earth which promotes physical and mental well-being. Horticulture is a part of each individual's daily life. It may be a profession, as it is for horticulture teachers and research workers, or it may be an occupation or vocation as it is for those who work in the production phases. It is strictly a business for the merchandiser, but it may be simply a source of exercise and health for the amateur gardener. People earn a living, relax, and survive through horticulture.

Fruits and vegetables occupy a prominent place in the daily diet of individuals. Well-balanced menus include items that originate in a horticulturist's greenhouse, orchard, or field, such as orange juice, apples, jelly, corn, pickles, and tomatoes. Approximately 30 percent of the food consumed in the world today is produced by horticulturists. Even flowers placed on the table are usually the result of some horticulturist's efforts.

Horticultural foods are excellent sources of energy and provide needed proteins, carbohydrates, vitamins, minerals, and bulk for the diet. Potatoes and sweet potatoes are high in starch. Fruits contain sugars, vitamins, minerals, and proteins. All are sources of carbohydrates. Certain vegetables and nuts—such as beans, peas, sweet corn, and pecans—are sources of protein, whereas avocados and most nuts are sources of fat. Spices, which were one of the principal reasons for early trade, and condiments (pepper, allspice, ginger, nutmeg, cloves) have been used for many centuries to add pleasure to eating.

Horticultural crops represent at least 20 percent of the total agricultural production on this continent. The annual income from the commercial production, processing, and utilization of horticultural crops is at least $15 billion. We cannot properly estimate the value of crops derived from home gardening. Similarly, the numerous business concerns catering to the hobbyist have a tremendous but unknown dollar value.

To obtain the maximum return on investments, the horticulture industry seeks ways to produce optimum yields. Factors that influence yield—which include geographical location, the growing structure and method of growth, the type of soil, the necessary pruning methods, the pest problems in a given area, the manner of propagation, the method of storage and marketing, and other individual factors—must be considered before a given plant is produced in an area. By making educated decisions, the horticultural crop producer should achieve optimum returns at a reasonable price to the consumer. Horticulturists use their knowledge to adapt to such changeable factors as soil and amount of available water, and they use a variety of plants to achieve their goals. In the production and storage of fruit, vegetable, flower, and nursery crops, commercial horticulturists utilize the latest technology to increase efficiency and improve the quality of their products. In landscape design, commercial horticulturists plan outdoor spaces for maximum utilization of the environment and landscape.

Improved production practices, including mechanization of many aspects of production, harvesting, and handling, have all emerged from the horticulturist's search for optimum yields. The consumer's constant search for quality and higher nutritive value will force the industry to seek new and still better methods to meet the demand. The development of new products creates a demand for more and different horticultural products. Roadside markets and pick-your-own operations, which involve the sale of fruits, vegetables, flowers, and plants, are a multimillion dollar industry. From a business standpoint, this type of enterprise is especially attractive to low-income families because it requires very little capital investment to begin. Also labor costs are less in this phase of the industry, and it is a source of recreation for the outdoor-oriented population of today's urban society. For the same reasons, such direct-marketing operations offer a new horticultural graduate the opportunity to get started in business quite readily.

Public awareness of the environment has created a surge of interest in home gardening and urban horticulture and has led to the increased use of landscaping to modify the functional and aesthetic aspects of the surroundings. People in every walk of life are becoming more interested in plants, in the environment, and in quality food for better health. Whether it be for homes, businesses, or recreational facilities, this awareness has created a demand for horticultural products that make a more natural, pleasing, and functional environment. Almost every home or business has potted plants, shrubs, and trees in the landscape, and sometimes fruit and vegetable gardens. The horticulture industry and related industries play an important role in preserving and enhancing the beauty and productivity of the environment. Gardening—whether backyard or balcony—fills a heretofore unmet need. The growing of plants is both therapeutic and recreational in nature.

Horticulture is for the individual; and in order to be easy for the individual, horticulture is a service industry and a supply industry. Horticulture is an industry directly related to survival and beauty. Horticulture can be the mechanization of the production and handling of the million containers of plants in a wholesale nursery in Pennsylvania or the hundreds of hectares of tomatoes grown in California for catsup. But it can also be one beautiful rose or the succulent perfection of sweet corn grown in someone's garden. Thus horticulture is both a vocation and an avocation, and these two aspects cannot be separated.

COMMERCIAL BRANCHES OF HORTICULTURE

There are five basic branches of horticulture today. Probably the oldest of these is *pomology*, fruit production. The other areas are *olericulture*, vegetable production; *floriculture*, flower production; *nursery culture*, nursery crop production; and *landscape design*, landscaping. These divisions can be broken down into different commercial branches such as the production and marketing of seed, nursery stocks, specialty and greenhouse crops, and pharmaceutical crops; market gardening; private gardening (as a gardener for a private estate); operation of trial grounds; arboriculture (growing of and caring for trees for specimen purposes, growing Christmas trees, tree surgery and tree moving); education (including teaching, research, and extension); turf production and management; landscape design; advertising; photography; special promotions (such as flower shows and exhibits); processing and storage; and operation of support industries, such as the manufacture, sale, and service of equipment, machinery, growing structures, and pesticides and other chemicals.

Pomology

The study of fruit production is called *pomology*; workers in the field are called *pomologists*. The horticulturist defines a *fruit* as an expanded and ripened ovary with attached and subtending reproductive structures. Generally a fruit is considered to be the edible, fleshy portion of a woody herbaceous or perennial plant whose development is closely associated with the flower. In some cases the ovary may be only a small portion of the fruit.

Pomology was the first branch of horticulture to develop in America. Fruit plantings were integral parts of most early homesteads as they supplied fresh fruit for use in fall and winter as well as fruit for cider, wine, and vinegar. The first nursery in America was a fruit tree nursery started by the Prince family in 1730 on Long Island in Flushing, New York. Commercial fruit production is of more recent origin, starting primarily after the mid-1800s when long-distance shipping became feasible.

Fruit crops are grown from Maine to Hawaii and from Canada to the tip of Florida. They include the deciduous tree fruits which are divided into pome fruits (pear, apple, and quince) and drupe or stone fruits (peach, cherry, plum, apricot, and almond); the small fruits (blueberry, blackberry, raspberry, grape, strawberry, currant, goose-berry, and cranberry); the nut tree fruits (pecan, filbert, walnut, chestnut, and macadamia nut); and the evergreen or tropical fruits, (lemon, orange, tangerine, lime, and grapefruit as well as banana, mango, date, avocado, pineapple, and papaya).

Olericulture

The study of vegetable production is called *olericulture*; workers in the field are called *olericulturists*. A *vegetable* is the edible portion of an herbaceous garden plant. There is no clear-cut delineation between fruits and vegetables, however; and some crops, such as tomatoes and melons, are classed as vegetable crops but are valued horticulturally for their production of fruits. Other vegetable crops—such as lettuce, celery, potatoes, and carrots—are valued for their edible leaf, stem, or root organs instead of their fruits. In Europe tomatoes and melons are considered fruits, but in the United States they are generally referred to as vegetables. Several vegetables, such as corn, beans, sweet pota-toes, Irish potatoes, peppers, and tomatoes, are of New World origin; however, most vegetables originated in the Old World.

Vegetables are used either fresh or processed. Onions, tomatoes, and potatoes enter world trade in the unprocessed form, but most vegetables must be processed before shipment because of their bulk or perishable nature. Fresh vegetables therefore, do not enter world trade as extensively as some of the fruit crops. Vegetable production is classified as either market garden or truck crop. *Market gardening* refers to the growing of an assortment of vegetables for local or roadside markets and *truck crop production* refers to large-scale production of a limited number of crops for wholesale markets and shipping.

Floriculture

The study of growing, marketing, and arranging flowers and foliage plants is called *floriculture*; workers in the field are called *floriculturists*.

Flowers and other ornamental plants have always played an important part in our lives. They have been used for generations to express joy or sympathy. Today flowers encompass every social function, being extensively used in arrangements, in corsages, and as pot plants.

Commercial flower growing in the United States is thought to have had its inception in the early part of the nineteenth century in the vicinity of Philadelphia, Pennsylvania. At first the flower products were grown outdoors, and they were of poor quality and limited variety, and available only for a small part of the year. In time, greenhouse production of flowers developed to meet the increasing and year-round demand, and retail stores began to sell the output. Gradually people became accustomed to buying from stores selling only flowers, and the florist business developed with a separation of the production and marketing operations.

Commercial floriculture today includes the production of plants and flowers for wholesale and retail sales. Commercial floriculture utilizes both the field and greenhouse for cultivation of its crops. Field production is necessarily confined to geographical areas or seasons where the climate is mild. (As an example, various cut flowers and foliage plants are grown extensively in the field in California.) Greenhouses are used throughout the world for flower production, but large greenhouse enterprises are located around the larger population centers in order to minimize shipping costs. Climate can also influence greenhouse production, as is shown by heavy concentrations of carnation production near Denver, Colorado, because of its high light levels.

Floricultural plants are classified as cut flowers, flowering pot plants, foliage plants, and bedding plants. Cut-flower production has dominated the floral industry. Roses, carnations, chrysanthemums, snapdragons, and orchids are grown for cut flowers predominantly in the greenhouse, whereas most gladioli, chrysanthemums, and straw flowers are grown outdoors during the summer or in suitable mild climates.

Flowering pot plants are sold as whole plants in bloom. These include the Easter lily, poinsettia, cyclamen, geranium, begonia, and hydrangea. Foliage plants are nonflowering plants grown in greenhouses or produced outside in Florida, Texas, and California. Examples are philodendron, ferns, and palms.

The interest in the environment during the last decade has resulted in an increased demand for bedding plants, which are mostly annual flowers used in the landscape. Therefore, the production of these bedding plants, such as petunias and marigolds, has developed into a major industry.

Nursery Culture

In the United States the production of fruit crops led to the development of the nursery industry. Although many nurseries do produce fruit trees, the great majority of the nursery business today is devoted to production of perennial ornamentals. A person who produces or distributes ornamental plants is called a *nursery grower*. A person who does research, teaching, or extension in this area is called an *ornamental horticulturist*.

The nursery trade involves the propagation, growing, and maintenance of young trees and shrubs and herbaceous annuals, biennials, and perennials other than bedding plants. The production of young fruit trees, some perennial vegetables (asparagus, rhubarb, and chives) and small fruit plants is an important part of this industry. Another important area is the production of young ornamental plants and turf for landscaping public and private buildings and residences, roadsides, and parks. The nursery industry also produces evergreen trees for use as Christmas trees.

Wholesale nurseries usually specialize in relatively few crops for maximum efficiency in production. These nurseries typically propagate their own stock and supply plant materials to retail nurseries, garden centers, landscaping companies, or florists. The retail nurseries may care for and develop the plants to a salable size, after which the plants are sold to the public. The retail nursery business frequently provides landscape design, planting, and maintenance services. There are also many mail-order nurseries. These nurseries usually grow a small portion of their catalog listings and buy the rest of their stock from wholesale houses.

Landscape Design

Landscape design is the profession concerned with the planning and planting of outdoor space to secure the most desirable relationship between land forms, structures, and plants to best meet man's needs for function and beauty. It is a part of horticulture because the essence of the landscape is living plant material. A person who is a landscape designer sells landscape design plans—which include general plans, construction plans, planting plans, and specifications—but does not sell plants.

Landscaping is not decorating the environment by planting trees and shrubs; it is the aesthetic and functional development of space. Plants and construction materials are used to enhance the site, not to detract from it. Landscape maintenance should be considered with the design, or the end result will be a breakdown in the desired environmental impact.

A landscape design encompasses the entire area, not only the garden area. The house, the utility, the comfort, and the aesthetic effect should all be considered in landscape design. The desires and space needs of the homeowner and family, coupled with their likes and dislikes, must be analyzed before a design is begun.

The landscape designer is educated in art and science related to landscaping. He or she must be competent in many engineering and architectural processes and techniques, and needs sensitivity in order to analyze the environment so that a work of art rather than a hodgepodge is created. Some landscapers try to decorate with their design; this detracts from the natural environment. The landscape designer should try to blend structures and personalities into the existing environment through good site planning and planting.

HORTICULTURE AND WORLD NUTRITION

Many of the underdeveloped nations of the world do not produce enough food to provide their populations with an adequate diet. Even certain developed nations, such as Japan, must import great quantities of food because they have inadequate land on which to grow their own.

Animals can provide a part of human nutritional needs, but animals depend on plants for their food supply. Therefore *all* animals, including human beings, are dependent upon the process of photosynthesis, which takes place only in green plants. Photosynthesis transforms light energy into the chemical energy which is necessary for life processes in green plants. Of the total energy available in the ecosphere, more than 99.9 percent is of solar origin. Since each progressive step in the food chain results in a reduction of the energy available for transfer, much energy must be converted to satisfy the physiological needs and comfort desires of the human population (Figure 1-2). To meet these demands, the farmers and gardeners of the world must understand the principles of crop growth in order to apply the most productive cultural practices to produce the highest possible yields of high-quality crops.

Thomas Malthus (1776–1834), an English clergyman-scholar, devised a biological or natural theory of population growth from the historical and scientific evidence of his time. His theory proposed that over a period of time food production has a tendency to increase at an arithmetic rate (1, 2, 3, 4, . . .) whereas human population increases at a geometric rate (1, 2, 4, 8, . . .). Since human population growth ultimately is limited by the amount of food the world can produce, Malthus believed that the rate of human population growth would be

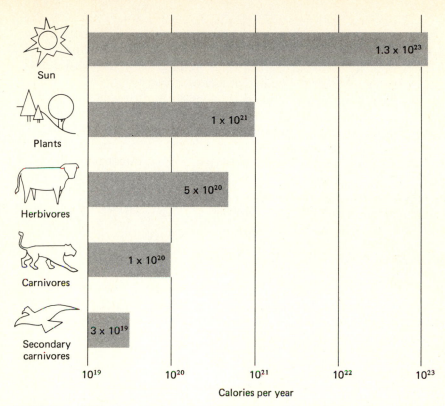

FIGURE 1-2 Utilization of solar energy decreases with each step along the food chain. These bars (on a logarithmic scale) show that plants use only 0.08 percent of energy reaching the atmosphere; plant eaters use only part of this fraction and flesh eaters even less. (From LaMont C. Cole, "The Ecosphere," *Scientific American*, 1958. Copyright © 1958 by Scientific American, Inc. All rights reserved.

brought into a proportional balance with food production by such natural checks as famine and pestilence. The twentieth-century advances in agricultural technology resulting in cultivars, improved nutrition of plants, increased use of irrigation, increased mechanization, and the development of agricultural chemicals have prevented the Malthusian law from taking effect or at least have delayed it. Unfortunately, in recent years it seems that these advances cannot equal the growing population, and there may again be a threat of mass starvation. Paul Ehrlich, in 1969, predicted on the basis of Malthus's theory that population growth is logarithmic. According to Ehrlich, at the rate population is currently expanding, there is a real and immediate threat of mass starvation. The population of the world is doubling approximately every 40 years, meaning that each day there

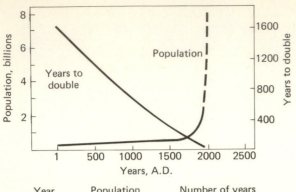

Year, A.D.	Population, billions	Number of years to double
1	0.25(?)	1,650(?)
1650	0.50	200
1850	1.1	80
1930	2.0	45
1975	4.0	35
2010	8.0	?

FIGURE 1-3 Estimated population of the world and the number of years required for it to double. [After Dorn, "World Population Growth," in Hauser (ed.), *The Population Dilemma,* The American Assembly, Columbia University, New York © 1963. Reprinted with permission of Prentice-Hall, Inc., Englewood Cliffs, N.J.]

are 200,000 new mouths to feed. Presently the world population stands at about 4 billion (Figure 1-3). Projections indicate that in the year 2000 the population will increase to 6 billion, even if Europe, the United States, and Japan maintain near-zero population growth.

Beyond the fact that the world may simply starve, an immediate concern is malnutrition. Inadequate amounts of protein in the diet of an expectant mother or lack of protein in infancy can result in permanent brain damage and a mental capacity much below normal. An individual suffering from malnutrition can expect a shorter life, and a person who is undernourished during infancy can expect stunted growth and mental retardation. Estimates indicate that in 1968 almost two-thirds of the world's 3.5 billion people were malnourished. Although the death of an undernourished person is usually attributed to some cause such as an infectious disease, many of these diseases generally are not fatal to individuals with an adequate diet.

Contrary to popular opinion, the caloric content of food does not present a complete picture of the suitability of the food for human nutrition (Figure 1-4). The caloric content measures only the *energy* provided by a given food. Frequently, the caloric intake of undernour-

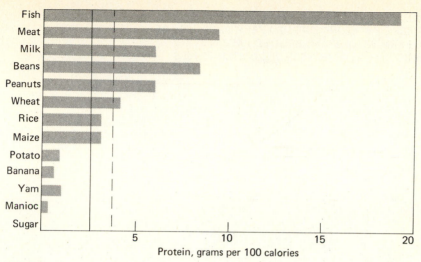

FIGURE 1-4 Protein-calorie ratios of various foods compared. It can be seen that some of the basic foods of tropical areas, such as banana, yam, and manoic, are protein-poor. The approximate adult dietary requirement of protein is shown by the solid line; that for children, by the dashed line. (From: Charles B. Heiser, Jr., *Seed to Civilization: The Story of Man's Food.* Freeman, San Francisco, Calif., © 1973.)

ished people is sufficient, but the nutritive content is inadequate. Proper human nutrition requires certain definite portions of vitamins, proteins, fats, and carbohydrates (organic compounds supplied by green plants), and minerals (inorganic compounds also supplied by green plants). Since animals incorporate into their bodies only 10 percent of the potential energy found in their plant food, the amount transferred to other animals that eat their bodies is not greater than 10 percent (see Figure 1-2).

It seems that plants have more potential than animals as a source of food; however, the protein in animals is also a great aid in human nutrition. This is especially true because animal products contain all the amino acids (building blocks of proteins) required by the human body. Plant proteins, in general, are deficient in one or more of these essential amino acids. If plants are to adequately provide all the protein necessary for the growth and development of the population, research must provide food crops of higher or more complete protein—such as Triticale, Opaque 2, grain and potato—and of more suitable composition.

Crop researchers and producers are continually increasing yields, but at the expense of energy. They develop and employ new methods of cultivation, fertilization, pest control, soil conservation, harvesting,

TABLE 1-1
NUMBER OF PERSONS SUPPLIED FARM PRODUCTS BY
ONE FARM WORKER AND RELATED FACTORS
DURING SPECIFIED YEARS IN THE UNITED STATES

YEAR	TOTAL PERSONS SUPPLIED PER FARM WORKER	TOTAL FARM EMPLOYMENT, MILLIONS	TOTAL U.S. POPULATION, MILLIONS
1820	4.12	2.4	9.6
1840	3.95	4.4	17.1
1860	4.53	7.3	31.5
1880	5.57	10.1	50.3
1900	6.95	12.8	76.1
1920	8.27	13.4	106.5
1940	10.69	11.0	132.1
1950	15.47	9.9	151.7
1960	25.85	7.1	179.9
1965	37.02	5.6	193.7
1977	57.00	4.2	217.0

Sources: M. S. Kipps, *Production of Field Crops*, 6th ed.,
McGraw-Hill, New York, 1970; U.S.D.A.

TABLE 1-2
ANNUAL AVERAGE RATES OF CHANGE (PERCENT PER YEAR) IN TOTAL
OUTPUTS, INPUTS, AND PRODUCTIVITY IN UNITED STATES AGRICULTURE,
1870–1971

ITEM	1870–1900	1900–1925	1925–1950	1950–1965	1965–1971
Farm output	2.9	0.9	1.5	1.9	2.3
Total inputs	1.9	1.1	0.3	−0.2	0.7
Total productivity	1.0	−0.2	1.2	2.1	1.6
Labor inputs[1]	1.6	0.5	−1.8	−4.3	−3.0
Labor productivity	1.3	0.4	3.3	6.6	5.4
Land inputs[2]	3.1	0.8	0.1	−0.8	0.5
Land productivity	−0.2	0.0	1.4	2.5	1.8

[1]Number of workers, 1870–1910; man-hour basis, 1910–1971.
[2]Cropland used for crops, including crop failure and cultivated summer fallow.
Sources: From U.S.D.A., *Changes in Farm Production and Efficiency*, Statistical Bulletin
233 (revised), Washington, D.C., June 1972; and D. D. Durost and G. T. Barton, *Changing
Sources of Farm Output*, U.S.D.A. Production Research Report no. 36, Washington, D.C.,
February 1960.

storing, and marketing. Consider the increase in production: in 1820 one farm worker produced food for 4 people; in 1965 one farm worker produced food for 37 people; and in 1977 one farm worker produced food for 57 people. There was a decrease in the number of farms in the United States from 5.6 million in 1950 to 2.8 million in 1977. A similar decline in the number of farm workers from 9.9 to 4.2 million people occurred during the same period (Table 1-1). It is obvious that technology has improved the productivity over the years, but along with higher yields have come a greater dependence on mechanization and greater use of energy. In the United States, with our highly mechanized techniques, it has been estimated that 5 to 10 cal of fossil fuel are required to produce 1 cal of food. In China, where most work is manual only 1 cal of fossil fuel (as opposed to human calories) is required to produce 50 cal of food. It would not, however, be feasible to meet the increasing needs of the world by doing away with mechanization and going back to manual labor. With manual labor, it would not be possible to maintain the present level of production, much less increase it.

Today, horticultural crops produce more food per hectare per man hour than ever before. Intensive cultivation and specialization of horticultural products have done much to increase the production of food for the world (Table 1-2). With increased population and less land for crop production, the intense culture of food crops will be of increasing significance if the world population is to be fed. Since horticultural crops require a minimum of land, the potential for these products will probably reach a peak higher than imagined.

The need to preserve and protect the environment and natural resources has long been recognized by horticulturists. Recent developments have brought the realization to the general public that, to survive, everyone must cooperate to preserve our vital assets and to preserve the land so that the increasing population can be accommodated and fed. With increasing growth in population there will be even less land for crops, but the demand will be greater for increased quantities of high-quality crops. The intensive cultivation of agricultural crops seems to be the promise of the future for our expanding population.

Horticulture must meet its share of the expanded food production necessitated by the increase in world population. Since horticultural products provide approximately 30 percent of the human food of the world and a considerable nutritional balance unavailable in other crops, the quality and nutritional content of the products grown has a sizable impact on the health and well-being of the world (see Table 1-3).

TABLE 1-3
PERCENTAGE OF TOTAL FOOD NUTRIENTS CONTRIBUTED BY MAJOR FOOD GROUPS

FOOD GROUP AND PERIOD 1973	FOOD ENERGY	PROTEIN	FAT	CARBOHYDRATE	CALCIUM	PHOSPHORUS	IRON	VITAMIN A VALUE	THIAMIN	RIBOFLAVIN	NIACIN	ASCORBIC ACID
	Percent	Percent	Percent	Percent	Percent	Percent	Percent	Percent	Percent	Percent	Percent	Percent
Meat (including pork fat cuts), poultry, and fish	19.9	41.2	34.2	.1	3.5	25.9	29.3	22.2	27.7	24.2	45.7	1.1
Eggs	2.0	5.3	3.0	.1	2.3	5.5	5.4	6.1	2.3	5.2	.1	.1
Dairy products, excluding butter	11.6	23.1	12.9	6.9	76.5	37.0	2.4	13.2	9.5	41.7	1.6	4.1
Fats and oils, including butter	17.9	.1	42.7	(1)	.4	.2	0	8.1	0	0	0	0
Citrus fruits	.9	.5	.1	1.9	.9	.7	.8	1.5	2.8	.5	.9	25.9
Other fruits	2.2	.6	.3	4.8	1.2	1.1	3.4	5.9	1.8	1.5	1.7	12.1
Potatoes and sweet potatoes	2.7	2.3	.1	5.3	.9	3.8	4.4	4.8	6.1	1.6	7.0	17.9
Dark green and deep yellow vegetables	.3	.4	(1)	.5	1.5	.6	1.6	20.7	.9	1.1	.7	8.3
Other vegetables, including tomatoes	2.5	3.2	.4	4.6	4.7	4.9	9.0	16.0	6.8	4.4	5.9	27.3
Dry beans and peas, nuts, soya flour	2.9	5.0	3.7	2.0	2.6	5.7	6.0	(1)	5.2	1.8	7.0	(1)
Flour and cereal products	19.3	17.8	1.3	35.0	3.3	12.4	28.9	.4	36.8	17.2	24.3	0
Sugar and other sweeteners	17.1	(1)	0	38.1	1.2	.3	6.2	0	(1)	(1)	(1)	(1)
Miscellaneous[2]	.7	.4	1.3	.6	1.0	1.8	2.5	2.3	.1	.7	5.0	3.6
Total[3]	100.0	100.0	100.0	100.0	100.0	100.0	100.0	100.0	100.0	100.0	100.0	100.0

[1]Less than 0.05 percent.
[2]Coffee and cocoa (chocolate liquor equivalent of cocoa beans) and fortification of products not assigned to a specific food group.
[3]Components may not add to total due to rounding.
Source: U.S. Department of Agriculture, *Agricultural Statistics*, 1974.

TOTAL PERCENTAGE CONTRIBUTION BY FIVE CATEGORIES OF HORTICULTURAL FOODS*

FOOD GROUP AND PERIOD	FOOD ENERGY	PROTEIN	FAT	CARBOHYDRATE	CALCIUM	PHOSPHORUS	IRON	VITAMIN A VALUE	RIBOFLAVIN	NIACIN	ASCORBIC ACID	THIAMIN
	%	%	%	%	%	%	%	%	%	%	%	%
Average 1957-59												
Horticultural foods	8.7	7.4	.9	17.1	9.3	11.2	20.5	50.8	18.5	17.5	93.6	18.5

*Total of citrus fruits, other fruits, potatoes and sweet potatoes, dark green and deep yellow vegetables and other vegetables including tomatoes.
Source: C. W. Basham, *Laboratory Activities in Horticulture*, Kendall/Hunt, Dubuque, Iowa, 1976.

BIBLIOGRAPHY

Basham, C. W.: *Laboratory Activities in Horticulture*, Kendall/Hunt, Dubuque, Iowa, 1976.

Chapman, S. R. and L. P. Carter: *Crop Production: Principles and Practices*, Freeman, San Francisco, 1976.

Christopher, E. P.: *Introductory Horticulture*, McGraw-Hill, New York, 1958.

Day, P. R. (ed.): *How Crops Grow: A Century Later*, Bulletin 708, Connecticut Agricultural Experiment Station, New Haven, Conn., 1969.

Edmond, J. B., T. L. Senn, F. S. Andrews, and R. G. Halfacre: *Fundamentals of Horticulture*, 4th ed., McGraw-Hill, New York, 1975.

Gardner, V. R.: *Basic Horticulture*, Macmillan, New York, 1951.

Hedrick, U. P.: *A History of Horticulture in America to 1860*, Oxford, New York, 1950.

Heiser, C. B.: *Seed to Civilization: The Story of Man's Food*, Freeman, San Francisco, 1973.

Hughes, H. D. and D. S. Metcalfe: *Crop Production*, 3d ed., Macmillan, New York, 1972.

Janick, J.: *Horticultural Science*, 2d ed., Freeman, San Francisco, 1972.

Janick, J., R. W. Schery, F. W. Woods, and V. W. Ruttan: *Plant Science*, 2d ed., Freeman, San Francisco, 1974.

Kipps, M. S.: *Production of Field Crops*, 6th ed., McGraw Hill, New York, 1970.

Pyke, M.: *Man and Food*, McGraw-Hill, New York, 1970.

Schilletter, J. C. and H. W. Richey: *Textbook of General Horticulture*, McGraw-Hill, New York, 1940.

Shoemaker, J. S.: *General Horticulture*, 2d ed., Lippincott, New York, 1956.

Shoemaker, J. S. and B. J. E. Teskey: *Practical Horticulture*, Wiley, New York, 1955.

Talbert, T. J.: *General Horticulture: Principles and Practices of Orchard, Small Fruit, and Garden Culture*, Lea & Febiger, Philadelphia, 1946.

Thompson, H. C. and W. C. Kelly: *Vegetable Crops*, 5th ed., McGraw-Hill, New York, 1959.

Webber, R.: *The Early Horticulturists*, Kelly, New York, 1968.

Zeven, A. C. and P. M. Zhukovsky: *Dictionary of Cultivated Plants and Their Centres of Diversity*, Centre for Agricultural Publishing and Documentation, Wageningen, Netherlands, 1975.

PART 1
Principles

CHAPTER 2
PLANT
CLASSIFICATION

The plant kingdom includes a wide diversity of plant types ranging in complexity from one-celled algae to large trees that often exceed 30 meters in height. Within this diverse plant kingdom, no two species are ever exactly alike. Some are very similar, and it is quite obvious from superficial observation that they are closely related. Others are so different from one another that few, if any, apparent bonds of relationship exist. Furthermore, these currently observed relationships among plants are not in a static condition because each of these plant groups is changing progressively with time through the process of evolution.

The naming of plants has always been a problem to botanists. For instance, dates are grown in Iran; coconuts come from Hawaii; and flower bulbs are grown and sold in the Netherlands. Each of these plants has a common name in the language of that location. To make matters worse, people in different areas of the same country sometimes use different names to designate a given plant. Therefore, each plant has a multiplicity of common names, which are quite indefinite. For an orderly system of classification, botanists must give to each group of plants a name that is recognized by botanists everywhere, regardless of the language they speak. This name is referred to as the scientific name or sometimes as the Latin name. Pronunciation of the scientific name may vary, but someone who knows scientific nomenclature can communicate with anyone from any country in the world regardless of national language. Scientific nomenclature is an international language, or at least a universal vocabulary.

To understand and utilize the various plants, it is essential to arrange them in an orderly system of classification and give each of them a name, a name that refers to this one and only this one group of plants. In 1886 De Candolle said in *The Origin of Cultivated Plants*, "Science can make no real progress without a regular system of nomenclature." Knowledge is easier to learn, understand, even classify and find (such as the missing elements in the periodic table) if it is related to associated knowledge. As scientists, human beings need unity, shape, and direction in order to learn a subject. The science that includes classification, nomenclature, and identification of plants is *plant taxonomy*, the oldest branch of botany. Botanists have worked for centuries on the classification of plants; and while much has been accomplished, much remains to be done.

The information base for taxonomy is being continually increased by additional research. Plant taxonomists work to identify the different species of plants and to classify each plant to show its true relationship to all other plant life. Taxonomists strive to group plants on the basis of anatomic, morphologic, physiologic, and genetic similarities and differences which are the expressions of the actual phylogenetic relationships. By analyzing these similarities and differences, a fairly orderly taxonomic system has been developed.

Taxonomy is vital to horticulture. The horticulturist identifies, names, and establishes relationships between and among plants through the use of taxonomy. Horticulturists across the world communicate in an exacting and precise manner through the use of taxonomy. And if used properly, there is no confusion among horticulturists as to what plant is being discussed. If a package of seed is mailed anywhere in the world, the recipient can tell exactly what to expect when the seed is planted and cultivated if it is correctly labeled. If a paper or popular article is written describing how to grow a specific plant, everyone in the world can understand that information.

CLASSIFICATION SYSTEMS

Over the centuries, many different systems have been devised for classifying plants. They generally fall into one of three categories: (1) artificial, (2) natural, or (3) phylogenetic. *Artificial systems* have been devised for convenience and are often based on arbitrary, variable, and superficial characteristics. The artificial system is based on the ultimate use of the plant being described. Agricultural plants are classified as to grain, timber, fruit, or medicinal plants, and ornamentals are classified into such groupings as herbs, greenhouse plants, foliage plants, garden perennials, shrubs, and trees. Since the artificial classifications serve chiefly for convenience in communication, they are used primarily as a nonscientific, practical system of identification rather than a system designed to indicate relationships among plants. This system uses convenient and readily observable characteristics of plants regardless of anatomic, morphologic, physiologic, genetic, or evolutionary relationships. But this classification system has so much overlap that other systems must be used for more exacting descriptions. For instance, azalea plants are used as flowering shrubs in landscaping, but they are also considered as flowering pot plants under the classification of a greenhouse crop.

The *natural system* of classification attempts to show relationships among plants through the use of selected morphological structures. The thesis underlying this system is that morphologically similar plants are closely related. The system represents an effort to reflect the order that exists in nature by utilizing all available knowledge.

By classifying plants according to their evolutionary pedigree, a *phylogenetic system* reflects genetic relationships between and among plants and establishes their progenitors. The chief limitation of this system is the limited knowledge of earlier plant forms. Through the use of cytogenetics, paleobotany, anatomy, and biochemistry, along with other sciences, taxonomists increase their knowledge and understanding of plants. In so doing, taxonomists can better determine the intricate relationships and interactions that exist within the plant kingdom as it exists for the living plants of today.

As plants evolve, taxonomy must follow. Plants cross-breed and mutate to create new forms, and existing species become extinct because their niche in the environment disappears. Plants also change as their environment changes or as they are moved from one area to another. For example, walnuts originally came from Persia and spread along the Mediterranean Sea, while at the same time spreading north into Germany. Walnuts were also brought to Spain, Chile, and California, and from the mountains of northern Europe to Canada. At present there are two distinct types or races of walnuts; each developed through the process of evolution. One is adapted to the colder weather and longer winter of northern Europe and Canada; the other has become adapted to the milder climate of the Mediterranean Sea and California. One can see that, after a period of time, two distinct species could evolve.

The systems of organized classifications have gradually shifted from purely artificial systems to natural or phylogenetic systems. Early systems were artificial and were based on the growth habits of plants; these systems were supplemented by a broadly adopted system based on the numerical aspects and sexual parts of the plant. This was later replaced by systems which established the natural morphological relationships as the focal point. The most recent systems use phylogenetic relationships to establish classifications.

Artificial Classification

Artificial Systems Based on Habit   The early Greek philosophers and medical practitioners were the first to leave written records of civilization's attempts to classify plants. The Greeks classified plants on the basis of vegetative characteristics, such as growth habits and leaf structure. Theophrastus (370–285 B.C.), a Greek naturalist, wrote *Historia plantarum*, the oldest existing botanical work. In it he classified 480 species of plants as herbs, undershrubs, shrubs, or trees. In his book, Theophrastus recognized and described families among flowering plants, such as the carrot family; he perceived relationships among the conifers, cereals, thistles, poplars, and birches; he recognized what is referred to today as genera in the sense of a group of species; and he applied Greek names to each species. For example, the thistles were grouped together because they had a similar flower shape. In recording his studies of seed germination and the growth of seedlings, Theophrastus found the root to be the first structure to appear from all seeds at the time of germination. He showed that not all roots are subterranean and that not all subterranean parts of plants are roots. Although he correlated the presence of one or two cotyledons with the occurrence of other characters, he did not make use of these characteristics in his classification system. He classified leaves and studied the arrangement of leaves on the stem. Many of Theophrastus' observations were so revolutionary for the time that they were not used in plant classification systems until 2000 years later; however, his work helped precipitate some of the later classification systems. Based on the greatness of all his early work, Theophrastus earned the title "father of botany."

Artificial Systems Based on Numerical Classifications   In 1737, *Genera plantarum*, by Carolus Linnaeus (1707–1778), introduced the era of systems based on numerical classification. Linnaeus, a Swedish naturalist, physician, botanist, and teacher, founded this type of artificial system and laid the groundwork for the natural system of plant classification which would follow in later years. Linnaeus's system arbitrarily attached special significance to the reproductive aspects of the plant and was known as the *sexual system*. He classified plants according to characteristics of flowers, such as the number of stamens and the number and organization of carpels. In 1753, Linnaeus published *Species plantarum*, which was the forerunner of the system of priority used in the present-day nomenclature of higher plants. The system of classification divided all plants into 24 classes. These primary divisions, or classes, were subdivided into orders. Approximately 7300 species were described, and the groups were arranged in an artificial arrangement of reproductive parts.

This work of Linnaeus was the foundation of a consistent and extensive use of the *binomial system* of nomenclature. Linnaeus divided the plant kingdom into major units called orders. He divided orders into classes, and each class was subdivided into generic parts. Each genus was then divided into species. The binomial or two-name system is comparable to the system of names that Americans use to name themselves. The plant's generic name is comparable to a person's family name, and the plant's species or specific name is like a person's given name. For example, in the Rogers family, the members of the family may be John Rogers, Mary Rogers, and Richard Rogers. In the genus *Acer* (maple), the species *Acer rubrum* (red maple), *Acer saccharum* (sugar maple), and *Acer platanoides* (Norway maple) are found.

Thus, new plants could easily be incorporated into such a system simply by adding new units at whatever level was required, or units that were found not to be separate could be combined. Since known plants were increasing by tremendous numbers during this period because of increased exploration, this system provided a simple classification which was very popular with botanists. Linnaeus, however, realized the limitations of the system in that it was more convenient than natural. His thinking was futuristic enough to indicate that increased knowledge of relationships among plants would someday permit a more natural system.

Linnaeus's greatness lay in his broad outlook and in his organizational ability. His system of classification was so revolutionary in its simplicity and workability that it earned him the designation "father of taxonomic botany." Much present-day work uses Linnaeus's basic fundamentals to establish classifications on the species level.

Natural Classification

Not until the latter part of the eighteenth century did other scientists realize that more was needed for plant classification and identification when naming plants than brief descriptions and illustrations. Between 1750 and 1800 there was a mass influx of plants into European botanical gardens. Plants such as African violets and orchids were introduced from Africa for the greenhouses of the aristocracy. Along with the increase in the flora came the realization that there were more differences in plants than merely the reproductive differences which Linnaeus had indicated. Awareness of the biological function of the sexual organs of a plant and a growing awareness of floral morphology led taxonomists to use these characteristics as a basis for a new system, referred to as the *natural system*, which was based on human understanding of nature during this period. Bernard de Jussieu, in 1759, rearranged plants in the garden at La Trianon, Versailles, France, using a system similar to Linnaeus's which divided flowering plants into groups on the basis of number of cotyledons, ovary position, presence or absence of petals, and fusion or distinctness of petals.

One obvious flaw in the approaches of this period was that they were all based on the premise that plants do not undergo evolutionary change. They failed to recognize the progressive changes that occur in all living things through the process of evolution. In this classification system, plants were treated as inanimate objects and were arbitrarily grouped into categories established by the scientist rather than into groups relating to the plant's geneology.

Phylogenic-Based Classification

Darwin's *Origin of Species by Means of Natural Selection or the Preservation of Favored Races in the Struggle for Life*, published in 1859, presented scientific facts based on the theory of gradual development and evolution of both plants and animals. His theory was based on three principles: (1) in nature there is overproduction and variation; (2) overproduction results in competition or struggle for existence; and (3) variation gives an opportunity for survival of the fittest. Darwin's theory of the origin of the species obliterated previous taxonomic theories, since its emphasis on variability and mutability of species was quite contrary to all the earlier long-held beliefs. The acceptance of Darwin's theory of evolution led to changes in the plant classification system. These changes led to the establishment of present-day taxonomic systems. The classification becomes a range from simple to complex with an emphasis on genetic and ancestral relationships.

In 1875, August Eichler was the first to present a plant classification system based on genetic relationships that was developed to go along with Darwin's concepts of evolution. This led to the publication of the 20-volume classic based on Eichler's theories, *Die natürlichen Pflanzenfamilien*, by Adolph Engler, Karl Prantl, and others, beginning in 1892. Engler and Prantl divided all plants with seeds into Gymnospermae and Angiospermae, with Angiospermae being subdivided into the classes Monocotyledoneae and Dicotyledoneae. However, they were not the first to do this. Under this system, subclasses were divided into orders, which were further divided into families. This is the backbone of the taxonomic system used today (see Table 2-1).

Plant Classification Today

The system of plant classification used today blends features of the natural and artificial systems. The family and specific categories are generally classified largely by natural and phylogenetic methods. This

TABLE 2-1

UNITS OF CLASSIFICATION LISTED IN ORDER FROM GREATEST TO LEAST MAGNITUDE, WITH COMPLETE CLASSIFICATION OF THE APPLE CULTIVAR 'JONATHAN' AS AN EXAMPLE

Kingdom—Plantae (plants)
  Division—Spermatophyta (plants that bear seeds)
    Subdivision—Angiospermae (seeds enclosed)
      Class—Dicotyledoneae (seeds usually with two cotyledons)
        Order—Rosales
          Family—Rosaceae
            Subfamily—Maloideae
              Tribe—none
                Genus—*Malus*
                  Species—*domestica*
                    Cultivar—'Jonathan'

results in a genus in which the species are similar in morphological features and are closely related genetically. Flowers and fruits are usually the basis for classification; but roots, leaves, and stems may also be used. Environmental variations do not affect the flower or fruit characteristics as much as they affect other parts. In addition, flowers have many distinct and discernible units. Phylogenetic studies establish the origins of each group of plants and its relationships to both its present-day relatives and also to those of long ago, including those that may now be very distantly related. In theory it should be possible to relate all present-day plants to that very first plant precursor which first began to photosynthesize.

Researchers continue to try to establish an even more exacting system of plant classification. Taxonomists use many methods to establish a logical classification system. Scientists realize that the ultimate classification system will never be devised because too many of the key pieces to the puzzles have become extinct. However, taxonomists continue to perfect the present system by delving into data on the relationships of individual plant groups. Studies of vascular and floral anatomy, embryology, and plant geography contribute precise knowledge needed for better classification. The more botanists learn about plants and the more they discover about natural relationships, the more changes will need to be made. All bases will be used in the future to develop as much as possible a classification system that will reflect the exact and true genetic-historical relationship among all plants within the plant kingdom. Among the areas taxonomists use for obtaining data for studying these possibilities and classifying plants are:

1. *Morphological*
   Traditional taxonomists use morphological characteristics to classify plants

into defined groups. They feel that the more similar two plants are in form and structure, the more closely related they are. In most cases, this is the only type of information available.

2. *Anatomical*

The anatomical taxonomists study cell types of the vascular system to establish evolutionary patterns. For example, a plant that has long fiber cells is believed to be more advanced in evolution than one with short fiber cells. Much work is also now being done on pollen anatomy.

3. *Embryological*

Embryological taxonomists use the morphology and anatomy of embryonic development to support classifications. Such characteristics as number and position of cells in the embryo sacs or the placement of the micropyle are used as bases of comparison.

4. *Biochemical*

Biochemical taxonomists use biological compounds such as sugars, amino acids, fats, oils, alkaloids, alcohols, terpenes, or phenols as additional characters to establish in which category a plant belongs. The basis for classification is that more complex compounds are believed to have evolved at only one time and in only one group of plants. Hybrids are classified by the presence of intermediate levels of the compounds which are present from both parental groups.

5. *Ecological*

Ecosystematics in botany isolates the natural biotic units resulting from natural barriers such as oceans or high mountain ranges. These are used to establish classification on the basis of divergent evolution, chiefly below the species level. Present-day researchers look at small units of plants within the species level, theorize and prove theories of evolution at these levels, and then extrapolate to larger and larger units.

6. *Numerical*

Numerical taxonomists take statistical data from divergent pieces of evidence, instead of only one or a few criteria, analyze them, and then classify the plant. They use the findings from all methods and present a composite analysis. Careful analysis has revealed that classification schemes arrived at numerically often correlate very closely with those arrived at by traditional means.

DEVELOPMENT OF SCIENTIFIC NOMENCLATURE

History and Background

Through the centuries, botanists have made repeated efforts to reach a common basis of botanical names through publications such as Linnaeus's *Genera plantarum* and *Species plantarum*. The scientific names of plants are based on the binomial system which Linnaeus presented in *Species plantarum* in 1753. Linnaeus was not the first to use binomial nomenclature but was the first to use it consistently.

The first organized effort toward standardization and legislation of nomenclatural practices was at the First International Botanical Congress in Paris in 1867. Up until this Congress, nomenclature was controlled by a few botanists of prestige and influence. The rules which were officially accepted at the Congress in 1867 were published by de Candolle in *Lois de la nomenclature botanique*. The Congress has met regularly since 1867, publishing the legislation agreed upon in various volumes of the International Code of Botanical Nomenclature. These rules apply only to scientific names.

The scientific names assigned to individual plants are usually derived from a descriptive trait of the plant, from the name of the place where the plant occurs, from some person that the taxonomist wishes to honor, or from a name in classical mythology. For example, *Magnolia grandiflora* is named because of its large or "grand" flowers; *Geranium carolinianum* was first named in the Carolinas; and *Rosa Wichuriana* was named in honor of Wichuray, a Russian botanist. Not until 1930 did an International Botanical Congress adopt rules for the determination and coining of plant names, the one area where new rules are made at each International Congress.

Organization of Nomenclature in Taxonomy

Convenience and uniformity have necessitated that a series of categories for the plant kingdom be adopted. The term *taxon* (plural, *taxa*) is used to distinguish the categories or taxonomic groups. The plant kingdom is divided into divisions, classes, orders, families, tribes, genera, and species, as shown in Table 2-1. Each of the taxa, except for the genus and species, is distinguished by having a name with a specific ending.

Horticultural crops are in the Spermatophyta, or seed plants, that *division* of the plant kingdom, which includes plants reproduced by seeds. Names of divisions of the plant kingdom all end in *-phyta*.

The division Spermatophyta is divided into two *subdivisions*—Gymnospermae and Angiospermae. Characteristics of Gymnospermae are: (1) plants are always woody, but vessels are not present in the vascular system; (2) seeds are produced on naked surfaces of scales; (3) scales are arranged in cones; and (4) flowers are absent. Examples of Gymnospermae are pines, spruces, and firs. The Angiospermae, like the Gymnospermae, are vascular plants characterized by having seeds, but they differ from the Gymnospermae in the following characteristics: (1) vessels are almost always present in vascular system; (2) flowers are present, each usually containing four whorls of floral organs (calyx, corolla, stamen, and pistil); and (3) seeds are enclosed in fruits formed by development of the pistil of the flower. Most horticultural crops such as lilies, grasses, geraniums, apples, potatoes, and orchids are in this subdivision.

The angiosperms are composed of two classes: (1) *Dicotyledoneae*—commonly called dicots—which contains approximately 200,000 species grouped in more than 250 families, including the broad-leaved trees, roses, peas, and sunflowers; and (2) *Monocotyledoneae*—commonly called monocots—which contains approximately 50,000 species grouped in more than 40 families, including the grasses, lilies, orchids, irises, and palms.

Dicots have the following characteristics:

1. Principal veins of the leaves branch out from the midrib or from its base, not parallel, forming a distinct network (Figure 2-1).

2. Flower parts are usually arranged in fours or fives (Figure 2-2).

3. The root system is characterized by a taproot.

4. The stem has vascular bundles in a single cylinder (Figure 2-3).

5. The cambium in woody species adds a new cylindrical layer (ring) of wood each year or each growing season (this is referred to as secondary growth).

6. The embryo has two cotyledonary leaves.

FIGURE 2-1 Surface view of cleared dicotyledon leaf of English ivy (*Hedera helix*) showing extensive pattern of venation. (Carolina Biological Supply Company.)

FIGURE 2-2   Dicotyledon flower of the peach (*Prunus persica*) showing the five-petal arrangement.

FIGURE 2-3   Diagram of cross section of vascular system of dicotyledon and monocotyledon stem. (*a*) Dicotyledon with continuous vascular system; (*b*) monocotyledon with discontinuous vascular system.

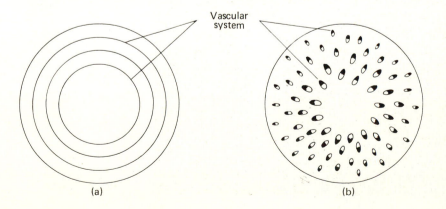

Vascular
system

(a)   (b)

FIGURE 2-4 Typical monocotyledon flower of the lily (*Lilium longiflorum*) showing flower parts in threes. (Courtesy of James Martin, Clemson University.)

Monocots have the following characteristics:

1. Principal veins of the leaves are parallel to each other.

2. Flower parts are usually in threes (Figure 2-4).

3. The root system is fibrous, without a taproot.

4. The stem has vascular bundles scattered irregularly through pithy tissue (Figure 2-3).

5. Stem and root are without a cambium and do not increase in girth by the formation of annual cylindrical layers of wood.

6. The embryo has one cotyledon.

Each of these classes is divided into *orders*. The Latin names of orders end in *-ales*, e.g., Rosales. An order is a group of closely related families with some common traits but some marked differences.

An order is divided into *families*. When the size of the family justifies it, and when the included genera may be naturally so grouped, the family is divided into *subfamilies*. The subfamily name is formed by adding *-oideae* to the stem of an included generic name. The family is the smallest of the major categories, but it usually represents a more natural unit than any of the higher categories. A family is a group of closely related genera or, in a few cases, a single genus with its name formed by adding *-aceae* to the stem of an included generic name—for example, *Rosaceae* from *Rosa* for the Rose family. The accent of the names of orders and families is placed on the first syllable of the termination *-ales* (pronounced a' les) for order and *-aceae* (pronounced a' se - e or "acey") for family. Thus it is Rosa'les (order) and Rosa'ceae (family). A *tribe* is a subdivision of a subfamily, and the tribe's name is formed by adding *-eae* to the stem of an included generic name. The Asteraceae (sunflower) family has 12 tribes whereas some subfamilies do not have any.

In the binomial system used today, the scientific name is represented by a Latinized form of the genus and species. The basic unit is the species which is a group of individuals of common ancestry with similar structure.

In writing the scientific name, the genus and species—such as *Viola canadensis*—are italicized if printed and underlined if handwritten or typed. An abbreviation of the name of the person who first gave this name to the plant is often written after the binomial, e.g. *Viola canadensis* L. Thus, the abbreviation L. following a plant name signifies that the botanist Linnaeus gave this name to the plant. If there were such a plant as *Enviola canadensis* L. Bailey, the name would indicate that Bailey had changed the name of *Viola canadensis* by Linnaeus to *Enviola canadensis*. Illustrations often accompany the original description; and the botanist, as a record, preserves in a herbarium a type specimen, which is the specimen on which the name was based. This plant specimen is dried and mounted on a sheet of herbarium paper. An *herbarium* is a collection of plant specimens that have been taxonomically classified, pressed, dried, and mounted on a sheet of herbarium paper. This specimen includes leaves, stem, flower buds, root, etc. A good herbarium normally contains several individual plants of each species. These individual plants represent the different stages of development and also the range of variation existing in the species. For the inexperienced person this approach may be misleading, since structural patterns and plant characteristics can be distorted by pressing and drying techniques. But for the experienced person it is an asset to have an accumulation of a large number of pressed specimens as an aid to plant identification. Should a dispute arise as to the application of a name, the type specimen can be compared with an herbarium specimen to determine if it is really different in important details from other species. Botanists engaged in taxonomic work usually have access to a large herbarium which they can use in comparing specimens to determine whether a new species really has been found.

The initial letter in the generic name is always capitalized, while in the specific name the initial letter is usually lowercase unless it honors a person. When the generic name cannot be mistaken or has already been referred to in a publication or paper, the genus can be abbreviated by using the initial letter and a period, *V. canadensis*. The names of the genera and species have diverse endings, with *-a, -ea, -ia, -i, -is, -um,* and *-us* being common. The International Code has established that only one group of plants can carry a particular genus name, and there can exist only one genus name per group of plants. For example, the genus name *Quercus* (oak) cannot also be used for any other group of plants and, if all the oaks are considered to be in a single group (genus), they must all carry the genus name *Quercus.* A species name, on the other hand, is often used more than once; for example the name *vulgaris* (common) is widely used—for example, *Phaseolus vulgaris* (bean), *Beta vulgaris* (beet) and *Lijaria vulgaris* (yellow toadflax). However, it can only be used once in any given genus.

The generic name is always a singular noun in the Latin nominative case. It may be descriptive, such as *Liriodendron* (lily or tulip tree); an aboriginal name of the plant, such as *Quercus*, which was the old Latin name for oak; or a name in honor of a person, such as *Linnaea* for Linnaeus. The species name may be an adjective with the same gender as the generic name, which usually indicates a distinguishing characteristic of the species, e.g., *Quercus alba*—white oak; a noun in opposition with the generic name, but agreement in gender is not necessary; a noun in the genitive singular, such as a species named in honor of a person, which may be either masculine or feminine and may or may not be capitalized, e.g., *Eugenia Hookeri*; or a plural genitive name indicating the species habitat, e.g., *Polygonum dumetorum*. Botanical Latin is discussed in more detail in Stearn's book of that title, cited in the bibliography for this chapter.

Horticulturists have used selective breeding in a given species in order to develop plants with specific characteristics for economic and scientific purposes. This has resulted in the recognition of cultivars. A *cultivar* is derived from a cultivated variety that has originated and persisted under cultivation, not necessarily referable to a botanical species, and is of botanical or horticultural importance, requiring a name. Cultivars are constantly being developed to improve the quality and yield of a crop and to permit wider climatic adaptation.

The term *cultivar* is used instead of *variety* to denote cultivated plants. This change in terminology resulted from the revised code of nomenclature adopted during the Fifteenth International Horticultural Congress in 1959, at Nice, France. Variety names used before 1959 will go unchanged, but cultivars named subsequent to that date are to be named in accordance with the provisions of the International Code. Cultivar names are now formed from not more than three words in a modern language and are usually distinguished typographically by the use of single quotation marks. Examples of cultivars are the 'Jonathan' apple and 'Seneca Chief' sweet corn, which are written as *Malus domestica* cv Jonathan and *Zea mays* cv Seneca Chief, thus making the nomenclatural system trinomial. The selections may vary from the original species in color, quality, size, time of harvest, yield, pest resistance, or a combination of these factors.

The cultivar may be quite distinctive outwardly, but it is usually a human creation or a selection from the wild. A cultivar may be a clone, abbreviated cl; a line; an assemblage; or a uniform group. A *clone* is composed of similar material derived from a single individual and propagated entirely by vegetative means (e.g., *Syringa vulgaris* cl 'Decaisine'). If a mutation has not taken place, the plants of a clone are genetically homogeneous (or identical) rather than carrying the genetic variation inherent in plants propagated by true seed. A *line* consists of plants of uniform appearance, which—because of its horticultural value—can be reproduced uniformly by seed (e.g., *Petunia* 'Rosy Morn'). An *assemblage* consists of individuals reproducing from seeds that show some genetic differences but have one or more characteristics by which the plants can be differentiated from other cultivars (e.g., *Phlox drummondi* 'Sternenzauber', a mixture of different color patterns, all characterized by the same starlike corolla). A *uniform group* is a first-generation hybrid (F_1) resulting from a cross of two other cultivars that can be reconstituted on each occasion by crossing two constant parents which are maintained either by inbreeding or as clones (e.g., *Allium sativum* 'Granex'). Hybrids are distinguished by an "x" between the generic or specific names, e.g., *Forsythia* x *intermedia*, or between the specific names of the two parents, e.g., *Daphne burkwoodii* x *retusa*.

The term *form* is used when a member or members of a population differ from other members to a degree not great enough to be called a cultivar. This results when an individual horticultural plant differs genetically. These deviations are considered only if they are beneficial to human beings and can be asexually propagated, thereby establishing a clone, to ensure that the desired characteristic can be retained. An example is the Starking 'Delicious' apple, which was a mutation from the original 'Delicious' cultivar.

Occasionally, one will find a plant with a generally recognized common name listed in two different publications under differing generic names, specific names, or both. Differences of opinion as to the proper botanical names for certain genera, species, and cultivars have always existed and will continue to be a problem despite the detailed rules established under the International Code of Botanical Nomenclature. One basic problem is the varying interpretation of the species. One taxonomist may believe that a group of differences are merely the *natural variations* to be found within a species, while another will proclaim that these variations are the distinguishing characteristics of two or more distinct species. These differences are usually resolved over the years by consensus or by more detailed study.

HOW PLANTS ARE PLACED IN TAXONOMIC CLASSIFICATIONS

There will always be variation with a species. These variations can be due to environmental differences in the amount of water, nutrients, or light a plant receives; to genetic differences which can occur randomly, giving a difference in color or size; or to hybridization, whether naturally occurring or induced by humans. Because of the variations, plants within a species can appear to be very different. Therefore, careful study of any plant is necessary to ensure that the plant is correctly identified either as a member of an established species and genus or as a new species. A plant is identified by determining if it is identical to or similar to an already classified plant. In other words, one determines if the taxon is similar or identical to another already established element.

Several methods are used in plant identification. The method most frequently used is to compare the plant to be identified with descriptions and illustrations from plant manuals which include a listing in appropriate order of the plant families of the principal taxonomic divisions. Some of the manuals cited in the bibliography which are most useful in identifying plants are Bailey's *Manual of Cultivated Plants* and *Hortus Third*, Rehder's *Manual of Cultivated Trees and Shrubs*, and Gray's *Manual of Botany*. Manuals describing the flora of specific states or regions are also useful, especially if the unknown plant occurs in a definite locality, in which case the local manual may be more authoritative. Regional manuals are easier to use since they contain only the plants for given regions and may provide more information than a flora covering a larger area. Plants can also be identified by comparing them with herbarium specimens. A third method of identification of plants is by the use of plant keys. A *plant key* is an analytical device whereby a choice between two or sometimes more contradictory propositions has to be made in each step. The worker selects the choice fitting the material and proceeds through the outline in this manner. In using a key, the user is presented with a sequence of choices, usually between two statements; i.e., the key is dichotomous, and by taking the series of correct choices one arrives at the name of the unknown plant. Keys based on successive choices between only two statements are used most frequently. In this type of key, each statement, called a *lead*, is identified by a letter or number; and each successive, subordinate *yoke* is indented directly under the preceding one. Sometimes these are bracketed instead of indented.

Plant identification with a key is an orderly process. For example, a rose may be placed first among the plants with buds, then among the seed plants having flowers, and so on from one subordinate group to another until it is finally associated with its nearest relatives. In the preparation of keys for woody plants, it is desirable to use vegetative characteristics to the greatest extent possible.

To illustrate the use of a plant key, suppose that some horticulturists are visiting the mountains in Tennessee and find an oak that they cannot immediately identify. The leaves are hairy with six lobes per leaf. They are narrow at the apex and do not have bristle tips on the lobes. The horticulturists know that the tree is an oak because it has alternate, simple leaves and has a fruit they identify as an acorn, a characteristic of the genus *Quercus* (oak). By using the observed plant characteristics, they can identify the plant through the use of a plant key, as shown in Table 2-2:

1. The two beginning "lines" or "leads" of the key, which begin at the extreme left margin and are numbered 1, make two opposing statements. The plant "fits" one, but not the other. The key is followed by going to the number indicated by the statement the plant fits. Since the leaf is not widest at the apex, the plant must fit the first lead.

2. The subordinate leads under the "1" are indented from the left margin and numbered "2." Again, a choice has to be made between two contrasting statements, which will direct the user to another pair of statements, indicated by a number appearing in the right-hand margin opposite the correct statement.

The user continues to follow the key in this manner, making choices between opposing statements until a name, instead of a number, appears in the right-hand column opposite the statement the plant fits. At this point, the specimen would be identified. In this case, the horticulturists would be led to *Quercus stellata*, the post oak.

Regardless of which method of identification is used, it is important

TABLE 2-2
A KEY TO THE SPECIES OF OAKS OCCURRING IN THE MOUNTAINS OF THE SOUTHEASTERN UNITED STATES

1. Leaves not widest at apex   2
    2. No lobes in leaves: *Quercus phellos* (willow oak)
    2. Lobes in leaves   3
        3. Bristle tips on leaves   4
            4. 3–5 lobes/leaf, hairy: *Quercus falcata* (Southern red oak)
            4. 5–9 lobes/leaf, smooth: *Quercus velutina* (black oak)
        3. No bristle tips on leaves   5
            5. 7–11 lobes/leaf, smooth: *Quercus alba* (white oak)
            5. 3–7 lobes/leaf, hairy: *Quercus stellata* (post oak)
1. Leaves widest at apex. *Quercus marilandica* (blackjack oak)

to become familiar with the characteristics and variations in form and size exhibited by the leaves, flowers, fruits, and seed. These morphological structures are visible to the eye; whereas, reproductive structures often have to be dissected or examined with a hand lens or dissecting microscope.

Leaves can be useful in identifying taxa. Care must be taken to note such characteristics as leaf arrangement, leaf type, and leaf margin.

Leaf arrangement refers to the arrangement of leaves on the stem. The leaves appear to be *alternate* on the two sides of the stem in some plants (e.g., holly). Some plants have two leaves directly across from each other at each node, giving an *opposite* arrangement (e.g., boxwood). In a few types of plants, three or more leaves appear at a node; these are described as *whorled* (e.g., barberry).

The leaf type, whether compound or simple, the leaf margin, whether serrate or entire, and the leaf texture, or feel, are useful in identifying plants. *Leaf type* refers to whether the leaf is simple or compound. A leaf whose blade is a single segment is *simple* (Figure 2-5). A maple leaf is simple but palmate. A leaf with the blade divided into several leaflets or sections is *compound*. A compound leaf with the leaflets arranged along both sides of the midrib is *pinnate* (like a feather), e.g., rose or walnut (Figure 2-6). A compound leaf with all the leaflets arising from one point at the end of the petiole is *palmately compound* (like the fingers on a human hand), e.g., buckeye or five-leaf ivy (Figure 2-6).

There is a great diversity in *leaf margins* (Figure 2-7), but there are only three basic types. Leaves with a *smooth* margin are entire and lack any indentations in the margin, e.g., boxwood. *Serrated* leaves have serrations or "teeth" along the margin of the leaf, e.g., beech (Figure 2-5). Some leaves are *lobed* and have very deep indentations, e.g., maple (Figure 2-5).

Leaf texture may be membranous or leathery; and the surface may be smooth, hairy (pubescent), or rough, like sandpaper (scabrous).

The Latin names for many plants are often descriptive of the plant characteristics; see Table 2-3 and Figures 2-8, 2-9, and 2-10.

Table 2-4 lists the most consistent identifying characteristics of plants. Other characteristics which may be used are leaf form, leaf apex, whether leaves are petioled (with a stalk) or sessile (without a stalk), and pith characteristics of twig or stem. Flowers and flower structures can also be used to identify the family and genus of plants; in fact, these structures are the best characteristics for this. Vegetative characteristics are generally useless in this regard. With a basic knowledge of flowers, natural classifications prove to be most helpful. In angiosperms, the basic patterns of flower and fruit structure commonly distinguish orders, families, and genera. Since reproductive

(a)

(b)

FIGURE 2-5 Leaf types: (*a*) Beech (*Fagus grandifolia*) twig with simple, pinnate slightly serrate leaves. Note the uniformity of the lateral veins which extend slightly beyond the blades' margins as short spines. (*b*) Simple palmately veined leaf of red maple (*Acer rubrum*); this is both serrate and lobed. (Carolina Biological Supply.)

(a)

(b)

FIGURE 2-6 Leaf types: (*a*) Rose (*Rosa* sp) leaf; an example of a pinnately compound condition. Venation is judged on the basis of the arrangement of leaflets on the petiole or midrib, not on the basis of the lesser venation within the individual leaflets. Note serration along margins of leaflets. (*b*) Buckeye (*Aesculus* sp) leaf. This exemplifies the palmately veined, compound leaf. The uniformly pinnate venation of each of the several leaflets sometimes results in incorrectly considering the entire leaf to be pinnate. (Carolina Biological Supply Company.)

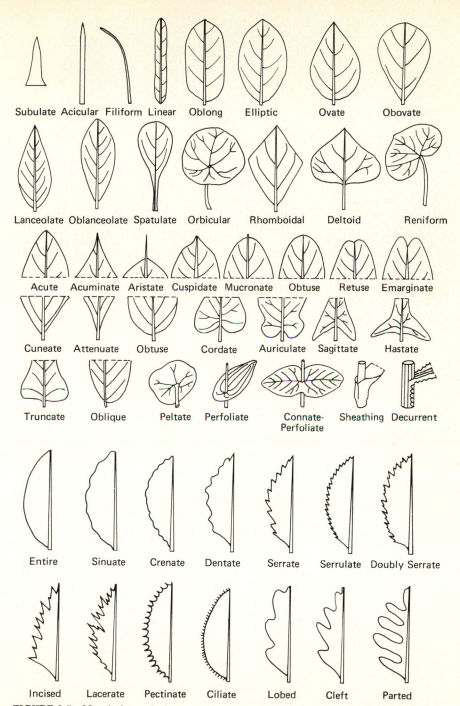

FIGURE 2-7 Morphological shapes of leaves used in plant identification. (Redrawn from *Hortus Third*, Macmillan, New York, 1976. Reprinted with permission of Macmillan Publishing Co., Inc. Copyright © 1976 by Cornell University for its L. H. Bailey Hortorium.)

TABLE 2-3
LATIN TERMS FOR PLANT CHARACTERISTICS

SEASONS OF THE YEAR

aestivalis—summer
autumnalis—autumn
hyemalis—winter
vernalis—spring

EMPHASIS, DEGREE, OR KIND

PREFIXES: SUFFIXES:
atro-—dark   *-bondus*—abundant
semper-—ever, always   *-escans*—resembling
sub-—somewhat   *-ferus*—bearing
                                   -issimus—very
                                   -oides—similar to
                                   -osus—with, bearing
                                   -ulus—somewhat

SIZE AND SHAPE (CHIEFLY AS PREFIXES IN COMPOUND WORDS)

alti-—tall   *gracilis-*—slender   *mega-*—large
angularis—angular   *grandi-*—large   *micro-*—small
angusti—narrow   *lati-*—wide   *-minimus*—very small
brevi—short   *longi-*—long   *ortho-*—straight
elongatus—elongated   *macro-*—large   *parvi-*—small
giganteus—huge   *maximus-*—very large   *tenui-*—slender

REGIONS OR HABITATS

aero—of or in air   *borealis*—northern   *rupestris*—rock-loving
agrarius—of fields   *montanus*—of mountains   *saxatilis*—rock-loving
agrestis—of fields   *occidentalis*—western   *sylvaticus*—of woods
alpinus—alpine   *oceanicus*—of the sea   *sylvestris*—of woods
aquaticus—aquatic   *orientalis*—eastern   *terrestris*—of earth
australis—southern   *riparius*—of river banks
Note: Place names, such as *americanus* and *germanicus,* can be derived.

CHARACTER, FORM, AND HABIT

acuminatus—tapering   *columnaris*—columnar   *divaricatus*—spreading
  (Figure 2-8)   *communis*—common,   *erectus*—upright
acutus—sharp-pointed   general   *esculentus*—edible (Fig-
alatus—winged   *cornutus*—horned   ure 2-9)
amabilis—lovely   *crenatus*—scalloped   *eximius*—distinguished,
annuus—annual   *cristatus*—crested   unusual
arborescens—woody,   *cultorum*—of cultivated   *filiferus*—having threads
  treelike   type   *flore-pleno*—double-
barbatus—barbed,   *cuspidatus*—stiff, pointed   flowered
  bearded   *deciduus*—deciduous   *floridus*—flowering
biennis—biennial   *decumbens*—bent down   *fragilis*—fragile
capitatus—headed   *densus*—dense   *fruticosus*—shrubby,
carnosus—fleshy   *dentatus*—toothed   bushy
coloratus—colored   *dissectus*—deep-cut   *glaber*—smooth

TABLE 2-3 (Continued)

CHARACTER, FORM, AND HABIT

glaucus—having a bloom
glomeratus—clustered
graveolens—heavy-scented
heterophyllus—with variously shaped leaves (Figure 2-10)
hortensis (hortorum)—garden type
hybridus—mixed, hybrid
incanus—hoary
laciniatus—torn
lanatus—woolly
lanceolatus—lancelike
marginatus—with margins
medius—intermediate
mollis—soft
muralis—of (for) walls
mutabilis—variable
nanus—dwarf

nudus—bare
nutans—nodding
odoratus—fragrant
officinalis—medicinal
paniculatus—panicle-shaped
pendulus—hanging, weeping
perennis—perennial
pleniflorus—double-flowered
plumosus—plumy, feathery
praecox—very early
procumbens—trailing
pumilus—dwarf
pungens—sharp-pointed
pyramidalis—pyramidal
racemosus—flower in racemes
radicans—rooting
recurvus—recurved

reflexus—reflexed
repens, reptans—creeping
rotundifolia—round foliage
scandens—climbing
serratus—sawtoothed
sessilis—stalkless
speciosus, spectabilis—showy
spicatus—with spikes
suaveolens—sweet-scented
suspensus—hanging, weeping
trivialis—common
unbellatus—bearing umbels
variegatus—variegated
vegetus—vigorous
vulgaris—common
zonatus—banded

PARTS OF PLANTS

andrus—stamen
anthus—flower
carpus—fruit
caulis—stem
florus—flower

folius—leaf
lobus—lobe
peas—foot
petalus—petal
podus—foot, stalk

phyllus—leaf (see Figure 2-10)
rhyzus—root
sepalus—sepal
spinus—spine
squamus—scale

COLOR

albus—white
argenteus—silver
aureus—golden
azureus—sky blue
caeruleus—dark blue
candicans—white
candidus—pure white
chromus—color
chryseus—yellow

citrinus—lemon yellow
coccineus—scarlet
coelestinus—sky blue
croceus—saffron
cyanus—blue
flavus—yellow
fulvus—tawny
incarnatus—flesh toned
lilacinus—lilac

luteus—yellow
niger—black
purpureus—purple
roseus—rosy
ruber—red
sanguineus—blood red
violaceus—violet
viren, viridus—green
xanthinus—yellow

NUMBERS AND QUANTITY

uniflorus, monophyllus—1
biflorus, diphyllus—2
trilobus—3
quadrifolius, tetraphyllus—4

quinqueflorus, pentanthus—5
hexaphyllus—6
heptaphyllus—7
octopetalus—8
enneaphyllus—9

decapetalus—10
centifolius—100
hetero-—various, (see Figure 2-10)
multi-—many
pauci-—few
poly-—many

FIGURE 2-8 *Musa acuminata* (*Musa x paradisiaca*—hybrid bananas). Giant treelike herb forming pseudostems up to 6 m. The leaves may be up to 3 m long; the yellow fruit is curved and forms a point at each end (*acuminata*—tapering gradually or abruptly from inwardly curved sides into a narrow point).

organs are less frequently affected by physical factors of the environment, they prove to be an ideal and stable basis for identifications. The deliniation of families and orders of flowering plants has been fairly well worked out by taxonomists. For example, a lily has three petals and an apple has five petals. The lily is a monocot and the apple is a dicot.

COMMON CLASSIFICATIONS

Some horticultural groupings are based on length and season of growth (annual, biennial, and perennial), form (tree or shrub), and use

FIGURE 2-9 *Lycopersicon esculentum* (tomato)—a perennial grown as a warm-season annual or in a greenhouse for its edible fruits (*esculentum*—edible). (Photograph, Joseph Harris Company, Rochester, New York.)

FIGURE 2-10 *Osmanthus heterophyllus* (holly osmanthus) —an upright evergreen shrub with an irregular shape. The leaves are variable in shape and size (*heterophyllus*—variously shaped leaves). (Courtesy of Benjamin T. Hendricks, Clemson University.)

TABLE 2-4
CONSTANCY OF IDENTIFYING CHARACTERISTICS, HIGHEST TO LOWEST*

Leaves needle (pine) or scalelike (juniper) versus leaves broad and laminate (magnolia)
Venation parallel (corn) versus venation reticulate (netted) (oak)
Leaves compound (mimosa) versus leaves simple (maple)
Plants erect trees (oak) or shrubs (blueberry) versus plants vinelike (English Ivy)
Stipules present (peas) versus stipules absent (petunia)
Leaves lobed or divided (maple) versus leaves neither lobed nor divided (camellia)
Leaves opposite (boxwood) or whorled (pachysandra) versus leaves alternate (holly)
Leaf margins entire (boxwood) versus leaf margins variously toothed (barberry)
Leaves persistent and coriaceous (podocarpus) versus leaves deciduous (dogwood)
Spines or prickles present (pyracantha) versus spines or prickles absent (ligustrum)
Leaf base tapering (hickory) versus leaf base broad (begonia)
Leaves with one main midvein (oak) versus leaves with three primary veins (maple)

*Order approximate.

(fruits, vegetables, or ornamentals). Horticulturists use the botanical classification in conjunction with an artificial but convenient system they have developed. This system often groups plants that are vastly different; its chief justification is the convenience of classification based on actual usage in gardening.

As the name indicates, annuals complete their life cycle in one year or less. These plants germinate from seed, produce vegetative growth, flowers, fruits, and seed, and die within one growing season. Examples of annuals are marigold, pea, bean, corn, squash, pumpkin, ageratum, and garden cress.

Biennial plants ordinarily require two years or at least part of two growing seasons with a dormant period between growth stages to complete their life cycle. The first year is used for building up reserves to allow the plant to flower and produce seed in the second year. Only vegetative growth is produced the first year, and the following spring or summer the plant produces flower, fruit, and seed, after which it dies. Examples of biennials are celery, sweet William, cabbage, carrot, and beet.

Perennials are plants which do not die after flowering but live from year to year. Perennials are further divided into herbaceous and woody perennials.

Herbaceous perennials have soft, succulent stems whose tops are killed back by frost in many temperate and colder climates, but the roots and crowns remain alive and send out top growth when favorable growing conditions return. Examples are rhubarb, asparagus, oriental poppy, chrysanthemum, and many cultivars of phlox.

Woody perennials contain woody fibers and are longer-lived and more durable. These perennials are classified according to form and habit of growth as trees, shrubs, or vines, although not all of them can be clearly classed in this manner. Trees are upright in growth, with stems or trunks forming the central axis or the main part of the framework. Shrubs do not have a central axis or predominant trunk but have a number of stems radiating from the root crown. The basic difference between shrubs and trees, then, is form, not size. Vines have stems that are too slender and flexible to support their branches and leaves in an erect position. Therefore, vines must grow on the ground or climb support objects, such as trees. Woody perennial plants are also classified into zones for hardiness.

The rules for artificial classification do not work consistently. Carrots, which are biennials, are often classed as annuals since they occupy the ground for only one growing season when they are used in vegetable production; however, if these same plants are grown for seed, they must remain for two seasons. Using special cultural treatment, many biennials can be forced into flowering, fruiting, or developing seeds in their first growing season. For example, young celery plants can be forced to "bolt" if exposed to low temperature when only a few weeks old—that is, to produce flowers and seed and then die before the end of the season. Eggplant and snapdragon are perennials and will produce a continuous succession of flowers, fruits, and seed; but they are grown most often as a single-season crop and are thus also called annuals.

Classification systems should be as natural as possible, but they should also be convenient for the user. A system of classification is of no value if it is impractical. Thus classifying plants by size, as is done by landscape designers, home gardeners, and nursery growers, may be a purely mechanical device, which groups diverse plants together under one heading; but the frequency and ease of use of the classification justifies to a certain degree its apparent disregard of botanical relationships. Also, classifying a plant as a fruit, a vegetable, or an ornamental is somewhat confusing, as it sometimes appears to overlook the botanical definition of a fruit. This system is based instead on the commodity use to which the plant is put. Certain plants are thought of as producing "fruits," and others "vegetables," because of the way they are grown, prepared, or eaten. Other plants are used for their aesthetic qualities as landscape ornamentals and not for their food value.

Horticulture is an old and venerable field of endeavor and as such contemporary horticulturists must put up with the "growing pains" and inaccuracies of the naming systems. It is like an old city: disorganized and ill-designed for modern life, but with comfortable old ways that "feel" right.

However, taxonomy and botanical systems of classification are gaining increased importance in the field of horticulture. The number of plants in horticulture is such that an orderly system of classifying and arranging is necessary to ensure accurate identification and classification of any given plant. The world has become too small for horticulturists to be using common names.

The taxonomic system of classification of plants serves as a common language by which horticulturists throughout the world can communicate with and understand one another. This common language is important in avoiding confusion and duplication. It provides each plant with one botanical name which belongs solely to it and to others of the same species. This system makes each plant unique, and it allows other members of the same species to be quickly identified.

BIBLIOGRAPHY

Bailey, L. H.: *How Plants Get Their Names*, Dover, New York, 1933.

Bailey, L. H. and E. Z. Bailey: *Hortus Third: A Concise Dictionary of Plants Cultivated in the United States and Canada*, Macmillan, New York, 1976.

Bailey, L. H.: *How Plants Get Their Names*, Dover, New York, 1933.

Bell, C. R.: *Plant Variation and Classification*, Wadsworth, Belmont, Calif., 1967.

Benson, L.: *Plant Classification*, Heath, Boston, 1959.

Benson, L.: *Plant Taxonomy, Methods and Principles*, Ronald, New York, 1962.

Davis, P. H. and V. H. Heywood: *Principles of Angiosperm Taxonomy*, Oliver & Boyd, Edinburgh, 1963.

Gray, A.: *Gray's Manual of Botany*, 8th ed., American Book, New York, 1950.

Heywood, V. H.: *Plant Taxonomy*, 2d ed., E. Arnold, London, 1976.

Lawrence, G. H. M.: *Taxonomy of Vascular Plants*, Macmillan, New York, 1951.

Porter, C. L.: *Taxonomy of Flowering Plants*, 2d ed., Freeman, San Francisco, 1967.

Radford, A. E., W. C. Dickison, J. R. Massey, and C. R. Bell: *Vascular Plant Systematics*, Harper & Row, New York, 1974.

Rehder, A.: *Manual of Cultivated Trees and Shrubs*, 2d ed., Macmillan, New York, 1940.

Stearn, W. T.: *Botanical Latin*, Hafner, New York, 1966.

Zielinski, Q. B.: *Modern Systematic Pomology*, Brown, Dubuque, Iowa, 1955.

CHAPTER 3
PLANT CELLS

 The vascular plant is composed of different tissues or groups of tissues, which are themselves composed of large numbers of similar cells. Cells are the smallest units of living matter capable of continued independent life and growth; each is a self-contained and at least a partially self-sufficient unit bounded by a cell wall. The life of the plant depends on the flow of energy from the sun through the living cells and back into the environment.

CELLS

Horticultural plants show a wide diversity in form, size, and structure. These structural and compositional differences are revealed when the pecan tree and watermelon or the maple tree and ivy are compared. In spite of these differences, all plants are composed of a basic structural and functional unit—the cell. Initially, all plant cells are basically similar in their physical organization and biochemical properties; but as these cells mature and differentiate, each cell develops into an individualized unit in form, function, and size. In a maple tree, for example, some cells manufacture food, some store food, some transport water or food, some give strength to the tree, some prevent water loss, some have the sole function of dividing to make new cells, some are large, some are small, some are long, and some are short.

History of the Study of the Cell

Robert Hooke first observed the plant cell as a structural unit in 1665, using a microscope with improvements he had developed himself. In his book *Micrographia: Or Some Physiological Descriptions of Minute Bodies Made by Magnifying Glasses*, he showed a sketch of a thin section of a piece of cork which had "little boxes or units distinct from one another . . . that perfectly enclosed air." He called these "little boxes" *cells*. He estimated that 16.4 cm³ of cork contained 1259 million cells. Hooke further concluded that plants are not homogenous but contain cells of various sizes, shapes, and appearances. It was not until the nineteenth century, after much study by other scientists of cell parts and functions, that the term *cell* gained common use. In 1838 Matthias Schleiden, a German botanist, and in 1839 Theodor Schwann, a zoologist, summarized the known cellular studies and formally proposed that the cell was the basic unit of all living things. Yet it took another 30 years of research before this theory was generally accepted by the scientific community.

Since Schleiden and Schwann, technology has enabled us to learn detailed information about cells and their component parts. Even though scientists have developed extremely complicated equipment in the fields of biochemistry, physics, and cytology, the functions and processes still are not totally understood today. Every cell contains several different types of very minute bodies that are collectively called organelles (Figure 3-1). The development of the electron microscope enabled scientists to study these organelles of the cell in detail. When one considers the complexity of the cell, it is not hard to understand why biologists took several centuries to discover the life processes that occur in the cell.

Size and Shape of Cells

Plants contain millions and even billions of cells, each of which is partially independent. A typical apple leaf, for example, is made of approximately 50 million cells. Plant cells are minute, ranging between 0.1 and 0.01 mm in diameter, and thus a microscope is required to study and examine them. Plant cells vary greatly in shape and size, which are generally related to function. The embryonic cells of the root tip are cubical in shape with angular corners. The food- or water-conductive cells are elongated cylinders. The cells that provide for translocation of foods in the plant range in length from 2 to 7 mm, whereas cells of which the supporting tissues are composed may be only 2 mm long.

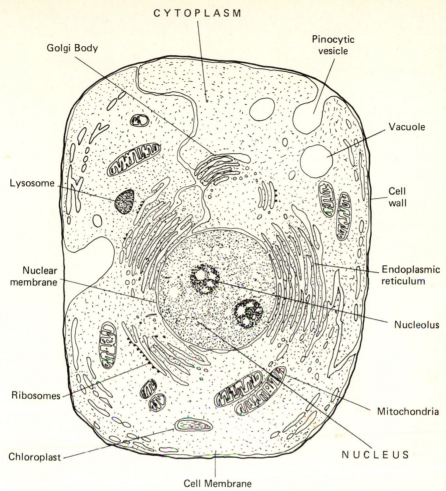

CYTOPLASM

Golgi Body

Pinocytic
vesicle

Vacuole

Lysosome

Cell
wall

Nuclear
membrane

Endoplasmic
reticulum

Nucleolus

Ribosomes

Mitochondria

Chloroplast

NUCLEUS

Cell Membrane

FIGURE 3-1 Cell and organelles reconstructed from electron micrographs. (Adapted from Jean Brachet, "The Living Cell," *Scientific American*, 1961. Copyright © 1961 by Scientific American, Inc. All rights reserved.)

Structure and Function of Cells

Plant cells, which are three-dimensional (Figure 3-2), are bounded by a membrane and, unlike animal cells, also by a rigid cell wall. The cell is made up of protoplasm, which in turn is composed of cytoplasm and organelles that carry on specific functions (Figure 3-1). Each of these cell components has a particular structural framework, chemical composition, and specialized function. One of the organelles—the cell nucleus—contains the chromosomes, which are the genetic blueprints for all cell functions.

All the activities of an organism ultimately result from the function and activity of single cells. Cells contain chemical elements and

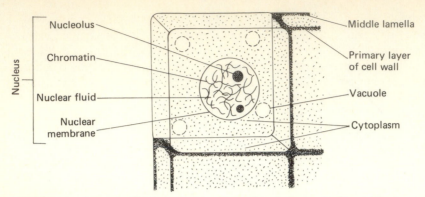

FIGURE 3-2 Diagram illustrating three-dimensional shape of cells and basic components of cell as seen with light microscope. (Courtesy of R. P. Ashworth, Clemson University.)

compounds that range from simple substances, such as water, oxygen, and carbon dioxide, to highly complex pigments, carbohydrates, proteins, and hormones. Cells reproduce, assimilate, respire, respond to changes in the environment, and absorb water and other materials from the immediate external environment. They are, however, mutually interdependent.

Lower plants, such as algae, may be composed of only one cell or of many cells, but the cells in these plants lack the diversity of size, shape, and function exhibited by cells of the higher plants. Lower forms of plants are composed of cells with similar structure. In more complex organisms, such as flowering plants, the cells show diversity in structure and function resulting from the process of *cell differentiation*—the changes a cell undergoes during growth, which result in a cell with a specialized form and function. Eventually the various differentiated cells create the entire plant.

Plant growth is defined as a permanent increase in volume, dry weight, or both. Cells increase in number by division and increase in size by enlargement. For cell division to occur, an increase in protoplasm must occur. This increase takes place biochemically through a series of events in which water, organic salts, and carbon dioxide are transformed into living material. In the cell, this process necessitates the uptake of water and minerals, the production of carbohydrates, and the formation of complex metabolites, such as proteins and fats. The energy required for growth is provided by cell respiration.

Cell Wall

The cell wall gives protection, support, and form to the plant cell, and is the basis for the overall support and form of the plant as a whole. The cell wall, which itself is nonliving, is produced by the living protoplasm.

It is composed of distinct layers, and its thickness varies greatly with age, type of cell, and function. The primary cell wall, which is formed during early stages of cell growth, is composed of layers of cellulose and hemicellulose. It is thin, elastic, plastic, and completely permeable and is stretched as a result of turgidity of vacuolar fluid and action of a cellulase which breaks down some cellulose, apparently followed or accompanied by synthesis of new cellulose to accommodate increase of cell size. Many cells of roots, fleshy stems, and leaves possess only this primary cell wall and depend on turgor pressure for rigidity.

A secondary cell wall is present in certain mature cells. This secondary wall is laid down on the inner surface of the primary wall after the cell has accomplished its maximum expansion or elongation. Typically, with the synthesis of the secondary cell wall, lignification occurs. This involves the deposition of a group of complex carbohydrates, collectively known as *lignin*, among the cellulose fibrils of the cell wall. As a result, such a lignified cell wall is impermeable, so that molecules of various substances are unable to enter or leave the cell. Cells with secondary walls are made rigid by cellulose microfibrils, lignin, and hemicellulose. These cells provide the mechanical support for the plant but can also exist in nonsupportive tissue.

Nonlignified cell walls in plants contain extremely small pores or pits through which cytoplasmic strands, *plasmodesmata*, extend. These strands provide living connections between cells. The cytoplasmic bridges facilitate the movement of materials such as water, minerals, photosynthates, and growth regulators from one cell to another. When lignification of the walls occurs, these pits are plugged and the plasmodesmata are severed.

The *middle lamella*, a viscous, jellylike substance, consisting of calcium and magnesium pectates, binds the primary cell wall of one cell to that of the adjacent cell. Chemically the pectic compound of the middle lamella consists of long chains of galacturonic acid or the calcium or magnesium salts of this acid. As fruits ripen, this long chain begins to break up. As it breaks up, the middle lamella is weakened, the cells begin to separate, and the fruit gets soft. Commercial preparations of pectin obtained from fruits are used in jellies to ensure "jelling."

Protoplast

The *protoplast* is the living unit inside the cell wall. It is composed of *protoplasm*, the living substance of the cell, and the plasma membrane, which is a differentially or selectively permeable, living, flexible membrane essentially composed of lipids and proteins. This membrane regulates the passage of water and organic and inorganic materials into and out of the protoplast.

Protoplasm is composed of water, protein, carbohydrates, fats, and inorganic salts. It is usually viscous, like raw egg white, but it may contain various pigments and be colored. Within the protoplasm, all functions necessary to sustain life, such as absorption, excretion, metabolism, growth, and reproduction, are carried on. If cells lose their protoplasm, they can serve only as conducting or supporting structures for the plant.

Cytoplasm   Enclosed by the plasma membrane is the cytoplasm, the living matter or physical substance of the cell. The cytoplasm consists of a fine structure involving an extensive and complex system of membranes and specialized bodies called *organelles*. Each cytoplasmic component is unique and has a specific structural framework, chemical composition, and function.

Endoplasmic Reticulum   The endoplasmic reticulum extends throughout entire units of cytoplasm (Figure 3-1). It functions in the transport of cell products, in serving as a surface for protein synthesis by the ribosomes, in the separation of enzymes and enzyme reactions, in support, and in the moving of cell membrane components into position, as in cell division. The endoplasmic reticulum is continuous with the nuclear membrane, the outer, bounding membrane of the nucleus.

Ribosomes   Floating in the cytoplasm and associated with the membranes of the endoplasmic reticulum are small, dense, globular particles called *ribosomes*. These ribosomes contain ribonucleic acid (RNA) and proteins. The synthesis of proteins, including the enzymes, takes place on these particles.

Plastids   Plastids, small cytoplasmic inclusions, are formed from proplastids, minute granules which are transmitted by the female reproductive cells. Plastids are organelles involved in food synthesis; in the storage of fats, starches, proteins, and various pigments; or in both. They are classified by their pigment. *Chloroplasts* (green), *chromoplasts* (yellow, red, and orange), and *leucoplasts* (colorless) originate from common proplastids; but they are not similar when mature and functional. For example, the chloroplasts of young fruits and developing petals may become the chromoplasts of the ripe fruit and mature flower. A leucoplast becomes a chloroplast upon being exposed to light.

The chloroplasts are dominant in the leaves and other green parts. These plastids are usually uniform in size and are ellipsoid or disklike in shape (Figure 3-3). Chloroplasts in a plant exposed to full sun are usually somewhat smaller than those in a similar plant in the shade. Chloroplasts are composed of proteins and lipids and contain the

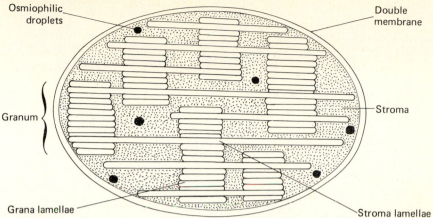

Osmiophilic droplets   Double membrane

Granum

Stroma

Grana lamellae   Stroma lamellae

FIGURE 3-3 Chloroplast showing a large number of disk-shaped grana lamella. Stacks of grana lamella make up a granum. (Courtesy of R. P. Ashworth, Clemson University.)

green chlorophylls a and b and the yellow or orange carotenoid pigments. The carotenoids are overshadowed by the chlorophyll and do not become evident until the chlorophyll has deteriorated. The color of dying leaves in autumn is created by the breakdown of chlorophyll, which is followed by the predominance of the carotenoids. The same process may be seen in the ripening of sweet peppers and tomatoes.

In each palisade cell of a green leaf there are 20 to 100 chloroplasts. In addition to chlorophyll, chloroplasts contain the enzyme energy compounds (adenosine triphosphate-ATP and adenosine diphosphate-ADP) and intermediate substances required for transfer of the sun's energy to chemical energy. This is accomplished by means of photosynthesis, which produces sugar from carbon dioxide and water. Each chloroplast has a large number of disc-shaped *grana lamellae* (Figure 3-3). Stacks of grana lamellae make up a granum. Chlorophyll, which occurs as molecules on the granum surface, receives light energy from the sun. The actual transformation of carbon dioxide to carbon-containing compounds occurs in the adjacent *stroma*, the granular portion of the chloroplast.

Yellow and orange flowers, fruits, and seeds lack green chlorophyll when mature. Instead, they owe their color to the chromoplasts. These variously shaped plastids synthesize and retain yellow (carotene), red (xanthophyll), and orange (carotenoids) carotenoid pigments, rather than the chlorophyll. Chromoplasts can originate directly from proplastids or from transformed chloroplasts in which the internal membrane structure and the chlorophylls have been destroyed.

Leucoplasts are nonpigmented organelles which vary in shape. These plastids are usually elastic and serve in the synthesis and storage of starch, oils, and proteins. Leucoplasts are located in colorless leaf cells (variegated leaves), stems, roots, and other storage organs. The underlying tissues of the plant that are not exposed to light, as well as tissues of the root systems, contain numerous leucoplasts.

Mitochondria The mitochondria are organelles which are much smaller than the plastids. They vary from spherical- to rod- or filament-shaped bodies. Each mitochondrion is made up of a stroma of proteins (including enzymes) and phospholipids and is bounded by a two-layered membrane. The mitochondria serve as the cell's "power-house" and are the site of the oxidation of organic molecules which release energy and of the conversion of the energy to molecules of adenosine triphosphate (ATP), the main chemical energy source for all cells.

Dictyosomes Golgi bodies, or dictyosomes, are groups of flat, disk-shaped sacs formed by cytoplasmic membranes which are often branched into a complex series of tubules. They develop numerous vesicles which serve as a source of plasma membrane materials. The Golgi bodies are also collection centers for complex carbohydrates and they probably have other functions.

Microtubules Microtubules are located in the cytoplasmic matrix of nondividing cells and in the spindle fibers of dividing cells. These may be involved in the growth of the cell wall, cell plate development, and mitosis. Microbodies, spherical organelles surrounded by a single membrane, serve an important role in glycolic acid metabolism associated with photosynthesis. Other microbodies have the enzymes necessary for the conversion of fats into carbohydrates during germination in many seeds.

Lysosomes These are single-membrane organelles in the cytoplasm. They contain hydrolytic enzymes capable of digesting other cellular particles.

Vacuoles In young meristematic cells, very small vacuoles are numerous, but the protoplasm fills most of the space within the cell wall. As the cells grow and mature, the numerous minute vacuoles in the cytoplasm fuse into a large vacuole which increases in size until it occupies most of the cellular space. In most mature cells, the nucleus, plastids, and inclusions are contained in the peripheral layer of cytoplasm with a large vacuole in the center. A vacuole is not a vacuum; it is filled with cell sap composed of a water solution of inorganic salts, various organic solutes, and crystals. It may also contain the anthocyanin pigments which are responsible for most of the red, blue, and purple colors of flowers, fruits, and autumn leaves. The red color of roses is a result of a concentration of pigments in the vacuoles of the flower petals' cells. Some cells also contain oil vacuoles. The vacuole is surrounded by a membrane called the *tonoplast.*

Nucleus   Cells of higher plants contain a true nucleus. This is proportionally small, usually spherical or disk-shaped, and surrounded by a nuclear membrane. In young cells the nucleus is centrally located, but as the cells mature, the vacuole fills most of the cell cavity, causing the cytoplasm with its nucleus to be pushed against the cell wall.

The nucleus is composed of a gelatinous nucleoplasm. It contains one or more spherical bodies, called *nucleoli*, composed of ribosomal RNA and proteins. A threadlike network of genetic material composed of DNA and proteins, referred to as *chromatin*, is also present. DNA of the chromosomes determines genetic information. When cell division begins, the chromatin mass shortens and thickens to form rod-shaped bodies called *chromosomes*. The nucleic acids of the nucleus control the synthesis of specific proteins on the ribosomes; this in turn controls the metabolic activities of the cell and thus determines morphological characteristics of the organism.

Cell Division and Growth

In flowering plants, the root tips and stem tips, as well as the cambium, are the regions where new cells form from preexisting ones. The DNA in the chromosomes of the parent cell contains genetic information which is transmitted during cell division to the daughter cells.

Mitosis   Cell division is the method by which cells duplicate themselves in plants and is a part of the process of basic growth. In this complex process the nucleus divides into two equal parts. Usually the cytoplasm then divides, a process known as *cytokinesis*. In some instances this phenomenon does not accompany nuclear division, so that rather than two daughter cells' being formed, each with a single nucleus, the original cell is multinucleate. This division is usually associated with and follows nuclear division or mitosis.

Once normal cell division begins, it continues without interruption. Mitosis has been divided arbitrarily into different stages, or phases, for convenience in studying the process. These stages are prophase, metaphase, anaphase, telophase, and interphase (Figures 3-4 and 3-5). During *prophase*, the first recognizable stage, the chromatin threads contract into short, thickened chromosome units. Each chromosome is made of two parallel identical longitudinal halves known as chromatids. For all organisms of a given species, the chromosome number is constant. Accompanying this contraction of the chromatin threads, the

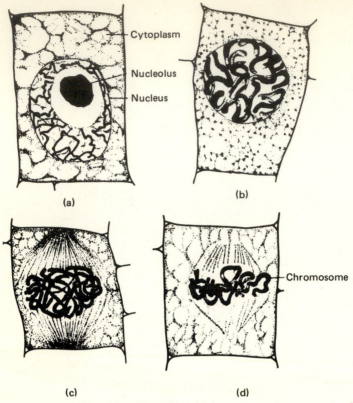

FIGURE 3-4 Mitosis in cells of root tip. *a–c, prophase stages; d, metaphase stages.* (*a*) Early prophase. (*b*) Medium prophase with chromosomes forming. (*c*) Later prophase; spindle becoming distinct and chromosomes grouping toward equator of spindle; nuclear membrane not evident. (*d*) Early metaphase; chromosomes definitely at equator of spindle; duplication of chromosomes can be seen in some instances. (From J. B. Hill, H. W. Popp, and A. R. Grove, *Botany*, 4th ed., McGraw-Hill, New York, 1967.)

nuclear membrane and the nucleoli disappear and the spindle fibers begin formation. These fibers first appear as clusters of short fibrils radiating from opposite positions in the cytoplasm.

Early in the second stage, *metaphase*, the spindle fiber apparatus becomes more evident. As metaphase proceeds, the chromosomes move to the center of the spindle, with the centromeres at the equatorial area of the spindle fiber apparatus and the arms of the chromosomes radiating outward. The *centromere* is that portion of the chromosome to which the spindle fiber is attached. *Anaphase*, the stage of movement, begins when the two chromatids of each chromosome separate. The separated chromatids, one from each chromo-

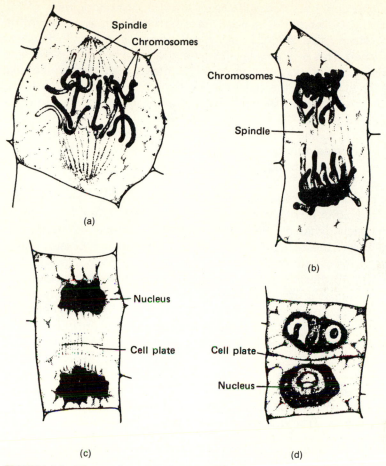

FIGURE 3-5 Mitosis (continued). *a*, anaphase; *b–d*, telophase. (*a*) Early anaphase; the duplicated chromosomes are passing from the equator toward the poles of the spindle. (*b*) Early telophase; chromosomes grouping at each pole of the spindle. (*c*) Somewhat later telophase; chromosomes rounding up at poles, preceding the formation of the nuclear membrane; cell plate forming on spindle at equator. (*d*) Two new cells formed. (From J. B. Hill, H. W. Popp, and A. R. Grove, *Botany*, 4th ed., McGraw-Hill, New York, 1967.)

some, pass in opposite directions along the spindle fiber to the opposite sides of the cell.

During *telophase*, the last stage of mitosis, the chromatids—which may now be considered as valid chromosomes—complete their movement to opposite sides of the cell, where they become longer and thinner and form a spherical mass of tangled chromatin threads. A nuclear membrane forms around each such cluster of chromatin, thereby creating two daughter nuclei within the cytoplasm of the parent cell. One or more nucleoli become visible in each daughter nucleus. During this stage, a cell plate develops and cytokinesis is accomplished. This cell plate is a partition that begins in the center and continues as a dividing layer across the center of the parent cell until it reaches the side walls, thereby creating two daughter cells. Following this there is a gradual restoration of the interphase condition in each daughter cell.

Interphase is the stage between mitotic divisions. During interphase, many biosynthetic processes take place within the nucleus. Among these is the duplication of the genetic material in preparation for mitosis.

The plasma membrane and intercellular substances are not easily distinguishable in newly formed cells, but new walls develop early. In cells in which division is not occurring, the nucleus has a distinct nuclear membrane separating the nucleus from the cytoplasm. Within the nucleus, the chromatin is granular and one or more nucleoli can be seen. After cell division has occurred, the daughter cells undergo a period of metabolic activity and increase in size. The increase in number and size of individual cells results in growth. When cells that have been carrying on cell division achieve a certain biochemical status, they cease dividing and undergo a process of elongation and differentiation.

Cell Enlargement and Differentiation Cell maturation or differentiation is the growth of specialized cells, which begins while the cell is enlarging. The ultimate appearance and function of each cell depend on the type of differentiation the cell undergoes. As a result, certain living cells of a plant may have thin walls and function in food storage, e.g., cells of potato tuber or carrot root; other cells may develop into specialized conducting components; while still others may develop thick, hard, tough cell walls and furnish strength and rigidity to the plant. At any given time, many cells in a mature, living plant may be dead but still may function in supporting the plant. Examples of nonliving but functional cells are the water-conducting components.

Some cells do not achieve full size immediately following cell division. In the young apple or tomato fruit, for example, cell division takes place rapidly within the first few weeks after pollination. Present in the small green fruit are most of the cells that will be found in the mature fruit. Weeks later, the fruit will continue growth, not by cell division, but by expansion of the small cells which have already been formed (Figure 3-6). The same situation occurs in leaves: a leaf at a very early stage contains most of the cells that will be found in a mature leaf.

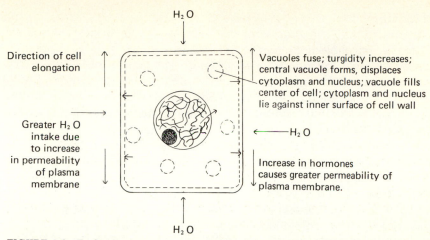

H_2O

Direction of cell elongation

Greater H_2O intake due to increase in permeability of plasma membrane

Vacuoles fuse; turgidity increases; central vacuole forms, displaces cytoplasm and nucleus; vacuole fills center of cell; cytoplasm and nucleus lie against inner surface of cell wall

H_2O

Increase in hormones causes greater permeability of plasma membrane.

H_2O

FIGURE 3-6 Early events in the elongation of a cell. Primary layer of cell wall is elastic. Cytoplasm produces additional cellulose fibrils to compensate for increase in cell length. (Courtesy of R. P. Ashworth, Clemson University.)

In a multicellular plant, cells undergo many changes as they mature. Cell differentiation during or after enlargement is necessary to form the highly specialized cells with respect to size, shape, formation of the secondary wall, and cytoplasmic content. Some cells complete all structural changes in a few days, whereas others remain relatively undifferentiated for a long time. Some remain active for years; others die in a few hours. Dead or alive, cells are part of the living plant. Many factors determine the differentiation and survival time of an individual cell. Root cap cells are sloughed off as the root grows downward in the soil; root hairs formed from epidermal cells are also of short duration. Cells that become adapted for the support or transport of water and minerals in solution also die, but they become a permanent and functional part of the plant. This change occurs after the secondary cell wall is formed from cellulose and lignin and after the ultimate death of the protoplasm. These specialized cells become tubes or vessels in an arrangement aligned parallel to the plant axis. The ends of these aligned cells are perforated, or they may disappear, producing continuous water-conducting vessels many cells in length.

Usually differentiation in cells is irreversible. Living cells, however, may at times change their function. For example, apical cells of a stem may initiate leaf primordia for a period of time, after which they may assume the function of floral initials and give rise to various parts of a flower. Also, some cells may become cork cambium cells which, by divisions, produce the cork of woody plants; and cells around a wound may revert to a physiological status in which they divide and form callus tissue.

TISSUES

From a morphological standpoint, a *tissue* is an organized group of cells with similar origin and function. Diversity may be found in cellular form and function within the tissue, but the cells that form the tissue are continuous and furnish some portion of the structural basis of the plant. Horticulturists need to know the characteristics and locations of specific cells and tissues because these determine the texture of the plant part involved and hence its quality. Cultural practices can be used to modify these tissues or to increase the size or number of a given type of cell. An undesirable characteristic in celery, for example, is stringiness, which is associated with the early production of thick-walled cells. Stringiness can be reduced by use of specific cultural practices or by cultivar selection. This type of selection requires expertise in plant morphology and anatomy.

Specialized Tissues

Various plant tissues are specialized for growth, the synthesis of food, storage of food, protection, support, and the conduction of water and mineral elements. Tissues that have thickened, strong cell walls which serve the plant for support or strength are often used as lumber, wood, or plant fibers. Tissues with thin, soft cell walls, such as those in fruits, leaves, and fleshy stems and roots, are consumed directly by human beings.

Cell types can be classified as either meristematic (cells capable of dividing) or permanent. Permanent cells are those in which growth and differentiation have been completed. Tissues composed entirely of one cell type may be considered as *simple* whereas tissues made up of two or more cell types may be designated *compound.* The basic permanent cell types of higher plants are: *parenchyma, collenchyma, sclerenchyma* (Figure 3-7), and the conducting components known as *tracheids* (Figure 3-8), *vessels* (Figure 3-9), *sieve,* and *companion* cells (Figure 3-10).

Parenchyma cells may form a simple tissue, such as the pith of stems or the edible portion of cantaloupes. On the other hand, parenchyma cells may be intermixed with various conducting components as well as sclerenchyma cells to form compound tissues commonly known as *xylem* and *phloem.* Xylem consists of trachieds, vessels, fibers (a type of sclerenchyma), and scattered parenchyma cells; phloem is made up of sieve cells, companion cells, fibers, and scattered parenchyma cells.

Meristem   Meristem cells function in cell division and elongation. These cells occur in undeveloped embryos, in the tips of roots and

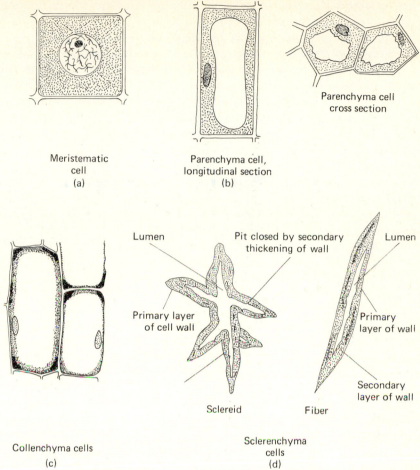

Meristematic
cell
(a)

Parenchyma cell,
longitudinal section
(b)

Parenchyma cell
cross section

Lumen

Pit closed by secondary
thickening of wall

Lumen

Primary layer
of cell wall

Primary
layer of wall

Secondary
layer of wall

Sclereid

Fiber

Collenchyma cells
(c)

Sclerenchyma
cells
(d)

FIGURE 3-7 Basic plant cell types. (*a*) Meristematic cell: possesses primary cell wall; carries on mitosis; occurs in root and stem tips, leaf and flower primordia. (*b*) Parenchyma cell: most abundant cell in plants; many modifications; only primary layer of cell wall present. (*c*) Collenchyma: some thickening on inner surface of angles of primary cell wall; no lignification in thickening. (*d*) Sclerenchyma-sclereid and fibers: secondary layer of cell wall, primary layer of cell wall, and middle lamella become lignified; lignification results in death of protoplast. (Courtesy of R. P. Ashworth, Clemson University.)

stems, and in leaf and floral primordia. *Apical meristems* of roots and stems are responsible for longitudinal or primary growth in the plant. The cells are generally small, are isodiametric in form, contain dense protoplasm with small vacuoles, and have thin walls which consist only of the primary layer (see Figure 3-7*a*). A spherical nucleus can be distinguished at the center of each meristematic cell. Older meristem cells undergo elongation and differentiation to form the permanent primary tissues of the plant.

The vascular cambium, a cylinder of fusiform or somewhat elongated meristematic cells within the dicotyledon stem and root, is responsible for an increase in the circumference or secondary growth of the plant. Meristematic tissues just above the nodes of grasses (monocotyledon) and at the bases of grass leaves are intercalary in that they cause the elongation of internodes and leaves of such plants. This meristem within the leaves allows for repeated mowing of lawn grasses without death of the plant.

Parenchyma Parenchyma cells are usually thin-walled and have large central vacuoles which force the cytoplasm and nuclei against the inner surface of their walls (Figure 3-7b). They may be primary or secondary in origin; i.e., they may have their origin from either the apical meristem or the vascular cambium of a stem or root. The cells are generally incapable of cell division. Their plasticity allows for a rapid increase in size in developing fruits, such as cherries and peaches. The primary cell walls of parenchyma cells are sometimes augmented and strengthened by lignified secondary walls. Parenchyma cells are commonly polyhedral (many-sided) in shape and contain protoplasm.

Parenchyma cells occur in all areas of the plant; i.e., in the pith, cortex, pith rays, xylem, and phloem of stems and roots as well as in leaves, floral parts, seeds, and fruits. The epidermis is also composed of modified parenchyma cells, the outer walls of which are impregnated with cutin. Nectaries occurring at the bases of flowers consist of parenchyma cells that secrete substances, such as sugar and oils, which attract insects. *Laticifer cells*, which are found in some plants, are parenchyma cells which are specialized to synthesize and store latex. Parenchyma tissues function in food storage, photosynthesis, the healing of wounds, and the formation of adventitious structures. Parenchyma cells, known as *chlorenchyma cells* when they contain chloroplasts, compose the palisade and spongy meosophyll cells of leaves. These cells are the principal site for photosynthesis. Since parenchyma cells retain the latent ability to divide, they play an important role in the healing and regeneration of wounds. They also initiate adventitious roots on stem cuttings. In addition, they play a role in the movement of water and the transport of food substances by diffusion and osmosis within the plant.

Collenchyma Collenchyma cells are actually a modification of parenchyma cells which are functional during primary growth in a plant (Figure 3-7c).

There are three forms of collenchyma: *lamellar collenchyma* with thickenings mainly on tangential walls, *angular collenchyma* with thickenings in the angles or corners, and *lacunar collenchyma* with numerous intercellular spaces and the most prominent thickenings located next to these spaces. Typically, collenchyma cells are living, have a high water content, and contain a central vacuole. These cells lend mechanical support to the young plant, yet allow it to yield slightly under stress and strain. Collenchyma occurs in the midrib and veins of leaves, as well as in the cortex of some stems during primary growth. Evidence of collenchyma tissue is found in the petioles of herbaceous vegetables, such as celery and rhubarb. They provide needed support for these plants; however, they are responsible for the stringiness of celery, which is objectionable to the consumer.

Sclerenchyma Sclerenchyma cells usually have no protoplast at maturity. They may develop in any or all parts of the primary and secondary plant body. They may be formed from meristem cells, by the cambium cells, or by modification of parenchyma or collenchyma cells. The principal characteristic of sclerenchyma cells is their possession of thick, usually *lignified*, walls. *Lignin* is a complex organic polymer which provides rigidity, toughness, and strength to these walls.

Sclerenchyma cells are diverse in shape and size, but two general types are definable—fibers and sclereids. *Fiber cells* (Figure 3-7*d*) are slender and elongated with tapering ends, which overlap and are often fused with one another. Usually their secondary walls are lignified. The lumen, or cavity in the fiber, is typically void of protoplast at maturity. These components hold stems erect and enable them to withstand stresses and to bear large loads of flowers and fruits. *Sclereids* (Figure 3-7*d*) are variable in shape. The term *stone cells* is often used synonymously for certain near-isodiametric sclereids which are unbranched and are without uniform or extreme form. The firm parts of seeds, nuts, and hard fruits are usually composed of stone cells of various types. They give hardness and mechanical protection when they are in mass, as in the shells of walnuts and the pit of peaches. In the fleshy part of pears, the scattered occurrence of groups of these cells lowers the quality of particular cultivars of pear. "Grit" or "stone" cells are inhibited primarily by picking the fruit before the pear is fully ripe. Off the tree, the pear ripens without forming excessive numbers of these cells. Commercial cultivars of pears grown today have fewer stone cells than certain other cultivars, such as the 'Kieffer.'

Conducting Tissue

Conductive or vascular tissue is made up of cellular components which are specialized for the translocation of substances throughout all parts of the flowering plant. This tissue is in turn made up of the *xylem* (from the Greek word *xylos*, meaning "wood") and the *phloem*. Each of these may be considered as a *compound* tissue, since each is composed of several types of cells. Each is continuous throughout the plant. Normally, components of the xylem conduct water and minerals in solution from the roots upward, whereas components of the phloem conduct food in solution from the point of its synthesis to various areas of the plant. Xylem and phloem formed by differentiation of meristem cells of the root or stem apex is said to be *primary*, whereas xylem and phloem in dicotyledon plants that is produced by cellular divisions of the vascular cambium is *secondary*. However, in the majority of monocotyledons there is no cambium, no vascular cylinder, and no secondary growth. All parts of such plants are primary in origin. Vascular bundles of these plants are scattered throughout the stem.

Xylem   The xylem is composed of tracheids, vessels, fibers, and parenchyma cells. Tracheids and vessels are nonliving when mature. *Tracheids*, which are the fundamental cell type in the xylem, occur in longitudinal rows and are elongated with tapered ends (Figure 3-8). In cross section, the tracheid is typically angular, though some rounded forms occur. Their walls have various arrangements of thickened, lignified materials and contain pits, perforations, or both, which permit ready diffusion of water and minerals in solution into adjacent cells. When longitudinally adjacent tracheids are completely fused and their end walls are broken down so that one long passageway is open, a *vessel* is formed (Figure 3-9). The vessel is not a single cell but a long continuous tube through which water and minerals may move with even less interference than through longitudinal rows of tracheids. When young, and before their longitudinal fusing, cells which form vessels contain protoplasts which lay down their lignified, thickened walls. Vessels, therefore, provide some support as well as functioning in conduction. They may contain pits, perforations, or both. The diameter of vessels is greater than that of tracheids; consequently vessels carry on most of the conduction.

Xylem fibers are elongated, strengthening cells with thickened walls. Their ends are pointed. At maturity, they do not contain protoplasm. They differ from tracheids chiefly in their thicker walls and reduced number of pits. The scattered *parenchyma cells* in the xylem are similar to parenchyma in other areas of the plant. They function in food storage. In some instances, their walls may thicken and become lignified.

Phloem   The phloem consists of sieve tubes (which are longitudinal rows of sieve cells), companion cells, fibers, and parenchyma cells

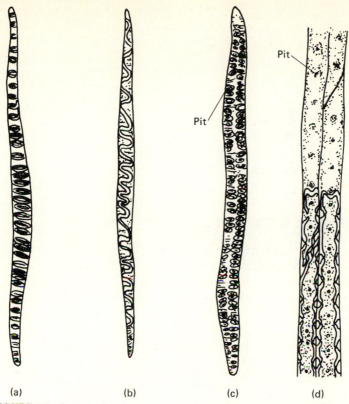

(a)   (b)   (c)   (d)

FIGURE 3-8 Longitudinal views of tracheids of the xylem tissue. (*a*) annular (ringed) tracheid; (*b*) helical (spiraled) tracheid; (*c*) scalariform tracheid; (*d*) pitted tracheid.

(Figure 3-10). *Sieve cells*, the main component of the phloem, are elongated, slender cells with thin, nonlignified walls. In dicotyledons they occur in rows parallel with the axis of the plant. Here, each cell of a row (i.e., of a sieve tube) is often designated as a *sieve element*. One or more small, slender *companion cells* are associated with the sieve cell (sieve element) in the phloem of angiosperms. These cells are living and in some way may be associated with the activities of sieve cells. Each sieve element has a large central vacuole that is basically surrounded by a thin layer of cytoplasm which is against the inner surface of the cell wall. To some extent, however, the cytoplasm and vacuolar fluid may be intermixed. Although living, they do not contain a discernible nucleus when "mature." Perforations in the cell wall allow strands of cytoplasm to extend from one sieve element to another. Callose, an accumulation of proteinaceous fibrils gradually

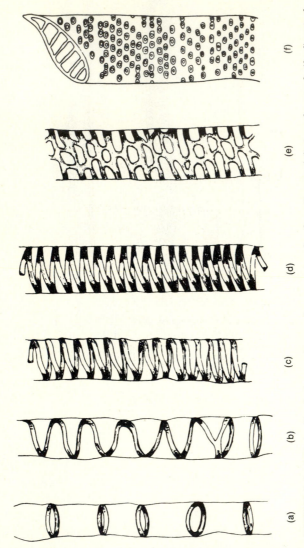

(a)   (b)   (c)   (d)   (e)   (f)

FIGURE 3-9 Longitudinal views of vessel segments of xylem tissue. (*a*) ringed vessel; (*b–d*) spiral vessels; (*e*) scalariform vessel; (*f*) pitted vessel with scalariform end wall. *a–d*, from the stem of the balsam plant (*Impatiens sultani*); *e*, from the stem of the sunflower (*Helianthus annuus*); *f*, from the stem of the alder (*Alnus*). (Drawings by Florence Brown; from J. B. Hill, H. W. Popp, and A. R. Grove, *Botany*, 4th ed., McGraw-Hill, New York, 1967.)

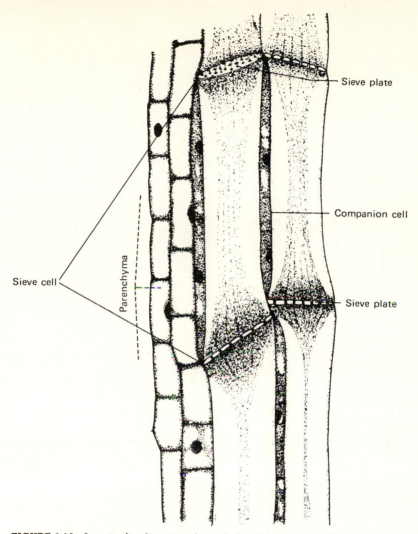

FIGURE 3-10 Longitudinal section through the phloem tissue of a stem, showing the cellular structure. (Adapted from J. B. Hill, H. W. Popp, and A. R. Grove, *Botany*, 4th ed., McGraw-Hill, New York, 1967.)

forms over these perforations. This interrupts the transport of soluble food and is followed by the death of the cell. *Phloem fibers* are nonliving, thick-walled, elongated cells which appear in groups within the phloem. Phloem parenchyma cells are living storage cells that are scattered among other cells of the phloem. They provide a means of lateral diffusion of substances in addition to possibly storing limited quantities of insoluble food.

BIBLIOGRAPHY

Eames, A. J. and L. H. MacDaniels: *An Introduction to Plant Anatomy*, 2d ed., McGraw-Hill, New York, 1947.

Esau, K.: *Plant Anatomy*, 2d ed., Wiley, New York, 1965.

Haynes, J. D.: *Botany: An Introductory Survey of the Plant Kingdom*, Wiley, New York, 1975.

Hill, J. B., H. W. Popp, and A. R. Grove, Jr.: *Botany*, 4th ed., McGraw-Hill, New York, 1967.

Troughton, J. and L. A. Donaldson: *Probing Plant Structure*, McGraw-Hill, New York, 1972.

Wilson, G. B.: *Cell Division and Mitotic Cycle*, Reinhold, New York, 1966.

Wilson, G. B. and J. H. Morrison: *Cytology*, Reinhold, New York, 1961.

CHAPTER 4
VEGETATIVE GROWTH AND DEVELOPMENT

 Most plants of the spermatophytes are composed of a vertical, cylindrical axis which has a variety and complexity of lateral appendages (Figure 4-1). The aboveground portion of the plant axis is most often the stem, and usually the part below ground is the root. Roots, stems, and leaves are vegetative parts. Each of these parts differs from the other in form and function, but all contribute to the total life and activity of the plant.

Vegetative parts are composed of a combination of different kinds of tissues. The monocotyledonous plant body develops solely from primary tissues, whereas the gymnosperm and dicotyledonous plant bodies develop from primary and secondary tissues.

Vegetative parts of plants may be used by the horticultural industry for various economic benefits.

ROOTS

The root is that part of the plant which anchors the plant, absorbs water and minerals in solution, and often stores food. Ordinarily it is without external buds, leaves, nodes, and internodes. Roots show great variation in form and structure. The root usually develops beneath the soil surface. However, some roots develop above ground, such as the

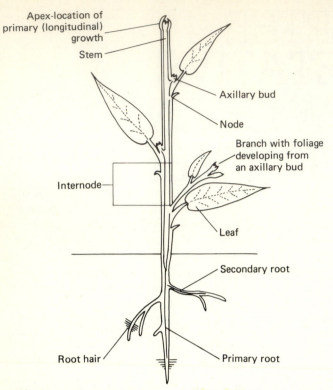

Apex-location of primary (longitudinal) growth

Stem

Axillary bud

Node

Branch with foliage developing from an axillary bud

Internode

Leaf

Secondary root

Root hair

Primary root

FIGURE 4-1 Longitudinal section of primary plant body of dicotyledon with basic lateral appendages. (Adapted from A. J. Eames and L. H. MacDaniels, *An Introduction to Plant Anatomy*, 2d ed., McGraw-Hill, New York, 1947.)

adventitious roots of the corn plant where prop or aerial roots extend from the stem and provide added support to the plant. In other instances, aerial roots may serve to anchor the plant or to attach climbing stems to some substratum, as in ivy and trumpet vine. Some plants, such as the orchids, produce only aerial roots. The above-ground roots of the bald cypress (*Taxodium distichum*), which develop in an aquatic environment, provide more adequate aeration for the root system.

The fundamental differences between roots and stems are based on external and internal structural adaptations. Roots possess a root cap and root hairs, distinct structures found on no other part of the plant. Roots usually do not form buds, but under certain conditions some species, such as sweet potato and blackberry, form adventitious buds which give rise to stems. This characteristic allows these plants to be propagated by root cuttings. The root typically has an internal core of vascular tissue and does not have a pith, whereas stems have their vascular tissue arranged in bundles surrounding a central pith.

Kinds of Roots and Their Functions

Roots can be classified as either primary, lateral or secondary, or adventitious roots. The primary root develops from the radicle or the hypocotyl of the germinating seed, or from both. It grows downward into the soil and branches repeatedly to produce lateral or secondary roots. The primary root usually grows vertically, whereas secondary roots ordinarily grow laterally or somewhat horizontally. If the primary root persists and maintains its dominance, it is referred to as a *taproot*. Plants with taproots are beet, carrot, turnip, hickory, pecan, parsley, camellia, and conifers (Figure 4-2). Since taproots penetrate deeply, plants characterized by taproots usually depend on cumulative annual rainfall for growth.

When primary and lateral roots develop more or less equally and have a limited quantity of cortex, they form a *fibrous root system* (Figure 4-2). This most often occurs when the primary root ceases to elongate. Numerous lateral and spreading roots of equal importance develop. These give rise to many secondary and tertiary roots, which

FIGURE 4-2   Root systems. (*a*) taproot—pecan;
(*b*) storage taproot—carrot; (*c*) fibrous—strawberry.

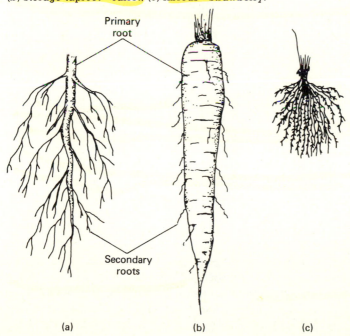

Primary
root

Secondary
roots

(a)   (b)   (c) constitute the "feeder roots" of the plant. These lateral roots remain small in diameter owing to very slow cambial activity. Most shrubs have this type of fibrous woody root system. Fibrous roots do not penetrate as deeply into the soil as do taproots. *Adventitious roots* are those which arise from a stem rather than from the branching of another root.

The root's functions are the absorption and conduction of water and dissolved minerals from the soil to the stem or trunk, the storage of food, synthesis of hormones, and the anchorage of the plant in the soil. *Fleshy* or *storage roots* often develop as part of biennial and perennial plants. These roots accumulate and store a rich supply of reserve food for the plants. Over the centuries people have cultivated many varieties of these plants for their own food. Turnip, beet, radish, parsnip, and carrot are examples of plants in which the taproot is developed into the storage structure (Figure 4-2). In the dahlia and sweet potato, branch roots develop into storage structures. The function of food storage in such roots as carrot, parsnip, beet, rutabaga, sweet potato, radish, turnip, and asparagus is most important, whereas with trees and shrubs it is of limited significance.

Structure of Roots

Roots are cylindrical in form and are usually colorless. In longitudinal aspect, they may be separated into basic zones: the root cap, the meristematic region, the region of cell elongation and differentiation, the zone of primary tissues, and the zone of secondary tissues (Figure 4-3). The latter two are generally referred to as the mature region.

Cells of the meristem region are isodiametric and are arranged in rows parallel with the axis of the root; they carry on numerous cell divisions. Cells formed at the apex or outer portion of this region differentiate into layers of the root cap. As the root grows downward, the outer cells of the cap are worn away and are replaced by underlying cells which have more recently been formed by the meristem region.

Directly behind the zone of cell division is the region of cell elongation. It is not sharply separated from the region of cell division. Here, polarity of the plant is demonstrated and the longitudinal axis of the cells increases more than the transverse axis. The cell walls stretch and the plasma membrane becomes more permeable so that large quantities of water are absorbed. The combined effects of cell division in the meristem region and cell elongation in the zone of elongation result in the root becoming longer. This increase in root length is known as *primary growth* and permanent tissues which are formed are known as *primary tissues*. These are called primary tissues because they are developed directly from the apical meristem. As

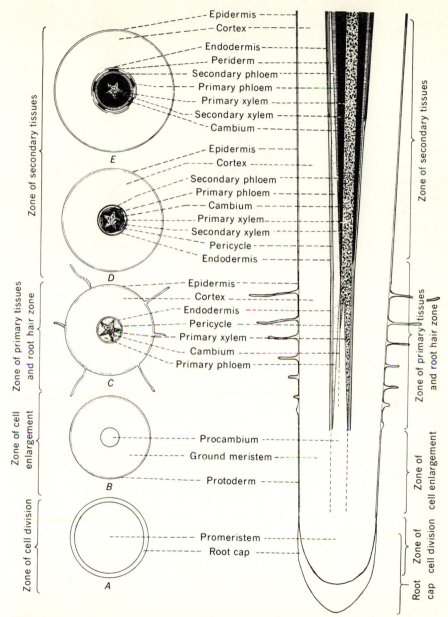

FIGURE 4-3 Diagrammatic representation of the origin and arrangement of primary and secondary tissues in a young root of a dicotyledonous plant; longitudinal section of root at right, and transverse sections at different distances from the tip (*a–e*) at left. (Drawing by Florence Brown; from J. B. Hill, H. W. Popp, and A. R. Grove, *Botany*, 4th ed., McGraw-Hill, New York, 1967.)

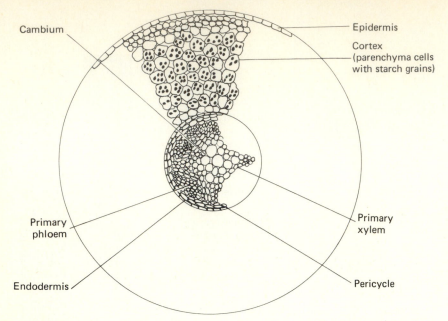

Cambium

Epidermis

Cortex
(parenchyma cells
with starch grains)

Primary
phloem

Primary
xylem

Endodermis

Pericycle

FIGURE 4-4 Cross section of Ranunculus root showing primary structures before secondary growth is initiated.

these cells mature and become a part of the mature region of the root, they are differentiated into epidermis, cortex, endodermis, pericycle, and vascular tissues (Figure 4-4).

The region of primary tissues, or zone of water and mineral absorption, develops after maturation of the cells. Water and minerals are also absorbed in the meristem and in the region of elongation to some degree.

Root Hairs   The outer walls of epidermal cells in the upper part of the zone of elongation and the lowermost part of the mature zone of the root form outward protrusions called root hairs, which are responsible for most absorption by the plant (Figure 4-5). *Root hairs* provide much more absorption area than the exterior of the root cylinder alone and have more permeable membranes than those of the epidermal cells of the older portions of the root. New root hairs are formed continuously as the roots extend into the soil. Root hairs in most plants live and function for only 2 to 3 days. An exception is the honey locust in which the root hairs can live for several months or even years.

The capacity of the plant to develop new root hairs rapidly is important in determining success in transplanting. When plants are removed from the ground, the root hairs are usually broken or damaged causing the plant to have less than its normal capacity to

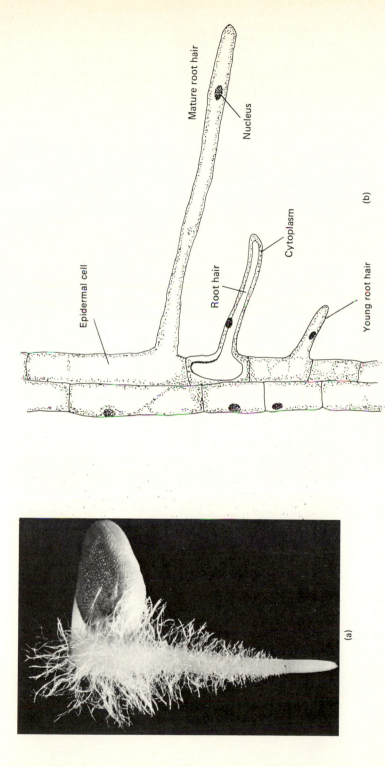

FIGURE 4-5 Root hairs. (*a*) Radish (*Raphanus*) seedling with root hairs. When very young the root may produce these hairs along its entire length rather than only at its tip. (Carolina Biological Supply Company.) (*b*) Developing and mature root hair. Cytoplasm streams into hair and lies against the inner surface of its wall. The nucleus is carried in cytoplasm from cell into hair; ultimately it may become located nearer the tip of the hair. The central areas of both epidermal cell and hair are filled with vacuolar fluid.

absorb water and minerals in solution. Tomato plants develop new root hairs from old root tissue and are easy to transplant successfully. Squash and cucumber seldom develop root hairs from old roots but form them only on newly developed roots; for this reason they are very difficult to transplant bare-root. As root hairs die and decay, the epidermal cells around and above them develop thicker walls and absorb only limited quantities of water and nutrients. Eventually, if the root carries on secondary growth (i.e., growth in diameter), the epidermis is cracked and broken, disintegrates, disappears, and cork or bark is formed. A few plants, such as Scotch pine and blueberries, do not have root hairs. Ninety to ninety-five percent of the water absorbed by a root enters through the meristem region, region of cell elongation, and root hair zone. Primary tissues have usually only partly differentiated at the level of the root hairs. Above the root hairs, little water is ordinarily absorbed by the root.

Transmission of Water and Solutes   Plants have developed a system for the conduction of water and solutes from the region of absorption by the root hairs in the root to the stem and leaves. This system basically consists of lateral diffusion of water through the cortical cells, the endodermis, and the pericycle of the root and thence up the xylem components. Water moves as a result of a gradient of concentration through cortical cells from the root hair zone to the endodermis. The *endodermis* is a single layer of cells between the cortex and pericycle in the roots. The *pericycle*, which is not found in stems, is a thin layer of parenchyma cells that separates the endodermis from the vascular components in roots. Once water has diffused along the concentration gradient from root hairs through the cortex, endodermis, and pericycle to the xylem, it moves upward through xylem components as a result of transpirational pull and, to some degree, of root pressure. (See chapter 8.)

Lateral roots   These roots are generally initiated at a level above the root hair zone where the epidermis no longer functions in absorption. Lateral roots are endogenous; that is, the lateral-root meristems are formed in the pericycle tissues of the mother root, and the lateral root appears externally only after its growth is well begun. In angiosperms and gymnosperms, the meristems arise in the pericycle just beneath the endodermis.

In the formation of lateral roots, several adjacent cells of the pericycle reassume meristematic activity, dividing to produce a new root apex. As this grows, it pushes outward through the surrounding tissues—the endodermis, the cortex, and the epidermis—to finally emerge outside of the root.

The primary portion of the lateral root, just as with the parent root, is composed of tissues differentiated from cells of the apical meristem. The primary part of the root is characterized by the areas or regions mentioned above.

Roots of gymnosperms and dicotyledons Roots of these plants possess a vascular cambium and carry on secondary growth. This occurs in the mature region of the root. Secondary xylem is laid down inside the vascular cambium and outside the primary xylem as the vascular cambium cylinder is formed; secondary phloem is laid down outside the vascular cambium and inside the primary phloem, pushing the primary phloem farther out. Gradually the primary phloem and the endodermis are stretched and crushed as a result of this pressure from underlying developing tissues.

Roots and Mycorrhiza In some plants, absorption of water and minerals is aided by mycorrhiza ("fungus root"), a fungus living in a mutualistic relationship with the roots of the vascular plant. Mutualism is a type of symbiosis in which both parties are benefited. In forest trees, it is often the secondary roots arising from the main root that are associated with these fungi which are also in intimate contact with the soil particles. Mycorrhizae are often useful in soils of high organic matter. Mycorrhizae appear to enhance water absorption. Fungi digest organic matter, making nitrogen and other nutrients available to the plant. In other cases, the mycorrhizae allow the plant to grow under unfavorable soil conditions. It is believed the fungus provides the plants with food and growth promoting substances of which they may be deficient.

Effect of the Environment on Roots Roots are sensitive to environmental influences. Depth of penetration may vary from several centimeters to 10 m, depending upon the plant. Lateral extension is generally much greater than depth of penetration. Roots of the beet or carrot become branched and malformed if they are grown where there is a high water table or in rocky, impervious soil. In well-prepared seed beds they develop straight, symmetrical, unbranched roots. Tomatoes, however, have a well-branched root system in well-prepared beds.

STEMS AND STEM ADAPTATIONS

The stem is the part of the axis of a plant which develops from the epicotyl of the embryo or from a bud of an already existing stem or root (Figure 4-6). *Buds* are apical meristems in a dormant condition in perennial plants. The stem is composed of nodes and internodes, as compared to roots which have neither, and bears leaves and buds.

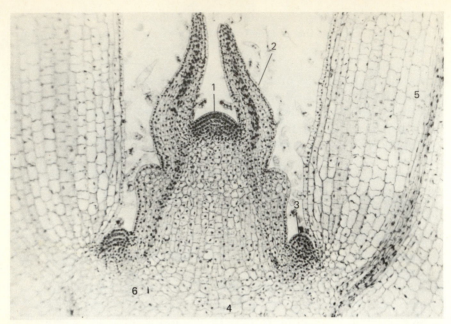

FIGURE 4-6 Longitudinal sections of shoot apex of coleus (*Coleus blumei*). Apex with older leaf primordia (embryonic leaves) and older axillary bud primordia. 1, apical meristem; 2, leaf primordia; 3, axillary bud primordia; 4, region of cell elongation in stem; 5, older embryonic leaves; 6, provascular strand (procambium strand). Leaves of coleus are opposite; thus the uniformity in origin of pairs of leaf primordia from the shoot apex. (Carolina Biological Supply Company.)

Flowers may be formed terminally or laterally at the nodes (Figure 4-7). Stems may have adventitious buds arising from their internodes. Regardless of the form or size of a stem, all stems perform the following basic functions: physical support of leaves and reproductive structures; conduction of food, water, and minerals in solution to and from the leaves; and food storage. Stem tissues are basically similar to root tissues, but the manner in which they are arranged in the stem differs from that of roots. Stems have nodes where branching occurs and internally have their fundamental vascular structures—i.e., their vascular bundles of the xylem and phloem—arranged cylindrically; whereas roots lack nodes, and their vascular tissues typically form a single central, longitudinal core.

Regions of the Stem

Stems develop from apical meristems or buds. Growth in the length of stems results from activity of the meristematic regions and the region of elongation and involves an increase in cell number followed by an increase in cell size. The stem is characterized by three regions similar

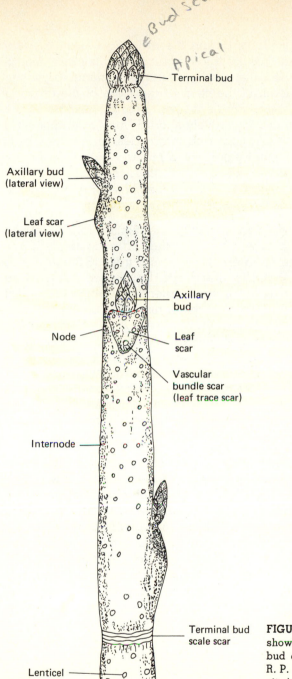

Bud Scales

Apical

Terminal bud

Axillary bud
(lateral view)

Leaf scar
(lateral view)

Axillary
bud

Node

Leaf
scar

Vascular
bundle scar
(leaf trace scar)

Internode

Terminal bud
scale scar

Lenticel

FIGURE 4-7 Diagram of stem showing node, internode, and bud arrangements. (Courtesy of R. P. Ashworth, Clemson University.)

to regions of the root: the regions of cell division, cell elongation, and cell differentiation. The region of *cell division* occurs within the buds of woody stems as they break dormancy, as well as in the tips of growing herbaceous and woody plants during the growing season. Annual herbaceous stems do not produce true buds. Instead, their meristematic regions once established after seed germination continue growth to some extent until the death of the stem at the close of the growing season. The area of *cell elongation* is the portion of the stem below the meristematic area; it may be relatively short or may extend several centimeters through several nodes and internodes. In this area, cells increase longitudinally, thereby producing primary growth in length. In the lower portion of this region, the *cell differentiation* takes place, and the permanent parts of the stem are formed: the epidermis, cortex, vascular bundles, vascular cambium, pith rays, and pith.

Makeup of the Stem   Stems exhibit wide variations in form, size, and structure. In length they may vary from less than 2.5 cm to nearly 122 m. In thickness they vary from hairlike structures to trees 18 or more m in circumference. Morphologically, stems can be divided into several parts. The *trunk* is the axis of the stem system; a *branch* is a lateral portion of the tree that originates from the trunk or from another branch and gives rise to shoots, twigs, and leaves. A *shoot* is a stem, one-year-old or less, that possesses leaves. A *twig* is a stem, one-year-old or less, without leaves. A scarred (bud-scale scar) circle on a twig shows the end of one season's growth and the beginning of the next. These bud-scale scars encircle the stem and remain as scars when the scales surrounding the terminal bud are shed as growth starts in the spring. *Spurs* are stems with short internodes, usually from older wood that bear leaves, fruit, or both.

Causes of Diversity of Form   The stems of plants exhibit a diversity of form because of competition between plants, because of the way they are pruned, or because of the very nature of the plant. Orchard trees have a stem or trunk that is pruned to branch lower than an oak or maple being used as a shade tree. The stem of the lily or strawberry is very short. Common asparagus has flattened, green stems and branches, that resemble and perform the functions of leaves, whereas the true leaves are reduced to small inconspicuous bracts. The stem of squash grows along the ground while grapes and pole beans usually have supplementary structures, derivatives of leaves or stems called tendrils, for attaching to a supplementary support. In most cacti the stem tissue is mostly of parenchyma which forms a storage area, with the outer layer of these cells containing chloroplasts where photosynthesis occurs; the spines are modified leaves. The stem of kohlrabi is enlarged and bulbous, with stored food in its cells.

Buds

Buds are undeveloped and unelongated stems composed of a very short axis of meristem cells from which arise embryonic leaves, lateral buds, and, sometimes, flower parts. Within the axils of the embryonic leaves are axillary bud primordia, from which a lateral branch may later develop (Figure 4-6). The central axis in the bud is often broader than it is long with the nodes extremely close together. Buds are diverse in size, shape, and appearance. They range in size from microscopic structures to terminal buds of the cabbage head weighing up to 4 to 5 kg. Buds in their early development are made up of undifferentiated meristematic cells.

The buds of temperate zone trees and shrubs typically develop a protective outer layer of small, leathery, modified leaves—the bud scales (Figure 4-8). Kalmia, tropical plants, most herbs, certain viburnums, carnation, black locust, and cabbage have naked buds. Bracts may replace scales in the bud structure as in chrysanthemum and other composite flowers; or the sepals may serve the same purpose as in rose and tomato.

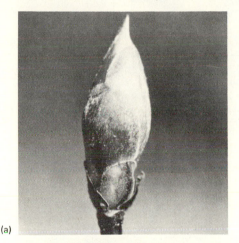
(a)

(c)

(b)

FIGURE 4-8 Bud structure: Stages in the opening of terminal foliage buds of hickory (*Carya* sp). (*a*) Closed bud with large pubescent (hairy) bud scales. (*b*) Bud scales opening. (*c*) Bud open, with young compound leaves expanding; bud scales have either dropped from bud or curled backward at base of expanded bud. (Carolina Biological Supply Company.)

Classification of Buds   Buds can be classified in different ways; one way is based on the types of tissues within the bud. A *mixed bud* contains primordia for both leaves and flowers. A *simple bud* contains either leaf or flower primordia, but not both. Apple trees have simple vegetative buds, but mixed flower buds (Figure 4-9). Peach, plum, and

(a)

(b)

FIGURE 4-9 Vegetative and reproductive development of a mixed bud of an apple tree from dormant bud stage to leaf and fruiting. (*a*) dormant bud; (*b*) leaf and flower primordia developing; (*c*) full flower stage; (*d*) mature leaves and developing fruit from the flower. (Courtesy of Benjamin T. Hendricks, Clemson University.)

(c)

(d)

cherry trees bear only simple buds—either vegetative or reproductive. Leaf buds are typically thinner and more pointed than either mixed or flower buds, which are rounded and somewhat plump. *Compound buds* occur on some plants such as grapes.

According to their position on the stem and manner of origin, buds may be classified as terminal, axillary, accessory, or adventitious. *Terminal buds* are at the tips of stems and are responsible for terminal growth; they are usually larger and more vigorous than other buds (Figure 4-7). *Axillary buds* are borne laterally on the stem in the axils of leaves. *Accessory* or *supernumerary buds* are lateral buds occurring at the base of a terminal bud or in an axil at the right or left of the axillary bud. In the silver maple and peach, accessory buds can be produced on both sides of the axillary bud. In the peach there are all combinations of flower and leaf bud arrangements. *Adventitious buds* originate at places on the plant other than nodes or stem apices, such as on the internodes of stems and sweet potato "slips." Blackberries can be propagated by root cuttings since their roots form adventitious buds. Adventitious buds may form watersprouts on the stems or suckers on the roots.

Market Value of Buds Buds from certain plants are economically important horticultural products. Opening flower buds of broccoli, for example, are consumed (Figure 4-10). In this case, portions of the stem

FIGURE 4-10 The flower buds of broccoli are horticultural food products. Portions of the stem as well as small leaves associated with the flower buds are eaten. (Photography, Joseph Harris Company, Rochester, N.Y.)

as well as small leaves associated with the flower buds are eaten. Cabbage and head lettuce are examples of unusually large terminal buds. In the case of globe artichoke, the fleshy basal portions of the bracts of the flower bud are eaten. Succulent axillary buds of Brussels sprouts become the edible part of the plant. Sexual reproduction is accomplished by the formation and maturing of flower buds, whereas asexual, vegetative propagation is accomplished in many instances by the development of foliage buds. Orchid meristeming (taking the minibud) has cut the cost of propagation tremendously. In this process, meristematic tissue is cultured on agar under controlled environmental conditions to propagate the plant.

Internal Structure of the Stem

The internal structure of the stem varies, depending upon whether the stem is an angiosperm or a gymnosperm. Variations are also found between monocots and dicots; and within dicots, there are differences between herbaceous (Figure 4-11) and woody plants (Figure 4-12). There are greater variations in structure between dicots and monocots than between dicots and gymnosperms. On the basis of the manner in which the portions of a dicotyledon stem are formed, such a stem is said to be composed of a *primary part* (sometimes referred to as the *primary plant body*) and a *secondary part*. The primary portion is formed by progressive differentiation of older cells of the apical meristem area as the stem carries on longitudinal or primary growth. As such differentiation occurs, a cylindrical pattern of *vascular bundles* (bundles of conducting components) surrounding a pith is formed. Areas of cells between adjacent bundles form so-called *pith rays*. Surrounding the cylindrical pattern of bundles is the *cortex* which is in turn surrounded by a peripheral layer of cells, the *epidermis* (Figure 4-13). Extending tangentially through each vascular bundle and transversing each pith ray is a layer (or thin region) of cells which maintain their potential to divide. Cells of this layer (or region) form a cylindrical layer known as the *vascular cambium*. This cambium, after all primary growth has been completed at a given level in the dicotyledon stem, produces the secondary portion of the stem.

Characteristically, primary growth is the only type of growth which occurs in monocotyledon stems, since they are without a vascular cambium. Their vascular bundles occur at random throughout the soft, parenchymatous tissue of the stem. Increase in circumference is accomplished in such stems by an enlargement of parenchymatous, pith cells formed by the meristem. This is brought about, as the stem becomes older, by the gradual increase in water concentration and turgidity that builds up in such cells. Examples of monocotyledon stems are: banana, lily, corn, asparagus, and date palm.

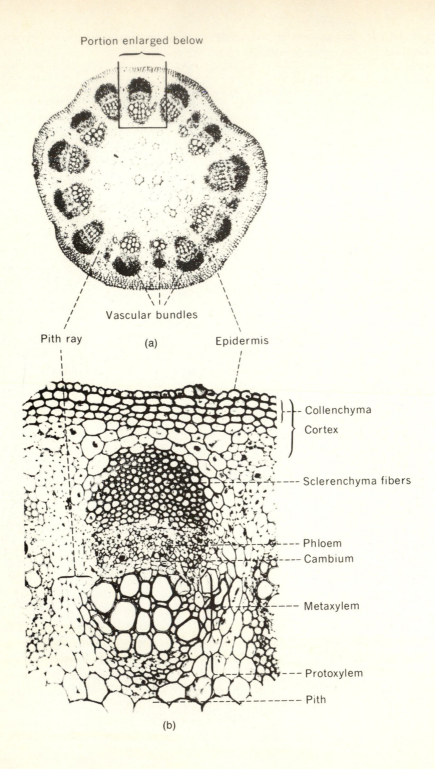

Portion enlarged below

Vascular bundles

Pith ray   (a)   Epidermis

Collenchyma

Cortex

Sclerenchyma fibers

Phloem

Cambium

Metaxylem

Protoxylem

Pith

(b)

Stems of annual, dicotyledon vegetables and ornamental flowering plants are herbaceous. Similarly, the stems of many dicotyledon perennial plants are annual and herbaceous, while their root systems are perennial. Characteristically, such herbaceous stems have a large amount of pith and cortex and rather wide pith rays. Their vascular bundles, arranged in a cylindrical pattern as described above, are comparatively small. Although most of the tissues formed in such stems

FIGURE 4-12 *Below:* Cross section of 2-year-old woody dicotyledonous plant, tulip tree (*Liriodendron tulipifera*). (Carolina Biological Supply Company.)

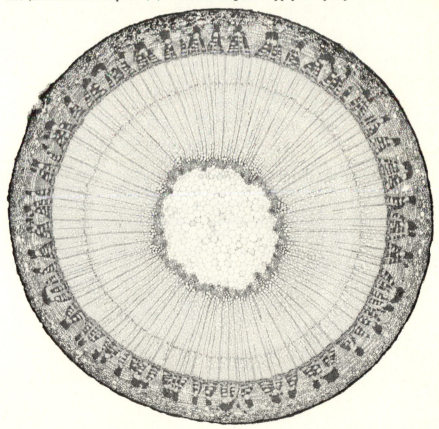

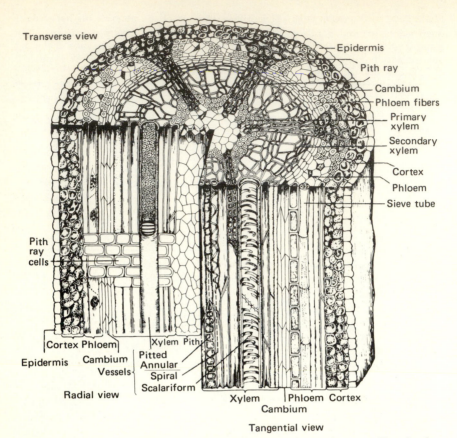

Transverse view

Epidermis
Pith ray
Cambium
Phloem fibers
Primary xylem
Secondary xylem
Cortex
Phloem
Sieve tube

Pith ray cells

Cortex Phloem   Xylem Pith
Epidermis   Cambium   Pitted
            Vessels   Annular
                       Spiral
Radial view Scalariform   Xylem Phloem Cortex
                                Cambium

Tangential view

FIGURE 4-13 Semidiagrammatic representation of a 1-year-old stem of Liriodendron, in transverse, radial, and tangential views. (Drawing by Edna S. Fox; from J. B. Hill, H. W. Popp, and A. R. Grove, *Botany*, 4th ed., McGraw-Hill, New York, 1967.)

are the result of primary growth, the stems do possess a vascular cambium which produces a limited amount of secondary growth during the middle to late part of the growing season. Since such stems die at the end of the single growing season, this secondary growth results in the formation of only a small percentage of the entire cellular volume of the stem. In biennials, however, secondary growth occurs throughout their entire second growing season. Because of this, their stems approach the structure of woody stems more closely than do stems of herbaceous annuals. Thus, the main structural difference between herbaceous, dicotyledon stems which tend to remain soft and those stems which become semiwoody or entirely woody is the quantity and degree of development of secondary tissues.

Gymnosperm and dicotyledon trees and shrubs have woody stems (Figure 4-13). From center to circumference, the parts of a one-year-old woody stem are: pith, composed of parenchyma cells; a cylindrical pattern of vascular bundles in which there is a cylindrical vascular cambium; narrow pith rays which radiate from the pith and extend between adjacent vascular bundles; a cortex; and an epidermis. Because of the narrowness of these pith rays (usually being one or possibly two layers of cells in thickness), the vascular bundles are much closer together than they are in herbaceous dicot stems; also, the vascular bundles represent a much greater percentage of the total volume of the woody stem than of the herbaceous one. In the woody stem all the tissues from the vascular cambium to the outer circumference make up the *bark*. In the young woody stem this includes the phloem of the vascular bundles, the cortex, and the epidermis. As the stem becomes older and increases in circumference (as a result of secondary growth produced by the vascular cambium), cells of the cortex and epidermis are stretched, gradually crushed, and eventually sloughed off. When this occurs, a cylindrical layer of cells of the cortex or of the phloem forms, which develops (or reinitiates) the capacity to divide. This layer becomes designated as the *cork cambium*. Its cells most frequently produce cells on their outer surface which develop waxes (suberin) in their walls and form the cork. Thus, in such older stems the bark consists of phloem, cork cambium, and cork.

The basic increase in stem diameter is due to the activity of the vascular cambium which, by successive cell division, produces secondary xylem (wood) on its inner surface and secondary phloem on its outer surface. This results in the formation of annual layers (growth rings) of xylem (wood). However, as new secondary phloem is formed on the outer face of the vascular cambium, the outer, older phloem is stretched and crushed. In this manner, the newly formed phloem merely replaces the older degenerate phloem, so that no annual layers of phloem are produced. Cambial activity varies seasonally. It is most active during spring and early summer. During this period of active cell division the bark "slips" easily; that is, it may be readily separated from the xylem. At this time the propagation procedure known as "budding" is carried on (Chapter 12).

The xylem or wood is divided into the outer sapwood and the inner heartwood. Heartwood consists of older annual layers, the cells of which contain deposits of resins and gums so that it gives strength to the plant. Sapwood is made up of the outer few annual layers of the woody stem and is that part of the xylem which functions in water and mineral transport. Progressively, year by year the inner layers of sapwood change into heartwood. This rate of change from sapwood to heartwood occurs more rapidly in slower growing stems than in faster growing ones.

Stem Adaptations

Many structures of the plant are modified stems. These adaptations function in storage and regeneration. *Offshoots* are short horizontal stems which occur in whorls or near whorls at the crown of stems. They bear fleshy buds or leafy rosettes such as the artichoke. A *stolon* is an aboveground stem which reclines or becomes prostrate (i.e., is horizontal) and may form roots at the nodes that may come in contact with the ground. Examples of stolons are the black raspberry and *Forsythia.* A *runner* is a slender stolon with elongated internodes; these internodes root at the nodes which touch the ground, as with strawberry. *Corms,* such as gladiolus and crocus, are short, fleshy, underground stems with few nodes and very short internodes (Figure 4-14*a*). *Rhizomes* are

FIGURE 4-14 Modified underground stems. (*a*) corm—gladiolus; (*b*) rhizome—Kohleria; (*c*) tuber—potato; (*d*) bulb—tulip. (*a, c,* and *d* redrawn from Hudson T. Hartmann and Dale E. Kester, *Plant Propagation: Principles and Practices*, Prentice-Hall, Englewood cliffs, N.J., © 1975. Reprinted by permission of Prentice-Hall, Inc.)

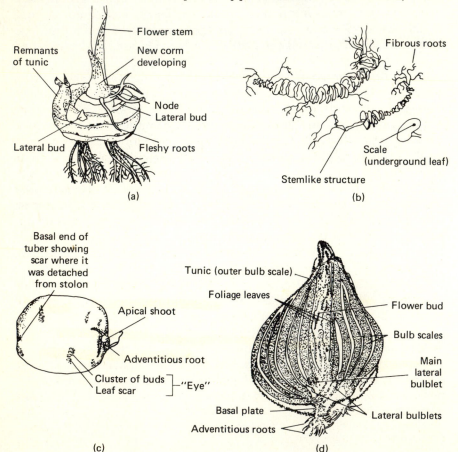

horizontal stems that grow partly or entirely underground. They are often thickened and serve as storage organs (Figure 4-14*b*). Rhizomes usually produce adventitious roots at their nodes and sometimes on their internodes as well; examples are: iris, many ferns, orchid, potato, banana, chrysanthemum, asparagus, and kohleria (Figure 4-14*b*). A *tuber* is a greatly enlarged fleshy underground stem, such as the Irish potato. The "eyes" of the potato occur at its nodes and consist of a leaf scar and axillary bud or buds (Figure 4-14*c*). A *bulb* is a budlike structure consisting of a small stem with closely crowded fleshy or papery leaves or leaf bases. Axillary bud primordia may occur in the axils of the leaves of some bulbs, such as the lily, onion, and tulip (Figure 4-14*d*).

Other aerial stem modifications are the *cladophyll* of asparagus, a leaflike structure which may bear flowers, fruits, and temporary leaves; *thorns*, which may be simple, as in osage orange, or branched, as in honey locust; and *spines*, which may arise as thorns or from leaves, crowns, or shortened stems. They may produce rosettes of leaves.

LEAVES AND LEAF MODIFICATIONS

Leaves are lateral outgrowths of stems which have developed special structural adaptations for photosynthesis. A very important difference between leaves and stems is their potential for growth. Stems are characterized by *indeterminate growth*; i.e., potentially their growth is limitless. Leaves, however, are characterized by *determinate growth*. After a period of multiplication and enlargement, leaf cells stop dividing and enlarging. With the achievement of maturity, the leaf functions for one or several seasons and then falls from the plant. The fundamental plant processes of photosynthesis and transpiration take place in the leaf, although they may also occur in other plant areas, such as in herbaceous and very young woody stems. Respiration also occurs in cells of the leaf as well as in all other living cells of the plant. Practically all the energy which is utilized by both plants and, ultimately, animals is manufactured in leaves by the process of photosynthesis.

Basic Makeup of the Leaf

The leaf consists of the *blade*, usually the flattened, green, expanded portion; and the *petiole*, or leaf stalk (Figure 4-15). *Stipules*, leaflike appendages often found on either side of the base of the petiole, may subtend the leaf but are not actual parts of the leaf. They are often flattened and leaflike, but may appear as tendrils or thorns on the stem. With some plants, they remain attached to the stem as long as the

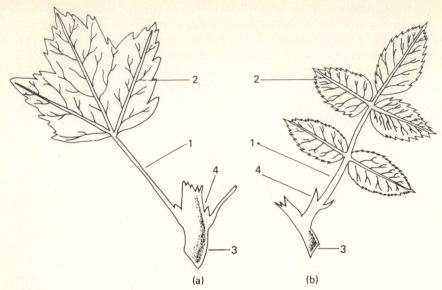

FIGURE 4-15 Parts of leaf. (*a*) Simple palmate leaf: 1, petiole; 2, blade; 3, stem; 4, axillary bud. (*b*) Compound pinnate rose leaf: 1, petiole; 2, leaflet (all leaflets make up the blade); 3, stem; 4, photosynthetic stipules.

petiole and blade remain, whereas with other plants they abscise soon after the leaf unfolds. The primary function of the blade is the manufacture of food. The petiole holds the blade out for maximum exposure to light and provides conducting tissue for transport of foods, water, and nutrients between the stem and leaf blade. In some plants, such as celery and rhubarb, the petiole has a large quantity of parenchymatous tissue that serves for storage. In some *Philodendrons* the petiole and associated tissues make up the broad, flat portion of the leaf with only a small segment constituting the blade. Sometimes the petiole is not present and the leaf is said to be *sessile*.

Types of Leaves

Leaves of dicotyledons have netted venation and are of two basic types: pinnately veined leaves and palmately veined leaves. *Pinnately veined leaves* have secondary veins extending laterally from a single midrib, as in apple, beech, and rose leaves. *Palmately veined leaves* have several large veins radiating into the blade from the petiole at the level where petiole and blade join, as in sycamore, maple, and buckeye leaves. Pinnate and palmate venations lend a degree of resistance to blade tear. Leaves with *parallel venation* are those in which there are several rather large veins that are essentially parallel to each other and are not connected by lateral veins. This type of venation is a characteristic of monocots, such as corn and other grasses. In banana and calla lily, the parallel veins run laterally from the midrib.

Simple leaves are those with blades consisting of one unit; such blades may have various types of marginal indentations. *Compound* leaves have blades divided into several leaflets which grow from the rachis (leaf axis). Both simple and compound leaves may be either palmately or pinnately veined. Maple leaves are simple palmate (Figure 4-15*a*), oak leaves are simple pinnate, roses have compound pinnate leaves (Figure 4-15*b*), and the buckeye is compound palmate.

Phyllotaxy of a plant refers to the arrangement of the leaves on the stem. When two leaves arise from opposite sides of the same node, as in dogwood and honeysuckle, they are said to be *opposite*. *Alternate* leaf arrangement, as in the rose and walnut, is characterized by one leaf per node; these are on different sides along the stem. In the *whorled* arrangement, three or more leaves develop at a node, as in the barberry. Other leaf characteristics that are used in plant identification are leaf shapes, leaf apices, leaf bases, and leaf margins.

Anatomical Structure

Anatomical structures of leaves may vary in accordance with the environment in which they grow; however, a "typical" dicotyledonous leaf is presented in cross section in Figure 4-16 and a "typical" monocotyledonous leaf showing parallel veins is presented in Figure 4-17. The *epidermis*, which is usually one-cell-layer thick, is a protective layer of cells composing the upper and lower surfaces of the leaf. It is covered with a waxy layer called the *cuticle* which reduces the loss of water from the leaf. Cells of the epidermis may be thicker on the side exposed to the sun. Small openings, known as *stomata* (singular, *stoma*), occur in the epidermis. They may occur in the upper and lower epidermis or, as is frequently the situation, in the lower epidermis only. Stomata are formed by specialized cells known as *guard cells*. These cells are sensitive both to light and to the relative humidity within the leaf and in the surrounding atmosphere. As a result of changes in their internal turgor pressure, guard cells open or close the stomata. An increase of this pressure causes the guard cells to open the stomata, and a decrease results in their closing. Through the stomata CO_2, O_2, and water vapor are exchanged with the atmosphere (Figure 4-18). The rate of gas exchange and water loss is affected by the number of stomata per unit of surface area as well as

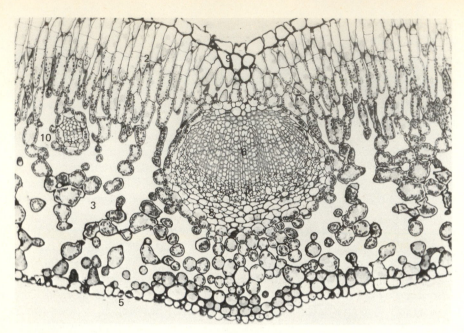

FIGURE 4-16 Leaf structure. Cross section of dicotyledon leaf (privet, *Ligustrum vulgare*) showing details of midvein and blade tissue on either side of midvein: 1, upper epidermis; 2, closely packed, multiple-layered palisade cells with numerous chloroplasts; 3, loosely packed spongy cells with chloroplasts (notice the large amount of intercellular space); 4, lower epidermis; 5, pair of guard cells and nearly closed stoma in lower epidermis; 6, radiating rows of xylem components of midvein; 7, phloem of midvein; 8, parenchyma cells associated with midvein (in some leaves, cells of this area develop into fibers with strikingly thick walls); 9, sclerenchyma cells just below upper epidermis and just above lower epidermis in midvein region; 10, branched vein enclosed in bundle sheath. Section cut 1.5μ thick. (Carolina Biological Supply Company.)

FIGURE 4-17 Leaf structure. Cross section of a monocotyledon (corn, *Zea mays*) leaf with large central midvein and smaller parallel veins. Cells of spongy mesophyll contain numerous chloroplasts. Pairs of guard cells with their stomata are in both upper end lower epidermal layers. Internal to each stoma is an enlarged area of intercellular space. Between both upper and lower epidermises and the midvein is a band of sclerenchyma cells. 1, upper epidermis; 2, lower epidermis; 3, spongy mesophyll; 4, guard cells and stoma; 5, midvein; 6, small vein paralleling midvein; 7, sclerenchyma cells. (Carolina Biological Supply Company.)

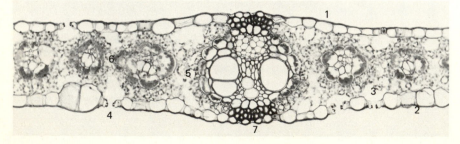

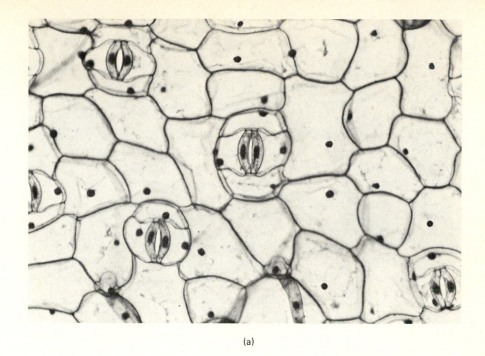

(a)

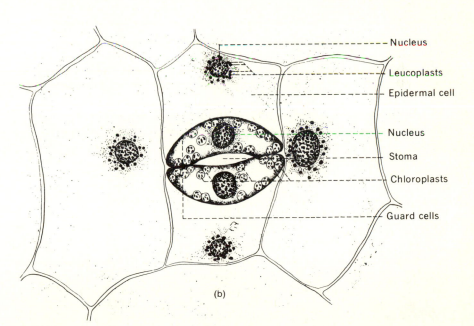

- - Nucleus

- Leucoplasts

- Epidermal cell

- Nucleus

- Stoma

- Chloroplasts

- Guard cells

(b)

FIGURE 4-18   (*a*) Surface view of epidermis from a leaf of *Tradescantia*. Guard cells are bordered by subsidiary cells. (Carolina Biological Supply Company.) (*b*) Surface view of a stoma from a leaf of *Tradescantia*. (Drawing by Elsie M. McDougle; from J. B. Hill, H. W. Popp, and A. R. Grove, *Botany*, 4th ed., McGraw-Hill, New York, 1967.)

by morphological features such as the presence of leaf hairs or trichomes, the thickness of the cuticle, or a sunken position of the stomata.

Between the upper and lower epidermal layers of the leaf is the *mesophyll*, a tissue consisting of palisade parenchyma cells and spongy parenchyma cells. Palisade cells are elongated and generally form the upper part of the mesophyll. They contain many chloroplasts with chlorophyll so that most of the photosynthetic activity within the leaf occurs in them. Spongy cells are present between the palisade and lower epidermis. These are loosely arranged, thin-walled, irregular cells of various sizes. Intercellular space occurs among them. Normally this space is saturated with water vapor when the leaf is fully turgid. Chloroplasts are present in limited numbers in spongy cells; thus some photosynthesis occurs in them. Small vascular bundles, or veins, occur through the spongy layers of the blade. These bundles are composed of xylem and phloem and extend from the midrib or main veins of the blade. Often they are encased in a *bundle sheath* composed of parenchyma cells or sclarified cells. Water, minerals in solution, and food in solution diffuse between the components of the small veins and the mesophyll cells.

Leaf Modifications

There are many kinds of leaf modifications from the usual foliage, photosynthetic leaves. Rhizomes have *scales* which are actually modified leaves. *Storage leaves* are found in bulbous plants, such as lily and onion. *Succulent leaves* are water storage structures characteristic of plants in arid and semiarid regions. Many *tendrils* are slender, twining leaf modifications used for support, as in the terminal leaflets of the English pea, grape, and cucumber (Figure 4-19a). Some tendrils are stipule modifications. *Spines* are sharp-pointed, woody structures, usually modified from a leaf or part of a leaf (Figure 4-19b). *Bracts* which are usually small, pointed, green modified leaves subtend many flowers or inflorescences and may appear to be a part of the flower. The sepals, petals, stamens, and pistils of flowers have evolved from leaves into reproductive units.

FIGURE 4-19 Leaf modifications. (*a*) Leaves of grape (*Vitis rotundifolia*) modified into tendrils. Leaves of the garden pea (*Pisum sativum*) also have modifications of their terminal leaflets into tendrils. (*b*) Leaves modified into spines on the barberry (*Berberis* sp) bush. Axillary buds are in axils of stem and leaves. In addition to these, the barberry stem also forms small photosynthetic leaves. (Carolina Biological Supply Company.)

(a)

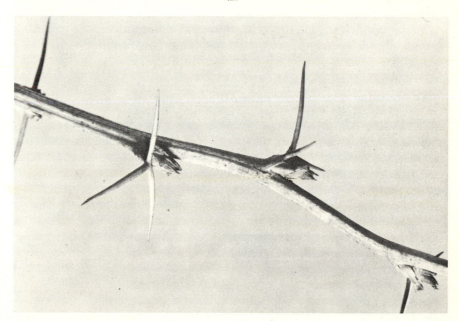

(b)

TISSUE SYSTEMS

Cells that form the various parts of the plant can be grouped in terms of function and structure into three so-called *tissue systems*. These are the dermal, the vascular, and the fundamental tissue systems.

Dermal System

The outermost layer of cells, the epidermis, forms the dermal system of floral parts, leaves, fruits, seeds, and of stems and roots until they begin secondary growth. The epidermal cells are specialized parenchyma cells; the shapes of these cells are very irregular because of the plasticity of the cell walls and the tension during the growth of the plant part. The epidermis usually is one-cell thick, but in some instances it is composed of several layers of cells and is then known as a *multiple epidermis* (examples include rubber plant and peperomia). Walls of epidermal cells are often impregnated with *cutin*, a group of waxy substances deposited both within the cell walls and on the outer surface of the walls. Cutin minimizes water loss. In roots, this deposition occurs in the epidermal cells above the root hair zone.

The epidermis is replaced in the older portions of roots and stems in which secondary growth occurs by the *periderm* which consists of cork (phellem), cork cambium (phellogen), and in some instances underlying layers of cells (phelloderm). The cork is composed of cells with suberized walls formed by the cork cambium. These cells are nonliving; their waxy suberin waterproofs the cell walls. The cork cells may also serve to protect, to some degree, the underlying tissues of the plant.

Vascular System

As was stated previously, the vascular system is composed of the xylem and phloem, which conduct water, mineral salts in solution, and foods in solution. The xylem and phloem also provide some support and strength. In mature perennial plants, the vascular system includes most of the root and stem. The arrangement of the vascular system varies with different kinds of plants and with different parts. In a dicotyledonous stem, the system consists of a number of vascular bundles which form a cylindrical pattern. Typically, the inner part of each bundle is composed of xylem; the outer portion consists of phloem. In monocotyledonous plants the vascular bundles occur longitudinally at random throughout the parenchymatous, fundamental system of the plant. In young roots, the vascular system makes up the central portion of the root which is surrounded by the pericycle and endodermis. The pericycle is made up of one or more layers of cells from which lateral roots arise. Collectively the vascular tissues and associated fundamental tissues of the root and stem are called the *stele*. The epidermis is associated with, but is not part of, the stele.

Fundamental System

All tissues of the plant encased by the epidermis, other than those of the vascular system, make up the fundamental system. In young roots, this system consists of the cortex; and in the young stem it consists of the cortex, pith, and pith rays. The cortex, as a result of extensive secondary growth in older woody stems as well as woody roots, is stretched, crushed, and gradually destroyed. Starch accumulates in cells of the cortex as long as it is present in the plant part. Basically, the cortex is composed of parenchyma cells, but collenchyma and sclerenchyma cells may also be present.

BIBLIOGRAPHY

Bold, H. C.: *Morphology of Plants*, 3d ed., Harper & Row, New York, 1973.
Eames, A. J. and L. H. MacDaniels: *An Introduction to Plant Anatomy*, 2d ed., McGraw-Hill, New York, 1947.
Esau, K.: *Plant Anatomy*, 2d ed., Wiley, New York, 1965.
Haynes, J. D.: *Botany: An Introductory Survey of the Plant Kingdom*, Wiley, New York, 1975.
Hill, A. F.: *Economic Botany*, 2d ed., McGraw-Hill, New York, 1952.
Hill, J. B., H. W. Popp, and A. R. Grove, Jr.: *Botany*, 4th ed., McGraw-Hill, New York, 1967.
Swingle, D. B.: *A Textbook of Systematic Botany*, 3d ed., McGraw-Hill, New York, 1946.
Tortora, G. J., D. R. Cicero, and H. I. Parish: *Plant Form and Function*, Macmillan, New York, 1970.
Troughton, J. and L. A. Donaldson: *Probing Plant Structure*, McGraw-Hill, New York, 1972.
Wilson, C. L., W. E. Loomis, and T. A. Steeves: *Botany*, 5th ed., Holt, New York, 1971.

CHAPTER 5
REPRODUCTIVE
DEVELOPMENT

All seed-producing plants have a life cycle in which there is both a sexual and an asexual phase. Angiosperms (flowering plants) produce seeds and fruits as a result of such a life cycle. This is of interest to the horticulturist not only as a reproductive process but also because in many instances the flower, fruit, or seed is the aesthetic or economic part for which the plant is grown. In addition, many angiosperms propagate themselves vegetatively by bulbs, tubers, corms, or stolons. Human beings, in their efforts to produce plants for their own welfare, have utilized these natural means and have devised several other means of vegetative propagation, such as budding and grafting.

FLOWERS

Structure

The complete angiosperm flower consists of four groups of parts: *sepals* (collectively comprising the calyx); *petals* (collectively comprising the corolla); *stamens*; and *pistil* or *pistils*. Sepals and petals comprise the two outermost whorls of the flower, and collectively are designated as the *perianth*. They are not of direct reproductive significance, but serve to protect the inner reproductive structures; in

FIGURE 5-1 Bisected complete dicotyledon flower of the strawberry. Sepals, petals, and stamens are attached to the receptacle around the ovary. 1, receptacle; 2, pistil; 3, filament; 4, petal; 5, sepal; 6, flower stalk; 7, anther; 8, stamen. (Photograph by Fred D. Cochran, North Carolina State University.)

addition, petals often attract insects either by their color or by their nectar, and thus facilitate pollination.

The *flower* is a shoot of determinate growth with modified leaves that is supported by a short stem, the *pedicel*. The receptacle is the enlarged apex of the pedicel (Figure 5-1) where the floral parts arise. It is sometimes called the *torus*, especially when it becomes conspicuously enlarged as in a strawberry or raspberry. *Bracts* (modified leaves) may occur at the base of the receptacle. They are of little concern with many plants, but with certain ornamentals, such as dogwood, hydrangea, and poinsettia, they are very important, since they make up the showy portions of the plant. The *sepals* usually form the outermost whorl of the flower (Figure 5-2).

The *stamen* is the male reproductive organ and consists of an *anther*, or pollen sac, supported by a long, slender stalk, the *filament*. Within the anther the meiotic process produces pollen grains which give rise to male gametes (haploid sperm cells).

The *pistil* is the female reproductive organ. The *stigma* of the pistil is the pollen-receiving site; the *style* connects the stigma to the *ovary* (the enlarged basal portion that serves as the site of seed formation), which contains the *ovules*. Within each ovule, a female gamete is formed. The pistil may be simple or compound. A simple pistil consists of a single carpel, a specialized leaf which has become modified for

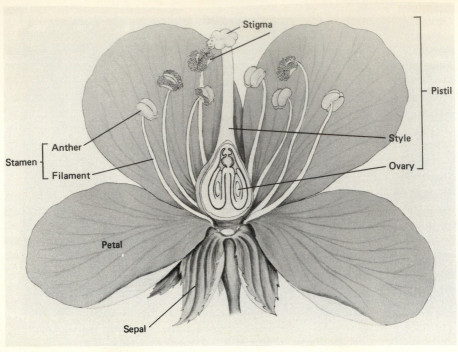

Stigma

Pistil

Style

Stamen — Anther
— Filament

Ovary

Petal

Sepal

(a)

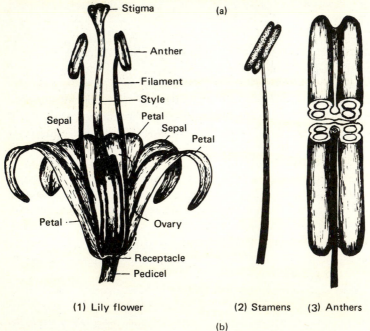

Stigma

Anther

Filament

Style

Sepal

Petal

Sepal

Petal

Petal

Ovary

Receptacle

Pedicel

(1) Lily flower (2) Stamens (3) Anthers

(b)

FIGURE 5-2 (*a*) Complete dicot flower with radial symmetry. (Carolina Biological Supply Company.) (*b*) Parts of the flower: (1) flower of lily with one sepal and four stamens removed; (2) view of stamen showing attachment of filament to anther; (3) an immature anther cut in half showing the four pollen sacs and the connective. (From J. B. Hill, H. W. Popp, and A. R. Grove, *Botany*, 4th ed., McGraw-Hill, New York, 1967.)

seed bearing. The ovary of a simple pistil has only one cavity or locule, e.g., pea. Compound pistils are composed of two or more carpels and have two or more locules in their ovary. The lily has a compound pistil with three locules.

The young anther is made up of four lobes with four chambers known as *microsporangia*. Within each microsporangium, diploid microspore mother cells are found. As the anther proceeds in growth, each microspore mother cell nucleus undergoes meiosis, as a result of which four haploid cells are formed. These are *microspores*. The four microspores, which collectively are known as a *tetrad*, later develop into four *pollen grains*. As the pollen grain matures its nucleus divides by mitosis to form two nuclei, which are designated as the *tube nucleus* and the *generative nucleus*. The tube nucleus is one of the nuclei of a pollen grain thought to influence the growth and development of the pollen tube. The generative nucleus is the nucleus of a pollen grain which by division forms the sperms. These two nuclei within the pollen grains are not separated by a cell wall, but are suspended in the cytoplasm of the pollen grain. They comprise the male gametophyte generation in the life cycle of the angiosperm. Pollen grains of different species have walls with various characteristic spines, plates, or ridges. Pollen dissemination takes place as lobes of the anther dehisce (split). Pollen is usually disseminated by wind or insects.

The ovary of the pistil of the flower is a structure which occurs only in angiosperms. It encases one or more cavities (*locules*). When young, one or more ovules (*megasporangia*) are initiated in each locule. In its early stages each ovule is a small mass of diploid cells. One of the innermost of these cells functions as a megaspore mother cell, since it divides by meiosis to form four haploid megaspores. Three of these megaspores deteriorate, leaving one functional megaspore within the small ovule. As the flower matures, the ovule increases in size and its functional megaspore strikingly enlarges to form a cavity within the ovule. This cavity is known as the *embryo sac*. As the embryo sac is formed, the nucleus of the functional megaspore divides and redivides within the sac, forming a group of haploid nuclei. Each nucleus is associated with a small quantity of cytoplasm. There haploid nuclei and associated quantities of cytoplasm comprise the female gameto-phyte phase in the life cycle of the angiosperm. In certain types of angiosperms eight of these nuclei occur within the embryo sac. One of these is a functional female gamete (megagamete).

Two other nuclei are designated as *polar nuclei*. The remaining nuclei are known either as *synergid* or as *antipodal* nuclei; these are without known function in the modern flower and ultimately deterio-rate. The wall of the mature ovule surrounds the embryo sac and is designated as the *integument*. It is several layers of cells in thickness. A small hole or passageway (micropyle), extends through the integu-ment; it is located near where the ovule is attached to the ovary wall.

Pollination and Fertilization

Transfer of pollen from the anther to the stigma of the pistil is *pollination*. Once pollination is accomplished, the wall of the pollen grain breaks down and the cytoplasm associated with its tube nucleus forms a *pollen tube* which grows downward through stigma and style, into the ovary and thence into an ovule (Figure 5-3). It grows through the micropyle of the ovule and into its embryo sac. As this pollen tube grows downward, both the tube nucleus and the generative nucleus with its cytoplasm are swept from the interior of the pollen grain to a location near the tip of the tube. Before or by the time the pollen tube penetrates the embryo sac of the ovule, the generative cell will have divided by mitosis into two haploid male gametes.

Once within the embryo sac, the end of the pollen tube deteriorates and the two male gametes are swept by cytoplasmic currents

FIGURE 5-3 With the occurrence of fertilization within the ovule (or ovules) of a flower, the following events occur: (1) The zygote develops (by cell division and cell differentiation) into an *embryo* (immature plant). (2) The ovule undergoes a period of growth and development and becomes a mature, functional *seed*. (3) The ovary containing the fertilized ovule (or ovules) undergoes a period of growth and development and becomes a mature *fruit*. (Courtesy of R. P. Ashworth, Clemson University.)

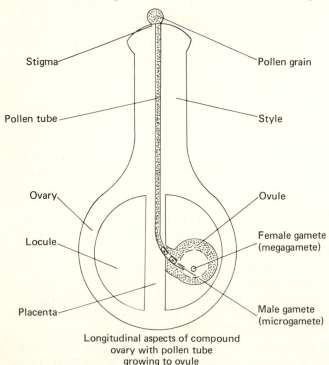

Stigma

Pollen grain

Pollen tube

Style

Ovary

Ovule

Locule

Female gamete (megagamete)

Placenta

Male gamete (microgamete)

Longitudinal aspects of compound ovary with pollen tube growing to ovule

from the pollen tube into the sac. One male gamete unites with the female gamete in the sac to form a diploid *zygote* (a zygote is a fertilized egg). This is the beginning of a new sporophyte phase. The second male gamete unites with the polar nuclei to form a polyploid (with more than two complete complements of chromosomes) *primary endosperm cell*. A union of the two male gametes with the female gamete and the polar nuclei is known as *double fertilization*; in almost all instances it is necessary for the ovule to continue its development. Once this occurs, however, the ovule undergoes various changes to ultimately form a seed. During these changes the zygote, by mitosis, develops into a *sporophytic embryo*; the primary endosperm cell, by mitosis, develops into a mass of food-storing cells known as the *endosperm*; and the *integument* becomes the seed coat.

In the horticultural sense, *self-pollination* of asexually propagated plants (those genetically alike because of vegetative propagation) is the pollination of a flower of one cultivar with the pollen from a flower of the same cultivar. All plants of asexually propagated plants are members of the same clone; this means that should a cross-pollination requirement exist, it would have to come from a member of a different clone and not a different plant of the same clone. *Cross-pollination* is the transfer of pollen to another plant of a different line. Cross-pollination in asexually propagated plants involves two distinct cultivars. Since 'Delicious' apples are essentially self-sterile, a pollinizing cultivar must be included in a 'Delicious' orchard. 'Golden Delicious' is a good pollenizer, since it flowers at the same time and has abundant pollen. Certain cultivars of apple, especially the triploid cultivars, have sterile pollen and therefore require a pollinizing cultivar. Self-pollination of sexually propagated plants must involve pollen and the stigma of the same flower or flowers on the same plant or flowers of the same inbred line. *Self-fertility* is the ability of a plant to set viable seed or fruit with pollen from the same cultivar. *Self-sterility* is the lack of this ability. Intersterile cultivars cannot set viable seed with pollen from each other.

If the male gamete formed by a pollen grain and the female egg nuclei within the ovule of a given flower or plant do not mature at the same time, self-pollination cannot occur; this results in the same condition as that in the imperfect male and female flowers of pecan. Therefore it is necessary in planting a pecan orchard to provide a group of cultivars that can pollinate one another. Flowers of many plants may open at about the same time (as with the peach, apple, and quince), or they may open progressively over a long period of time (as with the tomato, melon, citrus, cucumber, gladiolus, or snapdragon). Flowering thus extends over a considerable portion of the season; the harvest is also over a long period (Figure 5-4).

Pollen germination is stimulated by the presence of certain inorganic substances, such as manganese sulfate and calcium and boron

FIGURE 5-4 Lemon plant, showing fruit and flower present at same time.

salts, and probably also some organic substances. In the lily, boron is concentrated in the style and stigma. There is evidence that substances in the style chemically attract the growth of the pollen tube. If these substances are limiting, the growth of the pollen tube may be so slow that fertilization is delayed; this may result in the abscission of the flower. Following fertilization, the zygote, by cell division, grows into an embryo and the remaining parts of the ovule grow into a seed. At the same time, fruit development occurs and abscission, which normally takes place with unfertilized flowers, is prevented by complex metabolic processes which are not well understood.

Flower Types

Flowers of many horticultural plants are objects of beauty (Figure 2-4). Flowers of all fruit crops and of certain vegetables are necessary for the production of the edible product (Figure 2-2).

A *complete flower* is composed of a short axis or receptacle from which arise four sets of floral parts: sepals, petals, stamens, and pistils (Figure 5-2). Some examples of plants with complete flowers are: bean, tomato, pea, okra, pepper, eggplant, apple, peach (Figure 2-2), plum, pear, labrusca grape, raspberry, blackberry, most cultivars of strawberry (Figure 5-1), almond, gooseberry, avocado, citrus fruits, and most ornamental shrubs and trees. An *incomplete flower* lacks one or more of the four sets of floral parts. A *perfect flower* has both the pistil or pistils and the stamen, but it may lack sepals, petals, or both. An *imperfect flower* lacks either the stamen or the pistil. Complete flowers are perfect flowers, but incomplete flowers may be either perfect or imperfect. In *pistillate flowers*, only the pistils are present; there are no stamens. *Staminate flowers* have only the stamens; they lack the pistils. Sometimes neither the stamens nor the pistils are present. In this case, the flower is said to be *sterile* and is nonfunctional.

Ovules are borne within the locule or locules of the ovary. If an ovary has but one locule, it is said to be a *simple ovary*. If, however, it has two or more locules, it is said to be *compound*. The tissue of the ovary to which an ovule (or ovules) is attached is the *placenta*. In simple ovaries the single or several ovules are attached to the inner surface of the ovary wall. In the compound ovary, ovules may be attached to the inner surface of the ovary wall or to a *central axis* that separates the locules of the ovary. If ovules arise from the inner surface of the ovary wall, the ovary is said to have *marginal* or *parietal placentation*, whereas if they arise from the axis of the compound ovary, the ovary has *axial placentation*.

Classifications   Flowers are classified as hypogynous, perigynous, or epigynous, depending on how the different floral parts are attached to the receptacle or ovary. If the sepals, petals, and stamens are attached to the receptacle below the ovary, the ovary is said to be superior and the flower is *hypogynous* (Figure 5-5). If the receptacle is extended to

FIGURE 5-5 Hypogynous flower of tulip, bisected, showing superior ovary. Sepals, petals, and stamens are attached to the receptacle below the ovary. (Carolina Biological Supply Company.)

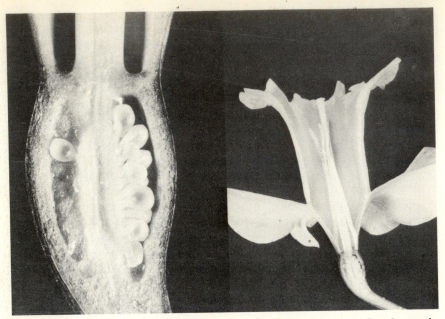

FIGURE 5-6 Epigynous flower of *Narcissus*, showing inferior ovary. Sepals, petals, and stamens are attached to the receptacle above the ovary. (Carolina Biological Supply Company.)

form a cuplike structure around a portion of ovary, the flower is *perigynous* (Figure 5-1). Again, the ovary is superior. The hypanthium may be inconspicuous or very prominent, but it must be present. If the perianth and stamens are attached above the ovary, the ovary is inferior and the flower is *epigynous* (Figure 5-6). Epigyny is considered to be a more advanced state in the evolutionary progression than either perigyny or hypogyny.

If the flower may be divided into similar parts by cutting it in more than one longitudinal plane and all the floral parts of each group are alike in size and shape, the flower is *radially symmetrical* and is said to be regular or *actinomorphic*; examples are tomato, magnolia, petunia, and peach. If the flower may be divided into two similar parts by division along one longitudinal plane only, the flower is *bilaterally symmetrical* and is said to be *irregular* or *zygomorphic*, e.g., orchid, garden pea and other legumes, snapdragon, mints, and violets. Irregular flowers are usually insect-pollinated, especially by the honeybee. Irregular symmetry is considered a more advanced evolutionary state than regular symmetry.

Sexual expression in plants is based on whether one or both of the sex organs are in the same flower. Perfect flowers have both male and female reproductive parts. Plants which produce staminate and pistillate flowers may be either monoecious or dioecious. In *monoecious*

(a)

(b)

FIGURE 5-7 A monoecious plant, sweet corn, showing (*a*) staminate and (*b*) pistillate flowers. (Carolina Biological Supply Company.)

species, both staminate or "male" flowers and pistillate or "female" flowers are borne on the same plant, as in sweet corn (Figure 5-7), oak, squash, pecan, filbert, cucumber, muskmellon, pumpkin, and watermelon. *Dioecious* species produce staminate and pistillate flowers on

FIGURE 5-8 Papaya fruit produced on pistillate plant with staminate flower on adjacent plant.

separate plants, as in asparagus, date palm, papaya (Figure 5-8), holly, spinach, and certain cultivars of grapes. *Andromonoecious* species, such as muskmelon, contain perfect as well as imperfect staminate flowers on the same plant. *Gynomonoecious* species, such as cucumber, contain perfect as well as imperfect pistillate flowers on the same plant. Flowers with large numbers of floral parts, such as numerous stamens or numerous pistils, are considered primitive. Reduction in floral parts is considered an advancement in evolution.

Inflorescences   Flowers may occur singly (e.g., tulip) or in closely associated clusters known as *inflorescences* (e.g., snapdragon). The inflorescence is supported by a main branch called a *peduncle*. Single flowers of the inflorescence branch off the peduncle. These single flowers may be connected to the peduncle by short stalks, called *pedicels*, or may be sessile and connected directly to the peduncle. A bract may occur at the axil of each flower or flowering branch.

The flower clusters of an inflorescence are classified on the basis of the way the individual flowers are attached to the stem (Figure 5-9). A *raceme* has stalked flowers on pedicels which are approximately equal in length on a single floral axis (hyacinth, snapdragon). In the *corymb*, a type of raceme, the pedicels of the lower flowers are longer than the pedicels of the upper flowers, resulting in a flat-topped

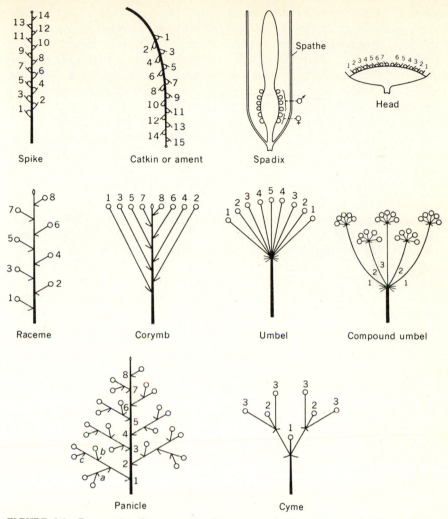

FIGURE 5-9 Types of inflorescences, diagrammatically represented. Flowers are shown by small circles, and bracts by short, slightly curved lines. Figures indicate the usual sequence of opening of flowers, number 1 opening first. (From J. B. Hill, H. W. Popp, and A. R. Grove, *Botany*, 4th ed., McGraw-Hill, New York, 1967.)

inflorescence (candytuft). A *panicle* can be either a cluster of racemes or corymbs and is distinctly branched. The *spike* is an indeterminate raceme of sessile flowers attached to the floral axis with the oldest flowers at the base (wheat, rye, plantain). A *catkin* is a type of spike that has unisexual flowers with a perianth. They are usually soft and pendulous and are found only on woody plants, such as oak and pecan. A *spikelet* is a unit of arrangement found in grasses. A *spadix* is a fleshy spike with very small male flowers above and female flowers below embedded in the spike axis, such as calla. The head, or *capitulum*, is a globose or disc inflorescence with a very short axis and sessile flowers (dandelion, sunflowers, clover). In the *umbel* the pedicels arise from a common point and are about equal in length (onion). A *compound umbel* is a series of simple umbels arising from the same point on the main axis (carrot). A *monochasium* has a peduncle bearing a terminal flower with a lateral flower producing branches below it. A *dichasium* has a peduncle with a terminal flower and a pair of lateral branches below each bearing lateral branch. These types are *indeterminate* inflorescences, since the flowers open from the bottom to the top or from the outer margins to the center. This blooming pattern permits further elongation of the inflorescence after flowering has started.

A second type of inflorescence is *determinate*, where the terminal flower is the first to form, preventing further elongation of the florescence. There are three basic types of determinate inflorescences—cyme, glomerule, and thyrse. A *cyme* is a broad, more or less flat-topped inflorescence, such as dogwood, in which the central flowers bloom first. The pedicels may be either alternate or opposite around the central axis. A cymelike inflorescence with a dense head is a *glomerule* (mint). A *thyrse* is a compact, condensed cyme or panicle (blackberries).

ALTERNATION OF GENERATIONS

Sporophyte and Gametophyte Phases

Two phases or generations occur in the life cycle of angiosperms. The asexual phase produces *spores*—a unicellular or few-celled structure of many types and forms—and is designated as the *sporophyte phase*. The sexual phase is the *gametophyte phase* and produces male and female sex cells (*gametes*). The sporophytic phase is conspicuous, consisting of roots, stems, leaves, and flowers. In contrast, the gametophytic phase is greatly reduced in activity and is composed of relatively few cells which occur in the tissues of the flower and are parasitic on the sporophytic phase. The gametophytic phase is of two entities: the male gametophyte (*microgametophyte*) and the female gametophyte (*megagametophyte*). Nuclei of the sporophytic phase are diploid (2N), whereas the gametophytic nuclei are haploid (1N).

Meiosis

The transition of the diploid chromosome status of the sporophytic phase to the haploid condition of the gametophytic phase is accomplished by a particular type of cell division known as *meiosis*. Chromosomes in sporophytic cells are in homologous pairs. During meiosis, daughter cells are formed in such a way that each such cell receives *one* chromosome of *each pair* of homologous chromosomes present in the parent sporophyte cell. Each daughter cell is thus haploid. However, before new, complete cell wall material separates the daughter cells into discrete entities (i.e., before cytokinesis is completed), each daughter cell divides by the process of mitosis. As a result, four haploid granddaughter cells are formed from the original diploid sporophytic cell. Within the life cycle of the angiosperm the particular sporophytic cells which divide by meiosis are known as *microspore mother cells* or *megaspore mother cells*. By meiosis each microspore mother cell forms four haploid *microspores*, whereas each megaspore mother cell forms four haploid *megaspores*. By mitotic division each microspore forms a haploid male gametophyte phase, and each functional megaspore produces a haploid female gametophyte phase.

Meiosis is the basis for all life. It is the process that produces the gametophytic phase in both the male and female reproductive parts of the flower. Meiosis includes one reduction division of chromosome number followed by one duplication division (mitotic division). After two mitotic divisions in the male and three divisions in the female, the gametes in the anthers and ovaries originate in which their chromosome number is reduced by one-half. In a normal plant each cell contains the same number of chromosomes. In many plants each cell contains two complete sets of chromosomes. This is referred to as the *diploid* or 2N condition for that organism. During meiosis the chromosome number for both the male and female reproductive cells or generative cells is reduced to half the normal number or to 1N, the *haploid* number of chromosomes for that organism.

Meiosis is essentially the same process in the anthers and ovaries. It takes place in two stages and results in four daughter cells, each cell with the haploid chromosome in both the male and female flower. In the female flower three of the daughter cells usually abort, whereas in the male flower all four daughter cells function in pollination and fertilization.

When one of these haploid gametes from the male parent fuses with one haploid sex cell from the female parent, fertilization has occurred. The single-celled structure that results after fusion of the two nuclei becomes the 2N zygote. Each diploid cell consists of chromosomes exactly like those contributed by both parents to the original zygote. Each chromosome given to a zygote from one parent has a compliment of chromosomes from the opposite parent. It is these two complements of chromosomes that pair in meiosis at the metaphase plate during future divisions.

For convenience, the process of meiosis (Figure 5-10) is considered as having distinct phases, even though it is actually a continuous sequence of events.

Stage 1 *Prophase* In the prophase the chromosomes become greatly elongated. They undergo pairing (synapsis), during which two chromosomes become intimately associated, aligning themselves side by side. Pairing is not indiscriminate, but always occurs between a chromosome derived from one parent and the corresponding chromosome derived from the other parent. The pairing relationship exists between homologous chromosomes or those with the same gene arrangement or morphology. This means that they are similar in appearance and bear corresponding genes with the same characteristics. By the time the chromosomes have paired, each has already duplicated itself longitudinally throughout its length and consists of two chromatids (as in the prophase of mitosis). The chromosomes thicken and shorten until it can be seen that each pair of chromosomes consists of four chromatids. The chromosome arms as well as centromeres have replicated exactly. There are, then, four arms and four centromeres per pair of chromosomes.

Metaphase Here the chromosomes move into the center of the cell along a metaphase plate. Their centromeres do not lie along the equator, as in mitosis, but appear to repel each other, so that the centromeres of the pairs lie toward the poles with their arms at the equator. At this stage the centromeres appear to repel each other even though the migrating chromatids may be in close contact.

Anaphase The paired chromosomes separate and move toward opposite poles along the spindle apparatus. The chromatids of each chromosome do not separate as they do in mitosis but the arms of the corresponding chromatids remain in close contact and migrate to the terminal ends of the chromatids as the centromeres move apart. The centromere splits at the metaphase stage of meiosis II and the chromatids move apart at this time.

Telophase Ordinarily nucleoli are not formed as they are in mitosis, but a nuclear membrane does develop. The chromosome number of the daughter nuclei has now been reduced to half that of the parent cell.

FIGURE 5-10 Stages of meiosis during the first and second meiotic divisions. Meiosis has been divided into artificial stages, or phases, for convenience in studying the process; however, the process is a continuous sequence of events.

(a) Meiosis I

(b) Meiosis II

Prophase I

(nuclear membrane disappears)

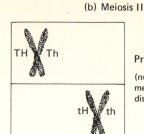

Prophase II

(nuclear membrane disappears)

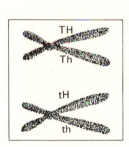

Metaphase I

(chromosomes have duplicated length-wise gene for gene and have migrated to metaphase plate; crossing over has occurred)

TH
Th
tH
th

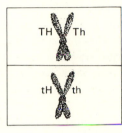

Metaphase II

(chromosomes appear at meta-phase II plate)

TH   Th

tH   th

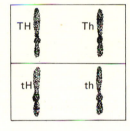

Anaphase II

(chromatids move to opposite poles)

TH   Th

tH   th

Anaphase I

(chromosomes move to opposite poles)

TH
Th

tH
th

Telophase II

(nuclear mem-brane reappears; 2 additional daughter cells; 4 types of gametes have been produced: TH Th tH th)

TH   Th

tH   th

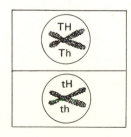

Telophase I

(2 daughter cells; nuclear membrane may reappear)

TH
Th

tH
th

Interphase

(nuclear membrane present)

Stage II *Prophase* In the prophase, nuclear membranes disappear.

Metaphase The chromosomes move to center.

Anaphase The centromeres of each pair of chromatids separate and the chromatids move toward opposite poles. When the chromatids have separated, they are again called chromosomes. Cell walls are produced.

Telophase The chromosomes reorganize into nuclei.

Interphase The interphase terminates the end of the mitotic (duplication) phase of meiosis and results in the reduction in number of chromosomes to one-half of the original number and eventually the production of the gametes.

Cytokinesis (cell division) during meiosis produces the microspores in the male flower and the functional megaspore in the female flower. Once these two structures are formed karyokinesis (nuclear division) produces the male gametes within the pollen grain and the eight nucleate embryo sac with the ovule.

Meiosis provides the means of maintaining a constant chromosome number from one generation to another. It also provides the means for recombination of genetic information or genes. During meiosis, homologous chromosomes pair or synapse and crossing-over may take place between chromatids, providing the basis of genetic exchange. Gametes are made up of recombinations of genetic material in new and often unique combinations. This recombination effect is often referred to as genetic variability that is common sexual reproduction.

MATING SYSTEMS IN PLANTS

Plants are composed of cells made up of cytoplasm which houses the nucleus. Located in the nucleus are elongated bodies called *chromosomes* which contain *genes*, the basic units of hereditary material. Genes dictate the characteristics of individuals; that is, they determine the potential appearance and abilities that can be expressed in an individual within a given environmental status.

The union of male and female gametes (fertilization) is an event in which the haploid chromosome complement of one gamete augments a similar chromosome complement of the second gamete, thereby forming a cell (the zygote) in which the diploid or paired status (sporophytic status) of the chromosomes is restored. During the process of meiosis some daughter cells receive at random one chromosome of each diploid pair of chromosomes, while other daughter cells receive the other chromosome of each pair. This results in random distribution of chromosomes in the various spores produced in meiosis and in the gametophytic individuals produced by these spores. It follows that there is random distribution of chromosomes in gametes; as a result, sporophytic individuals of a given species have random pairing of their chromosomes, so that individuals of the same species exhibit differing individual genetic characteristics.

Cells of plants from the same species have the same number of chromosome pairs. For example, the cells of the onion contain eight pairs of chromosomes, with one member of each pair derived from the male parent and the other from the female parent. As the plant grows, its cells divide by mitosis to produce daughter cells with the same chromosome complements (see Chapter 3).

Cells of the sporophytic flower have chromosomes in pairs (i.e., they are diploid). By meiosis (reduction division) spore mother cells within the flower produce daughter cells (spores) in which there is one chromosome of each pair of chromosomes which occurs in the parent 2N cell. These daughter cells are therefore haploid. Each spore develops by mitosis into a gametophytic phase, so that all cells of this phase, as well as the male and female gametes formed by this phase, are 1N.

For example, mature tomato fruits may be either red or orange as a result of genetic recombination during meiosis followed by random distribution of chromosomes. The difference is determined by a single pair of genes. Genes for red color may be designated with a T and those for orange color with a t (Table 5-1). The *homozygous* TT red-fruited plant will produce only male gametes and eggs for red fruit; the homozygous tt orange-fruited plant will produce only male gametes and eggs for orange fruit. Each gamete of the TT parent receives one gene for T, and each gamete of the tt strain receives one gene for t.

TABLE 5-1
CROSS BETWEEN
RED- AND ORANGE-FRUITED TOMATOES
RESULTS IN 3 RED-FRUITED PLANTS
TO 1 ORANGE-FRUITED PLANT

	T	t
T	TT	Tt
t	Tt	tt

T = red
t = orange

Both plants would be uniform with regard to fruit color, and progeny from either plant would show no segregation for fruit color; this would result in the establishment of two true breeding lines for fruit color. When a homozygous TT red plant and a homozygous tt orange plant are crossed, both a T gene and a t gene are present in the zygote and in all vegetative cells produced thereafter from the zygote. The resulting plant is *heterozygous* Tt and is referred to as a *monohybrid* (formed by one pair of genes). It will, however, produce only red fruit, since the T gene is dominant and the t gene is recessive (that is, t cannot express itself in the presence of T). A cross of this type may also be referred to as an F_1 *hybrid cross*; it is also designated as a *single cross* in the production of some hybrid seed. Hybrid plants possess the potential to produce male and female gametes with either red T or orange t genes.

If a monohybrid tomato plant that is heterozygous for color (Tt) is self-pollinated, the gametes produced can form three genetic combinations in the seedling progeny: TT, Tt, tt. Each of these is a different *genotype*; i.e., each has a different gene combination. The term *phenotype* refers to external appearance of the plant governed by its internal genotype. Thus individuals with the genotype of TT and Tt have a phenotype of red, whereas those with a genotype of tt have a phenotype of orange. The segregation ratio among the progeny obtained from a heterozygous red tomato plant are in the genetic ratio of one homozygous red TT, two heterozygous red Tt, and one homozygous orange tt. The phenotypic ratio of the above four plants would be three red to one orange. The two heterozygous plants would produce the same fruit color as the homozygous red plant, since the recessive t gene cannot be expressed in the presence of T. Subsequent self-pollination of the two heterozygous plants will yield the same genotypic and phenotypic ratios as the original self: 1:2:1 genotypic, 3:1 phenotypic. The two homozygous plants will breed true: one for red fruit only, the other for orange.

If a plant differs by two pairs of genes, it is known as a *dihybrid*. The two pairs of genes may occur as members on the same chromosome pair and in close proximity to each other in a linear arrangement along the chromosome arm. When two pairs of heterozygous genes appear on the same chromosome in a plant, it is possible to obtain nine different genotypes and four different phenotypes when the plant is self-pollinated. Chromosomes do not duplicate themselves during mitosis or meiosis. Their chromatids separate in anaphase in mitosis, but remain together during anaphase I of meiosis. The two parts are referred to as *chromatids*. In meiosis I, when the two chromosomes part (to form chromatids), an exchange of segments between nonsister chromatids takes place. Sister chromatids would comprise the two

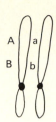
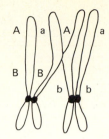
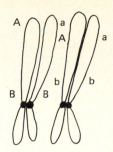

Both chromosomes duplicate themselves linearly before crossing over

Each chromatid becomes a chromosome in the male gametes and in the
egg cells

FIGURE 5-11 An exchange between nonsister chromatids known as *crossing-over.*

longitudinal pieces resulting from the duplication of one chromosome. The exchange between nonsister chromatids transfers genes from one chromatid to another. This phenomenon, referred to as *crossing-over,* results in a recombination of genes between the two original chromosomes that existed in the vegetative cells (Figure 5-11). Crossing-over may or may not occur between the two chromosomes of each of the chromosome pairs that exist within a plant cell. At the end of meiosis I, chromatid exchange and crossing-over are complete, and by the end of meiosis II, each sister chromatid has separated from the other and has developed into a functional chromosome.

In tomatoes another gene pair which controls the development of epidermal hairs on the stem may occur in the same chromosome pair with the genes for red and orange fruit color. Plants with hairy stems possess an H gene, whereas those with hairless stems possess an h gene. As was the case with the t gene in the presence of T, the h gene is unable to express itself in the presence of H. If a homozygous red-fruited, hairy-stem plant TTHH is crossed with a homozygous orange-fruited, hairless-stem plant tthh, all gametes formed by the TTHH parent will have TH, and all gametes formed by the tthh parent will have th. The resulting progeny will have a phenotype of red fruit and hairy stems and a genotypeTtHh (Table 5-2). This results in a heterozygous condition in which a dihybrid plant has been created by crossing two homozygous true breeding plants. Again, the plant could be referred to as an F_1 hybrid, but in this case it is heterozygous for two pairs of genes rather than one—thus it is a dihybrid. This would also be a type of single cross in hybrid seed production. Each homozygous plant used in the cross produces a single type of male gamete or egg cell with regard to genotype. For example, the T and H gene in the

TABLE 5-2

GENOTYPIC AND PHENOTYPIC RATIOS RESULTING FROM THE
SELF-POLLINATION OF A PLANT HETEROZYGOUS FOR TWO PAIRS OF GENES

FEMALE EGG CELLS	MALE GAMETES			
	TH	Th	tH	th
TH	TTHH* Red fruit Hairy stem	TTHh Red fruit Hairy stem	TtHH Red fruit Hairy stem	TtHh Red fruit Hairy stem
Th	TTHh Red fruit Hairy stem	TThh* Red fruit Hairless stem	TtHh Red fruit Hairy stem	Tthh Red fruit Hairless stem
tH	TtHH Red fruit Hairy stem	TtHh Red fruit Hairy stem	ttHH* Orange fruit Hairy stem	ttHh Orange fruit Hairy stem
th	TtHh Red fruit Hairy stem	Tthh Red fruit Hairless stem	ttHh Orange fruit Hairy stem	tthh* Orange fruit Hairless stem

*Denotes the four true breeding genotypes obtained from self-pollinating a dihybrid.

dihybrid was obtained through one chromosome coming from one parent, and the t and h gene through the other chromosome for that pair coming from the opposite parent. In this case it would be irrelevant which plant was used as the male or female parent in the cross, because the results would be the same.

If the heterozygous TtHh red-fruited, hairy-stem plant is self-pollinated and the T and H genes are completely dominant, the resulting seedling progeny appear in a 9:3:3:1 phenotypic ratio. When a dihybrid is selfed, there are four possible gametic combinations: TH, Th, tH, and th. In obtaining the 9:3:3:1 phenotype ratio, a large number of acts of fertilization must occur so that many new individuals can be produced by the random combination of male and female gametes. Limited numbers of such combinations fail to produce a clear-cut ratio.

The general genotypes and phenotypes for the 9:3:3:1 segregation would be nine red-fruited, hairy-stem plants T__H__, three red-fruited, hairless-stem plants T__hh, three orange-fruited, hairy-stem plants ttH__, and one orange-fruited, hairless-stem plant tthh. In making a genotypic as opposed to a phenotypic comparison, it must be remembered that TT and Tt both produce red fruit and HH and Hh both produce hairy stems. In the above designation T__ implies either TT or Tt, since at least one dominant gene is sufficient to give red. The HH and Hh genotypes would both produce hairy stems. Recessive genes have to be present in both chromosomes of a pair before a recessive characteristic can be expressed.

The progeny resulting from the self-pollination of an F_1 plant are called the *F_2 generation*; this refers to the second-progeny generation from the original cross between the two homozygous parents. By studying the genotypes and phenotypes of the progeny in Table 5-2, it can be seen that two additional phenotypes have been created from those of the original parents in just two generations. The three red-fruited, hairless-stem plants T__hh and the three orange-fruited, hairy-stem plants ttH__ were made possible through the chromosome-pairing relationships and crossing-over of nonsister chromatids during meiosis I. From the six plants involved in the new phenotypes only two would breed true with subsequent selfing. The genotype of the original parents that created the F_1 plant are also recovered in the F_2 generation: TTHH and tthh.

Another phenomenon occurring in genetics is *incomplete dominance*. With incomplete dominance the homozygous genotype shows a different phenotype from that of the heterozygous genotype. For example, with genes for red-colored fruit RR, the heterozygous genotype Rr would be intermediate in a color that would vary somewhere between red and orange rr fruit. A deep pink would probably result with the Rr genotype.

Plant breeders are especially interested in plant mutations. A *mutation* is a spontaneous change in the genetic makeup of the cell. A gene may change from a dominant allele to a recessive allele, chromosome materials may become rearranged, or segments of chromosomes may be lost or duplicated. The mutation may be dominant or recessive. The dominant mutation is evidenced immediately, whereas the recessive mutation is masked by dominant characters and will be expressed phenotypically only when an individual with the homozygous recessive genes is produced. Mutations can be induced artificially by treatment with ionizing irradiation with gamma rays, neutron bombardment, ultraviolet rays or x-rays, or treatment with certain chemicals. Mutations that occur as a result of external treatments are more likely to be undesirable and less stable; but in a few cases desirable traits, such as higher yields or resistance to disease, may result. These beneficial mutations may be used to strengthen a breeding program even though forced mutations rarely exceed those occurring naturally. The spur-type growth habit of apples appeared as a mutant.

A mutation which occurs in the apical meristem of a bud is termed a *point mutation* or a *bud sport* when the mutated portion of the meristem is involved in the differentiation of a lateral bud. The stem produced by the mutated bud is the only portion of the plant showing

(a) For one pair of chromosomes:

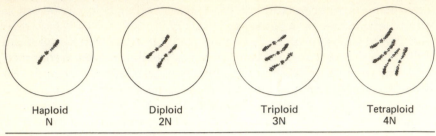

Haploid	Diploid	Triploid	Tetraploid
N	2N	3N	4N

(b) For four pairs of chromosomes:

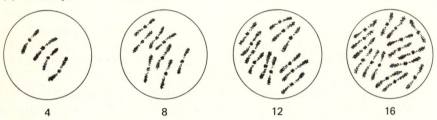

| 4 | 8 | 12 | 16 |

FIGURE 5-12 Plant breeders apply polyploidy where individuals have more than two chromosome sets in their somatic cells for a given pair of chromosomes.

variation. Bud sports can be propagated asexually and thereby establish new clones.

Plant breeders may also utilize polyploidy of cells to obtain or develop a new or desirable plant strain. *Polyploidy* is a condition in which individuals have more than two sets of chromosomes in their cells. It may result in the formation of larger or stronger plant parts, or in the occurrence of some other desirable factor. Normally plants are diploid, or 2N. However, 1N (haploid), 3N (triploid), and 4N (tetraploid) plants do occur (Figure 5-12). Often in plants meristem cells are 2N, but parenchyma and other cells are polyploid owing to failure of cytokinesis in late mitotic divisions of meristem cells. This is true in roots and possibly to a lesser degree in stems. Applications of colchicine and other chemicals, applications of heat, and wounding of the stem have been used to increase or change the number of chromosomes in plants. Colchicine is a chemical that prevents cell wall formation between daughter cells following cell division. Triploids are produced by crossing a tetraploid with a diploid, which was the technique used to produce the first seedless watermelons. The triploid is sterile and seedless owing to abnormal chromosome pairing that occurs during meiosis. Increased plant and fruit size and luxuriant growth often characterize polyploids. Although tetraploids occur more frequently in vegetable crops, tetraploid snapdragons and marigolds—which exhibit an increase in flower size—also occur.

Polyploidy is significant in plant breeding because it enables more genetic diversity in the plant kingdom. Polyploidy enables the plant breeder to change the chromosome number with greater ease than in diploids; however, care should be taken to avoid the unfavorable side effects, such as reduced fertility, that often occur in polyploid plants.

Through selection and hybridization, plant breeders develop new and superior plant characteristics. Uniformity and vigor may be obtained, early and total yields may be increased, and seedless fruit may be developed. Following hybridization, self-pollination and selection attempts to isolate plants with characteristics which may prove beneficial. Frequently, backcrossing an F_1 plant to one of the parents involved in the production of the F_1 may be the most rapid path for transferring desirable characteristics. The actual breeding methods are technical, individualistic, and dependent on the crop and objectives sought. Crops can be divided into four groups on the basis of pollination requirement or propagation method, and from these the specific breeding methods are determined: (1) those naturally self-pollinated, (2) those naturally cross-pollinated, (3) those often cross-pollinated, and (4) those vegetatively propagated.

In the peach, *Prunus persica*, some fruits have pubescent skin (peaches), and others have smooth skin (nectarines). If a peach is crossed with a nectarine, the F_1 produces all pubescent fruit provided the parent peach was homozygous (PP) for pubescence. When the F_1 from such a cross is self-pollinated, the resulting progeny segregate in the ratio of three peaches to one nectarine. Table 5-3 illustrates the characteristics for pubescence that plant breeders are able to control in peaches.

TABLE 5-3
CROSS OF PEACH AND NECTARINE
RESULTING IN 3 PUBESCENT-SKIN FRUIT
AND 1 SMOOTH-SKIN FRUIT

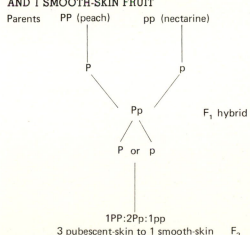

Parents   PP (peach)   pp (nectarine)

P   p

Pp F_1 hybrid

P or p

1PP:2Pp:1pp
3 pubescent-skin to 1 smooth-skin F_2

FRUIT

With many horticultural crops, the economically valuable portion of the plant is the fruit; therefore, since fruits are formed as a result of sexual reproduction, the development of these crops usually depends on pollination and fertilization. Only a few cultivars of fruits are seedless and thus do not require pollination and fertilization for the setting of fruit. The navel orange, certain cultivars of grapefruit, and Thompson seedless grapes do not require formation of seed for normal development of fruits.

Most seedless fruits are apomictic, and some are parthenocarpic. *Apomixis* is a form of reproduction in which new individuals are produced without nuclear or cellular fusion. The structures normally associated with sexual reproduction are involved, but no sexual fusion occurs. The embryo develops from an unfertilized egg or from tissues, such as the integument, which surround the embryo sac. In some forms of apomixis, pollination is required to stimulate apomictic development even though fertilization may not occur. A fruit that develops without fertilization is said to be *parthenocarpic*.

Botanically, a fruit includes a matured ovary plus those parts directly associated with the ovary. The ovary commonly enlarges to become the edible portion of the fruit. Since some fruits are consumed at an immature stage of development, it seems impractical from a horticultural standpoint to stress maturity as a basic part of such a definition. Usually, the ovary is considered to be a fruit when a rapid increase in growth occurs within it. Furthermore, a fruit may consist of not only the ovary but also the receptacle or other accessory parts of the flower.

Fruit development occurs in several phases: initiation of the ovary tissues; prepollination development of ovary; postpollination and post-fertilization growth; and ripening, maturity, and senescence of fruit (Figures 5-13 and 5-14). Cell division and enlargement are respon-

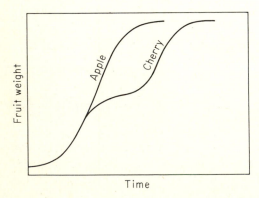

FIGURE 5-13 The growth curves for fruits are commonly of two types: sigmoid curves, as for apple, or double sigmoid curves, as for the cherry. (From A. C. Leopold and P. E. Kriedemann, *Plant Growth and Development*, 2d ed., McGraw-Hill, New York, 1975.)

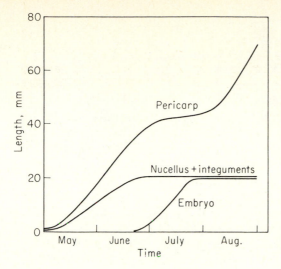

FIGURE 5-14 Comparison of the growth curves for various parts of the fruit of Elberta peach shows that the double growth curve is expressed by the pericarp tissues and that the time when the first growth period ends is approximately the time of completion of nucellar growth and commencement of embryo growth. (From A. C. Leopold and P. E. Kriedemann, *Plant Growth and Development*, 2d ed., McGraw-Hill, New York, 1975.)

sible for the increase in ovary size during prepollination development. In some very large fruits, cell division continues for some time after pollination and fertilization. The final size is a consequence of an increase both in the number and in the size of cells. It has been shown that pollen grains have relatively large amounts of auxins, which stimulate the ovary to develop into the fruit after pollination.

The importance of pollination in seed and fruit development has long been recognized. In commercial fruit production, it is recognized that not only pollination but also fertilization is necessary to obtain proper development in most fruits. Hormones produced by the growing tissues of the seed influence the ability of the fruit to continue growth and its ability to compete with other flower parts for food. An example of inability to compete is catfacing in strawberries where the ovary is destroyed by an insect. No seed is produced, and the fruit is flat-sided because auxins are not supplied by the seed to the receptacle for cell expansion.

The fruit wall, or pericarp, may have three distinct layers: the exocarp, the mesocarp, and the endocarp. Typically, the outer skinlike region of the fruit is the *exocarp*, the center portion of the fruit is the *mesocarp*, and the inner area is the *endocarp*; but variations do occur.

Classifications

Fruits can be classified into one of several main groups: simple (Figure 5-15), aggregate, or multiple (Figure 5-16). *Simple fruits*, which form from a single, ripened ovary, can be either fleshy or dry. *Fleshy fruits*, which include the berry, the pepo, the hesperidium, the drupe, and the pome, have a pericarp that is soft and fleshy at maturity (Figure 5-15). In *dry fruits*, the pericarp is often hard and brittle at maturity.

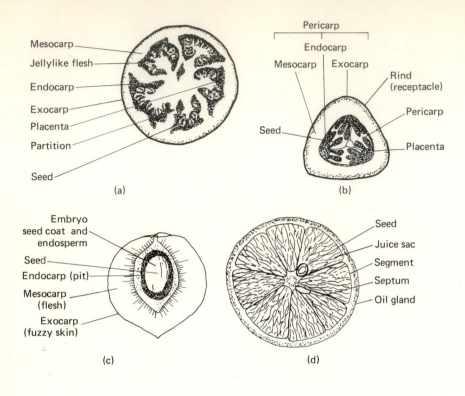

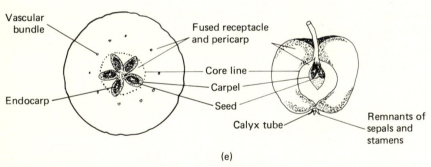

FIGURE 5-15 Simple fleshy fruits. (*a*) Berry (tomato); (*b*) pepo (cucumber); (*c*) drupe (peach); (*d*) hesperidium (orange); (*e*) pome (apple).

The entire pericarp of the berry is fleshy and is usually edible, as in the tomato or blueberry. *Pepos* are berries that have a hard rind around the fruit (cucumber or watermelon), whereas *hesperidiums* have a leathery rind (citrus fruits). The *drupe* has a thin exocarp, a mesocarp that is thick and fleshy, and an endocarp that is hard and stony (peach, plum, and olive). In the *pome* the portions produced by the pericarp are enclosed within fleshy parts that are derived from parts of the flower other than the ovary (apple, pear).

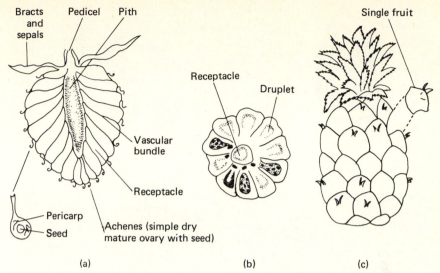

FIGURE 5-16   Aggregate fruits. (*a*) strawberry; (*b*) raspberry and multiple fruit; (*c*) pineapple.

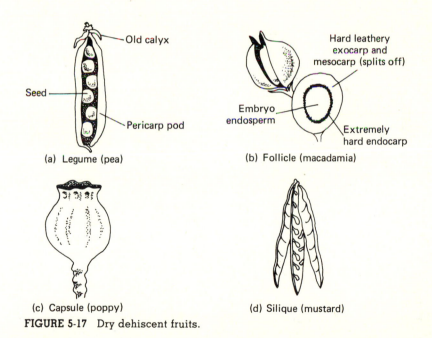

(a) Legume (pea)

(b) Follicle (macadamia)

(c) Capsule (poppy)

(d) Silique (mustard)

FIGURE 5-17   Dry dehiscent fruits.

The term *dehiscence* refers to the opening of the fruit at maturity, which allows the seeds to be disseminated. Dry fruits may be either *dehiscent*, where the carpel splits along definite seams at maturity, or *indehiscent*, where the fruit wall does not split at any certain point or

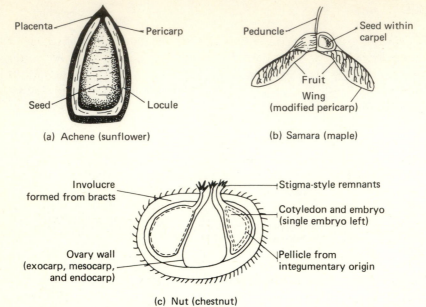

Placenta · Pericarp · Seed · Locule

(a) Achene (sunflower)

Peduncle · Seed within carpel · Fruit · Wing (modified pericarp)

(b) Samara (maple)

Involucre formed from bracts · Stigma-style remnants · Cotyledon and embryo (single embryo left) · Ovary wall (exocarp, mesocarp, and endocarp) · Pellicle from integumentary origin

(c) Nut (chestnut)

FIGURE 5-18 Dry indehiscent fruits.

seam at maturity. Many pulpy or nutlike fruits fall into this category. Their seed is released only by decay or by physically breaking the ovary wall. A dehiscent fruit may contain several to many seeds, whereas an indehiscent fruit usually has only one or two seeds. The legume (pea), follicle (macadamia), capsule (okra and poppy), and silique (mustard) are dehiscent fruits (Figure 5-17); the achene (sunflower), caryopsis (corn), samara (maple), and nut (chestnut) are indehiscent (Figure 5-18).

In the *legume*, one locule splits along two sutures (peas), while the *follicle* has one locule which splits along one suture (macadamia, milkweed). The *capsule* has two or more locules which split in various ways (lily, poppy, snapdragon, cotton). There are two fused locules in a *silique* which separate at maturity, leaving a persistent partition between them (cabbage or mustard).

The *achene*, an indehiscent dry fruit, has only one seed which is separable from the walls of the ovary, except where it is attached to the inside of the pericarp (sunflower). The *caryopsis*, or grain, has one seed which is completely fused to the inner surface of the pericarp (corn, oats). A *nut* is similar to the achene except that the pericarp is hard throughout (pecan, walnut). The *samara* has either one seed or two seeds, and the pericarp bears a flattened winglike outgrowth (maple, elm).

SEED

The seed of gymnosperms are borne on the surface of "scales" of ovulate (female) cones, whereas the seed of the angiosperms are borne within the ovaries that ripen into fruit. Even though seed of different species differ greatly in size and structure (Figure 5-19), they all consist of a plant embryo with associated stored food encased in a

FIGURE 5-19 Seed structure of representative species. (From Hudson T. Hartman and Dale E. Kester, *Plant Propagation: Principles and Practices*, Prentice-Hall, Englewood Cliffs, N.J., © 1975. Reprinted by permission of Prentice-Hall, Inc.)

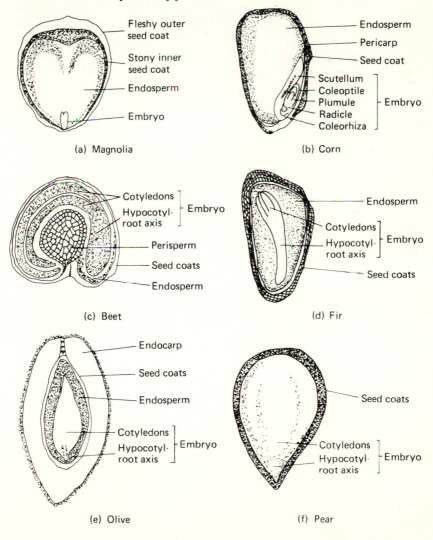

(a) Magnolia

Fleshy outer seed coat
Stony inner seed coat
Endosperm
Embryo

(b) Corn

Endosperm
Pericarp
Seed coat
Scutellum
Coleoptile
Plumule — Embryo
Radicle
Coleorhiza

(c) Beet

Cotyledons — Embryo
Hypocotyl-root axis
Perisperm
Seed coats
Endosperm

(d) Fir

Endosperm
Cotyledons — Embryo
Hypocotyl-root axis
Seed coats

(e) Olive

Endocarp
Seed coats
Endosperm
Cotyledons — Embryo
Hypocotyl-root axis

(f) Pear

Seed coats
Cotyledons — Embryo
Hypocotyl-root axis

protective seed coat. All seed has some stored food, even though it may be quite limited.

The miniature plant, or embryo, within the seed develops from the union of the male and female gametes and has three basic parts: (1) cotyledon or cotyledons, (2) hypocotyl, and (3) epicotyl. The lower portion of the hypocotyl is often designated as the *radicle*, since it develops into the primary root. The *hypocotyl* forms the portion of the stem up to the first node, whereas the *epicotyl* gives rise to all parts of the mature plant above the first node. The *cotyledon* is attached to the shoot at the first node. This is an embryonic leaf which may serve as a food-storing organ or may develop into a photosynthetic structure as the seed germinates. In members of Graminae the cotyledons neither store food nor become photosynthetic; rather, they absorb digested food from the adjacent endosperm of the seed as the embryo initiates germination.

The dormancy of seed is the inability to germinate because of genetically controlled internal conditions or unfavorable environmental conditions. There are two distinct types of dormancy in seed. *Rest*, or internal dormancy, is the inability of a seed to germinate as a result of conditions within the seed itself. These could include the presence of an undeveloped embryo within the seed where the embryo must complete its development before it can germinate (a requirement for low temperature) or the presence of certain inhibitory chemicals which must first be removed, usually by leaching, before germination can occur. The seed coat itself may be impermeable to water or oxygen or it may be too hard for the embryo to break through until the seed coat has been weakened. *External dormancy* is the inability to grow as a result of unfavorable external conditions such as moisture supply, temperature, oxygen, or light.

Seed as Food   The seed of many plants serve as an important source of food, such as the edible portion of nuts and peas. Oils extracted from certain seed, such as sunflowers, soybeans, corn, and cotton, are used as food products or in a variety of other products, including soaps, varnishes, and paints. Seed are used in beverages and medicines and as spices and condiments. Indeed the economic importance of some seed is such that the seed, rather than the plant, is brought to mind when the plant name is mentioned.

BIBLIOGRAPHY

Esau, K.: *Plant Anatomy*, 2d ed., Wiley, New York, 1965.
Fritsch, F. E. and E. Salisbury: *Plant Form and Function*, Bell, London, 1953.
Haynes, J. D.: *Botany: An Introductory Survey of the Plant Kingdom*, Wiley, New York, 1975.

Hill, J. B., H. W. Popp, and A. R. Grove, Jr: *Botany*, 4th ed., McGraw-Hill, New York, 1967.

Jensen, W. A. and F. B. Salisbury: *Botany: An Ecological Approach*, Wadsworth, Belmont, Calif., 1972.

Raven, P. H., R. F. Evert, and H. Curtis: *Biology of Plants*, 2d ed., Worth, New York, 1976.

Treshow, M.: *Environment and Plant Response*, McGraw-Hill, New York, 1970.

Weir, T. E., C. R. Stocking, and M. G. Barbour: *Botany: An Introduction to Plant Biology*, 5th ed., Wiley, New York, 1975.

Wilson, C. L., W. E. Loomis, and T. A. Steeves: *Botany*, 5th ed., Holt, New York, 1971.

CHAPTER 6
PLANT
METABOLISM

 The maintenance of life on earth is directly and totally dependent on a supply of energy. Energy in the form of radiation from the sun is "captured" by plants in the process of photosynthesis and is thus converted into chemical energy. Once fixed in a chemical form, this energy is available for plants and animals to maintain life. This chapter deals with the major metabolic systems involved in the fixation, transformation, and utilization of energy by plants.

PHOTOSYNTHESIS

Photosynthesis is the most important process on earth because it is the connecting link between solar energy and life on earth. On a global scale, vast quantities of CO_2 are fixed by plants; estimates usually range from 50 to 150×10^9 tons per year.

Pigments

The two major pigments which absorb radiant energy in photosynthesis are chlorophyll a and chlorophyll b, relatively large organic molecules. The empirical formula for the chlorophyll a molecule is $C_{55}H_{72}O_5N_4Mg$. The structural difference between the two types is

Chlorophyll a:

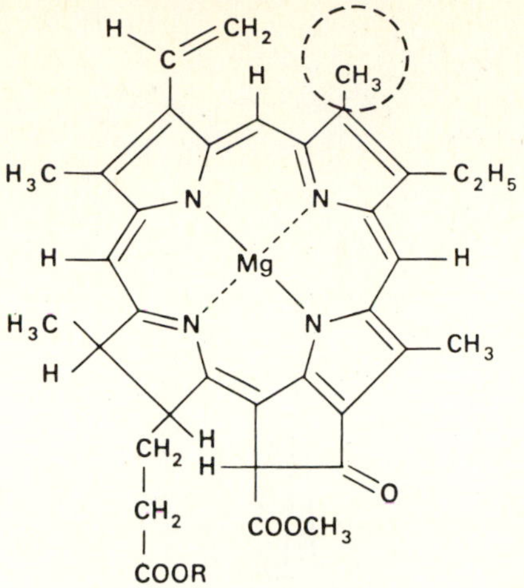

Chlorophyll b:

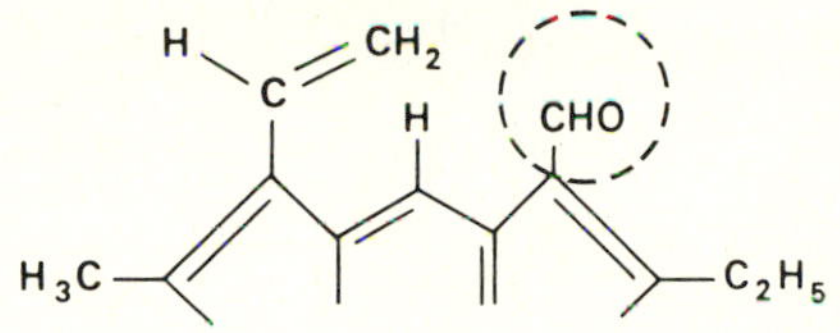

FIGURE 6-1 Chemical structure of chlorophyll a and distinguishing part of chlorophyll b (note dashed circles). The presence of a carbon atom is implied at each unlabeled junction of bonds. The residue R is a long-chain hydrocarbon, $C_{20}H_{39}$ or phytol. (Adapted from R. K. Clayton, *Light and Living Matter*, vol. 2, McGraw-Hill, New York, 1971.)

shown in Figure 6-1. Chlorophyll a is present in most plants in quantities about twice as great as chlorophyll b.

Radiant energy striking a leaf is absorbed by, reflected from, or transmitted through the leaf. The energy absorbed is converted to heat and reradiated, used in photosynthesis, or utilized in the vaporization of water. The *absorption spectrum* of a compound indicates its ability to absorb radiant energy of different wavelengths. The two types of chlorophyll have slightly different absorption spectra, but as can be seen in Figure 6-2, each has an absorption peak in the blue-violet range and a second peak in the red region of the spectrum. We "see" a leaf as green because the red and blue wavelengths are absorbed by chlorophyll, and the green is transmitted or reflected. Because it is not absorbed, radiation in the green portion of the spectrum is of no use in photosynthesis.

Chlorophyll occurs only in the *chloroplast*, the organelle in which photosynthesis is carried out. In most plants, light is required for the formation of chlorophyll. If grown in darkness, plants develop immature chloroplasts called proplastids which develop into mature chloroplasts after a few hours of illumination. The seed of certain plants, such as corn, will occasionally produce a seedling without the ability to

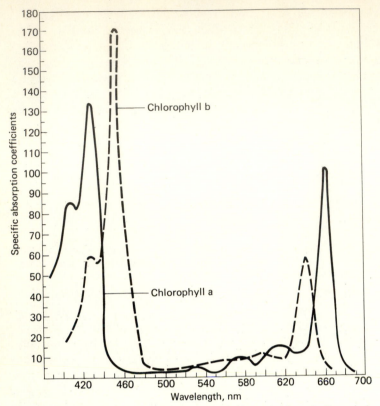

FIGURE 6-2 Absorption spectra of ether extracts of chlorophylls a and b. (After F. P. Zscheile and C. L. Comar, 1941, *Bot. Gaz.* 102:463–481.)

produce chlorophyll. Lacking chlorophyll and therefore the ability to carry out photosynthesis, these "albino" seedlings can survive in nature only until all the food reserves present in the seed have been utilized.

In addition to chlorophyll, chloroplasts also contain *carotenoids*, pigments ranging in color from yellow to red. These pigments, unlike chlorophyll, are not limited to the chloroplasts but may occur in other plastids called chromoplasts. The carotenoids consist of two groups, the carotenes and the xanthophylls. The *carotenes*, of which β-carotene is the most widely occurring, are made up only of carbon and hydrogen. The *xanthophylls* contain oxygen in addition to carbon and hydrogen. Carotenoids are not involved in photosynthesis as directly as chlorophyll, but are thought to affect photosynthesis in two ways. Although the mechanism is not well understood, carotenoids apparently can prevent the photooxidation of chlorophyll in bright light. Secondly, the carotenoids absorb light and pass the absorbed energy on to the chlorophyll molecule. In addition to the pigments, chloroplasts contain proteins, lipids, carbohydrates (such as starch), and various inorganic elements (such as iron, calcium, and potassium). It has been estimated that about one-half of the total leaf protein is in the chloroplasts.

Since *enzymes are proteins*, their importance in facilitating the many chemical reactions in the chloroplast cannot be overemphasized. Lipids are essential because they are vital components of membranes. Carbohydrates are an end product of photosynthesis and exist in various forms within chloroplasts.

Reactions of Photosynthesis

The following very simplified equation summarizes the complex series of reactions involved in photosynthesis:

$$\text{Light} + 6CO_2 + 12H_2O \xrightarrow[\text{chlorophyll}]{\text{enzymes}} C_6H_{12}O_6 + 6H_2O + 6O_2$$

This process is of tremendous significance because light energy is utilized to build a complex, high-energy molecule from simple molecules which contain little biologically useful energy.

Photosynthesis includes two separate and distinct sets of reactions. The photochemical reactions are initiated by light and are essentially unaffected by temperature or by oxygen or carbon dioxide concentrations. The second component of photosynthesis, called the *dark* or *Blackman reactions*, is very different because the reactions are strongly influenced by both temperature and the oxygen or carbon dioxide concentration, but can occur without light. As we shall see, however, the dark reactions use as reactants the various products of the light reactions.

Light Reactions   The first step in photosynthesis results in the splitting of water to form H+ ions, electrons, and O_2—a process called *photolysis*. The O_2 gas is released while the electrons and the H+ ions ultimately reduce NADP+ (nicotinamide adenine dinucleotide phosphate) (see Figure 6-3) to NADPH.

$$\text{Light} + 2H_2O + 2NADP^+ \xrightarrow[\text{chlorophyll}]{\text{enzymes}} O_2 + 2NADPH + 2H^+$$

Another part of the light reactions is called *photophosphorylation*. In this process some of the absorbed light energy is used to convert ADP (adenosine diphosphate) to ATP (adenosine triphosphate) (see Figure 6-3). The necessary reagents are H+, ADP, and inorganic phosphate, Pi.

$$\text{Light} + ADP + H^+ + Pi \rightarrow ATP + H_2O$$

(a)

The combined processes of photolysis, NADP+ reduction, and photophosphorylation are collectively called the light reactions and for our consideration may be combined into one equation:

$$\text{Light} + 2NADP^+ + 4ADP + 4Pi + 2H_2O \xrightarrow[\text{chlorophyll}]{\text{enzymes}} O_2 + 2NADPH_2 + 4ATP$$

Thus, as a result of the light reactions, light energy is converted to chemical energy in the form of $NADPH_2$ and ATP.

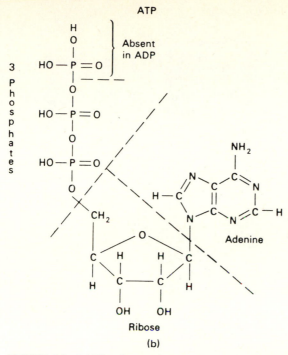

ATP

Absent in ADP

3 Phosphates

Adenine

Ribose

(b)

FIGURE 6-3 (*a*) The chemical structure of nicotinamide-adenine dinucleotide phosphate, NADP$^+$. Without the "odd" phosphate the molecule is NAD. The interconversion between the oxidized form, NADP$^+$, and the reduced form, NADPH, which takes place on the nicotinamide part is shown below. (*b*) The chemical structure of adenosine triphosphate (ATP). ATP is made of one molecule of adenine, one molecule of ribose (a 5-carbon sugar), and three molecules of phosphoric acid. With only two phosphoric acid residues, the molecule would be adenosine diphosphate (ADP). (From R. K. Clayton, *Light and Living Matter*, vol. 2, McGraw-Hill, New York, 1971.)

Dark Reactions   This very complex set of reactions is thought to involve the transfer of energy, H+ ions, and electrons to an intermediate which in turn bonds to CO_2 to produce a carbohydrate. Thus, ATP and NADPH$_2$ are high-energy "currencies" produced in the light reaction to provide energy for the dark reactions in which CO_2 is actually "fixed" or reduced.

The series of reactions leading to the fixation of CO_2 is known as photosynthetic *carbon reduction* or the Calvin-Benson cycle (Figure 6-4).

The cycle consists of several steps but can be summed up in three major phases:

1. Ribulose-1,5-diphosphate combines with CO_2 to form two molecules of 3-phosphoglyceric acid (PGA). This is the *carboxylation phase.*

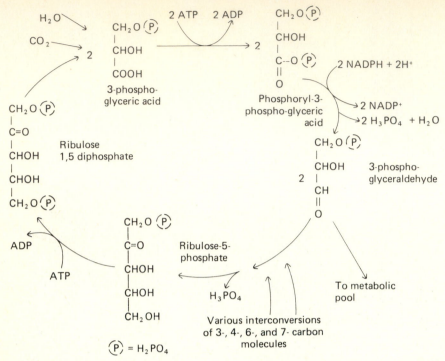

FIGURE 6-4 The carbon reduction or Calvin-Benson cycle.

2. PGA is reduced to 3-phosphoglyceraldehyde (a triose, or 3-carbon sugar) by $NADPH_2$. This is the *reductive phase.*

3. The 3-carbon sugars are interconverted to form 5-carbon sugars and ribulose-1,5-diphosphate, the CO_2 acceptor, is produced. This is the *regenerative phase.*

An overall reaction can be described by the following equation:

$$1RuDP + 6CO_2 + 18ATP + 12NADPH \rightarrow 1RuDP + glucose + 18ADP + 18Pi + 12NADP^+$$

In this reaction, 2 mol of NADPH and 3 mol of ATP are required for each mole of CO_2 fixed. In this cycle, the first product formed is the 3-carbon compound-phosphoglyceric acid; plants with only this mechanism of carbon fixation are called C_3 plants and constitute most of our crop and ornamental plants.

In recent years, a different mechanism of carbon fixation has been identified in certain plants. It is called the C_4, dicarboxylic acid or Hatch and Slack, pathway. Some of the C_4 plants identified to date include sugar cane, corn, sorghum, and certain species of *Amaranthus* and *Atriplex*. In this pathway, the first product after CO_2 is fixed is a 4-carbon compound, oxaloacetic acid (Figure 6-5a). The steps involved

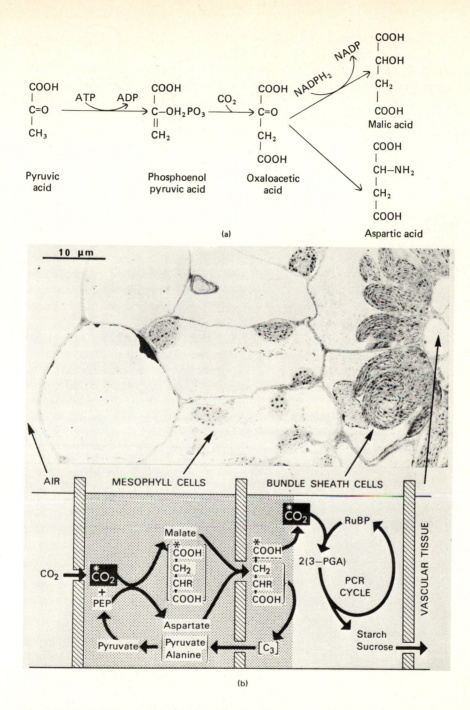

(a)

AIR   MESOPHYLL CELLS   BUNDLE SHEATH CELLS

(b)

FIGURE 6-5 (*a*) Fixation of carbon in the C_4, or Hatch and Slack, pathway. Phosphoenol pyruvic acid (PEP) combines with CO_2 to form oxaloacetic acid, which is converted (depending on species) to aspartic acid or malic acid. (*b*) The C_4 mechanism of carbon fixation depends on both leaf anatomy and biochemistry. (From M. D. Hatch, *Trends in Biochem. Sci.* **2**(9) 199–202, 1977. By permission of the International Union of Biochemistry)

145

are thought to include the phosphorylation of pyruvic acid to produce phosphoenol pyruvic acid (PEP) which in turn reacts with CO_2 to produce oxaloacetic acid. The oxaloacetic acid can either be reduced to form malic acid or transaminated (see page 163) to aspartic acid. In this method of CO_2 fixation, 5 mol of ATP plus 2 mol of NADPH per mole of CO_2 fixed apparently are needed, i.e., 2 additional mol of ATP as compared to the C_3 pathway. This apparent inefficiency in the C_4 system is presumably overcome by more photophosphorylation, more efficient collection of CO_2, and the absence of photorespiration (see page 153).

One can see in Table 6-1 that C_4 plants differ morphologically as well as biochemically from C_3 plants. As a rule, C_4 plants have two distinct types of chloroplasts located in two different kinds of cells. Around the vascular bundles or veins in leaves is a sheath of green parenchyma cells. These bundle sheath cells contain chloroplasts, which have starch grains, but lack grana. The other chloroplasts located in the mesophyll are smaller, contain no starch grains, but have numerous well-developed grana. The C_4 method of CO_2 fixation occurs in the mesophyll chloroplasts where CO_2 is fixed by the conversion of PEP into oxaloacetic acid and then into malic or aspartic acid. These acids then move to the bundle sheath chloroplasts where they are decarboxylated to pyruvic acid with the release of CO_2. This CO_2 is fixed to ribulose diphosphate in the C_3 pathway operating in the

TABLE 6-1
COMPARISON OF C_3 AND C_4 PLANTS

	C_3 (CALVIN CYCLE)	C_4 (HATCH AND SLACK PATHWAY)
Examples	Most crop plants, such as bean, tomato, apple, potato	Many tropical grasses; some crop plants, such as corn, sugar cane, sorghum
Maximum rates of net photosynthesis	6–30 $mgCO_2dm^{-2}h^{-1}$	50–100 $mgCO_2dm^{-2}h^{-1}$
Productivity (dry weight)	Low to high	High
First product of CO_2 fixation	Phosphoglyceric acid (C_3)	Oxaloacetic acid (C_4)
CO_2 acceptor	Ribulose 1,5 diphosphate (RuDP)	Phosphoenol pyruvate (PEP)
CO_2 compensation point	30–100 ppm	Near zero
Chloroplast location	Mesophyll	Mesophyll and bundle sheath
High temperature effect on photosynthesis	No effect or supresses CO_2 uptake	Increases CO_2 uptake
Irradiance for maximum photosynthesis	1000–3000 fc	10,000 fc +

bundle sheath chloroplasts. The pyruvic acid then returns to the mesophyll cells, giving the entire system a cyclic pattern (see Figure 6-5b).

Another feature of C_4 plants is that they are generally more efficient than C_3 plants in their use of water per unit of dry matter produced. Thus, in addition to having efficient methods of CO_2 fixation, minimal rates of photorespiration, and the ability to fix CO_2 present in very low concentrations, C_4 plants can also produce more dry matter per unit of water absorbed.

Factors Affecting Photosynthesis

Irradiance Under natural conditions, the process of photosynthesis is "driven" by the visible portion of the spectrum or the radiant energy between 400 and 700 nm wavelength, particularly the blue and red wavelengths. The level of natural irradiance varies with many factors including time of day, season, and cloudiness; but when the sun is at its zenith on a clear day in summer, irradiance levels are about 10,000 fc or 10.8 klx (see Chapter 9).

Photosynthetic rate is partially a function of level of irradiance. In complete darkness, no photosynthesis occurs; as irradiance is increased, photosynthesis increases until at the *light compensation* point, photosynthetic fixation of CO_2 exactly equals respiratory release of CO_2; thus there is no net movement of CO_2 into or out of the leaf. As irradiance is increased, net photosynthesis (gross photosynthetic CO_2 fixation-minus respiration) increases rapidly until the *light saturation* level is approached, at which point the photosynthetic rate levels off. Above the light saturation point, the CO_2 level is the factor limiting the net photosynthetic rate, whereas below this point, light is the limiting factor. The saturation level depends on many different factors including the species, CO_2 concentration, and the environment in which the plant developed. The striking difference in light response curves of C_3 and C_4 plants is apparent in Figure 6-6.

Single leaves of many crop plants are light-saturated at one-fourth to one-third of full sun, and some shade trees at one-fifth full sun; but C_4 plants are typically not light-saturated at twice the irradiance of full sun. It is also apparent from Figure 6-6 that the higher the level of light saturation, the greater the rate of net photosynthesis.

There is a strong effect of CO_2 level on photosynthetic rate. As the CO_2 level is increased above the atmospheric concentration of 300 ppm, the light level required for maximum photosynthesis increases, as does the rate of net photosynthesis at light saturation (Figure 6-7a). Although this is true for both C_3 and C_4 plants, the effect is most striking with C_3 plants. At levels of CO_2 much below normal (300 ppm), C_4 plants are much more efficient at fixing CO_2 than C_3 plants. One

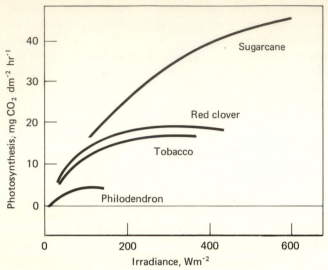

FIGURE 6-6 Light response of sugarcane (C_4 plant) and three C_3 plants. Note that sugarcane requires very high light for saturation, whereas the C_3 plants are saturated at much lower light levels. (Reproduced from J. D. Hesketh, and D. N. Moss, *Crop Sci.*, 3:107–110, 1963, by permission of the Crop Science Society of America.)

FIGURE 6-7 (*a*) Light response of sugar beet leaf at three CO_2 levels. As CO_2 level is raised, the light required for saturation also increases, as does the maximum rate of photosynthesis. (*b*) Photosynthetic response to increasing CO_2 levels at three levels of irradiance. At low light, photosynthesis is low and is saturated by low CO_2 levels; but at higher irradiances, more CO_2 is required for saturation, and photosynthetic rates are higher. (Adapted from P. Gaastra, *Meded. Landbouwhogeschool*, Wageningen, 59:1–68, 1959)

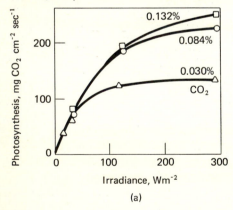

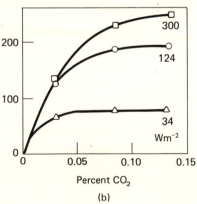

(a)   (b)

measure of this effect is the *CO$_2$ compensation point*, the minimum CO$_2$ concentration at which the plant is able to pick up CO$_2$ from its environment. In C$_3$ plants, this is usually at about 50 ppm; in C$_4$ plants the CO$_2$ compensation point is close to zero. This difference provides one means of determining whether a plant is the C$_3$ or C$_4$ type. If C$_3$ and C$_4$ plants are enclosed in an illuminated, airtight chamber, the CO$_2$ level will decrease as it is fixed in photosynthesis. Because the C$_4$ plant can still fix CO$_2$ when the concentration is below the CO$_2$ compensation point for the C$_3$ plant, the C$_3$ plant will ultimately die from its inability to photosynthesize.

The environment in which a plant or leaf develops can influence not only the light saturation level and the maximum rate of net photosynthesis but leaf morphology as well. A leaf developing under high irradiance is thicker, often as the result of larger and more numerous layers of palisade mesophyll cells. Associated with increased leaf thickness is a greater fresh and dry weight per unit area, specific leaf weight (SLW), usually expressed as mg cm^{-2}. Recent research has indicated a correlation between net photosynthesis and SLW; as SLW increased, net photosynthesis also tended to increase (Figure 6-8).

FIGURE 6-8 Relationship between the specific leaf weight (SLW) and net photosynthesis (Pn) of apple leaves. As SLW increases, Pn tends to increase. The various lines represent the relationship in different samples. (From J. A. Barden, *J. Amer. Soc. Hort. Sci.* 102:391–394, 1977.)

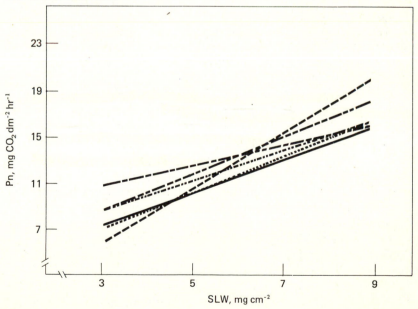

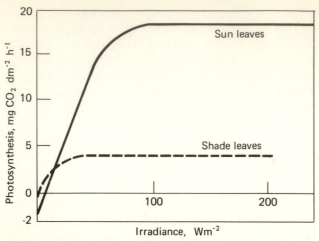

FIGURE 6-9 Sun leaves usually show higher rates of photosynthesis and dark respiration and greater light compensation points than shade leaves. The data are means for several species of sun and shade plants. (Adapted from R. H. Bohning and C. A. Burnside, *Amer. Jour. Bot.* **43**:557–561, 1956.)

At one time it was believed that species of plants could be divided into sun and shade types. It was proposed that this distinction could be made on the basis of light-saturation curves with "sun" plants being light-saturated at 2000 to 4000 fc and "shade" plants being saturated at 500 fc or less. More recent research has shown, however, that the distinction is not clear. Many sun species can be made to respond like shade species and vice versa by growing them under a different light regime. In other words a "sun" plant grown under low irradiance responds much like a "shade" plant (Figure 6-9).

Leaf Age Photosynthetic rates of young leaves are usually low but increase as the leaf approaches full expansion. After reaching maturity, subsequent photosynthetic activity varies widely with both species and environment. With evergreen species, such as citrus and conifers, leaves have been found to decline gradually in photosynthetic potential over a period of three years. In some deciduous species, those on which the leaf remains only one season, photosynthetic rates increase as a leaf reaches maturity and in some cases remain fairly stable for several weeks. As the season goes on, rates may decline slowly over a period of several weeks or rather rapidly, right before abscission. The latter situation is common in species which shed leaves relatively early or often in the growing season. The rapidity with which the photosynthetic potential of a particular leaf declines with age is influenced by several factors including the nutritional status of the plant and the degree of shading. If a mineral element is deficient, the leaf aging process may be accelerated, especially if it is an element which is readily moved from old to young leaves. Increasingly dense shade can also accentuate the decline in photosynthetic potential.

Temperature   The light-dependent reactions of photosynthesis are little affected by temperature, but the dark reactions are very temperature-dependent. When we speak of photosynthesis, we are usually referring to net photosynthesis or gross photosynthesis minus respiration. As seen elsewhere in this chapter, both dark and photorespiration rates are strongly influenced by temperature. There is, therefore, a striking difference in the photosynthetic rate response by C_3 and C_4 plants. Plants exhibiting the C_3 type of CO_2 fixation show little photosynthetic response to temperature between 10 and 30°C, because—although gross photosynthesis may increase as temperature rises—photorespiration and dark respiration increases neutralize the increased photosynthesis. With C_4 plants, however, with low or zero photorespiration, photosynthesis increases markedly as temperature rises from 10 to 35°C (Figure 6-10). In this range, net photosynthesis by

FIGURE 6-10 C_4 plants usually show greater photosynthetic capacity and higher temperature optima than C_3 plants. These light-response curves compare elephant grass (*Pennisetum purpureum*), a C_4 plant, and *Vigna luteola* (a tropical legume, a C_3 plant). This advantage is lost at low temperatures. (From M. M. Ludlow and G. L. Wilson, *Austral. J. Biol. Sci.* **23**:449–470, 1971.)

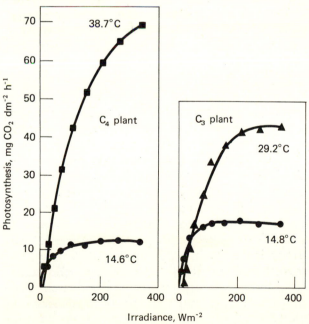

C_4 plants, such as corn and sugar cane, approximately doubles for each 10°C rise. The photosynthetic reactions of these species are said to have a Q_{10} of about 2.

Carbon Dioxide   Under many situations, CO_2 is the factor limiting photosynthesis. The normal atmospheric concentration of CO_2 is about 300 ppm which is well below CO_2 saturation (Figure 6.7b). There is a very strong interaction between CO_2 level and light level (Figure 6-7). At low irradiance, low levels of CO_2 will saturate the photosynthetic mechanism; but as light level increases, progressively higher concentrations of CO_2 are required to "use up" the $NADPH_2$ and ATP made in the light reactions. Another aspect of CO_2 levels is that within the canopy of a crop, CO_2 levels are often well below the general atmospheric level. This depletion depends on several factors including canopy density, wind speed, and irradiance.

It has been shown that, under certain circumstances, it is possible to increase yields in greenhouse crops by *CO_2 fertilization*. Although used commercially in some areas, widespread adoption of the procedure has not occurred. Some of the reasons for the lack of widespread use include the need for monitoring of CO_2 level, possible phytotoxicity, and the additional expense involved.

RESPIRATION

Plant and animal cells are continuously respiring. *Respiration* is in reality a group of processes in which various substrates are broken down with the release of energy which is then used for the wide array of vital processes. Substrates utilized in respiration include carbohydrates, fats, proteins, and organic acids. Using a hexose sugar as the substrate, respiration can be very simply expressed by the following equation:

$$C_6H_{12}O_6 + 6O_2 \rightarrow 6CO_2 + 6H_2O + energy$$

Although this equation is the reverse of that given for photosynthesis, it is somewhat of an oversimplification. Some of the enzymes and compounds involved are in fact the same, but respiration is not simply a reversal of photosynthesis. An important distinction is that the two processes are associated with different cell organelles. Photosynthesis occurs in the chloroplasts while respiration is primarily associated with the mitochondria.

A most important aspect of respiration is that, although it starts and ends with the same products as a fire might, each proceeds very differently. Rather than being an uncontrolled process, respiration is a controlled, complex, stepwise degradation of the substrate. A different enzyme catalyzes each reaction, and the chemical energy stored in the substrate is "trapped" or harvested during the process. The energy lost as heat can be readily measured, especially in very rapidly respiring tissues, such as germinating seeds. Respiration is, however, relatively efficient in the transfer of energy into other usable chemical forms.

Respiration occurs throughout the living regions of the plant whereas photosynthesis is limited to chlorophyll-containing, aerial plant organs. Horticulturists are vitally concerned with respiration of fruits and vegetables as they approach maturity as well as after harvest. Many of the storage treatments, such as low temperatures and controlled atmospheres, are aimed at minimizing respiration and thereby prolonging the life of produce. (see chapter 15).

Leaves respire in darkness; the rates can be detected quite readily by measuring CO_2 output. The question of respiration by illuminated leaves has, however, been very perplexing because CO_2 exchange by an illuminated leaf is a measure of net photosynthesis but gives no measure of either gross photosynthesis or respiration. It has recently been shown that respiration does occur in the light but that in many plants it involves two distinctly different types of respiration. Because of the distinct differences, the two types are commonly referred to as light respiration, or photorespiration, and dark respiration. *Dark respiration occurs in both darkness and light,* but rates in a typical leaf are relatively low, ranging from 0.3 to 1.5 mg CO_2 dm^{-2}h^{-1}. These rates are in the range of 5 to 10 percent of the net photosynthetic rate at light saturation. These rates vary not only with the plant species but with leaf age, previous irradiance, temperature, and CO_2 and O_2 levels as well. *Photorespiration* is a process which occurs in tissues of C_3 plants which are actively photosynthesizing. The substrates utilized come directly from photosynthetic products rather than storage materials as is the case in dark respiration. The process is distinctly different from dark respiration and is usually estimated to be several times greater than dark respiration when CO_2 output is considered. Photorespiration is very difficult to measure, not only because it is masked by photosynthesis but because CO_2 generated by photorespiration may be reabsorbed in photosynthesis and thus not detected by most measurement procedures.

The C_4 plants, such as corn and sugar cane, either have no photorespiration or have photorespiration which occurs at such low rates that it has not yet been detected. It is partially the absence of photorespiration in C_4 plants that is thought to make the plants so very productive. Also, since photorespiration is thought by some to be unnecessary, much emphasis is being placed on attempts to reduce or eliminate it from crop plants. If this were possible, yields might be increased 25 to 50 percent.

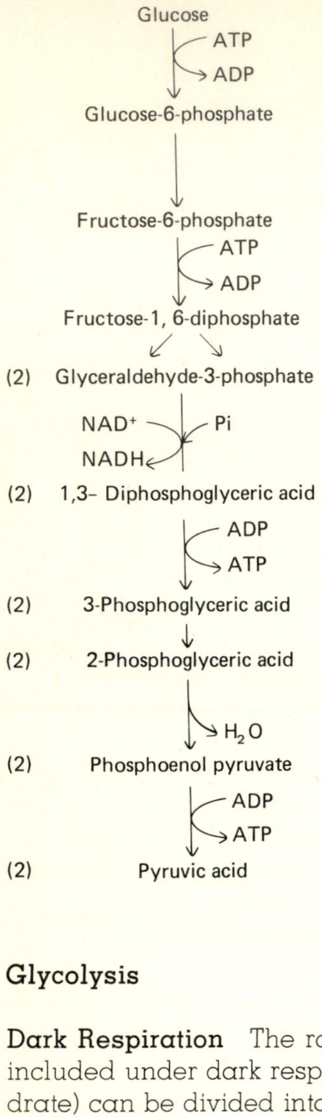

Glucose
├─ ATP
└─→ ADP
↓
Glucose-6-phosphate

↓

Fructose-6-phosphate
├─ ATP
└─→ ADP
↓
Fructose-1, 6-diphosphate
↙ ↘

(2)   Glyceraldehyde-3-phosphate

NAD⁺ ╲   ╱ Pi
      ╳
NADH ╱ ╲

(2)   1,3– Diphosphoglyceric acid
├─ ADP
└─→ ATP
↓
(2)   3-Phosphoglyceric acid
↓
(2)   2-Phosphoglyceric acid

└→ H_2O
(2)   Phosphoenol pyruvate
├─ ADP
└─→ ATP
↓
(2)   Pyruvic acid

FIGURE 6-11 In the process of glycolysis, a molecule of glucose goes through multiple intermediates and is ultimately converted to two pyruvic acid molecules. There is an input of two ATPs and an output of four ATPs, giving a net gain of two ATPs per molecule of glucose.

Glycolysis

Dark Respiration   The rather complex series of reactions which are included under dark respiration of glucose (a very common carbohydrate) can be divided into a few overall processes. The first of these is glycolysis and involves the degradation of carbohydrate to pyruvic acid, as shown in Figure 6-11. A summary equation for glycolysis which takes place in the presence or absence of O_2 is as follows:

$$C_6H_{12}O_6 + 2Pi + 2ADP + 2NAD \rightarrow 2CH_3 \cdot CO \cdot COOH + 2ATP + 2NADPH_2 + 2H_2O$$

The end result of the breakdown of 1 mol of hexose to 2 mol of pyruvic acid is 2 mol of ATP because, although 4 mol of ATP are generated, 2 were utilized. The 2 mol of NADH + H+ ultimately produce 3 mol of ATP each or an additional 6 mol. The net result, therefore, is 8 mol of ATP produced per mole of hexose broken down to pyruvic acid.

The end product of glycolysis—pyruvic acid—can be utilized in numerous metabolic processes. Under anaerobic conditions (without O_2), pyruvic acid is commonly converted to ethanol, acetaldehyde, or lactic acid. Under aerobic conditions (O_2 present), pyruvic acid is normally respired completely to yield CO_2 and H_2O. The series of steps in aerobic respiration (Figure 6-12) is called the Krebs cycle, tricarbox-

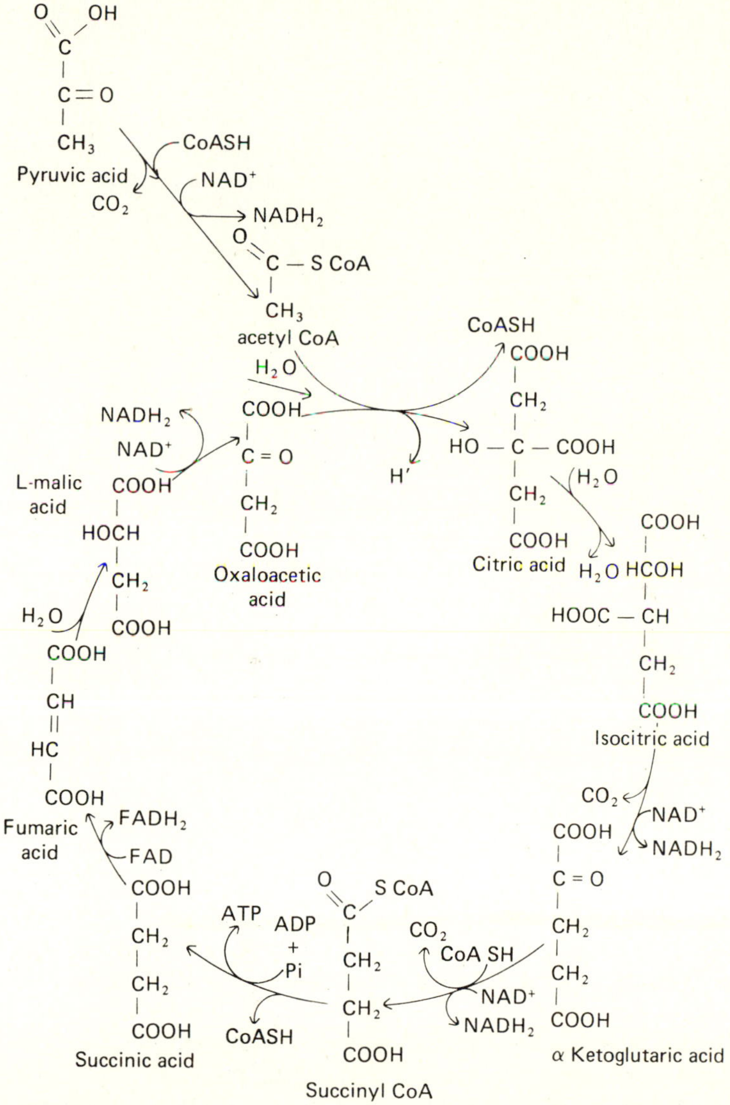

FIGURE 6-12 Krebs cycle. Each "turn" of the cycle produces two molecules of CO_2, three molecules of reduced NAD ($NADH_2$), one molecule of reduced FAD ($FADH_2$), and one molecule of ATP. See text for further details. (Adapted from Frank B. Salisbury and Cleon Ross, *Plant Physiology*, Wadsworth, Belmont, Calif. © 1969. Reprinted by permission of the publisher.)

ylic acid cycle, or citric acid cycle. In summary, one cycle releases 2 mol of CO_2 and produces 3 mol of reduced NAD ($NADH_2$), 1 mol of reduced FAD ($FADH_2$), and 1 mol of ATP. The energy in each mole of NADH can be transferred to ADP to form 3 mol of ATP; each mole of FADH can be used to convert 2 mol of ADP to ATP. Thus each mole of acetyl CoA releases energy enough to produce 12 mol of ATP per mole or the equivalent of 24 mol per mole of glucose. Adding this to the 8 + 6 (from 2 $NADH_2$) mol produced in glycolysis, a grand total of 38 mol of ATP are produced per mole of glucose.

Rather involved estimates indicate that each mole of ATP represents about 8 kcal of energy or a total of 8 × 38 or 304 kcal per mole of glucose oxidized. Since a mole of glucose is estimated to contain about 686 kcal of energy, the reactions described capture just under 50 percent of the total available energy.

As noted above, the energy captured in both glycolysis and the Krebs cycle is not directly transferred from NADH or FADH to ATP but is transferred through many intermediate compounds via the electron transport or cytochrome system. Although this system is quite involved and has yet to be fully worked out, it is well accepted at least in a general way. The interrelationships among glycolysis, the Krebs cycle, and the cytochrome system can be seen in Figure 6-13.

Other pathways for the oxidation of sugars have been found to exist. One of the most important is the pentose phosphate pathway or the hexose-monophosphate shunt.

COMPLEX COMPOUNDS

The initial products of photosynthesis are simple carbohydrates, but these serve as building blocks, carbon source, or energy supply for a wide diversity of more complex materials including polysaccharides, proteins, and lipids.

Carbohydrates

The term *carbohydrate* is used to include simple sugars or monosaccharides, disaccharides (that is, dimers of two simple sugars), and polysaccharides (that is, polymers of usually seven or more simple sugars).

Perhaps the most common monosaccharide is glucose, a 6-carbon sugar (i.e., a hexose) with the formula: $C_6H_{12}O_6$. As shown in Figure 6-14, the structural formula for glucose can be written in different ways—one depicting a straight chain configuration, the other a ring structure. Other hexoses include fructose and galactose, both having

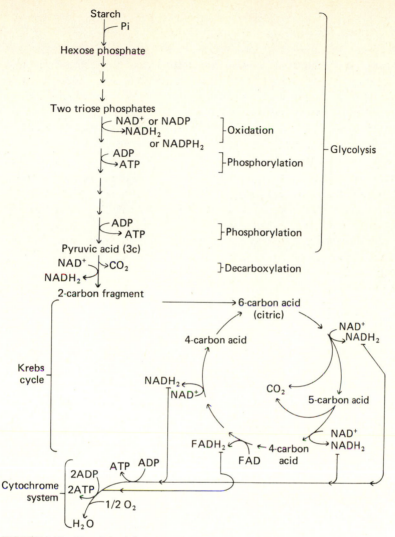

FIGURE 6-13 Simplified diagram showing the relationships of glycolysis, the Krebs cycle, and the cytochrome system. (Adapted from Frank B. Salisbury and Cleon Ross, *Plant Physiology*, Wadsworth, Belmont, Calif. © 1969. Reprinted by permission of the publisher.)

the same empirical formula as glucose but with different structural formulas (Figure 6-14).

Two monosaccharides may unite to form a disaccharide with the net release of one molecule of water. This type of reaction is called a *condensation reaction*. The reverse process is called *hydrolysis* since

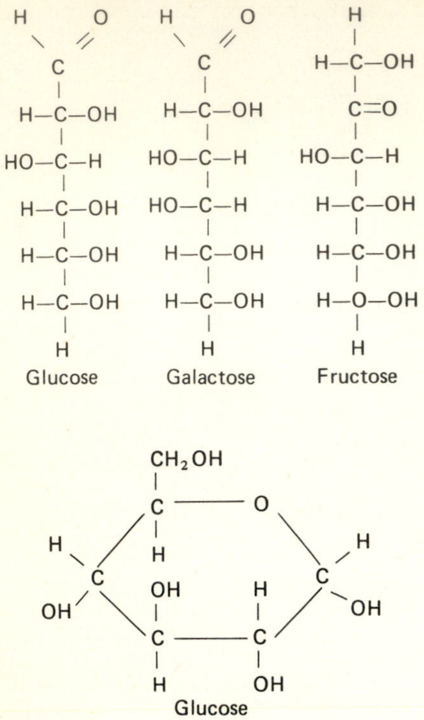

```
  H    O        H    O        H
   \  //         \  //        |
    C             C         H—C—OH
    |             |           |
  H—C—OH        H—C—OH       C=O
    |             |           |
 HO—C—H        HO—C—H      HO—C—H
    |             |           |
  H—C—OH        HO—C—H       H—C—OH
    |             |           |
  H—C—OH        H—C—OH       H—C—OH
    |             |           |
  H—C—OH        H—C—OH       H—O—OH
    |             |           |
    H             H           H
  Glucose      Galactose    Fructose
```

```
              CH₂OH
               |
               C ——— O
               |        \
   H  \        H         \  H
       C      OH      H    C
   OH /   \    |      |   / OH
           C ——— C
           |      |
           H     OH
             Glucose
```

FIGURE 6-14 Two ways of depicting the glucose molecule: *top,* as a straight chain or, *bottom,* as a ring structure. Galactose and fructose have the same molecular formula, $C_6H_{12}O_6$, as glucose and are said to be isomers of glucose.

the equivalent of one molecule of water is added during the reaction. For example, if a molecule of glucose is condensed with a molecule of fructose, the product is a molecule of sucrose. Sucrose is the carbohydrate obtained from sugar cane and sugar beets and is the common "sugar" used by everyone in many types of foods. In a similar manner two glucose units can unite to form maltose; a glucose and a galactose will combine to form lactose or milk sugar.

More complex carbohydrates are called polysaccharides because they are formed from multiple simple-sugar units. Two of the most common and important polysaccharides are starch and cellulose which are both made of glucose units but differ in the type of linkage between individual units (Figure 6-15).

The seemingly slight difference in the bonding between the sugar

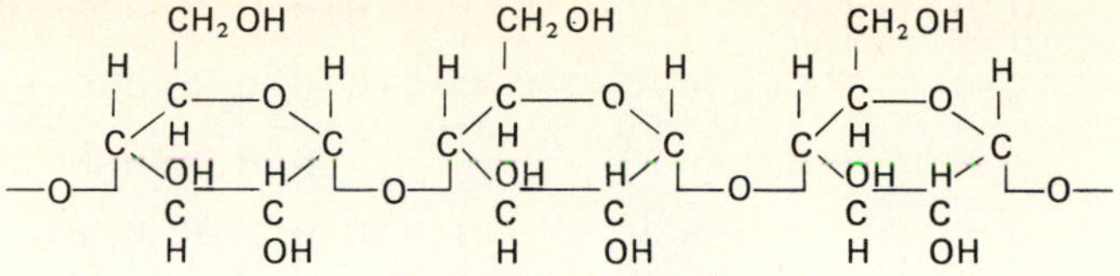

Portion of starch chain

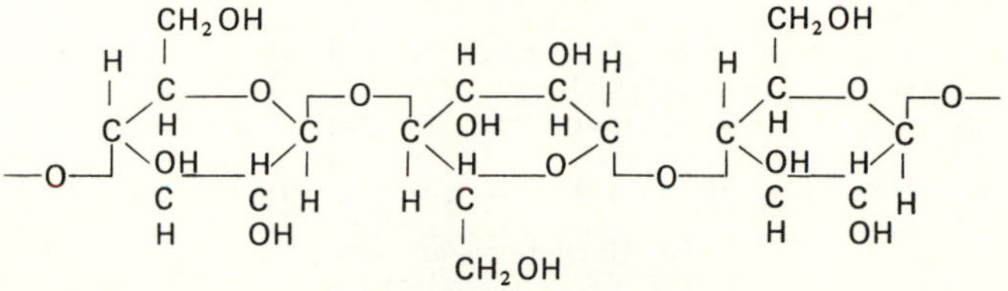

Portion of cellulose chain

FIGURE 6-15 **Polysaccharides are made of multiple units of simple sugars. Starch and cellulose are both made of glucose units but are joined by a different linkage.**

molecules makes a tremendous difference in the physical and chemical properties of the two polysaccharides. For example, starch is a good energy source because it is easily degraded by digestive enzymes, but few organisms can attack the cellulose structure to get at its sugar supplies. Cellulose would represent a tremendous source of carbohydrates—if we could get them out. Starch exists in many tissues of plants and serves as a storage material. Of major significance is the fact that, being insoluble, starch allows the storage of large quantities of carbohydrates in a form which does not influence the water potential of cell sap. Starch grains are very prevalent in storage organs, such as tubers, bulbs, and seeds, but are also often present in fruits, leaves, and stems. Although starch is insoluble, it can be readily broken down into soluble sugars within the plant through the action of amylase. Starch is also readily digestible by animals and so is a source of energy in many foods. Examples of foods with a relatively large portion of the food value in the form of starch are potato, cornmeal, wheat, and most other grains. Cellulose is also a high-molecular-weight polymer of glucose but is quite distinct from starch. Cellulose is largely a structural material and is a major component of cell walls. Because it is very resistant to degradation, cellulose does not reenter into cell metabolism after it has been formed; thus it is not considered a storage material. Cellulose is of no food value to human beings directly, as they cannot digest it.

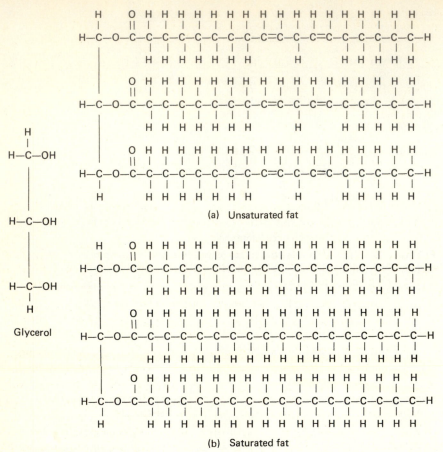

FIGURE 6-16 Simple fats and oils are made of one molecule of glycerol combined with three fatty acids. (*a*) Unsaturated fats have some double bonds, whereas (*b*) saturated fats have no double bonds.

Lipids

Another important group of compounds in the metabolism of plants is the lipids, which include three major types: (1) fats and oils, (2) phospholipids and glycolipids, and (3) waxes. Fats and oils are synthesized by the combination of one molecule of glycerol with three fatty acids (Figure 6-16); they are frequently called triglycerides. Since they contain mostly saturated fatty acids, fats are solid or semisolid at room temperatures while oils are liquid at comparable temperatures because oils contain unsaturated fatty acids (Figure 6-16). By now everyone is well aware of the controversy about saturated versus unsaturated fats. Fats and oils are primarily storage lipids and occur most abundantly in seeds which may contain up to 65 percent oil on a dry-weight basis (Table 6-2). Major commercial sources of vegetable

TABLE 6-2

OIL CONTENTS OF VARIOUS SEEDS

SPECIES	COMMON NAME	FAT AS PERCENT OF DRY WEIGHT
Cocos nucifera	Coconut	65
Helianthus annuus	Sunflower	50
Glycine max	Soybean	20
Zea mays	Corn	5
Triticum vulgare	Wheat	2

Source: J. Bonner, *Plant Biochem*, Academic Press, 1950, p. 354.

oils are the seed of corn, peanut, soybean, and cotton as well as the fruit of the olive. Fats and oils are high-energy materials containing more than twice the energy of sugar per gram.

The phospholipids are thought to be major constituents of membranes and function to regulate permeability. Although membrane structure and function are of paramount importance in the life of plants, this area of plant physiology is not well understood. As our understanding of membranes is gradually increased, many aspects of uptake and translocation of various substances should be clarified.

Waxes are present on the cuticle of leaves and fruits in varying amounts and are solids at normal temperatures. Waxes usually occur within the cuticle as well as exterior to it and serve as an effective barrier to water loss. The whitish, dusty "bloom" which is so typical of blueberries, grapes, plums, cabbage leaves, and certain apples is an example of wax. Although certain species of plants, such as the carnauba palm, yield commercial quantities of waxes, most plant waxes are of only indirect benefit to human life. Older leaves and leaves exposed to relatively severe environmental stresses, such as high illumination and low relative humidity, tend to extrude more wax. The wax layer and cuticle are of prime interest in attempts to increase foliar penetration of various chemicals such as nutrients and growth regulators.

Nitrogen Metabolism

Plants are able to absorb nitrogen in a variety of forms which include the nitrate ion, the ammonium ion, urea, and certain organic materials, such as amino acids. The most widely available source of nitrogen in the soil is nitrate, NO_3, because soil microorganisms rapidly oxidize reduced forms such as NH_3 to NO_3. Uptake of ammonium occurs under certain conditions, but excessive ammonium can be toxic if it is not converted into noninjurious compounds rapidly enough. Urea has been found to be a very useful fertilizer, especially when applied as a foliar spray. By applying the nitrogen source directly to the foliage, one overcomes the usual loss to weeds, leaching, and microorganisms, and response time is considerably shorter. Organic sources of nitrogen are decaying plant and animal materials in the soil, but these are not normally a major nitrogen source.

Regardless of the form in which nitrogen is absorbed by the plant, it must all finally be in the form of NH_3 to be utilized in the synthesis of nitrogenous compounds. The mechanism of nitrate reduction has not been completely worked out and certain intermediates between NO_3 and NH_3 are yet to be confirmed. It is well documented, however, that nitrate reductase is the enzyme involved in reducing NO_3 to NO_2, the first step in the pathway. Nitrogen reduction by the plant is an energy-requiring process and is thus dependent on respiratory energy release. The carbon compounds from photosynthesis are the carbon skeletons into which NH_3 is incorporated to form various nitrogenous compounds. Because of these interrelationships between carbohydrate and nitrogen metabolism, the carbohydrate status of the plant affects nitrogen utilization and vice versa.

Amino Acids

When nitrogen is incorporated into the carbon skeletons, the initial products are amino acids. The generalized structure of amino acids is given below:

$$R-\overset{\overset{\displaystyle H}{|}}{\underset{\underset{\displaystyle NH_2}{|}}{C}}-COOH$$

The "R" group varies among the twenty different amino acids identified in plants and may consist of various combinations of C, H, O, N and sometimes S molecules. Amino acids are formed by two different processes. The first is called *reductive amination* and is the formation of glutamic acid from α ketoglutaric acid and is catalyzed by glutamic acid dehydrogenase.

COOH		COOH	COOH	
|		|	|	
CH_2		CH_2	CH_2	
|		|	|	
CH_2	$+ NH_3 \longrightarrow$	CH_2 $\xrightarrow{\quad NADH_2 \quad}$	CH_2	$+ NAD^+$
|		|	|	
C=O		C=NH	CH$-NH_2$	
|		|	|	
COOH		COOH	COOH	
α ketoglutaric acid		α iminoglutaric acid	glutamic acid	

The reaction over the arrow is labeled "glutamic acid dehydrogenase".

The second and perhaps most important reaction in amino acid formation is *transamination*, which consists of the transfer of the amino group of one amino acid to a keto acid to form a new amino acid. The enzymes which catalyze these reactions are called *transaminases*. Glutamic acid is the most prevalent first amino acid formed and is also the major starting point for the transaminations which ultimately lead to the synthesis of the other amino acids.

Proteins

Of the many types of complex organic molecules in plants, some of the most important are the proteins. The vast array of chemical reactions within plants and animals proceed at normal rates only because of enzymes which act as organic catalysts. These enzymes are proteins, and most are very specific for a particular reaction or type of reaction. *Proteins* are made up of large numbers of amino acid molecules joined together in a type of linkage referred to as a peptide bond. This type of bond, depicted below, joins the carboxyl group of one amino acid with the amino group on the next:

$$
\begin{array}{c}
\diagdown \\
\ \ \ \ \ CH \\
\diagup
\end{array}
\ \
\begin{array}{c}
O \\
\parallel \\
C \\
\diagdown
\end{array}
\ \
\begin{array}{c}
N \\
\mid \\
H
\end{array}
\ \
\begin{array}{c}
R \\
\mid \\
CH \\
\diagdown
\end{array}
\ \
\begin{array}{c}
C \\
\parallel \\
O
\end{array}
\ \
\begin{array}{c}
H \\
\mid \\
N \\
\diagdown
\end{array}
\ \
\begin{array}{c}
CH \\
R
\end{array}
$$

Each protein molecule can consist of up to twenty different amino acids, each of which may occur repeatedly and in varying sequences. The order of amino acids in the protein molecule is not haphazard, but rather each protein has a very specific amino acid arrangement and structure. In addition to amino acid sequence (referred to as *primary structure*), protein molecules are often coiled in the shape of a helix (*secondary structure*) and in turn may be folded or looped (*tertiary structure*). It is currently thought that it is the secondary or tertiary structure (or both) which is lost or destroyed by denaturation of a protein. This denaturation can occur as the result of such factors as relatively high temperature or excessive changes in pH which can cause reversible or irreversible coagulation or precipitation.

Proteins which consist of amino acids only are called *simple proteins* and are further classified on the basis of solubility. Examples are albumins which are water-soluble, globulins which are relatively insoluble in water but are soluble in dilute salt solutions, and prolamines which are insoluble in water but soluble in water-ethanol mixtures. In the second group are the *conjugated proteins* which consist of a component in addition to the peptide. The part other than the amino acid is called the prosthetic group, and it is upon the basis of this part that they are classified; examples are lipoproteins and chromoproteins. Lipoproteins are complexes of proteins and lipids and are major structural components of membranes. Chromoproteins are combinations of a pigment and a protein, and most are enzymatic.

BIBLIOGRAPHY

Bidwell, R. G. S.: *Plant Physiology*, Macmillan, New York, 1974.

Devlin, R.: *Plant Physiology*, 3d ed., Van Nostrand, New York, 1975.

Lehninger, A. L.: *Biochemistry*, 2d ed., Worth, New York, 1975.

Leopold, A. C., and P. E. Kriedemann: *Plant Growth and Development*, 2d ed., McGraw-Hill, New York, 1975.

Meyer, B. S., D. B. Anderson, R. H. Bohning, and D. G. Fratianne: *Introduction to Plant Physiology*, Van Nostrand, New York, 1973.

Salisbury, F. B., and C. Ross: *Plant Physiology*, Wadsworth, Belmont, Calif., 1969.

Wilkins, M. B.: *The Physiology of Plant Growth and Development*, McGraw-Hill, London, 1969.

Zelitch, I.: *Photosynthesis, Photorespiration, and Plant Productivity*, Academic Press, New York, 1971.

PART 2
Plant Environment

CHAPTER 7
TEMPERATURE RELATIONS

Of the various environmental factors that affect plants, temperature is one of the most important. Plants are able to grow only within a rather narrow range of temperatures, although certain plants can survive at much lower and somewhat higher extremes. For most horticultural plants the optimum temperature for growth is between 15 and 35°C. Seed dried to the optimum moisture content can withstand very cold temperatures. Moderately cold temperatures can considerably extend the length of time during which seed will maintain viability.

The degree to which a plant can adapt to changing temperatures varies among species. The tomato, for example, cannot withstand temperatures below 0 to −1°C even when properly hardened off, whereas a properly hardened apple tree will commonly suffer no injury at −35°C. Certain fruits, such as the banana, will suffer "chilling injury" at 4°C.

BASIC ASPECTS OF HEAT AND TEMPERATURE

Before discussing the effects of temperature in further detail, let us go over a few basic definitions and concepts. *Temperature* is a qualitative term which gives us an indication of the *intensity* of heat in a body of matter, but does *not* give an indication of the *quantity* of heat present.

Heat

Heat is a form of energy that when transferred to a body of matter causes an increase in temperature or when removed causes a decrease in temperature, provided the matter does not change state during the process. The last phrase of the definition is necessary because changes of state, such as ice to water or water to water vapor, involve a transfer of heat without a change in temperature.

Units of heat most commonly used are the calorie, kilocalorie, and the British thermal unit (BTU). A calorie (cal) is the amount of heat required to raise the temperature of 1 g of water by 1°C. A kilocalorie (kcal) is 1000 times a calorie or the amount of heat required to change the temperature of 1 kg of water by 1°C. Under the English system, a BTU is the amount of heat required to change the temperature of 1 lb of water by 1°F. Thus, 1 BTU is equivalent to 253 cal.

Specific Heat   *Specific heat* is defined as the number of calories of heat required to change the temperature of 1 g of a substance by 1°C. A high specific heat implies a relatively small change in temperature in response to a given amount of heat energy.

From Table 7-1 it is apparent that water has a high specific heat which is lowered by one-half or more when found in the form of either ice or steam. The high specific heat of water is of major importance in horticulture. Since actively growing plants contain large amounts of water, its presence acts as a strong buffer against temperature change. Excessive tissue temperatures occasionally lead to "sun scald" on fruits and vegetables. This would be greatly accentuated if such tissues were not composed largely of water with its very high specific heat.

TABLE 7-1
SPECIFIC HEAT OF CERTAIN SUBSTANCES

SUBSTANCE	SPECIFIC HEAT, cal g^{-1} °C^{-1}
Water	1.00
Ice	0.50
Steam	0.48
Alcohol (ethyl)	0.58
Wood	0.42
Glass	0.20
Steel	0.11

TABLE 7-2
HEAT OF FUSION AND HEAT OF VAPORIZATION OF CERTAIN SUBSTANCES

SUBSTANCE	HEAT OF FUSION, cal g^{-1}	HEAT OF VAPORIZATION, cal g^{-1}
Alcohol (ethyl)	25.0	204
Oxygen	3.3	51
Water	80.0	540

Heat of Fusion   The heat of fusion is the amount of heat required to change 1 g of a substance at its melting point from the solid to the liquid state or vice versa.

Heat of Vaporization   The amount of heat required to change 1 g of a substance at its boiling point from the liquid state to the vapor state is called the heat of vaporization. The same amount of heat must be removed from 1 g of the substance in the vapor state to convert it to a liquid.

Just as water has a rather high specific heat (Table 7-1), water also has a relatively high heat of fusion and heat of vaporization compared with other substances (Table 7-2). These three physical properties make water a most unique substance, and all three properties are of major significance in horticulture.

Changes of State   The addition or removal of heat to a body of matter does not always lead to a change in its temperature. During a change of state no temperature change occurs until the entire mass has completed the change. The graph in Figure 7-1 is a plot of temperature versus time. Starting with 1 kg of ice at −100°C, heat is added at a constant rate of 100 kcal min^{-1}. Since the specific heat of ice is 0.5, each kcal of heat raises the temperature 2°C. Thus at 100 kcal min^{-1}, it takes 0.50 min to bring ice from −100 to 0°C. At 0°C the temperature rise ceases, and the temperature remains at 0°C until 80 kcal have been added—the heat of fusion of water. Thus, it takes 0.8 min at 100 kcal min^{-1} to convert 1 kg of ice at 0°C to water at 0°C. When all the ice is melted, the temperature again rises but now at a rate of 1°C per kcal. To raise the temperature of 1 kg water at 0 to 100°C takes 100 kcal or 1 min. When the temperature reaches 100°C, the boiling point of water, it remains constant until 540 kcal (the heat of vaporization) have been added. It therefore takes 5.4 min to convert all the water to steam. After this conversion is complete, the temperature rises at a slightly faster rate than for ice since the specific heat of steam is only 0.48.

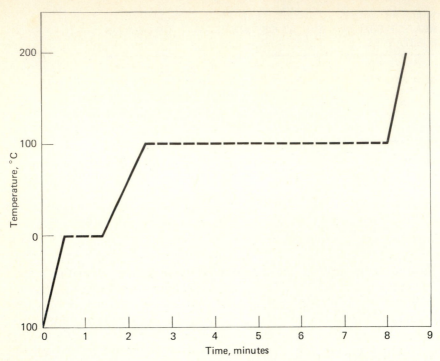

FIGURE 7-1 Plot of the temperature of 1 kg of ice starting at −100°C and with the addition of 100 kcal of heat per minute.

Heat Transfer

The transfer or movement of heat can occur through three different processes; in all three the net movement is always from the warmer to the cooler body.

Conduction is the flow of heat through a substance; the rate of flow is proportional to the cross-sectional area of the conductor and the temperature gradient and varies with the substance involved. For example, steel is a good conductor, wood is a relatively poor conductor, and air is a very poor conductor.

Convection involves the transfer of heat by a moving agent. When heat is added to a gas or a liquid, the density of the gas or liquid changes and a circulatory motion is established whereby heat is transferred. Thus, convection is dependent upon the establishment of convection currents. One can see convection currents in a pan of water being heated on a stove or in air over a very warm surface, such as black pavement exposed to direct sun.

Radiation is the transfer of energy with no connecting medium required. Radiant energy consists of electromagnetic waves traveling at the speed of light, 3×10^8 m sec^{-1}. Radiant energy covers the entire electromagnetic spectrum from gamma rays, x-rays, ultraviolet, visible light, infrared (heat), television, to radio waves. These differ only in wavelength, the range being from 10^{-15} to 10^5 m. Two important features of radiation are: (1) The predominant wavelength of the radiation becomes shorter as the temperature of the body increases. (2) The total radiation per unit area from all wavelengths is proportional to T^4 (T being the absolute temperature* of the radiating body).

The human eye responds to only a very limited portion of the electromagnetic spectrum. The diversity of wavelengths and characteristics of various portions of radiant energy, as well as the very small portion to which the human eye responds, is shown in Figure 9-1.

Temperature Measurement

In the past, scientists in the United States have used the Fahrenheit scale, in which 32° is the freezing point of water, and 212° is the boiling point. We are currently going through a transition period during which the United States is gradually converting to the metric system of weights and measures as well as to the Celsius (or centigrade) temperature scale. In the Celsius scale, the freezing point of water is 0°C and the boiling point is 100°C. The appendix of this book contains conversion tables.

Simple Thermometers A thermometer is a device for measuring temperature and is based on the principle that most substances *expand* as their temperature increases and *contract* when their temperature is decreased. A simple thermometer consists of a glass reservoir, capillary tube, and safety chamber. The reservoir and part of the capillary tube are filled with mercury or colored alcohol. The shape of the reservoir may vary from spherical to cylindrical, and the size of the tube in relation to the volume of the reservoir determines the "openness" of the scale. The remainder of the tube and the safety chamber are evacuated to allow for expansion of the liquid.

Simple thermometers can be very accurate if properly manufactured, but frequent checking is absolutely essential. One of the major problems with this type of thermometer is that the temperature scale is not on the tube of the thermometer. Thus, if the thermometer shifts in relation to the scale, all subsequent readings are inaccurate. Perhaps the easiest way to check the accuracy of a thermometer is to make an ice and water mixture in an insulated container and gently stir it with the thermometer. After a few minutes, the temperature will be 0°C, which is the melting point of ice or the freezing point of water. This

---

*Absolute (or Kelvin) temperature scale is based on "absolute zero," at which point matter contains no heat, and is equivalent to −273°C or −460.4°F. Ice melts at 273 K; water boils at 373 K. K = °C + 273 or °C = K − 273.

provides a reproducible reference temperature at which the thermometer can be checked and reset if necessary. This procedure is also suitable for standardizing the maximum, minimum, and maximum-minimum types described below.

Minimum Thermometers   This type of thermometer not only measures but also indicates the minimum temperature, and is commonly used in weather stations (Figure 7-2). A minimum thermometer contains alcohol in which a metal index is suspended within the tube (Figure 7-3). Because the index is free to move quite easily, the minimum thermometer is mounted horizontally, with the reservoir slightly lower than the tube. As the temperature decreases, the alcohol contracts, drawing the index down the tube by surface tension. When the temperature again starts to increase, the alcohol expands; but the

FIGURE 7-2 A minimum thermometer is mounted horizontally. The index lies within the liquid and is reset by tilting. (Photograph, Taylor Instrument, Sybron Corporation, Arden, N.C.)

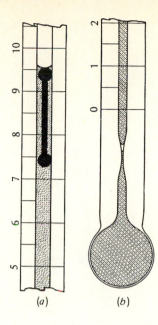

FIGURE 7-3 (a) Minimum thermometer showing index. (b) Maximum thermometer showing constriction in bore of tube. (From W. L. Donn, *Meteorology*, 4th ed., McGraw-Hill, New York, 1975.)

index remains at its lowest point since the alcohol merely bypasses the index. Thus the distal end of the index remains at the minimum temperature experienced. It is reset by tilting, thus causing the index to slide through the liquid until it is against the meniscus.

Maximum Thermometers Temperature-observing stations have a maximum thermometer in combination with the minimum type described above. A maximum thermometer contains mercury but differs from the simple type in that there is a constriction at the base of the stem. As the temperature increases, the mercury in the reservoir expands and is forced past the constriction. As the temperature decreases, however, the mercury cannot be drawn past the constriction (Figure 7-3). Thus, the top of the mercury column is at the maximum temperature experienced. To reset this type of thermometer, one must swing it around in a circle to force the mercury back past the constriction by centrifugal force. A similar type of thermometer, but with a limited range, is the clinical thermometer used in medicine.

Maximum-Minimum Thermometers This type of thermometer measures and records both maximum and minimum temperatures. It consists of a U-shaped tube with a safety chamber at one end (Figure 7-4). The middle part of the tube is filled with mercury, and creosote is added to both ends. The safety chamber is evacuated to allow for expansion. On top of each end of the mercury column is a metal index which is constructed so as to fit snugly enough inside the tube to allow passage of the creosote, but not the mercury. Thus, as the temperature

FIGURE 7-4 Maximum-minimum thermometer. The metal indices are at or above the mercury. The maximum and minimum temperatures are read at the lower end of the appropriate index. (Photograph, Taylor Instrument, Sybron Corporation, Arden, N.C.)

increases, the creosote and mercury expand. The mercury moves upward on the maximum side toward the safety chamber pushing the index immediately in front of the mercury. As the temperature decreases, the mercury moves back down the tube, leaving the *lower* end of the index at the maximum temperature reached. As this contraction occurs, the minimum index is pushed upward on the opposite end. Thus, the minimum temperature is read at the lower end of the second index. The temperature scales are reversed on the two sides of the thermometer so that, at any given time, the current temperature can be read on either side. The indices are reset to the top of the mercury columns by using a magnet.

The maximum-minimum thermometer is less expensive than using one each of the maximum and minimum types. If properly standardized, the maximum-minimum type can be sufficiently accurate for most purposes but can be subject to error since the scale is not on the tube as it is on the maximum and the minimum types.

Deformation Thermometers   In many situations it is desirable to have not only a record of maximum and minimum temperatures, but a continuous temperature record. An instrument which records temperatures is called a *thermograph* and usually utilizes a deformation thermometer. Deformation thermometers are also used in those thermometers with a round dial scale. The temperature-sensing element consists of a bimetallic strip in which two different metals are joined on a flat surface. The metals most often used are brass (copper-zinc alloy) and invar (nickel-steel alloy). Brass expands about 20 times as much as invar in response to a given increase in temperature. The actual shape of this strip can be straight, a spiral, or a crescent. One end of the strip is fixed. As the shape changes in response to temperature, this movement is transmitted by means of levers to a dial or in the case of a thermograph to a pen which records on a graph. The graph is attached to a rotating drum which is operated by a windup mechanism, batteries, or electricity. A thermograph is often combined with a humidity recorder, in which case the combination is called a *hygrothermograph* (Figure 7-5).

FIGURE 7-5   A hygrothermograph in which one pen records temperature as measured by a deformation thermometer and the second records relative humidity as measured by multiple strands of human hair. (Photograph, Weather Measure Corporation, Sacramento, Calif.)

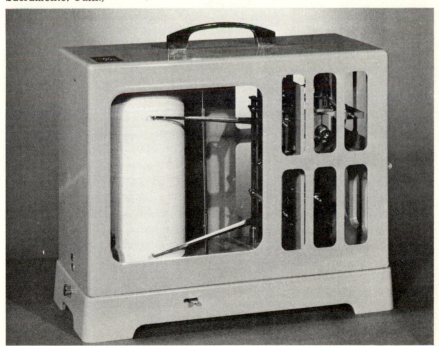

Bourdon Tube Thermometers   To record temperature with a thermograph, Bourdon thermometers are also often used. This type of thermometer utilizes a curved, chrome-plated phosphon bronze tube filled with a liquid, usually alcohol. As temperature increases, the liquid expands and bends the Bourdon tube which moves a link-and-lever assembly. This in turn moves the pen on the chart. As temperatures decrease, the reverse action occurs.

Electrical Thermometers   The two types of electrical thermometers are thermocouples and thermistors. *Thermistors* are mixtures of specially prepared metal oxides, which, when arranged in a circuit with platinum alloy wire leads, constitute semiconductors whose electrical resistance varies markedly with temperature. *Thermocouples* consist of two wires of dissimilar metals that are joined together at both ends. When the two junctions are at different temperatures, an electromotive force (emf) is generated and is proportional to the temperature difference. The reference junction is kept in an ice-and-water mixture in an insulated container to give a reference temperature of 0°C. When a meter or recorder is connected into the circuit, the voltage and, indirectly, the difference in temperature is measured.

Both types of electrical thermometers have their respective advantages. With thermistors, no reference temperature junction is necessary and longer lead wires are possible because of the larger current utilized. Thermocouples are less expensive, easy to make, and can be very small, being limited only by the sizes of wire available.

Location and Shielding of Thermometers   Not only must thermometers be accurate, they must be properly positioned and shielded to prevent errors. Under most situations we wish to determine air temperature; thus the thermometer must be shaded from direct radiation from the sun, the ground, and adjacent buildings. The thermometer should not be exposed to the sky at night as readings will be below air temperature because of excessive radiation from the thermometer to the cold sky. The typical weather observation station is a white, louvered structure which allows free air circulation but no direct penetration of radiation (Figure 7-6). For multiple thermometers used in crop areas, less elaborate thermometer shelters are used which provide protection primarily against the ground and sky.

For most purposes, thermometers are positioned at about 1.5 m above the ground. As will be seen later in this chapter, temperatures vary with increasing distance from the ground, but 1.5 m is used because it is both at eye level and in the general region of concern for crop production. For special applications, thermometers may be placed at various other heights above the ground.

FIGURE 7-6 Standard weather shelter for housing a maximum and a minimum thermometer. Note the double roof and louvered sides. (Photograph, Weather Measure Corporation, Sacramento, Calif.)

TEMPERATURE AS A PART OF CLIMATE

Whereas temperature is a major aspect of climate affecting horticultural crops, climate also includes a consideration of precipitation, humidity, sky conditions, wind, and atmospheric pressure. Although this chapter is devoted primarily to temperature, one must keep in mind that all the various aspects of climate must ultimately be considered as a whole.

Whereas *climate* refers to the average conditions over a long period, the term *weather* is used to describe the current and temporary atmospheric conditions. Climate and weather refer to the same basic factors, the only difference being the time span involved.

When discussing climate, we also need to be aware of the area involved. For example, *macroclimate* is used to describe the conditions over a relatively large area, perhaps a radius of 40 to 80 km. *Local climate* refers to a smaller area such as is used in comparing a valley and an adjacent mountain. The term *microclimate* is used to refer to the climate around a particular plant or leaf, or even that around a stomate.

Solar Radiation

Although a small amount of heat comes from the interior of the earth, essentially our only source of heat is the sun. To review briefly the relationship of the earth and sun, the diameter of the sun is approximately 1.4×10^6 km or about 100 times that of the earth. Its mean distance from earth is about 1.5×10^8 km. The surface temperature of the sun is estimated to be 6000°C, and it is constantly emitting energy in all directions. The earth and its atmosphere receive only about 4 ten-billionths of all the energy radiated by the sun. It is rather humbling to consider that all life on earth is sustained by such an infinitesimally small portion of the sun's energy output.

The total incoming solar energy reaching the outer edge of the earth's atmosphere averages 2 cal cm^{-2} min^{-1} or 2 langleys min^{-1}. This is referred to as the *solar constant*. It varies only slightly as the distance between the sun and earth changes.

The majority of solar energy is in the shortwave region of the spectrum. As noted previously, when the temperature of the radiating body increases, the predominant wavelength decreases. Of the incoming solar energy striking the outer atmosphere, about 10 percent is in the ultraviolet (UV) band, 40 percent is in the visible region, and 50 percent is in the infrared area.

Absorption As the solar energy penetrates the atmosphere, it is modified by several factors. Absorption by atmospheric components is of major consequence, especially in the case of UV. Oxygen and ozone absorb most of the UV—a factor of major importance to both plants and animals because neither can withstand very much ultraviolet radiation. Water vapor and carbon dioxide also absorb some energy in the infrared region.

Scattering by various particles in the atmosphere also alters incoming radiation. Large particles such as dust, smoke, and water droplets scatter all wavelengths nonselectively. White light is a mixture of all colors, and so the sky takes on a whitish color when there is a high concentration of such particles in the air. Smog and haze, which have become all too common, are one cause of nonselective scattering. Small particles, such as gas molecules, scatter the shorter wavelengths of visible light (blue) more than the longer wavelengths (red). For this reason, on a clear, dry day with little air pollution, the sky is very blue because of the preferential scattering of the blue portion of the spectrum. At sunset on such a day, the sky often has a reddish hue because the solar energy is passing through the equivalent of several depths of atmosphere and the shortwave (blue) energy is removed by scattering before the energy reaches our eyes.

Reflection   by clouds is a major factor affecting the amount of solar energy reaching the earth's surface. Clouds act as poor absorbers but very efficient reflectors of radiant energy. This is particularly apparent when one flies on an overcast day. As soon as the plane breaks through the cloud cover, the light above the clouds is very intense since most of the incoming energy is reflected. The proportion of energy reflected by clouds is often 50 to 90 percent depending on cloud thickness and density.

The total solar radiation striking the surface of the earth can be divided into two components: *direct* radiation and *diffuse* or sky radiation. The proportion of each is variable, depending on many factors such as cloudiness, latitude, angle of the sun, and altitude. On a clear day, most solar energy received is direct radiation, whereas on an overcast day most is diffuse radiation. Long-term averages indicate that of the total incoming solar radiation, about 19 percent penetrates directly, about 28 percent arrives as diffuse radiation, 19 percent is absorbed by the atmosphere, and 34 percent is reflected back into outer space.*

Outgoing Radiation

The earth is constantly radiating energy into the atmosphere. Because the surface of the earth is relatively cool compared to the sun, nearly all terrestrial radiation is of long wavelengths in the infrared region. Because of its wavelengths, the earth's radiation is very strongly absorbed and reflected by water vapor, ozone, CO_2, and clouds. Much of that absorbed by the atmosphere is reradiated back to earth. It is largely the result of this reflection and "counterradiation" that the drop in temperature at night is not more severe.

Greenhouse Effect

The net result of the atmosphere on incoming and outgoing radiation is termed the "greenhouse effect" because of the likeness between the effects of a pane of glass and those of the earth's atmosphere. Both

*G. T. Trewartha, *An Introduction to Climate*, 4th ed., McGraw-Hill, New York, 1968, p. 18.

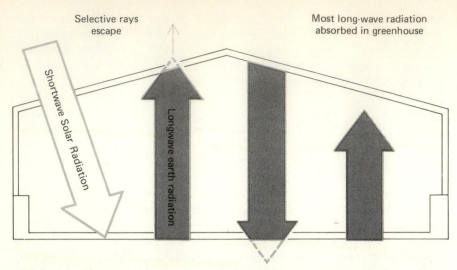

Selective rays
escape

Most long-wave radiation
absorbed in greenhouse

Shortwave Solar Radiation

Longwave earth radiation

FIGURE 7-7 The earth's atmosphere acts much like the glass in a greenhouse to produce the "greenhouse effect." The glass (and atmosphere) is relatively transparent to shortwave solar radiation, but relatively opaque to longwave radiation from the earth. (From G. T. Trewartha, *An Introduction to Climate*, McGraw-Hill, New York, 1968.)

allow the transmission of a high proportion of shortwave solar radiation but strongly inhibit the transmission of longwave terrestrial radiation (Figure 7-7).

FACTORS INFLUENCING TEMPERATURE

The temperatures typical for a particular location are affected by latitude, elevation, season, time of day, and local factors. Although all these factors interact, each will be presented individually.

Latitude

In Table 7-3 representative temperatures for several latitudes in the Northern Hemisphere are given. The temperature differences in Table 7-3 associated with latitude are caused largely by differences in insolation (radiation received from the sun) which is, in turn, dependent on both daylength and the angle of the sun.

The inclination of the earth's axis of rotation is $23\frac{1}{2}°$ to its plane of rotation around the sun (Figure 7-8). Only one-half of the earth is illuminated at any one time, and the portion illuminated is continually changing. At the spring and fall equinoxes, (March 21 and September 21, respectively), insolation reaches the earth from the North to the South Pole. Thus, all points on the earth have 12 h of daylight and 12 h

TABLE 7-3

TEMPERATURES (°C) AT FIVE LATITUDES IN THE NORTHERN HEMISPHERE

NORTH LATITUDE	MEAN ANNUAL	MEAN JANUARY	MEAN JULY	RANGE JANUARY–JULY
90°	−26°	−41°	− 1°	40°
60°	3°	−16°	14°	30°
30°	18°	14°	27°	13°
10°	26°	26°	27°	1°
0°	26°	27°	26°	1°

of darkness, and all regions of the *outer atmosphere* receive similar amounts of insolation. For two major reasons, however, insolation striking the earth's surface varies with latitude. The sun is directly over the equator; therefore, the solar angle decreases with increasing latitude. Thus, as one moves poleward, the same amount of radiant energy is spread over an increasing surface area. This is easily visualized by shining a flashlight on a dark wall at an angle of 90° and then decreasing the angle. As the angle of incidence decreases, the same amount of radiant energy is received by an increasing area, but intensity is decreased. The other factor is that as the angle of the sun decreases, the depth of atmosphere through which the insolation must pass increases. At 90°, the depth is 1 atmosphere; at 60°, −1.2; at 30°, −2.0; and at 10°, −5.7. As the depth of atmosphere increases, the amount of solar energy reaching the earth decreases owing to

FIGURE 7-8 The relative positions of the earth in relation to the sun at the summer and winter solstices and the fall and spring equinoxes. (From H. R. Byers, *General Meteorology*, 4th ed., McGraw-Hill, New York, 1974.)

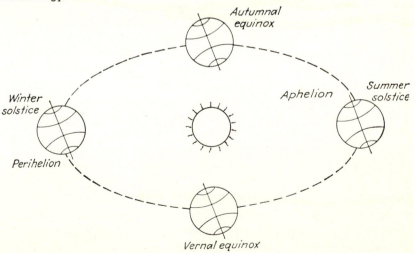

TABLE 7-4
EFFECTS OF LATITUDE AND SEASON ON DAYLENGTH

	DAYLENGTH AT SPECIFIED DATES			
LATITUDE	MARCH 21	JUNE 21	SEPTEMBER 21	DECEMBER 21
0°N (Equator)	12 h	12 h	12 h	12 h
10° (Caracas)	12 h	12 h 35 min	12 h	11 h 25 min
20° (Mexico City)	12 h	13 h 12 min	12 h	10 h 48 min
30° (Jacksonville, Fla)	12 h	13 h 56 min	12 h	10 h 4 min
40° (Columbus, Ohio)	12 h	14 h 52 min	12 h	9 h 8 min
50° (Winnipeg, Manitoba)	12 h	16 h 18 min	12 h	7 h 42 min
60° (Anchorage, Alaska)	12 h	18 h 27 min	12 h	5 h 33 min
70° —		2 months	12 h	0 h 0 min
80° —		4 months	12 h	0 h 0 min
90° —		6 months	12 h	0 h 0 min

Source: Adapted from A. Miller and J. C. Thompson, *Elements of Meteorology,* Merrill, Columbus, Ohio, 1970, p. 81.

absorption, reflection, and scattering by the atmosphere. These effects can be readily seen by comparing insolation at solar noon and in either the early morning or late afternoon when the angle of the sun is small.

Season

Seasonal effects on temperature are obviously very closely related to the influence of latitude. As shown in Table 7-3, when one moves poleward from the equator, temperatures not only decline but exhibit greatly increased seasonal variation. These seasonal effects result largely from the solar angle, discussed previously, and daylength (Table 7-4). Daylength at the equator is 12 h during the entire year, and solar angle is always relatively close to 90°.

As one moves poleward from the equator, summer daylength increases. Therefore, in spite of the low solar angle, total insolation is moderately high. In Table 7-3 the mean temperature difference between 0 and 90°N latitude in July is 27°C. During winter, far-northern latitudes have not only very low solar angles but very short days as well. The mean difference in January between 0 and 90°N latitude is 68°C.

Elevation

Temperatures decrease with elevation or altitude. Beyond the earth's atmosphere, where there is nothing to absorb radiation, temperatures are constant and very cold. On the surface of the earth, a zone of permanent snow exists above 4500 m in the tropics and above 3000 m in the temperate zones. For each increase in elevation of 100 m, mean temperatures decline about 0.6°C. A very striking example are two towns, each located 19 km from the equator: Belem, Brazil, at an elevation of about 10 m, has a mean annual temperature of 29°C; Quinto, Equador, at 2835 m, has a mean annual temperature of 13°C. Similar differences can be cited in many areas of the United States where sizable differences in elevation exist in close proximity. For example, Blacksburg, Virginia, at 640 m elevation, has a mean annual temperature about 4°C warmer than Mountain Lake, Virginia, 21 km northwest at an elevation of 1220 m. Similar differences between the latter locations are present both day and night throughout the year. The effects are very apparent in plant development in the spring, with plants at Mountain Lake being one or two weeks later in their seasonal foliation and flowering. Additional discussion of elevation effects is offered under "Air drainage."

If we compare two hypothetical sites—one at sea level and one at 1500 m elevation—we can readily explain the temperature differences on the basis of incoming and outgoing radiation. The important consideration here is the atmosphere and its effect on radiation. It is estimated that one-half of the total water vapor in the layer of atmosphere is in the lower 1800 m. This means that the lower 0.2 percent of the atmosphere contains more than 50 percent of the total water vapor in the atmosphere. Since water vapor is a much more efficient barrier to transmission of terrestrial rather than solar radiation, water vapor has major significance. Although a site at 1500 m receives slightly more insolation than one at sea level, the rate of radiant heat loss is much greater from the site at a higher elevation. Therefore, the net effect is a lower temperature at the higher elevation.

Time of Day

The effects of time of day are obvious to us all regardless of our location. At any given time, temperature is determined to a very large degree by the balance of incoming versus outgoing radiation. If we plot the energy received from the sun and that lost by the earth in relation to temperature, we can see the so-called "lag effect." It is apparent (Figure 7-9) that maximum temperatures occur not at the time of maximum insolation but rather when insolation has declined to where it equals radiational heat loss. The above statement holds true whether we are following daily or annual trends.

The maximum daily temperatures occur in midafternoon, and the daily minimum occurs just before sunrise. For the same basic reasons the annual maximum temperatures in the Northern Hemisphere occur in late July and August, well after the time of maximum insolation; the

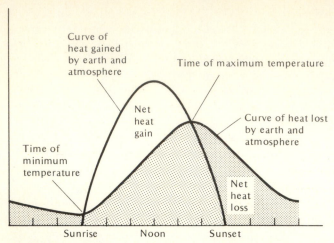

FIGURE 7-9 Variation of insolation and temperature during the day and night. Note that the time of maximum temperature lags behind the time of maximum insolation. Similarly, maximum annual temperatures lag behind maximum solar radiation received. (From W. L. Donn, *Meteorology*, 4th ed., McGraw-Hill, New York, 1975.)

annual minimum occurs in late winter, well after the period of minimal insolation.

Air Temperatures during the Day During the day, insolation passes through the atmosphere and strikes various solid surfaces on the earth's surface. As this insolation is absorbed, much of the radiant energy is converted to heat, thus elevating surface temperatures. Soon after sunrise, exposed surfaces exceed adjacent air temperatures and, by conduction, heat is transmitted to the air. Convection currents begin gradually, and heat is moved upward. However, throughout the day, the warmest air is close to the ground, and temperatures decline as distance from the earth's surface increases.

Air Temperatures at Night After sunset, the earth no longer receives solar radiation, but outgoing radiation continues throughout the night. The temperatures of many exposed surfaces, therefore, drop below that of the adjacent air. This is particularly true of efficient radiating surfaces, such as leaves, which have a large surface area in relation to their volume. As these surfaces become cooler than the surrounding air, heat is conducted from the air to the leaf—the reverse of the heat flow during the day. Since heat is radiated more rapidly than it is gained from the air, the flow of heat from the air to cold surfaces continues all night.

Temperature Inversions   As this conduction continues owing to continued radiant heat loss by solid surfaces, the air near the ground becomes cooler than the air above it. Since cool air is denser than the warm air above it, no convection currents are formed. If there is no wind to cause mixing within these lower levels of air, a *temperature inversion* results—that is, an inversion of the "normal" situation which is present during the day (Figure 7-10). Conditions which permit maximal radiant heat loss at night and therefore facilitate rapid cooling and temperature inversions are long nights, clear skies, cool dry air, and calm conditions. Long nights facilitate more complete heat loss through radiation. Since clouds are effective barriers to heat loss, temperatures drop much more slowly on cloudy nights. High levels of moisture in the air restrict heat loss as is readily seen by comparing the rapid rates of nighttime temperature decline in arid with those in

FIGURE 7-10   Diurnal variations in vertical distribution of temperature and of stability conditions, over a land surface. A, night and early morning; B, midday; C, evening. Arrows indicate direction of heat flow. (After G. T. Trewartha, *An Introduction to Climate*, McGraw-Hill, New York, 1968; and S. Petterson, *An Introduction to Meteorology*, 2d ed., McGraw-Hill, New York, 1958.)

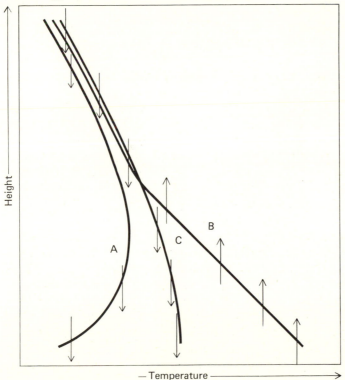

humid areas. As mentioned before, calm conditions are also necessary for a temperature inversion to develop.

Local Factors

Oceans and Continents Water absorbs heat to much deeper levels, has a higher thermal conductivity, and has a higher specific heat than soil; it therefore warms and cools more slowly than land masses. Temperatures vary less over oceans and other large bodies of water than over continents. The moderating effect of oceans on land is called the "oceanic effect" and is more apparent on the West than the East coast of the United States because of predominantly westerly winds. The effect of the Pacific Ocean is obvious when we examine Figure 7-11, which compares temperatures in San Francisco, exhibiting an

FIGURE 7-11 Monthly temperature means for St. Louis, Missouri, and San Francisco, California. The two cities had the same mean annual temperature, but very different climates.

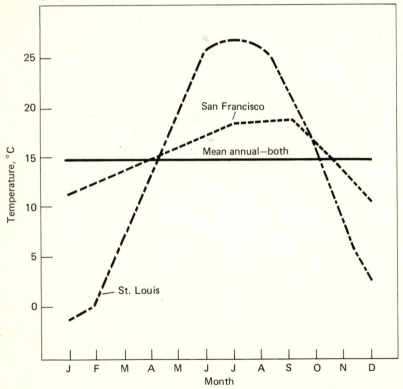

oceanic climate, with those of St. Louis, a continental climate. Both cities lie at about 38°N latitude. This graph also shows very clearly that mean annual temperature is a poor index of climate as both cities have a mean of about 13°C. The monthly means for San Francisco range from about 8 to 18°C, and those for St. Louis range from −1 to 26°C. The effects of these temperature differences on plants are extreme.

Similar effects, though to a lesser degree, are apparent on the eastern sides of the Great Lakes and have very important influences on the location of horticultural production enterprises. The moderating effect of Lake Michigan on temperatures east of the lake largely accounts for the concentration of fruit production in southwestern Michigan. The southeastern side of Lake Erie is a major grape-producing area for similar reasons.

Slope The direction in which a slope faces can alter temperatures sufficiently to affect some horticultural crops. Similar slopes would rank from warmest to coolest as follows: south, west, east, north. The effect is greatest in winter when the sun is low in the sky but is also apparent in the spring. The influence is particularly strong on soil temperature, and thus low-growing crops, such as strawberries, are more affected than apple trees, which are more influenced by air temperature. As with many of the factors described above, the reason for the temperature differences relate largely to insolation. The south slope receives more insolation than any of the other three because it is exposed to direct sun almost all day—even in winter when the sun is low in the sky. The west slope tends to be slightly warmer than an east slope because of warming by the air all morning. Little or no insolation is utilized in the melting of ice or evaporation of water as is true on an east slope.

Air Drainage As previously described, during the night the air near the ground is colder and therefore denser than warm air above. Because of its greater density, cold air "flows" downhill and collects in low areas from which it cannot drain. This phenomenon is referred to as *cold air drainage* and leads to the formation of "frost pockets," or depressions, which collect cold air. It is in these areas that plants are most subject to freeze damage.

Horticultural crops subject to freeze damage should be planted on sites selected on the basis of good air drainage. It is not uncommon for apple growers to plant orchards on hillsides with less than ideal soil, strictly to have good elevation and associated air drainage (Figure 7-12). A further refinement is to plant early-blossoming cultivars at the top of a slope, with the latest-blossoming cultivars at the lower levels. An important consideration is the area of the site to be planted in relation to the area into which cold air will drain. The sides of a narrow

FIGURE 7-12 An apple orchard planted on a sloping hillside in western Maryland. The air drainage is excellent, but the soil is quite thin, leading to frequent missing trees.

valley would be much less desirable than the sides of a hill which rises up from a relatively large area into which cold air can drain.

Soils There are several opposing factors relating to soil types and the rate at which they warm up in the spring. These factors include surface color, mass per unit volume, rate of heat conduction, air space, and perhaps the major factor—water-holding capacity. Sandy soils with low water-holding capacity tend to warm up first in the spring and are often referred to as "early soils." Organic or "muck" soils, although dark-colored, tend to warm up slowly and thus are considered "late soils." Loam, silt, or clay soils are intermediate between sandy and organic soils.

HOW PLANTS ARE AFFECTED BY TEMPERATURE

When relating temperature to plant growth, the term cardinal temperatures is often used. *Cardinal temperatures* encompass the following: the *optimum*, at which a plant functions best; the *minimum*, below which a plant cannot grow; and the *maximum*, above which the plant cannot grow.

Cardinal temperatures vary greatly among species—even among temperate-zone horticultural crops—although the great extremes lie outside the traditional crop plants. The situation becomes very complex as one considers that the cardinal temperatures for a particular species vary not only with cultivar but with the stage of development, the tissue involved, the process of concern, the length of exposure, and environmental factors such as insolation and water stress. With many crops the daytime optimum is higher than the optimum temperature at night.

Summer Temperatures

With perennial plants we are quite limited in control of summer temperature. There are, however, certain things which can be controlled, such as exposure, the use of shade to reduce leaf temperature, and the use of mulch to modify root temperatures. Most important, however, is the selection of a particular plant or crop that is suitable for a specific location. For commercial production, we select crops which are well adapted to the area, avoiding those with which the chance for success is less than acceptable. Landscape plants are sometimes used in marginal areas because the owner is not dependent upon them for direct income and because such an owner can take precautions to prevent injury which would be impractical on a large scale.

Reducing Summer Temperatures   For crops of very high value, such as ornamentals, certain techniques are used to reduce the unfavorable effects of high summer temperatures. Shading is a common practice in the production of both nursery, flower, and foliage plants. One major type of shading is a "lath house," which consists of wood lath that can be spaced in order to provide varying amounts of shade. Snow fencing is often used with the wood strips running from north to south to provide a continually changing shade pattern as the sun moves from east to west. One problem with wood lath is that the supports necessary to hold the weight must be substantial.

A much lighter and therefore easier-to-support system is saran shade cloth (Figure 19-11). In recent years, saran shading has become very common, and it is available in widely varying densities of weave to provide from minimal to 95 percent shade.

It should be remembered that shading has little or no direct effect on air temperature; its effect is primarily on tissue temperature. By intercepting part of the insolation, the temperature of leaves and other tissues is markedly reduced.

In certain plants, excessive soil temperatures can be detrimental. When this is a problem in high-value crops or plants, mulching is commonly practiced. By providing a layer of 10 to 15 cm of a mulch on the soil surface, soil temperatures are suppressed considerably in the summer. Most mulches are organic in nature. The type used depends largely on cost and availability. Mulching materials include chopped corn cobs, straw, hay, wood chips, bark, leaves, sawdust, and—for the homeowner—grass clippings. All these materials are good insulators in that they absorb little heat and conduct it poorly. In landscapes, white stones are used occasionally. Mulches also have beneficial effects on soil moisture which are discussed elsewhere.

In greenhouses, the most widely used procedure for temperature control other than ventilation, is to apply a shading compound to the outer surface. This material is usually a white latex or calcium paint mixture sprayed on the glass which reflects some of the insolation, preventing its entry into the greenhouse. The shading compound gradually weathers during the summer, and any excess remaining in the fall can be washed off. In addition to shading, or where the reduced illumination is undesirable, fan-and-pad cooling systems are utilized. If exhaust fans are mounted on one side of a greenhouse and air is drawn into the other side of the house through damp pads, considerable cooling is accomplished—especially in low-humidity areas. As the water evaporates, it absorbs tremendous quantities of heat and thus is an effective cooling mechanism.

With crops in the field, increasing use is being made of overhead irrigation for *evaporative cooling* in addition to adding water to the soil. This procedure is often called "air conditioning." The plant, the soil, and the surrounding air are all cooled by means of the vaporization of water. Because leaf temperatures are lowered, respiration is reduced. If temperatures are excessive, water stress and the resultant closing of stomates can also be alleviated and thereby photosynthesis is increased. The reduction in tissue temperature and water stress has also been found to be effective in increasing fruit-set in tomato and bean; reducing abscission of young fruit in grapes, apricots, and citrus; and increasing quality in various crops. If the irrigation is being used strictly for the purpose of cooling and reducing water stress, the system is cycled on and off to add only enough water to keep the plant and soil surfaces wet. Obviously the same system can be run continuously when it is necessary to add sufficient water to raise the moisture level of the soil.

Temperature Requirements

For annual crops, we are concerned primarily with temperatures during the growing season. Two major classifications are used in describing flowers and vegetable crops which are grown as annuals.

One of these classifications is based on the minimal temperature which can be withstood and categorizes plants as *hardy, half-hardy,* or *tender.*

Hardy plants can withstand minimum temperatures of approximately −4 to −2°C. Examples are English peas, spinach, kale, turnip, cabbage, and pansy. These are usually planted in early spring, three or more weeks before the average last freeze for the area. In Southern areas, hardy crops may be planted in the fall.

Half-hardy crops can survive minimum temperatures of −1 to 0°C and are planted two to three weeks before the average last freeze date. Examples include carrots, beets, lettuce, and celery.

Tender crops are those which cannot withstand 0°C. These include beans, corn, squash, melons, cucumber, tomato, and many annual flowers. These are not usually planted until after the danger of freezing temperatures is past unless provision is made for protection by the use of hot caps or row covers.

There is a direct correlation between these categories of hardiness and the temperatures at which the seed will germinate. Hardy species will germinate at relatively cool soil temperatures and can therefore be planted quite early. Half-hardy types need somewhat higher soil temperatures for germination, whereas the tender class needs considerably higher temperatures. It is impractical to seed crops such as tomato, pepper, and eggplant directly in the field if the goal is to harvest for the early market. To facilitate early harvest, you should transplant young plants to the field and thereby avoid the problem of seed germination in cool soil. The use of hot caps is helpful, not only to reduce freeze injury, but to provide higher temperatures to accelerate early plant growth and to ease the transition to the more adverse environment of the field.

A second classification scheme involves the growing-season temperatures required for optimal growth. The two classes are *cool-season* and *warm-season* crops and are used particularly in relation to vegetable crops. Those requiring cool temperatures of 18 to 24°C include spinach, lettuce, cabbage, Irish potato, and English peas. Examples of warm-season (25 to 35°C) vegetables are tomato, eggplant, corn, melons, cucumbers, and sweet potato. Obviously, there are others which are intermediate. It is apparent that there is a correlation between the two classification systems in that the hardy crops thrive in cool weather whereas the tender types need warm growing seasons.

In spite of the distinctions made previously, a particular crop can be grown in areas not normally considered ideal. For example, the Irish potato, a cool-season crop, is best-suited to only the Northern tier of states in the United States. Production statistics show, however, that potatoes are grown all the way to Florida. This is accomplished primarily by the time of planting and cultivar selection. In the Northeast, long-season cultivars are planted in the spring and harvested in the fall. Farther south, shorter-season cultivars are planted earlier and harvested before the warmest part of the growing season occurs. In Florida, potatoes are grown during the winter. Although yields are generally lower in the more Southern areas, decreased transportation costs somewhat offset the disadvantage as does the marketing of freshly dug potatoes in the "off" season.

Another example of adjusting a major vegetable crop to temperatures is commercial head lettuce production. Most head lettuce comes from California and Arizona. Since lettuce is a cool-season crop, much of the summer production is in cool, irrigated regions such as those around Salinas, California. Winter production is concentrated more in southern California and, particularly, Arizona. Since these areas which are suited for winter lettuce production have summers too warm for lettuce, warm-season crops, such as cantaloupe and honeydew melons, are grown in the summer.

Spinach thrives under cool temperatures but is grown very widely in the South during the cooler parts of the year. It may be seeded in late summer and harvested in the fall, or seeded in late winter for spring harvest. In Southern states, such as Texas, spinach is grown during the winter.

Although it is practical to grow cool-season crops in warm areas, the reverse is much less feasible except by the use of greenhouses.

Length of Growing Season

Probably the oldest and simplest systems for assessing the crop-production potential of an area is the length of the growing season. This is expressed as the average number of consecutive frost-free days or the period from the last frost in the spring to the first in the fall. Growing-season length varies from more than 320 days in subtropical areas to 365 days in Hawaii, 100 to 120 days in Northern sections of the United States, to only 80 to 90 days in the upper, interior regions of Alaska. Although the length of the growing season decreases as one moves northward, the increased daylength in summer partially compensates for the short season by accelerating crop growth.

Within a given state or area, the length of the growing season is markedly affected by elevation and proximity to large bodies of water. For example, in coastal Virginia, the average number of frost-free days is 225, whereas in the more mountainous Western regions it is only 165 days. Differences in California are even greater with the range being from 320 to 340 on the Coast to 100 days or less in the interior mountains. Part of the difference in California is also due to latitude.

The length of the growing season is often cited to indicate what *cannot* be grown but does not tell us what *can* be grown. For example, even though the growing season would be sufficiently long, if temperatures are not warm enough for the crop in question, its production is not practical.

Temperature Summation

For many decades researchers have attempted to correlate temperatures with crop development. If this were possible, it would be feasible not only to predict the maturation date for a crop but to schedule successive plantings of the same crop to give a continuous, orderly harvest. Large vegetable processors often contract for the production of such crops as beans, peas, and corn for their processing plants. For peak use of their facilities, the processors need a continuous supply of each crop. Because of increasing temperatures as the season progresses, weekly plantings of corn or other vegetables will mature at less-than-weekly intervals. Perhaps the most widely used method of temperature summation is the *"degree day."* For each crop, a base temperature is selected below which crop growth and development is minimal. For each day this base temperature is subtracted from the maximum for that day. For example, with English peas, a base temperature of 40°F is used; for a day with a maximum of 50°F, 10 degree days would be accumulated. If the daily maximum is lower than the base temperature, no degree days are added or subtracted. By using long-term averages, we can approximate the number of degree days needed to reach maturity. An average figure of 2000 degree days is common for many cultivars of peas. For warm-season crops, such as sweet corn, a base temperature of 50°F is used. Unfortunately the system of degree days is not very accurate because it ignores the nonlinear response of plants to temperature as well as other important factors, such as illumination, daylength, nutrition, disease and insect problems, and water relations.

FREEZES AND FROSTS

Through the years, the terms freeze and frost have been used interchangeably by many people, especially in agriculture. In this section we will consider various terms and usages, but strongly recommend the use of radiational freeze and advective freeze as being the most descriptive and useful. Meteorologists recommend the term *freeze* for any situation where the average temperature over an area drops below 0°C.

Radiational Freezes

If the freeze is associated with calm conditions, radiational cooling, and a temperature inversion, it is called a *radiational freeze*. The likelihood and severity of a radiational freeze is directly dependent on the factors which encourage radiant heat loss, as described earlier under Temperature Inversions.

The occurrence of the first really cold night of the autumn season is effectively described in the following quotation from John Cole:

> You can sense it in the stillness of the October evening, the chill clarity of the starry sky. It's a cruel beauty, this windless brilliance; it allows the killing frost to do his work without moderation from the clouds or disturbance by the wind. The way to the garden is quite without defenses and the cold moves in flowing close to the ground on invisible rivers of malice. Newspapers spread about, tarps hastily pulled over the tomatoes, smudge pots lit, hay heaped here and there—none of it works against the merciless advance of a true killing frost. When you inspect the scene of the crime on the white, still morning after, there is a finality so evident that hope for survivors never even flickers.

> The garden is gone.*

Advective Freezes

When freezing temperatures occur as the result of the invasion of a large cold air mass, the term *advective freeze* is used. It is common in much of the United States for large, dry, cold air masses to move southward out of northern Canada and cause rather sudden, drastic temperature drops. These air masses are frequently accompanied by windy conditions. Although radiant heat loss occurs during an advective freeze, the situation is quite different from a radiational freeze. No temperature inversion is present; much of the heat loss is directly to the cold air by conduction; and freeze damage control is much more difficult.

Frost versus Freeze

Rather than using the terms radiational freeze and advective freeze, horticulturists have traditionally used frost and freeze, respectively. *Frost* has thus indicated temperatures of 0°C or below associated with a temperature inversion, with or without the deposit of moisture. A *white frost* occurs when the dewpoint is *above* the minimum tempera-

*Permission granted by: John N. Cole, Editor, *Maine Times*, Topsham, Maine 04086.

ture and is obvious from the deposit of ice or "frost" on exposed surfaces. When you look out on the first cold fall morning and see ice on tender plants you are immediately aware that freezing injury has occurred. A *black frost* results when the dewpoint is *below* the minimum temperature. The term black frost is used because no ice is deposited, but tender plants turn black the following day as the result of freezing injury. By late morning on a day following a black frost, injury is very obvious on plants such as tomato, eggplant, squash, and bean. The initial symptoms are only a water-soaked appearance by sunrise, but by midmorning injured tissues turn black.

In this system of terminology, an advective freeze is referred to as a *freeze* and, to review, is associated with a cold air mass, windy conditions, and the absence of a temperature inversion.

In spite of the traditional usage of frost and freeze, we recommend the use of the terms advective freeze and radiational freeze to clearly differentiate between these very distinctly different phenomena. As will become evident in the following sections on freeze damage control, a clear distinction between the two is essential.

FREEZE DAMAGE CONTROL

At the outset, it should be emphasized that *the best method of freeze damage control is good site selection*. Even today, it is all too prevalent for growers to invest thousands of dollars in freeze protection systems for a site which should never have been planted. Time and effort spent in assessing the normal temperatures experienced in an area, *before* setting out an orchard or other long-term crop, is time and effort well spent. This can be accomplished by strategically placing accurate minimum thermometers in the early spring and keeping records of the temperatures at several locations. Sometimes, information can be gained by talking to old-timers in an area. They may well remember a time when your proposed planting site was in crops and how comparably productive it was.

Radiational

To prevent or minimize freeze damage to crops during a radiational freeze, you may use several effective methods. With some of these techniques, such as flooding, expense is small; but others require large investments in either labor or equipment. The various possibilities are considered in the following paragraphs.

Reduce Outgoing Radiation Hot caps consist of small "tents" of

FIGURE 7-13 A field of tomatoes with each plant covered by a hot cap made of translucent paper. The hot cap acts as a greenhouse providing protection against intense sun, wind, and freezing temperatures. (U.S. Department of Agriculture photo.)

translucent waxed paper which are placed over tender crops such as early tomatoes, peppers, and eggplants. These crops are transplanted in the field before the average last freeze date (Figure 7-13). Hot caps are put on at the time of transplanting as they also serve to reduce the stress of intense sun and wind on young plants which are not yet adapted to field conditions. As the plants become better adapted to their new environment, the hot caps may be removed but put back if a freeze is imminent, or they may be torn open to an increasing degree with time. The shock of going from a sheltered environment to outside conditions is thereby eased, and as the season progresses, the likelihood of a freeze also declines.

In certain areas of the country, such as California, large, clear polyethylene *row covers* are used. These are much like hot caps but cover entire rows rather than individual plants. The sides or ends of these covers can be raised during warm days and lowered at night. Because of the expense involved, row covers and, to a lesser degree, hot caps are used primarily on high-value crops.

With both hot caps and row covers, the covering is an effective barrier to radiant heat loss. Radiation from the plant and soil is absorbed or reflected by the cover. Part of that which is absorbed is reradiated back down, and part is radiated outward and lost. The net effect with either type of cover is higher temperatures and relative humidities during both day and night.

Cranberries are grown in low-lying areas often referred to as bogs. These sites are constructed so that they can be *flooded* during freeze conditions and drained after the danger is past. Most horticultural crops will not withstand extended periods of submersion, but cranberries show no ill effects from several hours under water.

A very common method to control freeze damage to strawberry fields is to cover the field with straw (Figure 7-14). The cost of labor today makes this method very expensive because the straw cannot be left on the plants during the day. Thus, it may have to be applied and removed several times during the blossoming period.

A more recent development for low-growing crops in particular has been the use of foams. By mixing foaming and stabilizing agents with water, scientists have made possible the generation of foams which can be applied by machine and will last through a night. Soon after sunrise, the foam dissipates with no residue.

FIGURE 7-14 A mulched field of strawberries in the early spring. Since the covering is not very thick, removal will be unnecessary. The mulch will help keep the berries clean.

FIGURE 7-15 Artificial fog being generated with the goal of providing protection against a radiational freeze. (Photograph, Mee Industries, Inc., Rosemead, Los Angeles.)

Another recent development in protection from radiational freeze has been artificial fog (Figure 7-15). By generating very small droplets of water containing a protein stabilizer to slow evaporation, one can produce a persistent fog. Even with a temperature inversion, some air is usually in motion so that the fog is generated upwind from the area to be protected. The layer of fog floating over the crop reduces radiant heat loss much like clouds.

Add Heat One of the most commonly used techniques in controlling freeze damage has been the generation of heat. Through the years a wide variety of materials has been burned, but increasingly stringent air pollution regulations have drastically changed the situation. For example, the burning of used automobile tires does generate considerable heat, but the accompanying foul-smelling, long-lasting smoke is no longer tolerated in most areas of the country.

For many years the standard heater used was a 5-gal can of fuel oil. These so-called *"smudge pots"* were set out in the area to be heated, filled with oil, and covered with a tight-fitting lid. When necessary to fire the pots, the lids were removed and enough gasoline added and lighted to ignite the oil. Considerable heat was generated; but because smudge pots do not burn very hot, considerable smoke was also produced. Smudge pots are therefore also banned in areas which have laws controlling air pollution. It is important to realize that smoke particles in the air are of essentially no value in reducing radiant heat loss. A heavy smoke layer can also delay the warming of the area after sunrise because smoke particles absorb some solar radiation. More efficient oil burners have been used including the "hotstack" and "return-stack" (Figure 7-16). In these oil burners the flame is confined within the stack; therefore they are hotter and burn oil much more efficiently and with less smoke than smudge pots. More recent introductions have been coke bricks and wax-impregnated blocks. The latter are moderately expensive and have the drawback that once ignited, they cannot be extinguished and reused. They do,

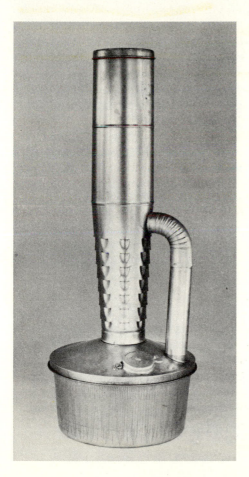

FIGURE 7-16 A modern, clean-burning return-stack orchard heater. (Photograph, Scheu Products Company, Inc., Upland, Calif.)

however, provide the grower with protection for minimal capital investment.

To avoid air pollution and also to drastically reduce labor input, most modern heating systems involve permanent installation of pipelines to carry oil or gas (natural or propane) to individual burners. These systems are expensive to install, but are much more efficient in their use of labor. Not only can these systems be controlled from a central valve, but some can even be ignited electrically from a control center. In some of these systems the heaters are disconnected and taken from the orchard after the danger of a freeze is past. The actual heaters vary greatly from system to system, some being very elaborate, others being homemade from locally available materials. Many oil systems of this type use a furnace nozzle to vaporize the oil and thereby obtain very clean burning.

Utilize a Temperature Inversion An entirely different principle for freeze protection is the use of large fans, often referred to as wind machines (Figure 7-17). These rotating fans are mounted on towers, at a height of 9 to 12 m above the ground. Various models are available including electric-, gasoline-, or diesel-powered. Some have a single fan; others have two fans facing in opposite directions. The basis for using wind machines is the mixing of warm air at the top of a temperature inversion with colder air below. *Without a temperature inversion, wind machines are totally ineffective*. The degree of protection provided depends on the size of the inversion—usually expressed as the temperature difference between about 1.2 and 15 m altitude. With a large inversion of 4 to 6°C the temperature at crop level may be raised 2 to 3°C whereas with an inversion of only 2 to 3°C temperature increase will be minimal. The effect is greatest close to the wind machine and decreases with distance. The initial investment is large, but labor requirements are very small. With the recent rise in petroleum prices, wind machines are becoming increasingly popular due to the greatly reduced fuel requirements compared to heaters.

A recent innovation, similar in principle to wind machines is the use of helicopters. In certain areas it is possible to contract for helicopters to fly at low altitudes over an area and provide mixing of air much like that of wind machines. The cost per hour is quite high, but there is no capital investment, and they are used only when needed. By use of thermistors, the pilot can seek out and fly in the warmest layer of the inversion and thereby produce maximal temperature increase at crop level. This added flexibility is an advantage over wind machines which are fixed at a particular height.

For maximal protection with either wind machines or helicopters, heaters are sometimes used. Thus, growers are not only utilizing the warm air within the inversion but adding heat as well.

FIGURE 7-17 A wind machine in a citrus grove. During a temperature inversion, the large fan mixes warm air above trees with cold air near the ground to elevate temperatures around trees. (Photograph, SSP Agricultural Equipment, Inc., Burbank, Calif.)

Use Overhead Irrigation Under the proper conditions, overhead irrigation is a very effective system for freeze protection. As water freezes, the heat of fusion (80 cal g^{-1}) is liberated, and in this case much of the heat is absorbed by the plant and surrounding air. For this procedure to be of value, enough water must be added to maintain a constant film of water over whatever ice has formed. As long as liquid water is present, the temperature of the plant and ice will remain at about 0°C, which is not an injurious temperature for most crops. Such systems apply water continuously during the danger period, starting with a low rate of application and increasing the rate as necessary. The continued formation of icicles is an indication that sufficient water is being applied. If, on the other hand, the water freezes immediately on contact, the temperature is below 0°C, and damage is likely to occur. An average rate of irrigation would be about .25 cm h^{-1}. For this system to be practical, one must consider these requirements: a plentiful water supply, sufficient piping and nozzles to irrigate the entire area, and good soil drainage to avoid waterlogged soils. This system is most effective under calm conditions since even a moderate wind will accelerate evaporation. It must be remembered that the heat of vaporization of water is 540 cal^{-1}g^{-1}; thus, for every gram that evaporates almost 7 g must freeze to "break even." This technique, therefore, is more effective in humid than in arid climates.

It is recommended that once irrigation is started it be continued all night and long enough after sunrise for the ice to melt. If the system is shut off too soon, vaporization may pull the temperature below freezing and thus cause serious damage. Another consideration is that evergreen trees, such as citrus, cannot support the tremendous ice load which can develop; irrigation, therefore, is limited to low-growing crops or deciduous fruit trees which have very limited leaf surface during the freeze danger period.

Advective

With the windy conditions that accompany an advective freeze, most of the above protective measures are either much less effective or totally useless. Heaters will do very little good, wind machines and helicopters are of no value, and overhead irrigation would be much less effective. Hot caps, row covers, and straw mulch will help, but perhaps the only completely effective method would be flooding as used in cranberry bogs.

Delay Crop Development   A new and entirely different approach to spring freeze damage control has been developed in recent years. By overhead irrigation during the early spring, the development of deciduous tree fruit crops can be delayed. Research in various parts of the United States indicates that the date of bloom of tree fruits, such as apple, may be delayed by as much as two weeks, thus drastically reducing the likelihood of freeze damage. By turning the irrigation system on when tissue temperatures rise above 7°C, the heat of vaporization maintains tissue temperatures well below unirrigated trees. It is through this suppression of tissue temperatures that the bloom period is delayed. Although still in the experimental stages, this technique may well hold considerable promise for those growers with the facilities for overhead irrigation.

Thermometers

One of the most critical decisions facing growers is when to put a freeze protection system into operation. Not only must they know the critical temperature for each crop at the current stage of development, but they must have an accurate indication of the temperature. Unfortunately, it is not unheard of for a person to invest several thousand dollars in a protection system and base the decision of when to start it on a 79-cent thermometer. *The absolute necessity for properly standardized, dependable thermometers cannot be overemphasized.*

Although tissue temperatures can be measured, decisions are usually based on air temperature at crop level. A thermometer should be enclosed in a cover which shields it against radiation from the ground as well as from exposure to the cold sky. This can be accomplished in a covered white shelter which is louvered on all sides and has a bottom (Figure 7-6). Less elaborate covers are available as well.

Normally, the freeze control system is started when the temperature reaches a degree or two above the critical temperature to allow for the time lag before the temperature is stabilized or raised.

To avoid the necessity for continually checking thermometers during the night, you may install a "frost alarm" which will activate an alarm. Certain systems such as wind machines can also be triggered to start automatically by a thermostat in the area to be protected. Since freezing conditions often occur for two to three consecutive nights, modern conveniences and safety features can be very worthwhile investments. To save a crop for two nights only to lose it the third by falling asleep can be a devastating blow.

Winter Temperature

Minimum temperatures during the fall and winter have profound effects on the survival of perennial plants. Many tropical and subtropical species cannot withstand any freezing temperatures, and their distribution is limited to areas with very mild winters. Examples of such crops in the United States are avocado, pineapple, mango, and many others which are grown largely in California, Florida, and Hawaii. Citrus species are slightly hardier but are limited commercially to areas which experience only infrequent light freezes.

To aid in plant selection, plant-hardiness maps are used (Figure 7-18). These maps are based on long-term averages, and we are able to predict with some degree of accuracy what minimal temperatures can be expected in a particular area. By knowing the minimum temperatures that a plant can withstand, one can decide on the feasibility of its use.

Unfortunately, even long-term averages are not always dependable. Certain horticultural industries have been profoundly affected by one abnormally cold period. An outstanding case in point is the 'Baldwin' cultivar of apple. Until the winter of 1933–1934, 'Baldwin' was the major cultivar grown in New York and New England. During that uniquely cold winter great numbers of 'Baldwin' trees were killed, and this cultivar has never regained its popularity.

The resistance of a plant to environmental stress is very dependent upon the growth status of the plant at the time the stress occurs. This is particularly true of low temperatures but is apparent with high temperature and drought stresses as well. In fact, a plant that has developed resistance to one type of environmental stress is also resistant to other stresses. When a young greenhouse-grown cabbage plant is "hardened off" by exposure to cooler temperature, temporary water stress, and reduced nutritional levels, it becomes more tolerant of both heat and cold as well as drought. The apparent changes which occur during the "hardening" of such a plant are: slowed growth rate, greater cuticle thickness, and loss of succulence in both leaf and stem tissues.

Hardening As growth slows in the late summer and fall, the utilization of photosynthate by new growth gradually ceases. There is, therefore, an accumulation of carbohydrates, pectins, nucleic acids, and various proteins in the various tissues of the plant. In many plants a conversion of starches to sugars also takes place as the temperature declines. Another phenomenon associated with these changes is an increase in protoplasmic viscosity. These processes are reversible, and a period of warm weather in late fall or winter will cause the plant to lose hardiness—as does the onset of spring.

It has recently been learned that much of the difference in cold hardiness is due to the ability of some plants to avoid actual ice formation within plant tissue. About 80 percent of deciduous trees and shrubs (including many important horticultural plants) apparently have supercooled water in the living xylem at subfreezing temperatures as low as $-40°C$. Also, the floral primordia in *Rhododendron*, azaleas, *Vaccinium*, and many *Prunus* species survive by supercooling, although not often to $-40°C$. As research on cold hardiness continues and as our understanding of the phenomenon increases, we will hopefully be able to decrease the severity of cold injury.

FIGURE 7-18 Plant hardiness map indicating average minimum temperatures for 10 zones. (U.S. Department of Agriculture map.)

Factors Affecting the Degree of Cold Hardiness The amount of foliage in relation to the size of the plant is of major importance in determining the degree of hardiness. Leaf injury by insects or diseases and late-summer pruning tend to make plants less hardy because of reduced leaf area. With fruit plants, an excessively heavy crop can leave a tree with too few reserves for the development of maximum hardiness. Any factor which shortens the normal growing season can have a similar effect. The influence of nutrition is also apparent in many cases. Excessive nitrogen levels in the late summer and fall delay the normal cessation of growth and thereby the onset of dormancy. Such plants are particularly susceptible to early-fall freeze damage. Poor nutrition during the growing season can also reduce the degree of hardiness developed, particularly to severe midwinter freezes.

In trees in particular, the buds and other tissues in close proximity to the leaves develop maximum hardiness first; the crotches and trunk are the last to harden. This is apparent in that an early cold snap in the fall will injure the trunk and major crotches but not the rest of the tree. By midwinter, however, the trunk may withstand colder temperatures than the buds.

Dormancy

With perennial plants the development of resistance to environmental stress is much more complex than with annual plants. This should be obvious when one considers the vast range of environments in which perennial plants are able to survive. A concept of prime importance in understanding the physiology of plant survival is *dormancy*, a term used to describe a state of inactive growth due to either internal or external factors.

A plant develops hardiness during the summer and fall but does not reach maximum hardiness until exposed to progressively colder temperatures. Under normal conditions, growth tapers off in late summer, leaves are shed in the fall, and as the temperature and daylength decline, hardiness increases. Thus dormancy is a pre-requisite for the development of maximum hardiness in deciduous plants.

The entrance and exit from dormancy are gradual processes and vary with different portions of a plant. For example, the axillary buds on a current year's apple shoot are fully dormant in mid- to late summer when the leaves are still fully active and while fruit growth is actively proceeding. Gradually, as the fall proceeds, the entire above-ground portion of the plant becomes dormant. The roots, however, do not become dormant but rather continue to grow when the soil temperature is above a minimum of about 4°C. Partially because of the roots' lack of dormancy and partially because they are not exposed to as low temperatures as the top of a plant, the cold tolerance of the roots is considerably less.

During midwinter, the entire aboveground portion of temperate-zone perennial plants exhibit dormancy. However, because certain parts of a plant, such as buds or seeds, may be dormant while other tissues of the same plant are actively growing, it is desirable to consider the dormancy of different tissues separately.

Bud Dormancy   During the active vegetative growth of shoots on a deciduous perennial plant, buds are formed in the axils of the leaves. Most of these axillary buds do not develop into shoots during the early summer because of apical dominance. This is readily shown by removing the shoot apex, thus destroying apical dominance and allowing the axillary buds to grow. By late summer, however, removal of the apex does not lead to the growth of shoots from the axillary buds, even though environmental conditions are suitable. It is, therefore, apparent that there has been an internal physiological change within the buds which prevents their growth.

Bud dormancy is brought about by a variety of environmental factors including short photoperiods, low temperatures as occur in the fall, and severe drought. In some species, growth is resumed as soon as the environment is altered. In others, however, growth is not initiated until bud dormancy has been broken, usually by exposure to cold temperatures. This necessity for exposure to cold temperatures has been termed the "chilling requirement" and is usually expressed as the number of hours below 7°C required to break bud dormancy. Among deciduous fruit trees the chilling requirement is of major importance in their commercial production. The southern limit of peach production has been set by a chilling requirement of 600 to 1000 h for most cultivars. The breeding of peach cultivars with chilling requirements of 250 h or less has recently extended commercial peach production into Florida, where there is insufficient chilling to break dormancy of previously available cultivars.

Until the chilling requirement is met, a plant will not grow and develop normally. In many flowering trees, such as the peach, flower buds have a slightly shorter chilling requirement than vegetative buds. After a mild winter the trees may flower normally, but vegetative bud break is very sporadic. If the winter is very mild, even the flower buds will open unevenly, with the result that the bloom period is greatly extended.

Part of the chilling requirement can be satisfied by special treatments. Exposure to high but sublethal temperatures and to certain chemicals, such as oils, ethylene chlorohydrin, and others, will induce buds to break if the majority of the necessary chilling has been received. Such treatments may have commercial application in regions of the world where the growing season is suitable but the winters are too mild. Examples are areas south of the Mediterannean as well as the Southern United States.

Extreme environmental stress will sometimes break bud dormancy, even prior to any significant amount of chilling. A striking example seen by one of the authors was an apple orchard in which the grass had burned in late summer. As the apples were maturing in October, the lower branches were in full bloom. The very high temperatures brought about by the fire had broken the dormancy of both flower and vegetative buds which would normally have not developed until they had received several hundred hours of chilling.

Some seed also goes through a dormant state, and the requirements to break seed dormancy are varied. With many, there is a direct relationship between the requirements for breaking dormancy and the survival of the young seedling. Apple and peach seed must be exposed to an extended period of damp cold before it will germinate. This low-temperature requirement has led to the practice of storing such seed at 2 to 4°C under moist conditions and is termed *stratification*. This requirement for germination effectively prevents germination in the fall or early winter and thereby increases chances of seedling survival.

The seed of many weeds, the sweet potato, and a variety of other species must have the seed coat weakened or ruptured before germination will occur. In such species, the seed coat restricts O_2 or water penetration or prevents emergence of the embryo. Various mechanical treatments are used, such as soaking in concentrated sulfuric acid or using mechanical abrasion. Such treatments are called *scarification*.

Certain seed requires light for germination to begin. The need of lettuce seed for light has been extensively studied, and it is now well known that red light induces germination, but far-red light is inhibitory. This light response is repeatedly reversible, and the end result depends on the type of light to which the seed was last exposed. Certain other seed germinate best in darkness.

Tomato seed will not germinate until removed from the fruit, washed, and dried. On the other hand, grapefruit seed germinate readily while still in the fruit as is often observed upon cutting open a grapefruit.

The complete understanding of dormancy evades us, but it is apparent that dormancy is a major contributing factor in the development of cold hardiness by perennial plants.

VERNALIZATION

In certain plants temperature plays a very vital role in flower initiation and development. This is particularly true in biennial plants which typically produce only vegetative growth the first year. After exposure to an extended cold period during the winter (vernalization), flowering occurs in the second year. Many of the plants would remain vegetative indefinitely without a cold period. After such a biennial plant has reached a certain minimal size, it is possible to induce flowering the first year by exposure to cold temperatures followed by proper temperatures and photoperiod.

One of the earliest uses of the mechanism of vernalization was in the "conversion" of winter wheat to spring wheat. Winter wheat is planted in late fall; it germinates before the onset of freezing temperatures and is thus exposed to the winter temperatures as a young seedling. This exposure to cold temperatures vernalizes the wheat; it then flowers in the spring and is ready for harvest sometime between June and August. If winter wheat is planted in the spring, it remains in a short vegetative state and does not flower for long periods of time. As the name implies, spring wheat is planted in the spring and is harvested in the fall as no vernalization is required.

With some plants the effect of vernalization is quantitative rather than an absolute requirement. Crops which have a requirement for low temperatures for flower induction include beet, celery, and cabbage, whereas other crops, such as lettuce and radish, will flower more readily after exposure to cold. In some responsive species, vernalization can occur with moistened seed, some others will respond only in the seedling stage, and still others will respond in either stage. An additional aspect of vernalization is that the effect is localized in the meristems, and the stimulus is not mobile as is the photoperiodic stimulus. Of interest also is the fact that plants can be readily devernalized by exposure to relatively high temperatures. For example, vernalized grains can be devernalized by 1 day's exposure to 35°C.

Types of Winter Injury

There are various types of winter injury other than damage from extreme cold.

Winter Desiccation An important type of damage, especially to ornamentals, is winter desiccation. Although often associated with cold temperatures, the injury actually results from desiccation, or drying out, of tissues—in particular the leaves. By midwinter the soil becomes quite cold, and in more Northern areas may freeze to a depth which includes a sizable part of the root system. If a warm period occurs with sunny days and windy conditions, evergreen plants transpire quite rapidly. If the ground is frozen or even cold, the roots are unable to obtain enough water to meet the transpirational demand. Winter desiccation is particularly severe on evergreens in areas where they are not totally adapted. A prime example among landscape plants is Chinese holly. It also occurs on certain deciduous plants such as raspberry, which has rather fleshy canes that lose water rapidly.

Sheltered locations and windbreaks are effective means of minimizing the severity of this problem. Because they lower wind speed, evergreen hedges, snow fencing, and planting on the leeward side of a building drastically reduce transpiration rates. Mulching is also effective not only because a mulch reduces the loss of moisture from the soil but also because it reduces the cooling of the soil and may prevent freezing of the soil. Watering when the ground is not frozen also helps to minimize winter desiccation by ensuring adequate soil moisture. The application of antidesiccant sprays has been moderately effective in reducing the transpiration rate. There are several materials of this type consisting primarily of plastic or wax emulsions which form an artificial coating on the leaves. Because most are not persistent enough to last all winter, two to three applications are required.

Actual Freezing Injury With a sudden temperature drop, injury is much more severe than if the drop were over a period of days. With very rapid freezing, cell sap may freeze within the cell; its protoplasm becomes disorganized, proteins are denatured, and death results.

Normally, temperatures decline somewhat slowly, and the results are quite different. The purest water is between the cells in the intercellular spaces, and because of its low solute content, it freezes first. As this happens, water is withdrawn from the cells, making the contents of the cell more concentrated and therefore more resistant to the release of water.

There are at least two theories as to the cause of the death of cells if enough water is withdrawn. One is dehydration—the cell sap becomes so concentrated owing to the loss of water that disorganization of the contents of the cell results. The other theory is that as the intercellular ice masses continue to enlarge, the cells are crushed or the walls punctured by the ice crystals.

Frost Heaving A problem with small perennial plants, particularly those which are not fully established after transplanting, is frost heaving. Water seeps into crevices in the soil, and as it freezes exerts upward pressure. As this occurs repeatedly, plants can be pushed up several inches above their original position, exposing roots to drying winds as well as extreme cold.

Root Injury As mentioned previously, the roots of a plant do not become fully dormant as do the aboveground portions. When soil temperatures get very cold, root injury may result. Root injury is perhaps more common in the midsection of the United States than in far-northern areas where a snow cover provides considerable insulation.

Mulching is a common protective measure where practical. Rootstock selection for cold hardiness is also widely practiced in fruit production.

Ice Damage Freezing rain can be disastrous to many types of plants. When tissue temperatures are below freezing, or when the falling rain is supercooled, the rain freezes on contact, and considerable thickness of ice can result.

With most plant materials, ice damage consists of limb breakage. However, with certain fine-textured trees, such as white birch, the ice load not only causes limb breakage but may bend the trees over to the ground. If the ice does not melt rapidly, such trees are often permanently deformed.

Bark Splitting With apple trees in particular, bark splitting is an all-too-common occurrence (Figure 7-19). It is typically associated not with extreme cold, but rather with a moderately cold period in the fall. With young apple trees, growers try to maximize growth in the early years to develop the bearing surface rapidly. Occasionally, a dry summer followed by a wet fall will make excess nitrogen available in the soil, thereby delaying the cessation of growth and subsequent hardening off. Such trees are subject to injury by temperatures in the range of −7 to −10°C in November, especially if the temperature drop is rather sudden. The usual symptom is splitting of the bark on the trunk, which is the last portion of the tree to harden off. After the bark splits, it curls back and exposes the cambium beneath. If the loose bark is tacked down soon enough to prevent drying of the cambium, injury is often minimal. The problem, however, is that the injury is often not detected soon enough to prevent considerable cambial injury.

Southwest Injury Tree trunks exposed to afternoon sun in midwinter suffer a unique type of injury known as southwest injury or sun scald. In winter, when air temperatures are low, exposed tree trunks absorb insolation, and bark temperatures exceed air temperatures by several degrees. The east and southeast sides of the trunk cool gradually as the

FIGURE 7-19 Bark splitting and death on the trunk of a young apple tree. Freezing injury occurred over a year before the time when the photograph was taken.

sun moves. The southwest side, however, is warmed by the afternoon sun; when the sun sets, tissue temperatures drop very rapidly to air temperature. Southwest injury is the result of this rapid freezing and shows up as cracking and often in the death of the bark on the southwest side of the trunk.

It has become a standard practice in areas where this type of injury is prevalent to protect tree trunks. With ornamentals, the trunk is often wrapped with burlap or plastic tree guards. On a larger scale, such as in orchards, trunks are painted with white latex paint to reflect part of the insolation and to thereby suppress the temperature of the trunk.

BIBLIOGRAPHY

Ballard, J. K., and E. L. Proebsting: *Frost and Frost Control in Washington Orchards*, Washington State University Extension Bulletin 634, 1972.

Daubenmire, R. F.: *Plants and Environment*, 3d ed., Wiley, New York, 1974.

Gardner, V. R.: *Principles of Horticultural Production*, Michigan State University Press, East Lansing, 1966.

Gerber, J. F., and J. D. Martsolf: *Protecting Citrus from Cold Damage*, University of Florida Extension, Circ. 287, Gainesville, 1966.

Miller, A., and J. C. Thompson: *Elements of Meteorology*, Merrill, Columbus, Ohio, 1970.

Neuberger, H., and J. Cahir: *Principles of Climatology*, Holt, New York, 1969.

Puffer, R. E., and F. M. Turrell: *Frost Protection in Citrus*, University of California Agric. Extension, AXT 108, Berkeley, 1967.

Trewartha, G. T.: *An Introduction to Climate*, 4th ed., McGraw-Hill, New York, 1968.

Trewartha, G. T., A. H. Robinson, and E. H. Hammond: *Fundamentals of Physical Geography*, 2d ed., McGraw-Hill, New York, 1968.

Valli, V. J.: *Basic Principles of Freeze Occurrence and the Prevention of Freeze Damage to Crops*, Spot Heaters, Inc., Sunnyside, Washington, 1971.

CHAPTER 8
WATER
RELATIONS

In the production of horticultural crops, the importance of sufficient moisture cannot be overemphasized. Within a particular temperature zone, the availability of water is perhaps the most important factor in determining which plants can or cannot be grown. The development of vast areas of agricultural production around the world depends upon water supplied by irrigation projects, and it is probably upon such projects that future generations of the world's inhabitants will depend for food.

Water performs many vital functions within plants. It is a necessary constituent of all living plant cells and tissues. Water serves as the *solvent* in which a wide variety of materials such as nutrients from the soil, carbon dioxide from the air, and elaborated materials move within the plant from the site of uptake or production to the site of utilization. Water is a major *reagent* in a wide diversity of chemical reactions, including photosynthesis. The growth of young expanding cells, the functioning of the stomates, and normal plant turgidity are all directly dependent upon sufficient water. As described in Chapter 7, the physical properties of water are of major consequence in the temperature relations of plants. First, as 1 g of water is transpired, 540 cal of heat are absorbed; thus transpiration serves to dissipate heat and thereby suppress leaf temperatures. Second, water has a uniquely high specific heat, and because of their high percentage of water, plant tissues are resistant to rapid temperature change.

ATMOSPHERIC MOISTURE

Before we discuss water relations of plants per se, the various aspects of atmospheric moisture will be presented. Atmospheric moisture can be expressed in various ways, each of which has inherent advantages and disadvantages.

Methods of Expression

The weight of water vapor per unit volume of air is called the *absolute humidity* and is expressed in grams per cubic meter. As air temperature increases or decreases, its volume changes if pressure is constant. Therefore, even with no change in moisture content, absolute humidity will vary with temperature, only because of changes in volume of the air. The *specific humidity* is defined as the weight of water vapor per unit weight of air and is expressed in grams per kilogram. Since specific humidity is expressed on a weight/weight basis, temperature or pressure changes have no effect. The amount of water vapor present in the air as a percent of the amount at saturation at the same temperature and pressure is called the *relative humidity*.

The capacity of air to hold water vapor is very temperature-dependent. As shown in Table 8-1, the capacity of air to hold moisture increases more than seven times as the temperature increases from 0 to 30°C. It is very apparent that on a warm summer day the atmosphere can hold many more times the amount of water vapor than can be held on a cold day in winter; a similar comparison can be made between tropical and polar regions. Another important aspect in Table 8-1 is that the higher the temperature, the greater the change in water-holding capacity per 10°C temperature change. For example, as temperature is raised from −10 to 0°C, the moisture-holding capacity increases by 2.01 g kg^{-1}; with a change from 10 to 20°C the water-

TABLE 8-1
SPECIFIC HUMIDITY OF SATURATED AIR AT VARIOUS TEMPERATURES*

TEMPERATURE, °C	SPECIFIC HUMIDITY, g kg^{-1}	CHANGE PER 10°C
−30	0.32	—
−20	0.78	0.46
−10	1.79	1.01
0	3.80	2.01
10	7.67	3.87
20	14.70	7.03
30	26.90	12.20
40	47.30	20.40

*Data for an atmospheric pressure of 1000 mbar.

TABLE 8-2
**EFFECT OF INCREASING TEMPERATURE ON RELATIVE HUMIDITY
WITH A CONSTANT SPECIFIC HUMIDITY**

TEMPERATURE, °C	SPECIFIC HUMIDITY, g kg⁻¹	RELATIVE HUMIDITY, %
0	3.80	100
5	3.80	70
10	3.80	50
15	3.80	36
20	3.80	26
25	3.80	19
30	3.80	14
35	3.80	11

holding capacity increases by 7.03 g kg^{-1}, or more than three times as much. An even more striking example is that as temperature rises from 20 to 30°C, moisture-holding capacity *increases* by 12.20 g kg^{-1}, which is far more than the *total* amount which can be held at 10°C.

The relationship between relative humidity and specific humidity is shown in Table 8-2. If we start with air at 0°C and 100 percent relative humidity and increase the temperature to 30°C, the relative humidity drops to 14 percent, with no change in composition.

It is this relationship between temperature and relative humidity which accounts for the diurnal variation in relative humidity. Let us assume that in midafternoon on a particular day, the temperature is 20°C and the relative humidity is 26 percent. If the temperature falls to 0°C by midnight, the relative humidity would then be 100 percent, with no modification of actual water content. Although relative humidity is the most widely used method of expressing atmospheric moisture, the previous example shows it to be meaningless when comparing moisture content at different temperatures.

Probably the most meaningful and useful method of expressing atmospheric moisture is *vapor pressure*. Water vapor pressure is that part of the total atmospheric pressure due to water vapor, and like atmospheric pressure it is expressed in millimeters of mercury (mm Hg) or millibars; 1 mm Hg is equivalent to 1.32 millibars. Since relative humidity is readily measured (by techniques to be discussed later in this chapter) and is an expression of the relative saturation of the atmosphere, we can calculate the actual vapor pressure by multiplying the saturation vapor pressure by the relative humidity. For example, in Table 8-3 the saturation vapor pressure at 30°C is 31.82; at 50 percent relative humidity, the vapor pressure is 15.91 or 31.82 × 0.50. When comparing the vapor pressure at a relative humidity less than 100 percent with the saturation vapor pressure, the term most often used is the *vapor-pressure gradient*. The vapor-pressure gradient allows us to relate different situations on a comparable basis; this is

TABLE 8-3
VAPOR PRESSURE OF THE ATMOSPHERE AT VARIOUS TEMPERATURES AND RELATIVE HUMIDITIES

TEMPERATURE, °C	ACTUAL VAPOR PRESSURE (mm Hg) AT INDICATED RELATIVE HUMIDITY										
	0	10%	20%	30%	40%	50%	60%	70%	80%	90%	100%
0	0	0.46	0.92	1.37	1.83	2.29	2.75	3.21	3.66	4.12	4.58
5	0	0.65	1.31	1.96	2.62	3.27	3.92	4.58	5.23	5.89	6.54
10	0	0.92	1.84	2.76	3.68	4.60	5.53	6.45	7.37	8.29	9.21
15	0	1.28	2.56	3.84	5.12	6.40	7.67	8.95	10.23	11.51	12.79
20	0	1.75	3.51	5.26	7.02	8.77	10.52	12.28	14.03	15.79	17.54
25	0	2.38	4.75	7.13	9.50	11.88	14.26	16.63	19.01	21.38	23.76
30	0	3.18	6.36	9.55	12.73	15.91	19.09	22.27	25.46	28.64	31.82
35	0	4.22	8.44	12.65	16.87	21.09	25.31	29.53	33.74	37.96	42.18
40	0	5.53	11.06	16.60	22.13	27.66	33.19	38.72	44.25	49.79	55.32

not possible with any of the other parameters described. For example, the vapor-pressure gradient at 30°C and 40 percent relative humidity is $31.82 - 12.73 = 19.09$; at 20°C and 40 percent relative humidity, the vapor pressure gradient is $17.54 - 7.02 = 10.52$. Although the relative humidity is 40 percent in each case, the relative rate of evaporation from a water surface would be 19.09/10.52 or 1.8 times as rapid at 30°C as at 20°C.

A very important fact to remember is that with a constant amount of water vapor present, vapor pressure *does not change* with temperature. Obviously, however, the vapor-pressure gradient changes drastically with temperature even though the moisture content and vapor pressure are constant.

The relative humidity within the stomatal cavity of a leaf is assumed to be 100 percent. Thus, to find the vapor pressure of water in a leaf or other succulent plant tissue, we use the column under 100 percent relative humidity. If a leaf is at 30°C, its water vapor pressure is 31.82. If the atmosphere is at 30°C and 50 percent relative humidity, its vapor pressure is 15.91. The vapor-pressure gradient would be, therefore, $31.82 - 15.91 = 15.91$. If however, the leaf temperature is 35°C owing to the absorption of direct insolation, its vapor pressure is 42.18. With the same air temperature and relative humidity as in the previous example (30°C and 50 percent relative humidity) the vapor-pressure gradient would be $42.18 - 15.91 = 26.27$. The relative rates of transpiration would be 26.27:15.91, or 1.6:1. This temperature difference would be quite reasonable for comparable leaves—if one is in the shade and the other in the sun.

Measurement of Atmospheric Moisture

The most commonly used instrument to measure water vapor content of a mass of air is the *psychrometer*. A typical type consists of two similar thermometers—one a "dry-bulb," the other a "wet-bulb." The wet-bulb is covered with clean muslin which is thoroughly wetted with clean (and preferably distilled) water prior to and during use. A "sling psychrometer" has the two thermometers mounted side by side and by means of a handle is rapidly swung through the air (Figure 8-1). If the relative humidity is below 100 percent, water evaporates from the wet-bulb—the lower the humidity, the more rapid the evaporation. As the water evaporates, it takes up approximately 540 cal g^{-1} (heat of vaporization), and the temperature of the wet-bulb is lowered. For an accurate determination, the psychrometer is swung until the temperature of the wet-bulb no longer decreases. The principle involved is that as the wet-bulb cools by evaporation, the vapor pressure of the water in the muslin wick decreases (see Table 8-3). When the temperature of the wet-bulb no longer declines, the vapor pressure of the water in the wick and the water in the atmosphere are equal. By knowing the dry-bulb temperature and the wet-bulb depression (the difference between the wet- and dry-bulb temperatures), one can readily obtain

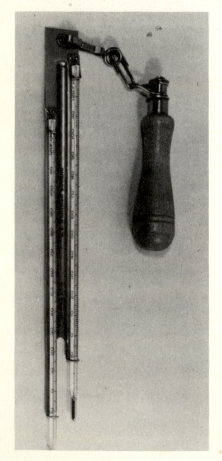

FIGURE 8-1 A sling psychrometer consists of wet- and dry-bulb thermometers. The thermometers are swung, and with the dry-bulb temperature and the wet-bulb depression (the difference between the two temperatures), one can use psychometric tables to look up the relative humidity. (From H. R. Byers, *General Meteorology*, 4th ed., McGraw-Hill, New York, 1974.)

the relative humidity from tables. It is then a simple matter to look up the vapor pressure in the appropriate tables and to calculate vapor-pressure gradients.

Other psychrometers are available which have been developed for particular situations where a sling psychrometer is too cumbersome. Models with battery-powered fans to blow the air over the two thermometers are advantageous in small areas. It is also possible to devise wet and dry thermistors or thermocouples for measurements in very confined locations. The major criteria that must be met are comparable temperature-sensing mechanisms and relatively rapid air circulation.

For a continuous record of atmospheric moisture levels, *hygrometers* are used (see Figure 7-5). Probably the oldest type is made from human hair (this type is still widely used). Several strands of hair are attached so that their expansion or contraction moves a pen on a recording chart by means of levers. As humidity increases, hair expands owing to the absorption of moisture.

A concept of major importance in understanding atmospheric moisture is *dew point*. The dew point is the *temperature* at which the relative humidity reaches 100 percent and the vapor-pressure gradient is zero. Particularly during nights with extensive radiational heat loss (see Chapter 7), many surfaces become cooler than the air temperature. Surface layers of air are subsequently cooled by conduction, and if the dew point is reached, condensation occurs. If the dew point is above 0°C, water is deposited as *dew*; if the dew point is below 0°C, water is deposited as ice. The value of dew or frost to the water needs of plants is minimal because of the very small quantities of water involved.

The formation of fog also occurs as the result of the temperature dropping to the dew point. In the case of *fog*, however, the condensation of water into droplets occurs *within* the lower layers of the atmosphere rather than on solid surfaces as with dew and frost. *Hygroscopic* ("water-loving") *particles* such as salt and smoke in the atmosphere serve as *condensation nuclei* on which water condenses. When the droplets become large enough, they scatter light and thereby reduce visibility. Since fog formation is dependent on condensation nuclei, it is most common in regions such as the Pacific and North Atlantic coasts and the Appalachian Highlands. Appreciable rainfall seldom results from fog because it is a phenomenon only in surface layers of air and is usually dissipated by evaporation soon after sunrise.

PRECIPITATION

Precipitation is usually described as drizzle, rain, snow, or hail. *Drizzle* and *rain* differ only in particle size; drizzle is less than 0.5 mm in diameter, whereas rain is larger than 0.5 mm. *Freezing rain* is either supercooled rain or rain very close to 0°C which freezes on contact with cold surfaces. *Snow* consists of loose ice crystals formed as the ice particles pass downward through a cloud.

As is obvious to us all, the formation of clouds does not necessarily lead to precipitation. Clouds can float through the atmosphere indefinitely if the condensed particles are small enough. Precipitation occurs only when droplets become so large that they can no longer remain suspended. One might assume that raindrops result from condensation occurring over a period of time starting with the very minute condensation nuclei. Although the mechanism is not well understood, meterologists indicate that continued condensation cannot account for the rapidity with which particle size can increase. Another possible mechanism is that as a droplet comes under the influence of gravity, it falls through the cloud and coalesces with other droplets in its path.

Precipitation is measured in various types of receptacles commonly called *rain gages* (Figure 8-2). The simple ones collect rain in a tube. More refined types are needed to collect freezing rain, snow, or hail. The ratio of snow depth to rainfall equivalent is usually about 10 or 12 to 1 but may vary considerably depending upon the moisture content of the snow.

Precipitation Cycle

Evaporation   Four major processes are involved in the precipitation cycle. Obviously the first of these is evaporation, the great majority of which occurs from the oceans although sizable quantities of water vapor also originate from transpiration by plants and evaporation from land masses and inland bodies of water. The evaporation rate depends upon the temperature of the air and the evaporating surface, air movement, the vapor-pressure gradient of the air, and the availability of water for evaporation. For example, in an arid area little evaporation occurs because of the unavailability of moisture even though potential evaporation is very high. Estimates of the earth's evaporation indicate that about 80 percent of the total evaporation occurs from the zone between 35°N and 35°S latitude.

Transportation   The second step in the precipitation cycle, and equally important, is transportation or *advection* of moisture-laden air masses. Tremendous quantities of heat are carried with the water vapor. As water evaporates it absorbs about 540 cal g^{-1} latent heat, which is released when condensation occurs. This latent heat transfer in the movement of water vapor has tremendous effects on the heat balance of areas of the earth's surface.

FIGURE 8-2 Rain gages are used to measure rainfall. They should be placed in locations away from buildings and trees which could obstruct normal rainfall. (Taylor Instrument, Sybron Corporation, Arden, N.C.)

Condensation In the formation of dew or frost on cold surfaces, condensation begins to occur at the dew point. In the atmosphere, however, there are fewer solid surfaces on which condensation can form. If the air were completely devoid of foreign particles such as dust, condensation would be very slow to occur. Normally, however, there are enough particles such as dust, smoke, and salt to serve as nuclei for condensation. Some particles such as salt from coastal areas and certain organic particles are hygroscopic and may actually remove moisture from the air at relative humidities below 100 percent.

To have sufficient condensation for precipitation to occur, large air masses must be cooled to the dew point or below. The only significant process by which this cooling occurs is through the rising of air masses. As air or any other gas rises, it expands because of decreased pressure upon it. As a result of this expansion, its temperature decreases. When

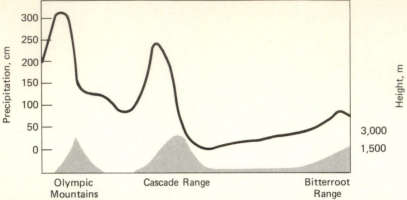

FIGURE 8-3 Effect of two mountain ranges on rainfall patterns in Washington and Idaho. (Adapted from H. Neuberger and J. Cahir: *Principles of Climatology—A Manual in Earth Science.* Copyright © 1969 by Holt, Rinehart and Winston. By permission of Holt, Rinehart and Winston.)

no heat is exchanged with surrounding sources, the temperature change is termed *adiabatic*. As an air mass descends, pressure increases and temperature rises. An air mass rises until it reaches a portion of the atmosphere of similar temperature. If condensation occurs, clouds are formed, much as fog is formed in surface layers. As air masses are forced to rise up and pass over mountain ranges, considerable adiabatic cooling occurs. The effects of mountain ranges in Washington and Idaho are shown in Figure 8-3.

Precipitation and Crop Production

Average annual precipitation around the world ranges from 0.05 cm in parts of Chile to more than 1182 cm in Cherrapunji, India.* Within the contiguous 48 states of the United States, the extremes are less than 5 cm in Death Valley, California, to more than 317 cm in northwestern Washington. The state of Hawaii provides some of the most amazing differences in precipitation in the entire world when the very small areas are considered. For example, the island of Kauai is approximately 47 km wide from ENE to WSW. The average annual precipitation varies from about 51 cm at low leeward elevations to 1181 cm on the windward peaks of the mountain (Table 8-4). As moisture-laden trade winds pass up over the island, adiabatic cooling and resultant condensation cause very heavy rainfall at high elevations. As these air masses move down the leeward side, they warm up and are relatively dry. Because of this, the islands have their "wet" and "dry" sides.

*From G. T. Trewartha, *An Introduction to Climate*, 4th ed., McGraw-Hill, 1968, p. 153; (after Landsberg).

TABLE 8-4

MEAN ANNUAL PRECIPITATION FOR SEVERAL LOCATIONS ON THE ISLAND OF KAUAI, HAWAII

LOCATION	DISTANCE FROM ENE COAST, km	ELEVATION, m	PRECIPITATION, cm
Anahola	2	55	124
Kaneha Resort	10	253	254
Hanalei Tunnel	18	371	444
Mt. Waialeale	26	1570	1181
Puehu Ridge	42	507	91
Hukipo	45	244	64
Kehaha	47	3	56

In relation to horticultural crop production, mean annual precipitation tells us little. To be of value, precipitation records must be on a monthly basis. For example, San Francisco, California, and Dodge City, Kansas, have essentially the same mean annual precipitation (Figure 8-4). Seasonal distribution is so different, however, that the potential for crop production is totally different. In San Francisco, the great majority (92 percent) of the total precipitation falls between November and April with the remaining 8 percent coming between May and October. In Dodge City, 73 percent falls between May and October with 27 percent falling from November through April. Thus without irrigation, crops in an area such as San Francisco are limited to winter annuals or very drought-tolerant perennials. Crop potential in the summer is obviously greater in an area such as Dodge City. An area in the humid East with heavier precipitation has even greater potential, not only because of relatively heavy rainfall but also because of rather uniform monthly distribution (see Figure 8-4).

In addition to the amount and monthly distribution of precipitation, the type of precipitation can also be of major consequence in horticulture. The amount and persistence of snowfall is of importance in winter survival of perennial crops in Northern areas because snow is a very efficient insulator. A continuous snow cover during the coldest parts of winter in the Northern tier of states allows survival of plants which would be killed in more Southern areas which lack snow. A striking example seen by one of the authors in New England was the survival of peach flower buds covered by snow during extremely cold winters. The only blossoms which opened the following spring were those on low branches covered by snow during the extremely cold period. Because of their greater cold tolerance, no effect of the snow cover was apparent on vegetative buds. If snow falls on cold but unfrozen ground, little or no soil freezing will occur if the snow depth is 0.3 to 0.6 m. Snow can thus provide a great deal of protection against cold injury to roots.

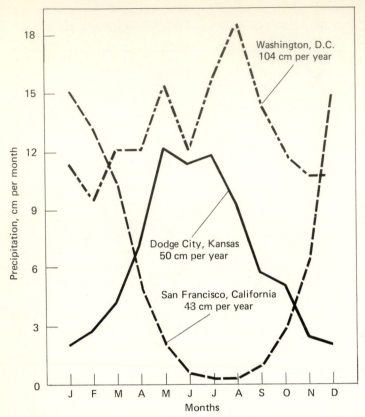

FIGURE 8-4 Precipitation by months at three locations in the United States. Note that San Francisco and Dodge City have similar total annual precipitation, but very different seasonal distribution.

Another important aspect of precipitation relates to the rate at which rainfall occurs as well as the total amount occurring during any given rainfall period. In a light shower in which the total accumulation is only 0.25 cm, very little rain penetrates the soil sufficiently to reach plant roots. Rather it is held on the surface of leaves and in the surface layers of soil, only to be lost rapidly to evaporation. Very rapid rainfall, as occurs in summer thunderstorms, often falls so quickly that a very large proportion is lost through runoff, particularly when the soil is dry prior to the rain. For maximum benefit, a slow "soaking" rain is best, as a high proportion is absorbed by the soil. The rate of infiltration is also dependent on soil type and any covering to protect the soil surface from compaction by rain drops. The coarser the soil type, the more rapidly the soil will soak up water. A soil with vegetation or a mulch will also take up water much more rapidly than a bare soil.

The occurrence of a hailstorm can bring total devastation to many horticultural crops in a matter of minutes. The possibility of hailstorms is of major importance and varies considerably over the country. Hail insurance is available from certain companies, and the cost for a particular location is based on long-term averages of the frequency and intensity of such storms for that area as well as on the crop involved.

WATER ABSORPTION AND MOVEMENT IN PLANTS

The water content of plant tissues varies widely, not with only the type of plant, but with the tissue and stage of development as well. Many properly dried seeds contain only about 5 percent moisture, whereas succulent fruits and vegetables consist of 85 to 95 percent water. Most other tissues have a water content intermediate between these two extremes, but actively growing plants are surprisingly high in water content. The vast majority of water is taken up but is readily lost by transpiration and must be replaced. Estimates of water absorbed by an actively growing squash plant are about ten times the fresh weight of the plant per day, or approximately the equivalent of its fresh weight for each daylight hour. To be able to absorb this quantity of water, the absorbing mechanism of a plant must be very efficient.

Absorption

Root System   The water needs of most plants are met almost exclusively by the root system in the soil. Notable exceptions are orchids, which are able to absorb sizable quantities through highly developed aerial roots. Certain other species can absorb water through modified leaf hairs, the base of leaves with bromeliads, or needles of certain conifers. In general, however, plants are essentially totally dependent on water uptake through roots in the soil. Although the type of root system varies greatly with species, plants have larger and more expansive root systems than is commonly realized.

Various researchers have estimated the expanse of the root systems of plants. The results of these studies provide us with some idea of the absorbing surface. For example, a squash vine with 140 leaves and a stem length of about 9 m had a total of 86 m of main and branch roots. This did not include root hairs, which usually increase the root system by 5 to 12 times. A winter rye plant has been estimated to have 14 billion root hairs with a surface of 1310 m². Another aspect of major consequence is the rapidity with which new roots and root hairs are produced. A rapidly growing squash vine has been estimated to produce up to 300 m of new roots and root hairs per day. It is quite easy to visualize the significance of new root growth and thereby soil exploration by comparing rapidly growing plants in a container with those in the field. A tomato plant in a 4-L pot may require watering twice a day, whereas a comparable plant in the field may go one to two weeks without rainfall. This difference is largely because the plant in the field continually grows into new soil and thereby obtains water, whereas the plant in the pot rapidly extracts the available moisture from the limited soil mass.

Root distribution In many plants, the root system is much larger in lateral spread than the aboveground parts. Others with tap roots are limited in spread but penetrate very deeply. In many cases it is possible to explain the adaptation of particular plants to their environment on the basis of their inherent rooting characteristics. For example, cacti generally have very extensive shallow roots which allow maximum benefit from infrequent light rains experienced in arid areas. Among crop plants there is considerable variation in average rooting depth. Shallow-rooted species include lettuce, strawberry, and radish; more deeply rooted types are tomato and corn. Many large perennial fruit trees such as peach, apple, and pear have about one-half their roots in the surface 60 cm of soil. Other tree crops, such as pecan and walnut, tend to be more deeply rooted.

Factors limiting root penetration Many factors can markedly affect the rooting depth of a particular species. Among these are a high water table, a hardpan, soil aeration, and soil compaction. Most plant roots are unable to grow or even survive when submerged, as when the water table rises into the root zone after a heavy rain. The *water table* is the upper edge of free water in the soil. The severity of this problem varies not only with species, but with the stage of plant growth. Although rice thrives with its roots submerged, many other plants cannot survive submergence for more than a very few days. Whereas apple trees are able to withstand flooded conditions for periods of two to four weeks in the dormant season, during the growing season such long periods would be very detrimental.

A widely fluctuating water table is also undesirable. When the level drops, roots may penetrate into these lower soil layers, only to be killed as the water table once again rises. A high water table in the spring and early summer prevents deep root growth and thereby may make a crop more drought-sensitive in the late summer than in an area with a uniformly deep water table. For crops of high cash value, a soil with a high water table can be drained by several systems, depending on the particular situation. In cranberry bogs and muck soils, the fields are usually trenched and the excess water is drained or pumped out of the ditches. In more typical situations, tiles or perforated pipe covered with coarse gravel can be laid in ditches at the appropriate depth and the water table lowered. The spacing of drainage lines depends on soil type; since lateral movement is more rapid in light-textured soils than in heavy-textured soils, drainage lines can be more widely spaced in light-textured soils.

When planning to set a field where a high water table is a potential problem, the field should be checked, preferably in the spring. This is readily done by digging holes and periodically examining the level of water in the holes. After a heavy rain the water table often rises, but if drainage is adequate, it soon drops to an acceptable level. If the level of free water stays at an undesirable height in the soil, the site should be either drained or not planted at all. Careful observation prior to planting can be a most effective prevention against financial disaster, particularly for a long-term crop such as an orchard.

A hardpan or impervious layer in the soil normally prevents the penetration of roots and can force a plant to be shallow-rooted. This effect is particularly apparent in dry seasons, in that plants on such sites are the first to suffer water deficiency. The effect is not very different from that described previously for a plant in a pot with a limited soil mass.

Little can be done where soil is underlaid with rock; but if a shallow hardpan is present, its effects can be reduced by subsoiling and thereby breaking up the restricting layer. A *subsoiler* is a heavy subsurface plow which lifts and thereby breaks up a hardpan. Subsoiling is most effective when the hardpan is relatively thin and when the soil beneath the hardpan is suitable for root growth. Soil compaction is often the result of heavy equipment being driven over fine-textured soil, especially when soil moisture is high. Obviously it is easier to avoid than to cure the problem. Subsoiling may also be helpful here, as with a hardpan.

Among horticultural crops, there is enough variability in both rooting depth and tolerance to "wet feet" that by judicious planning a producer can at least partially overcome the problems described previously. For example, a fruit grower knows that deciduous tree fruits rank in decreasing tolerance to wet feet as follows: pear, apple, peach, and cherry. Pear trees will grow and produce well on a wet site where cherry trees would die during the first year. Within some species there is also the possibility of rootstock selection. For example, with plums, the Myrobalan stock is more tolerant of excess water and poor aeration than the Marianna stock. In well-aerated soils, plums are also grown on peach rootstocks. If there is a hardpan at 45 cm, it is much more logical to grow a shallow-rooted crop, such as lettuce or radish, than a deep-rooted species, such as corn or tree fruits.

Since plant roots respire, they are utilizing oxygen and liberating carbon dioxide. When soil aeration is restricted because of flooding, compaction, or any other factor, oxygen is depleted and carbon dioxide accumulates. The tolerance to poor aeration varies with species, but the roots of most crop plants cannot survive at oxygen levels below 2 to 5 percent. Many roots die if the oxygen level is below 1 to 2 percent. For normal root growth, oxygen levels must be in the range of 5 to 15 percent, depending on species. It seems that low oxygen is more detrimental than high carbon dioxide, although the two generally occur simultaneously.

Water Uptake As mentioned previously, the root hairs greatly expand the root system of most plants. Water uptake is largely through the root hairs, which are ideally suited for this process, not only because of the large surface area but also because of the rapidity with which they are produced. As a root penetrates into the soil, root hairs are formed immediately behind the root cap and therefore are continually expanding into unexploited soil.

The root hairs grow into spaces among soil particles and are thus in intimate contact with the films of water surrounding these particles. Moisture in the soil is not "pure" water but rather contains a wide spectrum of dissolved substances, many of which are vital nutrients for plants. Some of these solutes originate from the soil particles, some from fertilizers, and some from decomposition of both plant and animal tissues. This soil solution is separated from the cell contents by a semipermeable membrane. This membrane serves not only to separate the two solutions but to control uptake of both water and solutes.

Cell sap is even more complex than the soil solution. In addition to many of the inorganic ions present in the soil, many simple and complex organic molecules are present. For a root cell to extract water from the soil, the concentration of solutes of the cell sap must exceed that of the soil solution. Water moves from a region of higher to lower water concentration by osmosis; it therefore moves into the root as long as solute concentration is higher in the root. If the reverse is the case, water may move out of the root into the soil solution. This occurs primarily when excess fertilizer is added or in the case of alkali soils and is sometimes referred to as "physiological drought." One satisfactory solution to this problem is application of large quantities of water to leach out the excess soluble salts. Although leaching is often practiced in potted plants in greenhouses, in field situations it may be difficult because of the quantity of water needed and because excellent subsoil drainage is necessary.

Water uptake by plants is also influenced by soil temperature. During midwinter in cold areas the ground may freeze, and water absorption from frozen ground is impossible. However, even when the soil is not frozen, cold temperatures may restrict water uptake.

This can be partially explained by realizing that the viscosity (resistance to flow) of water is twice as great at 0°C as at 25°C. Thus, at cold temperatures, water movement in the soil as well as in the plant is slower. This fact is reflected in the practice by florists and wholesalers of putting cut flowers in warm water to accelerate water uptake and thereby shorten the time necessary to restore turgidity after shipping.

In this section absorption—water uptake—has been briefly discussed. Below, the movement of water within a plant is described; the section headed "Transpiration" (page 230) describes its ultimate loss to the atmosphere. Although one of these three processes may occur independently from the others for short periods of time, in the normal situation all three are going on concurrently.

Water Movement in Plants

The xylem of a plant acts not only as the water-conducting system, but by being the connecting link between the transpiring and absorption mechanisms, it transmits the demand for increased or decreased water uptake. Water must pass through living cells between the epidermis of the root hair and the xylem, but once in the xylem its movement is relatively unrestricted until reaching the living cells of the leaf or other aerial organ. Thus water conduction in the xylem is "passive" in that the nonliving elements obviously cannot supply energy as living cells could. It must be remembered, however, that xylem is not strictly a "pipe"; it consists of individual cells with perforated end walls. The end walls offer little resistance to water flow, but do prevent large air bubbles from developing by confining small bubbles within individual cells; air blockage is thereby reduced.

Root Pressure   The forces behind the rise of water in tall plants, such as trees, have been a topic of discussion and debate since the days of the early plant physiologists. The phenomenon of *root pressure* has been part of this debate. As can be seen by the exudation of water from the cut stem of many plants with little or no transpirational surface, the roots can generate a positive pressure. This is particularly apparent in the "bleeding" of grapevines pruned in the spring. Leaves of certain plants exude liquid water by the process known as *guttation*. Most commonly at night following a day in which transpiration has been rapid, water is released through specialized structures called *hydathodes*. "Bleeding" and guttation indicate a positive pressure in the plant which can only reflect a force generated in living roots and is therefore attributed to root pressure. Under conditions of rapid transpiration, root pressure is probably not of major importance because, although the pressure can be significant, the rate of water movement is rather slow.

Transpirational Pull   It is widely accepted that the rise of water in tall plants is largely caused by what is often termed *transpirational pull*. As water is transpired by the cells within a stomatal cavity of a leaf, a water gradient or tension is established whereby water moves toward these cells. This tension is transmitted back to the xylem of the leaf vein, through the petiole, to the stem or trunk, and ultimately back to the roots. Thus under conditions of rapid transpiration, the entire plant is under a tension by which water is "pulled up" from the soil to the leaves. Although this tension is difficult to measure directly, an instrument called a *dendrometer* has been used to monitor expansion and contraction of tree trunks accurately. During periods of rapid transpiration, tree trunks can be shown to shrink. This observation provides evidence in support of the theory that tension exists within plants. This theory also depends on the cohesive forces (attraction of like particles) of the water molecules present in the xylem elements and the adhesive forces (attraction of unlike particles) between water molecules and the xylem elements. It should also be noted that absorption "lags" behind transpiration.

TRANSPIRATION

Leaves of crop plants have presumably evolved over time to provide maximum efficiency in their prime functions of capturing light energy and exchange of CO_2 and O_2 with the surrounding atmosphere. A leaf which is efficiently designed for photosynthesis is also a very effective transpirational unit. The "benefits" of transpiration to plants have been the subject of heated debate over the years, and the issue is yet to be completely settled. The proposed beneficial effects of the normally rapid rates of transpiration encountered have included (1) dissipation of heat by heat of vaporization and (2) increased absorption of nutrients. It is generally accepted that transpiration results in both of these effects, but many feel that these processes would proceed quite satisfactorily without transpiration and have therefore labeled transpiration as an unavoidable evil. It is considered "unavoidable" because of leaf morphology and an "evil" because of the tremendous loss of water and injuries which can result from tissue dehydration.

Stomates

Transpiration occurs largely through the stomates of leaves. The leaves of most plants have stomates on the lower surface, although certain species, especially grasses, have them on both upper and

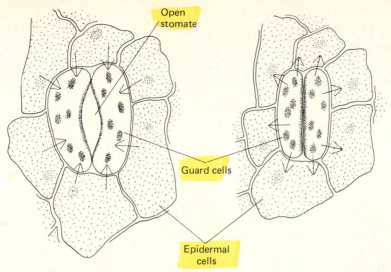

FIGURE 8-5 As water moves into guard cells, the cells become turgid, and the stomate opens. As guard cells lose turgidity because of loss of water, the stomate closes.

lower surfaces. The cuticle on leaves is a covering of cutin which is an effective barrier to water loss. The cuticle is also often covered with a waxy layer. Most water loss, therefore, is under control of stomatal opening and thus the guard cells. Guard cells are modified and highly specialized cells in the epidermal layer. The sides of the guard cells which surround the stomatal opening are usually visibly thickened. The opening of stomates is the result of increased turgor pressure in the guard cells in relation to surrounding cells, whereas closing is caused by decreased turgor pressure in the guard cells. Since the cell wall surrounding the pore is thickened, it does not expand as does the rest of the cell wall, thus causing the distortion necessary to open the stomate (Figure 8-5).

The number of stomates is usually very large but also shows wide variation with species and environment (Figure 8-6). Estimates range from 2000 to 100,000 cm^{-2}. Even when stomates are wide open, the combined area of the pores is usually less than 3 percent of the total leaf area. It has been estimated that water loss by this 1 to 3 percent of the surface area is approximately one-half of that from a free water surface equal to the area of the leaf. It is thus apparent that the stomatal pores are far more efficient in gas exchange than is indicated by their combined surface area.

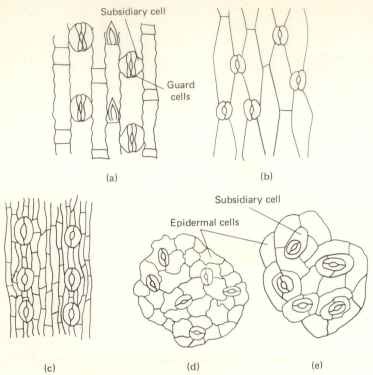

FIGURE 8-6 Stomatal arrangement in: (*a*) Graminae (corn), (*b*) Liliaceae (onion), (*c*) Gymnospermae (pine), (*d*) plants with elliptical stomata (broad bean), (*e*) most succulent plants (*Sedum spectabilis*). (From H. Meidner and T. A. Mansfield, *Physiology of Stomata*, McGraw-Hill, New York, 1968)

Transpiration Ratio

The efficiency with which plants utilize water is referred to as the *transpiration ratio* and is expressed as the ratio between the units of water absorbed by the plant and the units of dry matter produced. To determine dry matter, the plant is dried in an oven. The loss in weight during drying is the moisture content; the remainder is dry matter. If a plant weighed 4 kg before drying and 0.4 kg after drying, it would be said to have been 90 percent water and 10 percent dry matter. To calculate the transpiration ratio, one must know the amount of water absorbed by the plant. Let us assume that the total water uptake was determined to be 200 kg of water. The transpiration ratio for this plant is 200/0.4 = 500. Because the transpiration ratio is expressed as weight of water/weight of dry matter, the units must be the same but can be in terms of g/g, kg/kg, pounds/pounds, tons/tons, etc. Transpiration ratios have been shown to vary from about 30 to 3000, but as a general rule, crop plants fall in the range of 300 to 1000. To put this figure in terms of the amount of water per unit area, perhaps the most important consideration is the amount of dry matter produced. Although desert plants have a surprisingly high transpiration ratio (about 2000), they are able to survive through very slow growth and low dry weight accumulation per hectare. Estimates of cactus growth indicate an annual production of only 225 kg ha^{-1} per year. Using a transpiration ratio of 2000, the amount of water would be 225 × 2000 = 450,000 kg ha^{-1}. This amount of water would be the equivalent of 450,000 L ha^{-1} or 4.5 cm of water.

If, however, a crop of peach trees producing a total of 5600 kg of dry matter per hectare per year is considered, the water needs would be much greater. Using a water requirement of 500, such a hectare of peach trees would absorb 500 × 5600 = 2.8 million kg or 2.8 million L or 28 cm of water.

It must be kept in mind in calculating the transpiration ratio that only the water which is taken up by the plant is included. The proportion of total precipitation or irrigation which is ultimately absorbed by the crop plant varies from 0 percent, from very light showers which are intercepted by the foliage and lost to evaporation, to somewhat over 50 percent under ideal conditions. The remainder is lost through runoff, evaporation, uptake by competing vegetation, and percolation beyond the root zone. A rule of thumb is that about one-third of the total water falling on cropland is absorbed by the crop. Using this figure of one-third, the cactus would require about 4.5 × 3 or 13.5 cm of water, whereas the peaches would require 28 × 3 or 84 cm.

Soil moisture, nutrition, soil aeration, temperature, light level, atmospheric humidity, and wind speed can all influence the transpiration ratio. In general, the transpiration ratio is at its minimum when each factor is at its optimum. Thus, a crop ideally suited to its environment is most efficient in its utilization of water.

Factors Influencing Transpiration

Several rather diverse factors affect rates of transpiration. Among these are (1) irradiance, (2) temperature, (3) wind speed, (4) availability of water to the roots, (5) water vapor pressure of the air, and (6) a group of plant factors. Increasing irradiance usually causes stomates to open and thereby accelerates transpiration. An indirect effect is that as irradiance increases, leaf temperature rises and thereby the vapor-pressure gradient between the leaf and the air is also increased. The effect of increasing air temperature is similar, because the intercellular spaces inside a stomate remain at 100 percent relative humidity. The result is that the water vapor pressure increases, whereas the vapor pressure of the air does not change with temperature (see Table 8-3). Under calm conditions, the air surrounding a transpiring leaf gains increasing amounts of water vapor, and the vapor-pressure gradient is therefore reduced. As wind sweeps this moisture-laden air away and replaces it with relatively dry air, however, the vapor-pressure gradient is increased and therefore transpiration is accelerated. This effect is greatest as wind speed increases from 0 to 8 km per hour, after which the effect increases only slowly. A confounding effect is that if the irradiance is high, increasing wind speed tends to cool the leaf and thereby lower the vapor-pressure gradient and transpiration. When there is an abundance of soil moisture, transpiration is controlled largely by the aerial environment and transpiring surface, but when water availability to the roots is limited, water stress develops which can limit transpiration. As water stress develops within a plant, the stomates start to close, serving as a defense mechanism against tissue desiccation. The influence of the vapor-pressure gradient on water loss was described previously in this chapter. Obviously, the greater the gradient, the higher the transpiration rate. This effect is accentuated by leaf temperatures higher than air temperature, but is also very important in influencing transpiration rates in arid as opposed to humid climates.

The physical characteristics of plants can have a considerable effect on the rate of transpiration. Upright leaves intercept less radiation than horizontal leaves during midday and thereby tend to be cooler. Certain plants, such as pineapple, have sunken stomates and are less affected by wind speed. Many succulents open their stomates at night rather than during the day, thereby drastically reducing water loss.

Although we have discussed water absorption by roots and transpiration as dependent processes, each process can proceed independently for limited periods of time. Plant parts severed from the root system continue to transpire, as is very apparent with cut flowers. Obviously, if allowed to transpire without replacement of water, they desiccate very rapidly. If, however, the stems of cut flowers are placed in water, they are able to live for extended periods. With no roots, uptake of water is by transpirational pull. It has recently been discovered that a factor which limits the life of cut flowers is blockage of xylem elements. When added to water used to sustain cut flowers, recently tested chemicals can significantly extend flower life by delaying the development of xylem blockage.

Water uptake by roots without concurrent transpiration occurs commonly at night until any deficit which developed during the day is erased. Stump bleeding and guttation, mentioned previously, are indications of excess uptake of water.

Methods to Reduce Transpiration

Depending upon the particular situation, the suppression of transpiration by horticultural plants varies from an absolute necessity to a desirable but not totally necessary goal. For example, when softwood cuttings are being rooted, unless transpiration can be drastically reduced, the cuttings cannot survive. Cuttings are usually placed in beds which may be misted frequently, or the relative humidity is kept very high by means of polyethylene covers (see Chapter 12). In addition, new cuttings are usually shaded to reduce leaf temperature and thereby lower the water vapor-pressure gradient between the leaf and air. With dormant hardwood cuttings taken in winter without leaves, these precautions are not necessary because the cuttings usually have started to form roots by the time leaves appear. Newly rooted cuttings of ornamentals are taken from the propagation bed and grown for a period under lath or saran cloth shade to allow further root development prior to exposing the plants to the extreme transpirational demand that occurs in full sun and wind.

Windbreaks are often used to suppress transpiration. If plant material is used for the windbreak, major considerations in its selection are rapid growth, dense foliage, resistance to damage by winds, and the type of root system. For example, trees with widely spreading root systems should be avoided as windbreaks for orchards because they may compete with the adjoining trees for water and nutrients. For such situations, windbreak plants should be deep-rooted for support but have limited lateral spread. For a container-stock nursery, the rooting habit of the windbreak would be of much less concern.

Young plants are often started in a greenhouse or other sheltered location for transplanting to the field. This is done to obtain early production of vegetables such as tomatoes. Plants which are growing rapidly in a greenhouse tend to be very tender and are unable to withstand immediate transplanting to the field. Transplants require *hardening off* before being put in the field. Several methods are used to slow the growth rate and to cause the plant to become less subject to drying out. These include withholding nutrients, allowing incipient wilting to occur, and exposure to lower temperatures. This hardening off is done primarily to allow the plants to cope better with water stress when transplanted.

For many years, researchers have sought means of reducing transpiration other than by controlling the environment. Certain commercially available materials are effective under some situations. Most of these are wax or plastic materials which are diluted with water and sprayed on the foliage to form a film upon drying. These have been most widely used for the prevention of winter desiccation (see Chapter 7). Various researchers have evaluated the potential of these materials on actively growing crops and transplants, but in general they have not proven commercially practical. Although these transpiration suppressants reduce transpiration, the film which is formed may also suppress photosynthesis if the stomates are covered. Another problem is that the film tends to crack and peel, which reduces its effective life.

Another potential drawback is that as transpiration is suppressed markedly, leaf temperature rises and the resulting increase in the vapor-pressure gradient tends to nullify the transpiration decrease.

An entirely different approach to transpiration reduction has been explored in recent years. Certain chemicals have been discovered to cause stomatal closure, whereas others have the ability to reduce or prevent stomatal opening. When stomates partially close, transpiration is usually suppressed more than photosynthesis. If a chemical could be found which could control stomatal opening without adverse side effects, such as reducing photosynthesis, its value in agriculture could be tremendous. Not only would the control of transpiration conserve great quantities of water, but many additional benefits would occur if the occurrence of water stress could be reduced. For example, many horticultural crops develop disorders associated with even temporary water stress. Some of the more common problems attributed, at least partially, to water stress are blossom-end-rot of tomato, poor blossom set on grapes, and the shedding of young fruits in many species.

Whether or not chemical control of transpiration ever becomes commercially practical remains to be seen, but the potential value to agriculture makes the search most worthwhile.

MOISTURE LOSS DURING STORAGE AND MARKETING

With many horticultural products, control of water loss is of major importance. Fresh fruits, vegetables, and flowers are stored at cool temperatures and high relative humidities to minimize water loss. In most storage environments the relative humidity is kept between 90 and 95 percent to avoid condensation of water which accentuates disease problems, such as rots. The examples in Table 8-5 indicate the

TABLE 8-5
EFFECT OF THE TEMPERATURE OF FRUIT
ON RELATIVE WATER LOSS IN COLD STORAGE

Situation 1. Apples at 35°C, put in cold storage at 0°C, 90% relative humidity
  Vapor pressure of apples: 42.18 (Temperature 35°, relative humidity: 100%)
  Vapor pressure of air: 4.12 (Temperature 0°, relative humidity: 90%)
  Vapor-pressure gradient 38.06
Situation 2. Twenty-four hours later. Apples at 20°, storage: same as before
  Vapor pressure of apples: 17.54
  Vapor pressure of air: 4.12
  Vapor-pressure gradient 13.42
Situation 3. Seventy-two hours later. Apples at 0°, storage: same as before
  Vapor pressure of apples: 4.58
  Vapor pressure of air: 4.12
  Vapor-pressure gradient 0.46

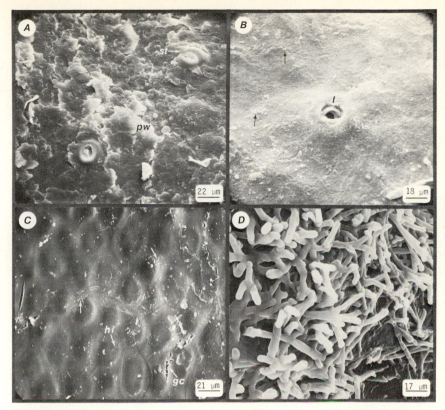

FIGURE 8-7 Scanning electron photomicrographs of some variations in fruit surfaces. A, Orange fruit surfaces are covered by platelet wax (pw) and have stomatal openings (st) which may be at least partially covered by wax. B, Mango fruit surfaces have a thin, slightly pebbly wax coating (black arrows) and lenticel openings (l) similar to apples. C, Tomato fruit have a very thin but pebbly epicuticular wax coating (white arrows) and no natural openings. Growth cracks (gc) or handling injuries (hi) are evident on the surface. D, Blueberry fruit have rodlet surface wax and occasionally an opening (not shown) heavily covered by rodlet wax is present. (Plate courtesy of L. G. Albrigo and P. Anderson, University of Florida.)

effect of the temperature of fruit on water loss under normal cold-storage conditions. (See Table 8-3 to find vapor pressure and vapor-pressure gradients.)

Relative water loss from the fruit in these situations would be as follows: 38.06:13.42:0.46 or 83:29:1. These ratios emphasize the advantage of rapid cooling as a means of minimizing water loss. Obviously excessive water loss can result in shriveling of fruits and fleshy vegetables as well as wilting of leafy vegetables. The surface wax on fruit surfaces is quite variable, depending on the species, but in most cases it provides considerable protection against water loss (Figure 8-7, above).

FIGURE 8-8 'Golden Delicious' apples are stored in boxes containing an unsealed polyethylene liner to retard moisture loss. (Photograph, U.S. Department of Agriculture.)

Polyethylene or other films which are relatively impervious to water vapor are used with some fruits (Figure 8-8). For example, 'Golden Delicious' apples are stored in containers lined with polyethylene because even at normally high relative humidity in cold storage, these apples lose enough water to shrivel. Most other apple cultivars have sufficient cuticle so that excessive water loss is not a major problem.

Many fresh vegetables are waxed to maintain their crispness. Waxed cucumbers have a somewhat greasy coating. A harder wax is used on turnips. Polyethylene bags are used with many fruits, not only for reducing water loss but for ease of handling and because the consumer can readily inspect the contents. These containers are commonly used with apples, citrus fruits, carrots, and lettuce. Many other products are packaged in overwraps for a similar purpose. When a semirigid base is used, such as cardboard or styrofoam with an overwrap, rather perishable items such as tomatoes, blueberries, cherries, and plums can be not only attractively displayed but also protected from mechanical damage and excess moisture loss.

Most polyethylene bags in which produce is sold are perforated with small holes to prevent the excess modification of O_2 and CO_2 levels by respiration. If, for example, apples are sealed tightly in polyethylene bags, respiration by the fruit will often lead to low O_2 and high CO_2. In this situation, anaerobic respiration occurs, leading to "off" flavor and breakdown of tissues. The perforations allow sufficient exchange of O_2 and CO_2, but water loss is reduced sufficiently to minimize weight loss and shriveling.

Waxing is also used on certain ornamental plants during marketing. Many rose bushes are marketed by digging the bush, putting some damp peat moss around the roots, and tying a polyethylene bag over the root system. Because of the rather fleshy nature of the stems, however, transpiration during storage, shipping, and marketing can lead to desiccation and death. A commonly used procedure is to dip the stems of the plant in melted wax to provide a reasonably continuous coating. As the bush is planted and growth starts, the wax gradually flakes off—but only after it has served its purpose.

IRRIGATION

In spite of efforts to suppress transpiration by crop plants, there is no substitute for an adequate supply of moisture. It appears that increasing dependence on irrigation is inevitable.

Although irrigation has been utilized for centuries for crop production in arid areas, its use is expanding in semiarid and humid climates as well. Food production must be increased to meet the needs of the world's rapidly expanding population. This will mean not only better management of land currently used in food production, but the utilization of vast areas of land not formerly used for agricultural production. Another aspect of this overall problem—at least in the United States—is the loss of prime agricultural land to urban sprawl, interstate highways, and modern suburban subdivisions which are so common today. Increasing real estate taxes are another serious problem for agricultural producers whose land is closed in upon by urban and suburban development. This problem is particularly severe with horticultural producers, who have commonly been located relatively close to large areas of population.

Determining the Feasibility of Irrigation

The shortage of water is the major factor limiting crop production in vast areas of the world. Much of the most productive land is totally dependent on irrigation; crop production in many other areas is greatly improved by irrigation. In an arid or semiarid climate the need for supplemental water is obvious, and without it crop production could not be even considered. In areas which normally receive marginal or sufficient precipitation, however, the economic feasibility of irrigation becomes considerably more complex. As discussed earlier in this chapter, seasonal distribution of precipitation is of prime importance. If the long-term weather records indicate that sufficient rainfall can generally be expected during the growing season, the decision to purchase irrigation equipment is often based on the frequency with which insufficient rainfall can be anticipated. If a serious drought has occurred on the average of once in 10 years, irrigation is much less practical than if a drought has occurred once every 3 years. In many cases, irrigation can be considered as insurance against dry periods which may reduce yield or, if extreme, may cause the loss of the entire crop. The unpredictability of weather makes this decision a gamble, and perhaps the best one can do is make decisions on the basis of long-term averages for the area.

In addition to the amount of precipitation which can be expected during the growing season, one also needs to know how rapidly this water will be depleted. Probably the most useful measure of water loss is called the *evapotranspirational potential*. This measure takes into account both evaporation from the soil and transpiration by the crop.

Several methods have been used to estimate the evapotranspirational potential, including (1) elaborate calculations of incoming, outgoing, and net radiation and (2) the use of lysimeters. A *lysimeter* is a large container of soil in which a crop is grown; the water loss is determined by weight loss of the entire container. Perhaps the most common method uses various available parameters such as temperature, relative humidity, wind speed, and cloudiness to calculate an approximation. The rate of evapotranspiration depends on a complex interaction of factors including soil type, crop, stage of development, and environment. An area in Texas with relatively high temperatures, low humidity, and intense sunshine has a higher rate of evapotranspiration than a comparable crop and soil in Illinois. Therefore, more water would be needed to produce the crop in Texas than Illinois.

Rooting Depth   There are rather striking differences among horticultural crops which affect the decision to install irrigation. Depth of rooting can markedly affect the length of time which the crop can thrive without the addition of water. Lettuce, strawberry, and radish plants are shallow-rooted and therefore deplete the root zone of available moisture more rapidly than deeply rooted crops, such as tomato and tree crops. Thus, irrigation is more necessary for the shallow-rooted than for the deep-rooted crops in areas with marginal rainfall.

FIGURE 8-9 Cracked 'Stayman' apple as the result of a dry summer followed by early fall rains. Internal pressure has caused the fruit to split.

Growth Pattern   The pattern of growth of a crop is of major importance. An entire crop of strawberries can be lost during a 3- to 4-week drought in the spring when the fruit are developing. A similar period of dry weather could reduce the yield of long-season crops such as apples, but the net effect would be much less drastic because of the long period of fruit development. Another example is the comparison of apples and peaches. The fruit growth curves are different. Apples enlarge at a more or less linear rate from full bloom to harvest. Peaches, however, have a rapid growth rate during the latter part of the season, and this is referred to as the "final swell." Inadequate soil moisture during the final swell of peaches is much more harmful than a comparable dry period occurring at any time during the growth of apples.

Certain crops also suffer when the availability of moisture fluctuates. As soil moisture is depleted, growth slows down. When water again becomes available, especially if it comes in large quantities, growth resumes rapidly. The result of this rapid growth can be seen as splitting such as that which occurs in 'Stayman' apples (Figure 8-9) and in cabbage. With both of these, the majority of the crop may be destroyed in a matter of a few hours.

Balance of Cost and Return   The potential returns from an irrigation system depend on the crop, how often the system would be needed, and the increased value to be realized. The costs are very variable, depending on the system chosen, the frequency of use, and the cost of water. The cost of water is very dependent on the location involved. In large irrigation areas in the Western United States, the cost of water varies from less than $10 to well over $100 per acre-foot of water. In humid regions some producers can use large lakes and rivers, whereas others must use deep wells, adding considerably to water costs. The high cost of water may well be a major impetus to the use of trickle rather than other types of irrigation.

Water Rights   Another concern faced by many agricultural producers considering installation of irrigation is water rights. The laws regulating the use of water vary widely in different areas of the United States but are of importance everywhere. The increasing demand being placed on our finite water supply will no doubt lead to many lawsuits. It is not an easy task to decide who should have access to water when so many are competing for it, including agricultural producers, cities, industries, and those interested primarily in recreational uses. With the current situation, it certainly behooves a person contemplating an irrigation installation to be sure of the water rights before, rather than after, purchasing the system.

Systems of Irrigation

The four basic irrigation systems used today are surface, aerial (sprinkler), trickle, and subsurface. Within each of these there is a variety of types designed to meet particular situations.

Surface Irrigation   The application of water directly to the soil surface is known as *surface irrigation*. A major advantage that it offers over sprinkler application is reduced evaporational loss, since the water does not pass through the air. This is one of the major reasons for the widespread use of surface irrigation in arid and semiarid regions. Another reason is that in such areas irrigation water is often supplied to the field in ditches or large pipes, so that no pumps are necessary to obtain the water initially. Because gravity flow is used for surface distribution, no power is required. Several commonly used systems of surface irrigation are described next.

Flood irrigation   *Flood irrigation* is used primarily with those crops which are tolerant of excess moisture. Since it usually involves the flooding of an entire field at once, it is limited to essentially *level* land. About the only horticultural crop irrigated by flooding is the cranberry, but rice is commonly grown using this method. With cranberries, flooding is also useful for freeze protection, insect control, and harvest.

FIGURE 8-10 Basin irrigation in a walnut orchard. Ridges are "raised" by machine and each elevation flooded separately. (Photograph, Diamond/Sunsweet, Inc., Stockton, Calif.)

Basin irrigation   A modified system of flooding called *basin irrigation* is used to some degree in orchards (Figure 8-10). The land must be relatively level, but the design can accommodate small differences in elevation. Ridges or levees are raised along contour lines, and water is introduced into individual "basins" which are essentially level. The levees may be permanent and seeded to grass or may be re-formed prior to each irrigation.

Furrow irrigation   Perhaps the most widely used type of surface irrigation for horticultural crops is *furrow irrigation*. Typically used with row crops, such as vegetables and sugar beets, it offers the advantages of more even distribution and more efficient use of water than flood or basin irrigation. The water is delivered to the field in a head ditch or gated pipe at one end of the field from which it flows into individual furrows (Figure 8-11). Depending on the planting system, furrows may exist between every two rows of plants, or two rows of the crop may be on the bed between furrows.

A disadvantage of furrow irrigation is that, in order to ensure that sufficient water reaches the far end of the field, the end closest to the

FIGURE 8-11 In furrow irrigation, water is siphoned into each furrow from a main ditch.

source receives excess water. Perhaps the greatest disadvantage today, however, is the high labor requirement. Furrow irrigation requires continual supervision and maintenance to ensure the application of the proper amount of water at the correct rate. If water is added too fast, erosion may occur, causing individual furrows to merge and form channels. If it is applied too slowly, the distant end of the furrow will receive insufficient water.

A real advantage of surface irrigation of crops is that, since the irrigation water does not wet the foliage, disease problems are minimal. Surface application is, however, of little or no value in freeze protection or "air conditioning" (see Chapter 7).

Aerial (Sprinkler) One of the most widely used types of irrigation today is sprinklers. Under this heading is a vast array of types, but all deliver water through the air to the crop and soil. The basic parts of any sprinkler system are pump, main delivery pipe, lateral pipes, and sprinklers. Some of the more common types are described below.

Permanent systems *Permanent systems* are installed with no expectation of moving them. In certain permanent systems the laterals are mounted on supports high enough above ground to allow movement of workers and equipment underneath. These were most widely used with oscillating, perforated steel pipes as the delivery mechanism. Most were installed prior to the mid-1940s, when the availability of lightweight aluminum pipe led to great advances in sprinkler irrigation technology.

Today, most permanent systems are installed with both the main line and laterals buried underground and only the risers and sprinklers above ground. The main lines and laterals are often made of PVC plastic (polyvinyl chloride), although some steel pipe is also used. These systems are often used for high-value crops such as orchards, vineyards, and nurseries. Other applications include lawns, golf courses, and landscapes where convenience, appearance, and low maintenance become more important than cost.

Some permanent installations are designed to irrigate only one section at a time rather than the entire area simultaneously. This enables the owner to utilize a smaller pump as well as a water source with lower rates of delivery. A similar situation exists if the system is to serve for both irrigation and "air conditioning" (see Chapter 7). For air conditioning, the sprinklers are often cycled so that each area receives 15 sec of irrigation per minute. Thus, only one-fourth of the sprinklers are operating at any one time. For freeze protection (see Chapter 7), the entire area to be protected must be irrigated continuously. This use, therefore, requires larger pumps and greater water delivery rates than use for irrigation only or irrigation plus air conditioning. Assuming an adequate water supply, however, the cost of additional pump capacity would be a small price to pay for the potential benefits of freeze protection. With tall crops, such as fruit trees, it is preferable to utilize permanently installed risers with the sprinklers above tree level because of the greater stability of permanent risers. Usually the risers on movable systems are shorter and deliver water at or below tree level, and so are not suitable for either air conditioning or freeze protection.

Major disadvantages of permanent systems are the very high initial cost and lack of flexibility; the chief advantage is the minimal labor requirement.

Portable pipe The availability of aluminum pipe with "quick-coupler" connections gave tremendous impetus to sprinkler irrigation. In many situations a farmer could not justify the expense of a permanent installation, but the use of a *portable pipe* system is practical. Assume that a farmer has 50 ha of land under cultivation and needs the potential to irrigate once every 10 days for 8 h. If only one setting is used per day, a system covering 5 ha is needed. If it is possible to irrigate two areas per day, only 2.5 ha would have to be done at one time but two moves per day would be required.

FIGURE 8-12 A solid set sprinkler system in operation in a potato field. (Photograph courtesy of Wade Rain, Portland, Ore.)

With the conventional portable pipe system, the laterals are spaced from 12 to 27 m apart, with spacing based on sprinkler size (Figure 8-12). Rotary sprinklers are driven by a combination of water pressure and a spring-operated lever (Figure 8-13). The force of the stream of water deflects the lever, which is rapidly returned by tension

FIGURE 8-13 Sprinkler nozzle which rotates 360° by means of a combination of water pressure and spring tension.

on the spring. Each deflection rotates the sprinkler slightly so that it gradually makes a complete revolution.

The major drawback of this system is the high labor requirement. Although aluminum pipe is relatively light, a great deal of hard work is involved in each move. The increasing cost of labor has led to the development of the systems described next, which have much lower labor requirements.

Self-propelled Self-propelled systems are available today in various types. One is the *power roll* system in which the lateral pipe is mounted as the axle in large, lightweight wheels spaced at 18 to 30 m (Figure 8-14). Sprinklers are mounted at appropriate intervals on the pipe and operate when the system is stationary and the sprinklers are upright. Either the main line is of flexible pipe or a coupling is made at proper places in a portable main line of hand-moved aluminum pipe. The system is "driven" to each new setting by a small air-cooled gasoline engine, mounted in the center, or in some cases at one end with a drive shaft going to the center, where the drive wheel is located. Many of the

FIGURE 8-14 A power-roll system with drive wheels in foreground. (Photograph courtesy of Wade Rain, Portland, Ore.)

FIGUER 8-15 Central pivot systems rotate around a central source of water, usually a well. (Photograph, Valmont Industries, Inc., Valley, Neb.)

power systems are up to 1500 m in length, and they may be considerably longer.

The *center-pivot* system is designed to operate with minimal labor, as it is completely mechanized. The pipe (usually steel) and sprinklers are mounted on A-frames and wheels or skids, and the entire system rotates around a swivel joint at the pivot point—often a deep well (Figure 8-15). The couplings are strong but flexible enough to allow for variation in terrain among the A-frame towers. It is fascinating to fly over an arid or semiarid region and see perfectly round, green sections in the otherwise brown landscape (Figure 8-16). These areas may range from 8 to 80 ha or more, depending on the length of the lateral pipe, which often is from 150 to 450 m long. The A-frames are about 30 m apart, and cables or trusses are used to support the pipe between the A-frames. Because each tower must travel at a different speed to keep the lateral pipe straight, each has a separate drive mechanism which can be adjusted individually. As distance from the center increases, the area covered by each section of the lateral pipe increases, so that the sprinkler size must be increased or the spacing between sprinklers decreased to ensure uniform water distribution. Some center-pivot systems are also de-

FIGURE 8-16 The fields irrigated by center-pivot systems are usually round. (Photograph. Valmont Industries, Inc., Valley, Neb.)

signed so that they can be towed from field to field by a tractor. To do this, the wheels are rotated 90° to lie parallel with the pipe, or, if skids are used, special wheels can be added. In addition to the fact that this system economizes on labor, the lateral pipe can be high enough to clear a crop up to 2.5 to 3 m in height. The power-roll type normally is designed to clear only about 1.2 m. One disadvantage is that the center-pivot system will not effectively irrigate corners of a square field. An attempt to alleviate this problem is to have large sprinklers at the end of the lateral pipe which operate only when the pipe is in a corner.

Perhaps a better approach for square fields is the use of *self-propelled side-move systems* with multiple sprinklers. These operate similarly to the center-pivot system in that each A-frame tower is powered but travels in a straight line rather than rotating around a center pivot. A water-winch is on one end, and water power from the main line drives the winch. A steel cable is solidly attached at one end of the field and is unwound at the start. As the sprinklers irrigate, the cable is gradually wound up by the winch, and the entire lateral pipe moves across the field. The main line is a flexible pipe which is gradually either straightened out or looped, depending on whether the lateral pipe is moving away from or toward the source of water. By turning the wheels on the tower parallel with the lateral pipe, one can move the lateral to the other side of the main line and irrigate the other side of the field.

Single-sprinkler or *gun* systems employ one large sprinkler rather

FIGURE 8-17 Traveling gun irrigation system which moves by a cable and winch. Water is supplied by flexible hose.

than multiple small sprinklers in the types described previously. At one setting, a gun sprinkler may irrigate from 0.4 to 2.4 ha depending on the sprinkler size and water pressure used. Although both hand-moved and tractor-pulled types are available, perhaps the most popular is the traveling-gun type which is self-propelled (Figure 8-17). The hand-moved and tractor-pulled models are moved from set to set in much the same manner as the portable pipe systems described earlier. The traveling gun is self-propelled and moves from one side of the field to the other while irrigating; this unit may be driven by water power or engine. A flexible hose is used to supply water much like that described for the self-propelled side-move type. With the main line in the center of the field, the traveling gun is started at one side with the supply line extended. It then moves to the center, looping the supply line, which is again extended as the gun moves to the opposite side of the field. The gun is moved over the next area, and the direction of travel is reversed.

Trickle, Drip, or Daily-Flow Irrigation Although it has received widespread publicity in recent years, the concept of trickle irrigation is not new. Automatic watering systems of this type have been used in greenhouses and nurseries for years. With container-grown plants, as in a greenhouse or nursery, the entire root system is brought to field capacity when water is applied to the container. Under field conditions, however, the entire soil mass is not irrigated. The basic goal with trickle irrigation is to add sufficient water to maintain about 25 percent of the root system at or near field capacity. Research has shown that if about 25 percent of the root system is adequately supplied with moisture, moisture stress throughout the plant can be greatly reduced. Somewhat less water is absorbed by the plant than if the entire root system had adequate water, but 25 percent of the root system can adequately meet the need for water for the entire plant. The amount of water lost by transpiration is slightly reduced, but the major savings in water are obtained by reducing evaporation from the surface of the soil, since only a limited amount of the surface is wet by trickle irrigation.

The basic components of a trickle irrigation system include the following: pump, pressure control valve, filter, pressure regulators, plastic main lines, headers, laterals, emitters, and flush valves. Various other components can be added, such as fertilizer metering equipment, flowmeters, pressure check points, electronic timers, and even soil moisture sensors.

In any particular application the system depends on the crop, plant spacing, soil types, evapotranspirational potential, and the manufacturer. All trickle systems, however, do have basic things in common. First, since they are operated daily for several hours on the same acreage, water application per hectare per hour is low, making pump capacity requirements per hectare low. Second, most trickle systems operate at pressures below 1 kg cm^{-2} (sprinkler systems are usually operated at 2.5 to 4.2 kg cm^{-2}).

Water is applied through outlets called *emitters*. Many types have been developed during the past 15 years, and improvements are being made continually. One of the problems with the early emitters was plugging by foreign particles. Improvements in emitters, filtering systems, and flushing valves have overcome some of the earlier problems, but clean water is a definite asset for trickle systems.

Pipes in trickle systems are made of black plastic because of its low cost, flexibility, and inhibitory effect on the growth of algae (algae would be a problem in clear plastic lines). Lateral lines are characteristically 1.2 cm in diameter and are fed by 5-cm main lines. The sizes of both can vary according to the number of laterals and the number of emitters per lateral.

Experiments in Michigan apple orchards indicate that a reasonable amount of water to add in July would be about 3600 L ha^{-1} per day. Since it is most effective to irrigate only about 12 h per day, an application rate of about 300 L ha^{-1} h^{-1} would supply sufficient water. Since 300 L h^{-1} is equivalent to only 5 L min^{-1}, a well with a capacity of only 30 L min^{-1} could irrigate 6 ha in 12 h, or if run for 24 h could irrigate 12 ha (6 during the day and another 6 at night). For larger acreages, bigger wells or other sources of water would be required. In any case, however, water demand per day is low compared with that of other irrigation systems, where large quantities of water are added at each irrigation.

A daily application rate of 3600 L ha^{-1} is less than 0.25 cm per day. Sprinkler systems range in rates of water application from 0.25 to 0.5 cm h^{-1}, indicating the much greater need for water per hour to operate a sprinkler system. It is therefore practical to consider the installation of trickle irrigation with wells which cannot deliver adequate water for sprinkler systems.

With large orchard trees 1 or 2 emitters are installed at each tree so that spacing between emitters may be from 3 to 6 m. At 6-m spacings, emitters might be designed to give 7.6 L h^{-1}; at 3 m, only 3.8 L h^{-1}. With row crops, perforated pipes are often used because of the great number of emitters which would be required.

Elevation differences in an area make trickle irrigation somewhat more complex. Owing to the relatively low pressure being used, small differences in elevation can affect the output of the emitter. This is usually overcome by adjusting the emitter to negate the effect. The adjustment system varies with the particular type of emitter used.

In some installations the laterals are buried to avoid problems of traffic destroying the lines as well as potential damage by rodents. In row crops, the entire system is removed at the end of the season; therefore, it is installed on the surface. In either case the emitters are usually above ground, although certain systems place the emitters underground. The potential problem here is that clogged emitters are harder to locate and to clean if necessary.

The potential water saving of the trickle system compared with other systems ranges from 20 to 50 percent. This saving is based largely on reduced surface evaporation. When water is added at a point source, such as an emitter, the area of wetted soil is balloon-shaped. Thus, the diameter of the soil wetted in the root zone is larger than that on the surface. The wetting pattern varies with soil type; in a sandy soil the pattern is more vertical, whereas in a loam or clay soil a more horizontal movement is common.

As technology advances, it appears that trickle irrigation will be more widely used. The advantages are many, particularly in regions where lack of plentiful and inexpensive water has been a major limiting factor in the installation of previously available irrigation systems.

Subsurface Irrigation The reason for subsurface irrigation is to elevate the water table to within 0.3 to 0.6 m of the root system of the crop. Water then moves upward by capillarity into the root zone. This system is much more limited in application than others because, for it to be considered practical, certain ideal conditions must be present. Subsurface irrigation is based on artificially raising the water table; to do this the water table must already be relatively near the surface, or an impervious layer must be present which will restrict downward movement of water. Because of these restrictions, only those sites which require drainage of excess water in the wet season are suitable for subsurface irrigation in the dry season. Fortunately, the system developed can serve both purposes: the drainage of excess water and the addition of needed water.

There are only two basic types. The first consists of digging open-ditch laterals which are connected to a head or main ditch. The second involves subsurface laterals of perforated pipe, jointed concrete, or drain tiles which are in turn connected to a main line. The desired level of water is maintained by pumping water to the desired level maintained by dams, or pumping excess water out.

BIBLIOGRAPHY

Black, J.D.F.: *Daily Flow Irrigation*, Department of Agriculture (Victoria, Australia), Leaflet H 191, 23 pp., 1971.

Bidwell, R.G.S.: *Plant Physiology*, Macmillan, New York, 1974.

Byers, H. R.: *General Meteorology*, 4th ed., McGraw-Hill, New York, 1974.

Daubenmire, R. F.:*Plants and Environment*, 3d ed., Wiley, New York, 1974.

Devlin, R. M.: *Plant Physiology*, 3d ed., Van Nostrand, New York, 1975.

Donn, W. L.:*Meteorology*, 4th ed., McGraw-Hill, New York, 1975.

Gardner, V. R.: *Principles of Horticultural Production*, Michigan State University Press, East Lansing, 1966.

Kenworthy, A. L.: *Trickle Irrigation—the concept and guidelines for use*, Michigan State University Research Report 165, 19 pp., 1972.

Kramer, P. J.: *Plant and Soil Water Relationships*, McGraw-Hill, New York, 1969.

Mather, J. R.: *Climatology: Fundamentals and Applications*, McGraw-Hill, New York, 1974.

Meyer, B. S., D. B. Anderson, R. H. Bohning, and D. G. Fratianne: *Introduction to Plant Physiology*, Van Nostrand, New York, 1973.

Miller, A., and J. C. Thompson: *Elements of Meteorology*, Merrill, Columbus, Ohio, 1970.

Neuberger, H., and J. Cahir: *Principles of Climatology*, Holt, New York, 1969.

Treshow, M.: *Environment and Plant Response*, McGraw-Hill, New York, 1970.

Trewartha, G. T.: *An Introduction to Climate*, 4th ed., McGraw-Hill, New York, 1968.

Trewartha, G. T., A. H. Robinson, and E. H. Hammond: *Fundamentals of Physical Geography*, 2d ed., McGraw-Hill, New York, 1968.

Turner, J. H., and C. L. Anderson: *1971, Planning for an Irrigation System*. American Association for Vocational Instructional Materials, Engineering Center, Athens, Georgia.

U. S. Dept. of Agriculture: "Water," *Yearbook of Agriculture*, 1955.

CHAPTER 9
LIGHT
RELATIONS

Radiation received from the sun is the source of essentially all the energy on earth. By the process of photosynthesis, green plants convert radiant forms of energy to chemical forms which can then be utilized by nonphotosynthetic organisms. The basic plant processes are described in Chapter 6 with emphasis on photosynthesis, which is directly driven by radiant energy. Many other effects of the light environment on plants will be described in this chapter. These include effects on seed germination, vegetative growth, flowering, and plant morphology.

RADIANT ENERGY

Before discussing the effects of light on plants, we should clarify our terminology and background information relating to radiant energy. *Radiation* is one of the forms in which energy is transferred and in a vacuum travels at the rate of 3×10^{10} cm sec^{-1}. Radiant energy behaves partially as if it were transmitted in waves and partially as if it were in discrete particles. The wave characteristic is used in classifying radiation on the basis of *wavelength* (the distance between successive crests or troughs) and *frequency* (the number of wave crests passing a point in a particular time span). The wavelength is inversely proportional to the frequency. The particulate nature of radiation is described in units of energy called light *quanta* or *photons*. This description fits the way in which light behaves when it is absorbed by matter.

Spectrum of Radiation

Radiant energy is depicted in Figure 9-1 below in what is called the *electromagnetic spectrum*, which ranges from the shortwave, high-frequency, cosmic rays to the longwave, low-frequency, radio waves.

Visible light, or the portion of the spectrum to which the human eye responds, extends from about 380 to about 760 nm. The response curve varies somewhat among individuals, but the standard visual sensitivity, as adopted by the International Commission on Illumination (C.I.E.), is given in Figure 9-2. Sensitivity peaks at 555 nm and drops off rapidly at both longer and shorter wavelengths. The visible portion of the spectrum also encompasses the wavelengths which are most active in photosynthesis and in the many effects of light on plant growth and development.

Of utmost importance to life on earth is the "atmospheric window," which allows the relatively unrestricted penetration of visible light

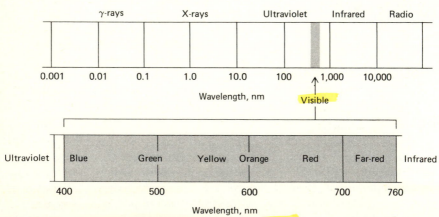

FIGURE 9-1 Electromagnetic spectrum of solar radiation. (From H. Smith, *Phytochrome and Photomorphogenesis*, McGraw-Hill, London, 1975. Copyright © 1975 McGraw-Hill Book Co., UK, Ltd. Reproduced by permission.)

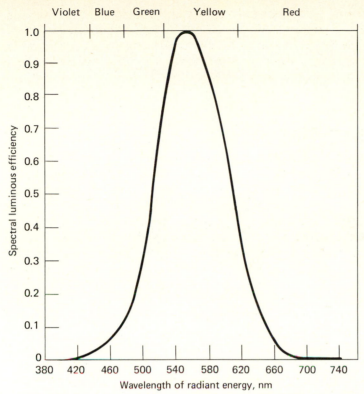

FIGURE 9-2 Standard (CIE) spectral luminous efficiency curve (for photopic vision), showing the relative capacity of radiant energy of various wavelengths to produce visual sensation. [From J. E. Kaufman (ed.), *IES Lighting Handbook*, 5th ed., Illuminating Engineering Society of North America, New York, 1972.]

through the earth's atmosphere (Figure 9-3). Ozone, CO_2, and water vapor are effective in reducing the penetration of specific wavelengths but freely transmit the visible region of the spectrum. Of the total radiation reaching the surface of the earth, the proportion which is in the visible wavelengths is about 50 percent. It might be considered coincidental that about one-half of the solar energy received can be used by various biological systems such as the vision of animals and photosynthesis of plants. It seems, however, much more plausible to speculate that through the process of natural selection, organisms have evolved which can utilize this available range of the total spectrum.

It should also be noted that the spectrum of radiation emitted by

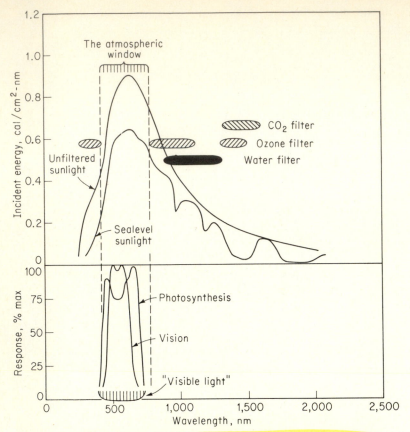

FIGURE 9-3 The radiation received by the earth from the sun is most abundant in the 400-to-700-nm range, which coincides with the wavelengths that are used in photosynthesis of plants and in vision in animals. (From A. C. Leopold and P. E. Kriedemann, *Plant Growth and Development*, 2d ed., McGraw-Hill, New York, 1975.)

the earth is very different from that received from the sun. The so-called "terrestrial radiation spectrum" is entirely in the infrared region, ranging from about 4000 to 100,000 nm. This major difference between the solar and terrestrial spectra is directly related to the difference in temperature of the radiating surface. As was noted in Chapter 7, the "greenhouse effect" of the atmosphere is due to the dissimilarity in the emission spectra of the earth and the sun and to the interaction of spectral quality and absorption, scattering, and reflection by the atmosphere.

Terminology

The level of radiant energy is not only measured variously but also expressed in various ways; the resulting confusion is apparent in many areas of endeavor today. One of the primary reasons for this unfortunate state of affairs is that many disciplines are interested in different parts of the electromagnetic spectrum. Perhaps the majority are concerned only with the visible portion of the spectrum as related to such everyday activities as conventional photography and the lighting of homes, offices, shopping areas, and streets.

We are all probably familiar with the concept of a "light meter" in determining proper exposures for photography. Although this kind of light meter measures visible light, it does not provide readings in specific units; it merely indicates the proper *f*-stops and shutter speeds. For lighting engineering, more refined light meters have been developed, and the most widely used unit of measurement is the *footcandle* (fc).

Originally, the footcandle was defined as the illuminance from a standard candle received on a 1-ft² surface that is everywhere 1 ft from the candle. This has been replaced with a definition which relates it to a more reproducible standard and is also defined as 1 lumen per square foot. The footcandle is being replaced by the metric unit *lux*, which is defined as illuminance of 1 lumen per square meter; thus 1 fc equals 10.76 lux.

To put the units footcandle and lux into perspective, a few examples are helpful. Full sunlight at noon in summer is normally about 10,000 fc or 108,000 lux; full moonlight is about $\frac{1}{50}$ fc; for reading about 20 fc is considered adequate.

It should be noted that footcandles and lux are units of illuminance, not intensity. *Intensity* refers to the *source* of the light and is expressed in candelas.

The footcandle and lux have been widely adopted by researchers in the plant sciences, presumably because of availability and low cost of the instruments. Unfortunately, the use of footcandle meters has caused considerable difficulty in interpreting data where sources of artificial light have been used. The footcandle meter is designed with a response curve which approximates that of the human eye (Figure 9-2), whereas the response curves for plant processes are very different (see Figure 6-2). Thus for data to be meaningful, a researcher should provide not only the illuminance but the spectral emission of the specific light source. Typical fluorescent-lamp spectra are presented in Figure 9-4, and these can be compared with the very different spectrum of an incandescent lamp (Figure 9-5).

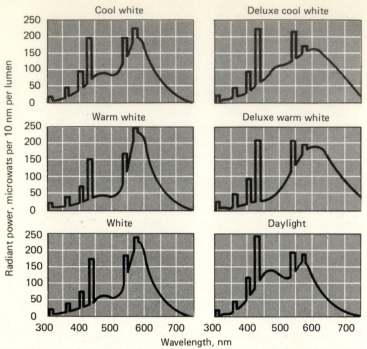

FIGURE 9-4 Spectral energy distribution curves for typical "white" fluorescent lamps. [From J. E. Kaufman (ed.), *IES Lighting Handbook*, 5th ed., Illuminating Engineering Society of North America, New York, 1972.]

FIGURE 9-5 Spectral energy distribution for incandescent lamps, including tungsten-halogen. [From J. E. Kaufman (ed.), *IES Lighting Handbook*, 5th ed., Illuminating Engineering Society of North America, New York, 1972.]

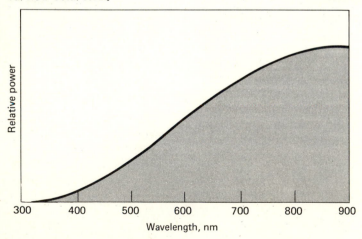

Instruments

Those instruments which are designed to measure visible light are termed *photometers* or simply *light meters*. These are standardized to respond much like the human eye, with peak responsiveness at about 550 nm. Therefore, since they are designed to match human vision, they are not ideal for use in plant studies. For very refined research, such as with the vitally important pigment phytochrome described later in this chapter, instruments called *spectroradiometers* are used. These instruments can measure the amount of radiant energy received at various wavelengths so that the radiation can be described in terms of quantity and quality. For example, if one is concerned with the light climate within the canopy of a plant as opposed to light in the open, a portable spectroradiometer could be used.

A *radiometer* measures not only visible light but also radiant energy of both shorter and longer wavelengths. Black surfaces absorb essentially all radiation which strikes them, whereas white or silvered surfaces reflect almost all the incoming radiation. Thus, the temperature difference between a black surface and a white surface would be a good measure of radiation received. Since infrared radiation has very strong heating effects, this part of the spectrum must be filtered out if only visible light is to be measured. Instruments employing this principle are the *pyrheliometer, pyranometer,* and *bolometer.*

The trend in agricultural research today is to express radiant energy in terms of photosynthetically active radiation (PAR). A quantum sensor is used, and the measurements are expressed in microeinsteins per square meter per second ($\mu Em^{-2}sec^{-1}$) in the wavelength range between 400 and 700 nm.

PHYTOCHROME

The major photoreceptive pigment involved in photomorphogenesis is *phytochrome.* This chromoprotein is widely present in plants and was initially isolated in 1959, although its presence had been proposed much earlier.

Phytochrome has two forms, which have different absorption peaks and are photoreversible. The red-absorbing form (Pr) has an absorption peak at 660 nm and is readily converted to the far-red-absorbing form (Pfr). In at least some plants, a reversion of Pfr to the Pr form takes place in the dark. This is much slower, however, than the conversions which occur in response to specific wavelengths of light. The Pfr form has its absorption peak at 730 nm and is converted to the

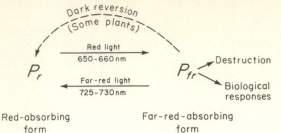

FIGURE 9-6 The red-absorbing form of phytochrome (Pr) is converted by red light to the far-red-absorbing form (Pfr). Pfr may be used to induce a biological response, may be destroyed, or may be slowly reconverted to the Pr form in darkness. Far-red light converts the Pfr form back to the Pr form. (From A. C. Leopold and P. E. Kriedemann, *Plant Growth and Development*, 2d ed., McGraw-Hill, New York, 1975.)

Pr form upon absorption of far-red radiant energy (see Figure 9-6, above). The active form is Pfr; the Pr form is not physiologically active.

Daylight (sunlight) tends to convert phytochrome from the Pr to the Pfr form, although the ratio may be approximately 50:50. The effect of artificial lighting depends on the spectral energy distribution of the source. Fluorescent lamps are quite deficient in the far-red wavelengths (see Figure 9-4 on page 260). For this reason, when fluorescent lighting is used for growing plants it is usually supplemented with some incandescent lamps to provide a better balance of wavelengths.

There is evidence that phytochrome occurs very widely in plants; in fact, it is thought by some to be universal in its occurrence. The distribution of phytochrome within a plant is very broad, with its presence having been confirmed in roots, stems, leaves, flowers, fruits, and seeds.

A most intriguing question relating to phytochrome is its mode of action. Research into its mode of action has paralleled work on pigment identification since the beginning.

Although not clearly shown as yet, there are theories relating to the probable effects of phytochrome on hormone systems, membrane permeability, and enzymatic systems. It seems that with the rapid progress made in our understanding of phytochrome and the major importance of this area of research, considerable information will be forthcoming over the next few years.

PHOTOPERIODISM

Photoperiodism can be defined as the developmental responses of plants to the relative lengths of the light and dark periods. It should be emphasized here that photoperiodic effects relate directly to the *timing* of the light and dark periods, whereas the total amount of light energy is unimportant as long as it exceeds the very low minimum level required.

Responses to Photoperiodism

The responses of plants to photoperiodic stimuli are numerous and very diverse. Some of the more important responses related to photoperiodism include: flowering, tuber and bulb formation, and bud dormancy.

Flowering The striking effect of photoperiod on flowering was discovered in 1920 by Garner and Allard while working with a mutant of tobacco called 'Maryland Mammoth'. They found that in the summer, this cultivar grew to a height of 3 to 5 m and did not flower in the Washington, D.C., area. During the winter months in the greenhouse, however, the same cultivar grew to a height of only 1 m, flowered, and set seed. They hypothesized that the response was due to daylength, and this was confirmed by experiment. Subsequent experiments by many researchers have shown that the effect of photoperiod on reproductive development is widespread among plant species. It also became apparent that not all plants respond the same, but that in fact there are three distinct groups: short-day, long-day, and day-neutral plants. Further description of the three groups follows:

1. *Short-day plants* will be induced to flower only when the daylength is below a certain critical length. If the critical daylength is exceeded, the plant remains in a vegetative state. The critical daylength varies among species and cultivars. Examples include chrysanthemum and poinsettia as well as 'Maryland Mammoth' tobacco and most cultivars of soybean.

2. *Long-day plants* will be induced to flower only when the critical daylength is exceeded. Until the daylength exceeds the critical length, vegetative growth continues. As with the short-day plants, the critical daylength varies both among and within species. Examples are spinach, beet, and radish.

3. *Day-neutral plants* will flower under any daylength and are exemplified by plants such as tomato, dandelion, rose, and African violet.

Unfortunately, there is considerable confusion about photoperiodism, and much of it is the result of inaccurate terminology. The short-day plants are not named because they require a daylength which is necessarily shorter than a long-day plant. It is rather because for a short-day plant to flower, the photoperiod must be *shorter* than the critical daylength. The long-day plant must have a photoperiod *longer* than the critical daylength. A widely studied short-day plant is cocklebur (*Xanthium*), which has a critical photoperiod of $15^1/_2$ h; as long as the daylength is less than $15^1/_2$ h, it will flower. A long-day plant which has been widely studied is henbane (*Hyoscyamus*); it has a critical photoperiod of 11 h, which must be exceeded for flowering to occur. It is apparent that at a daylength of 13 h both the short-day plant cocklebur and the long-day plant henbane will flower.

A further breakdown of the groups can be made on the basis of whether the photoperiodic requirement is qualitative or quantitative. Those which *must* have short days to flower have a *qualitative* requirement and are called *obligate* short-day plants. Others, in which short days will accelerate flowering but can be substituted for by temperature or other manipulation, are termed *quantitative* short-day plants. Similar types of obligate and quantitative long-day plants also exist.

The occurrence of a flowering hormone has been hypothesized since the 1880s, but was strongly proposed in the mid 1930s. The term *florigen* was proposed by the Russian plant physiologist Chailakhyan in 1936 and has been widely used since that time. In spite of intensive research efforts, the existence of florigen is still unproven, although the concept of a flower-inducing hormone has widespread acceptance.

Tuber and Bulb Formation   The induction of tuber and bulb formation has several similarities to the onset of flowering as previously described. The photoperiodic stimulus is received by the leaves and is apparently transmitted to the part which actually enlarges to develop into the storage organ, such as tubers (in the Irish potato and Jerusalem artichoke) and bulbs (in the onion). The mechanism of development of these storage organs is not well understood, and since we are concerned primarily with induction rather than development, this area will not be explored further here.

The formation of potato tubers is affected by photoperiod, but the degree of control is dependent on the cultivar and temperature. Some potato cultivars form tubers regardless of photoperiod but do so more readily under short days. Certain cultivars may form tubers under short days only. The stimulus has been shown to be perceived by the leaves and to move in a basipetal direction with little lateral movement. A particularly interesting observation is that although the tuberization response to photoperiod would make us classify the potato as a short-day plant, the flowering of the potato puts it in the long-day category. The acceleration of tuber formation in the potato by short days is nullified by a night interruption, another similarity to the flowering phenomenon described previously. Other similarities include a quantitative response to the number of short days and a gradual reversion to the noninduced state.

Bulb formation in the onion is encouraged by long photoperiods, the critical photoperiod ranging from 12 to 16 h, depending on the cultivar.

Bud Dormancy   The onset of bud dormancy in many species is, to at least a large degree, a response to the shortening days of late summer and fall. In response to the lengthening days of spring, dormancy is broken and a new season's growth begins. The initiation of growth in the spring is also associated with the advent of rising temperatures. Since buds will not break dormancy in the short days of winter, the plant is provided with protection against premature growth, such as might occur in response to a midwinter warm spell. In many deciduous species the initiation of growth in the late fall and early winter is prevented by the requirement for a minimum amount of hours of low temperatures to satisfy the chilling requirement. In the spring, however, no leaves are present, and the lengthening days are perceived by the buds themselves.

Lateral (axillary) bud formation usually occurs in the early summer on the newly developed shoots of temperate woody plants. In many species these buds do not grow during the season in which they are formed unless an unusual situation arises. Examples of factors which can cause buds to grow in the season of formation include defoliation and severe pruning. Under normal conditions these buds do not grow and by late summer have gone into a state of physiological dormancy or rest, at least partially because of the shortening days. Cooler temperatures which occur in late summer and early fall also contribute to the onset of bud dormancy. The photoperiodic cue leading to bud dormancy is thought to be received by the leaves in most plants. It thus seems plausible that some substance is formed in the leaves and is translocated to the buds—much like the flower-inducing substances involved in the flowering response. In most species, the onset of bud dormancy can be prevented or delayed by either an extension of daylength or a night interruption. Indications are that these reactions are the result of phytochrome, but this has not been conclusively proven as yet.

Most woody species of the temperate zone have a rest period or chilling requirement (see Chapter 7). In order for bud break to occur normally, a lengthy period of cool temperatures must occur. The optimum temperature for satisfying the chilling requirement is about 7°C and the number of hours required ranges from a few hundred to more than 1000. The onset of dormancy in late summer is due to a signal perceived by the leaves as the photoperiod decreases. The chilling requirement for buds to come out of dormancy is a characteristic of individual buds; each branch and even each bud is independent of others. It seems that there is no translocation of a substance involved in the exit from the dormant state.

Further Considerations on Photoperiodism

Night Interruption   When the photoperiodic phenomenon was initially discovered, the assumption was made that the critical aspect was length of the day. Subsequent research soon showed, however, that the length of the dark period was the determining factor in whether or not a plant flowered. An interruption of the night with a few minutes of light will prevent a short-day (long-night) plant from flowering and promote flowering of long-day ones. An interruption of the day with a few minutes of darkness has no effect at all. The night interruption has considerable horticultural significance. It is much more energy-efficient to break the night up by a short night interruption than to extend the daylength by artificial lighting in either the morning or the evening.

Another interesting aspect of the night interruption is shown in Figure 9-7. The red–far red phenomenon exhibited by phytochrome is apparent in the night-interruption data. A few minutes of red light prevents flowering; an exposure to far-red light does not. This response to either the red or the far-red light is repeatedly reversible.

FIGURE 9-7   The influence of a short exposure of red (R) or far-red (FR) light or of sequences of R and FR during the inductive night of a short-day plant.

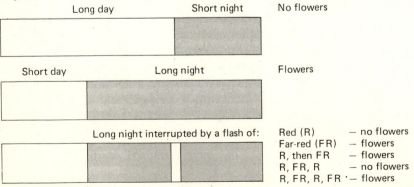

Long day	Short night	No flowers

Short day	Long night	Flowers

Long night interrupted by a flash of:

Red (R)	— no flowers
Far-red (FR)	— flowers
R, then FR	— flowers
R, FR, R	— no flowers
R, FR, R, FR	— flowers

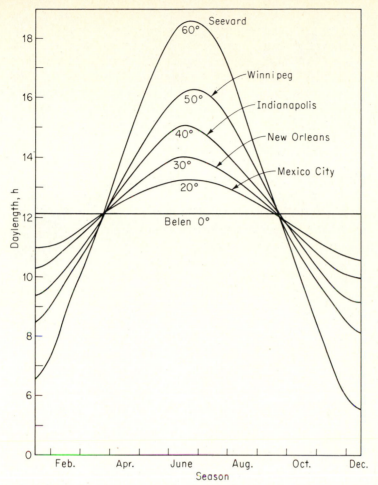

FIGURE 9-8 As latitude increases, the annual variation in daylength increases dramatically. (From A. C. Leopold and P. E. Kriedemann, *Plant Growth and Development*, 2d ed., McGraw-Hill, New York, 1975.)

Effect of Latitude The very striking effects of latitude on photoperiod are shown in Figure 9-8. Since the daylength is approximately 12 h all year round at the equator, there is only a minimal photoperiodic cue to which plants can respond. Obviously, the further poleward one goes, the greater the variation, with the extremes being found at the poles. In the middle latitudes there is a sizable change in daylength, the general range being from 15 h of daylight in the summer months down to about 9 h in midwinter. The changes in daylength with seasons provide a dependable cue to plants. It must be remembered that plants respond not only to the photoperiod of a particular day or week but (equally

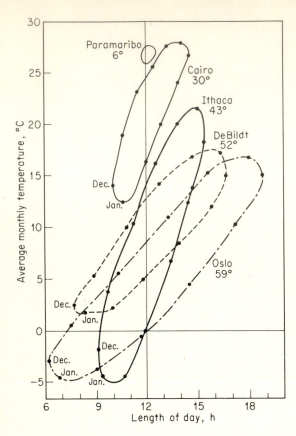

FIGURE 9-9 Seasonal changes in daylength and temperature can be plotted so that by connecting the 12 points the range becomes apparent. Paramaribo has minimal seasonal changes, whereas Oslo has extreme variation. (After J. H. A. Ferguson, *Euphytica* 6:97–105, 1957; reproduced from A. C. Leopold and P. E. Kriedemann, *Plant Growth and Development*, 2d ed., McGraw-Hill, New York, 1975.)

important) to whether the days are becoming progressively longer or shorter. The effect of photoperiod on a great many plant processes is a major factor in the natural distribution and adaptation of plants to various latitudes.

The interaction of temperature and photoperiod makes the situation somewhat more complex. The relationship between temperatures and photoperiod is depicted in Figure 9-9. It is interesting to contemplate the tremendous acclimation required of plants at Ithaca, New York, as opposed to Paramaribo, Surinam, as the seasons change. At least part of this ability to acclimate results from the cues received from photoperiod which cause the plant to initiate various responses.

Perception of the Photoperiodic Stimulus From many experiments it is well agreed that leaves are the organs which perceive the photoperiodic stimulus. This was originally established by J. C. Knott with spinach. With some plants the exposure of one leaf to the necessary photoperiod will cause the initiation of the flowering process. In other plants the degree of flowering increases with the proportion of the total leaf surface which is exposed to the proper daylength. Although the photoperiodic stimulus is received by the leaves, the initiation of flowering is at the apex, which in some cases may be at some sizable distance. Researchers have yet to isolate the flowering stimulus which is apparently formed in the leaves and translocated to the apex. It has, however, been quite well documented that the substance is translocated through the phloem.

SEED GERMINATION

The seed of certain species of plants require exposure to light before germination. This is a common phenomenon in weeds and offers such plants a competitive advantage in that they germinate only when near the soil surface or where the soil has been disturbed. A crop species which has been extensively studied is lettuce. If lettuce seed are allowed to absorb water in the dark, only a few will germinate. If the seed are exposed to red light (660 nm) after having imbibed water, seed germination will approach 100 percent. If, however, after exposure to red light the seed is exposed to far-red light (730 nm), the stimulatory effect of the earlier exposure is completely nullified. This effect can be reversed repeatedly, as was described earlier with the flowering response associated with phytochrome.

Although light stimulates the germination of lettuce seed, certain other species are inhibited by light. Examples are the American elm and some cucurbits. It should also be noted that many different kinds of seed are unresponsive to the presence or absence of light.

A further complication of the light effects is that there is a strong interaction with temperature. 'Grand Rapids' lettuce seed has been shown to require light at 20 to 30°C but will germinate in darkness at 10 to 20°C. At 35°C the seed will not germinate in either light or darkness. In some kinds of seed, alternating high and low temperatures can overcome a light requirement.

PHOTOTROPISM

When plants are irradiated unequally from two sides, they generally bend in the direction of the more intense light, a response termed *phototropism*. The bending of the plant causes the plant to become more evenly irradiated. Phototropism has held the interest of researchers for at least 100 years, but there are still unanswered questions. Much of the research to date has been done with the coleoptiles of grasses such as *Avena* (oat). The coleoptile is the leaf sheath—a tubular, leaflike structure, closed at the top, which emerges first from the soil. Early research showed that the receptive site on the coleoptile was the tip, which if removed made the coleoptile relatively unresponsive. It is now well accepted that the curvature is due to a higher concentration of auxin on the shaded side which leads to a much greater rate of cell elongation. In spite of considerable research, the exact mechanism leading to the differential auxin levels has yet to be agreed upon. The absorption spectrum for the perception of the phototropic stimulus indicates that the receptor pigment is probably either carotene or riboflavin and absorbs light in the blue part of the spectrum.

LIGHT RELATIONS IN THE CANOPY OF PERENNIAL PLANTS

The effects of the naturally occurring variations in light levels in the canopy of both annual and perennial plants have come under much more intense scrutiny in recent years. Among horticultural crops, the apple tree has been investigated most thoroughly, and some of the results of such investigation provide insight into the complexity of the light relations of plants.

Photosynthesis

It is well known that rates of photosynthesis increase with increasing irradiation levels until *light saturation* is reached. Above this irradiation, there is little or no response to increasing light. There is a direct relationship between the level of irradiation under which a leaf has developed and its light-response curve (Figure 9-10). In these data, leaves exposed to the sun were not light-saturated at 29 klx, whereas leaves grown in 80 percent shade (either continuous or intermittent) were light-saturated at 19.4 klx. Within the canopy of an apple tree there are various levels of irradiation from full sun on the periphery to very heavy levels of shade in the interior. Presumably, then, there is a very broad spectrum of photosynthetic potentials among the leaves on a given tree.

The following responses show the varied effects on leaves of decreasing light levels: thinner leaves due to fewer layers of palisade cells, lower dry weight per unit area, and lower rates of dark respiration. It is apparent that the so-called shade leaves adapt to shade but are likely much less productive than those leaves with better exposure to the available light.

The effects of shade within the canopy are also very apparent on fruit production and quality. Heavy shading can suppress flower bud formation, fruit set, and fruit size, presumably because of lowered leaf

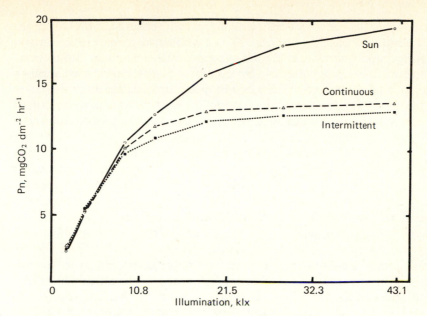

FIGURE 9-10 Net photosynthesis of apple leaves at various illuminations as influenced by 80 percent shade from continuous (shade cloth) or intermittent (slats) light. (From J. A. Barden, *J. Amer. Soc. Hort. Sci.*, **102**:391–394, 1977.)

efficiency. Direct effects on fruit color are also apparent. The side of an apple facing outward colors earlier and more intensely than the side facing inward. The red color of apples is due to the formation of anthocyanin pigments for which light is a necessity. The action spectrum for the coloration of apples indicates that phytochrome is involved.

The degree of shading which occurs within a tree canopy is partially under the control of the fruit grower. Control of tree size through selection of rootstock, interstock, and scion cultivar is effective in that the smaller the tree, the lower the proportion of the tree which is heavily shaded. Other techniques to improve the level of light in the interior of the trees include spreading of branches, adequate pruning during the dormant period, and, in particular, summer pruning.

Transpiration

The influence of level of light on transpiration is great and is both direct and indirect. It is well known that, with most plants, most transpiration takes place through the stomates. In most plants the stomates open in response to light and close in the dark, and so stomatal transpiration is directly dependent upon light. The exceptions to this rule are some of the succulents, such as cacti, which typically open stomates at night and close them during the day. The effectiveness of this characteristic in suppressing water loss by these succulents is obvious. The mechanism by which CO_2 is fixed is by necessity different from that in other plants and is called *crassulacean-type metabolism* because it is common in the family Crassulaceae as well as in other succulent plants. Most of the CO_2 fixation occurs in the dark and exists largely in the form of organic acids.

The indirect effect of light on transpiration is related to leaf temperature. The energy budget of a leaf is a complex interaction of several energy fluxes. A leaf receives energy from its surroundings as well as from solar radiation. They efficiently absorb photosynthetically active radiation but tend to transmit and reflect much of the infrared wavelengths which would have particularly strong heating effects. If it were not for this selective absorption, leaf temperature would probably reach lethal levels when exposed to intense solar radiation. Of the radiant energy absorbed, only small proportions are utilized in photosynthesis; much is converted to heat, and most is reradiated or is dissipated by conduction, convection, or transpirational cooling.

A leaf exposed to full midday sun may be 1 to 5°C warmer than air temperature, but this temperature differential depends on many factors, in particular the water status of the plant. If sufficient water is available to avoid moisture stress, heat dissipation by transpiration (heat of vaporization—see Chapter 7) is quite rapid, and leaf temperature may be near air temperature. If, however, the plant is under moisture stress and stomates close, leaf temperature may rise considerably above ambient temperature. Other major factors affecting leaf temperature include wind speed, orientation of the leaf, leaf shape (entire versus deeply dissected), and leaf thickness. It is apparent that leaf temperature is not constant but varies in response to a diversity of factors. Under artificial sources of light, spectral quality (see Figures 9-4 and 9-5) is a prime consideration.

Our understanding of plant responses to quantity, quality, and duration of light has improved dramatically in the last 25 years. The impact of research on photoperiod is particularly apparent in floriculture, where certain crops can be induced to flower as desired. As we better understand photosynthesis from both physiological and environmental aspects, we can, it is hoped, increase food production. With the rapidly rising costs of energy, increased efficiency in artificial lighting is an absolute necessity. For example, night interruption by a short period of artificial light can accomplish the same end result as several hours of daylight extension. Many areas remain imperfectly explored in which research is vitally needed if we are to understand and thus be able to manipulate plants for maximum benefit. Fortunately, much of this research is currently in progress. The next twenty years should be fully as exciting as the last twenty.

BIBLIOGRAPHY

Bickford, E. D., and S. Dunn: *Lighting for Plant Growth*, Kent State University Press, Kent, Ohio, 1972.

Canham, A. E. (ed.): *Symposium on Electricity and Artificial Light in Horticulture*, Technical Communication of the International Society for Horticultural Science, no. 22, The Hague, Netherlands, 1969.

Clayton, R. K.: *Light and Living Matter*, vol. 2, Biological Part, McGraw-Hill, New York, 1971.

Daubenmire, R. F.: *Plants and Environment*, 3d ed., Wiley, New York, 1974.

Kaufman, J. E. (ed.): *IES Lighting Handbook*, 5th ed., Illuminating Engineering Society, New York, 1972.

Leopold, A. C., and P. E. Kriedemann: *Plant Growth and Development*, 2d ed., McGraw-Hill, New York, 1975.

Reifsnyder, W. E., and H. W. Lull: *Radiant Energy in Relation to Forests*, U. S. Department of Agriculture Forest Service Tech. Bul. 1344, 1965.

Salisbury, F. B.: *The Flowering Process*, Permagon, London, 1963.

Smith, H.: *Phytochrome and Photomorphogenesis*, McGraw-Hill, Maidenhead, Berkshire, England, 1975.

Van der Veen, R., and G. Meijer: *Light and Plant Growth*, Phillips Technical Library, Eindhoven, Netherlands, 1959.

Vince-Prue, D.: *Photoperiodism in Plants*, McGraw-Hill, Maidenhead, Berkshire, England, 1975.

Zelitch, I.: *Photosynthesis, Photorespiration, and Plant Productivity*, Academic Press, New York, 1971.

CHAPTER 10
SOILS

A basic understanding of soils is of utmost importance to horticulturists. Crops can be grown hydroponically (roots growing in aerated water), but the problems and expense involved are prohibitive. Since we are still dependent on soil as a primary growing medium, our soils remain a vital natural resource.

Soil can be defined as the outer, weathered layer of the earth's crust which has the potential to support plant life. As a growing medium for plants, soil provides not only physical support, but water and nutrients as well. An aspect of soil which is often overlooked by the novice is that soil is a very variable, complex entity—both chemically and biologically—which undergoes continual change.

If a cut is made down through the soil, horizontal layers are usually apparent by differences in color, particle size, and general appearance (Figure 10-1). The layers or gradations from the surface soil to the bedrock are called *horizons*, and the whole group is called a soil *profile*. The *A horizon* or upper layer (often called *topsoil)* is often higher in organic matter and therefore usually darker in color than the layers below. The middle layer of the profile is called the *B horizon*; it usually contains less organic matter and more clay, is brighter in color than the A horizon, and is often referred to as the *subsoil*. The lowest layer in the profile is the *C horizon*; it extends from the lower limit of the B horizon down to the bedrock. This third layer is called the soil *parent*

FIGURE 10-1 A soil profile showing A, B, and C horizons. The lines between horizons are indicated by white dots. Note the relative similarity between the A and B horizons but the distinct difference between B and C. (Photograph, courtesy of William J. Edmonds.)

material, or the *substratum.* The gradations between horizons may be fairly distinct, or the layers may merge gradually with little or no line of demarcation. The degree with which the horizons can be distinguished is very dependent on the age of the soil, with older soils having more clearly defined horizons. Very often, distinct subhorizons are present which allow the more complete description of the soil profile. The total depth represented by the three horizons is often in the range of 1 m for temperate-zone soils, although it may range from a very few centimeters to many meters. The depth, type, and components of the profile are used in classifying soils into types and thus in predicting their value for crop production or other uses.

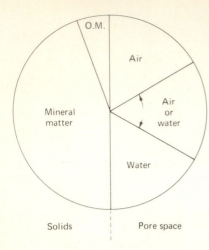

FIGURE 10-2 The relative volumes of solids, air, and water present in a typical A horizon. (From L. M. Thompson and F. R. Troeh, *Soils and Soil Fertility*, 4th ed., McGraw-Hill, New York, 1978.)

SOIL COMPONENTS

The four major components of a mineral soil are mineral particles, organic matter, water, and air. The mineral particles originate from the breakdown of parent material, and the organic matter comes from a broad spectrum of plant and animal life. These solid components are interspersed with openings of innumerable sizes and shapes which are called *pore spaces*. The pore spaces are occupied by water and air in proportions which vary widely. When water is added to a soil, the smaller pore spaces may remain full of water, whereas the larger ones tend to drain and thus fill with air. A "rule of thumb" for a silt loam soil is that it should have about 50 percent pore space which should be split about equally between air and water (Figure 10-2). It must be remembered, however, that the proportion of the pore space occupied by air and water is constantly changing as water is added or withdrawn from the soil.

Mineral Particles

Although it stands for one category, the term *mineral particle* includes an extremely broad spectrum of materials from both a physical and a chemical viewpoint. According to U.S.D.A. guidelines, mineral particles are classified as sand, silt, or clay as follows: *sand*, 0.05- to 2-mm diameter; *silt*, 0.002- to 0.05-mm diameter; and *clay*, less than 0.002-mm. Pieces from 2- to 76-mm diameter are *gravel*, and those over 76-mm diameter are *stones*. The larger particles, such as sand, are essentially inert from a chemical standpoint and are called *primary minerals*. The smaller particles, especially the clays, are much more weathered and are called *secondary minerals*. They are extremely active. Part of this activity relates to surface area, which is vast in clay particles because of their small size and plate-shaped structure. This aspect will be discussed further under cation exchange capacity.

Organic Matter

The organic fraction of a mineral topsoil constitutes from about 1 to 6 percent, but its importance far outweighs its proportion, as will be discussed later in this chapter. The organic component consists of both plant and animal residues which are in various stages of decomposition. Through the action of microorganisms and chemical reactions, decay is a continuous process which results in a dark-colored, amorphous organic material called *humus*. Its composition is not well documented, but it is quite resistant to further breakdown, although this does ultimately occur with the release of carbon dioxide, water, and various mineral components.

Water in the Soil

From the standpoint of crop production, the importance of water is obvious. There are very strong interactions of moisture in the soil with both the mineral particles and the organic matter. The moisture-holding capacity of a soil increases (1) as organic matter levels are increased and (2) as the size of mineral particles decreases. These two factors are also of prime importance in water infiltration and movement in soils. An important fact to keep in mind is that the water in soil is not pure water but is a solution containing a variety of dissolved substances, many of which are nutrients needed by plants. The composition of the soil solution is continually altered as water and chemical components are added or removed. Thus, the soil solution is an ever-changing constituent of the soil, which will be further described later in this chapter.

Air in the Soil

Roots must have oxygen in order to respire and function normally. Adequate aeration of the soil is therefore of prime importance in maintaining optimum growth and production. With sandy soils, the large pore spaces allow relatively free gas exchange between the air in the soil pore space and the atmosphere. In clay soils, however, poor aeration can be the result of a combination of very small pores and a higher proportion of the existing pore space being occupied by water. The normal concentrations of oxygen and carbon dioxide in the atmosphere are about 20 percent and 0.03 percent, respectively. Under conditions where gas exchange between the soil and atmosphere is limited, oxygen is depleted and carbon dioxide accumulates as the result of respiration by both roots and microorganisms. Both low oxygen and high carbon dioxide levels can be detrimental to roots.

Organisms in the Soil

In addition to the obvious living inhabitants such as plant roots and a wide variety of macroscopic animals, soils contain tremendous quantities of microscopic organisms, especially in the A horizon. Although not considered a component of soil, these living organisms are of utmost importance in the formation of soil, influencing soil properties, and in the degradation and synthesis of the organic fraction.

Animals The larger animals in the soil include rodents such as the woodchuck, prairie dog, ground squirrel, and mole. The activities of these animals are important in the movement, aeration, and drainage of soil, as well as in the incorporation of organic matter. Insects are numerous and are important in many ways. The serious effects of grubs, wireworms, and other insects of the soil will be covered in Chapter 14. Earthworms have long been recognized as major contributors to the fertility and productivity of soils. Tremendous quantities of soil are passed through the digestive tract of these animals, and not only is the organic matter partially broken down, but it is mixed with the soil. Certain elements are made more available by the digestive processes of the earthworm; most markedly affected is nitrogen. The holes dug by earthworms considerably improve drainage and aeration both of which are vitally important characteristics of a productive soil. Earthworms are most numerous in soils which are moist but well-drained and high in organic matter. In midsummer when soil becomes dry and warm, earthworms burrow to deeper levels in the soil and return to surface layers when moisture is replenished.

Other organisms of the soil are protozoa and nematodes. Although the single-celled protozoa are present in extremely high numbers and are represented by more than two hundred species, little is known about them and their possible interactions with plants. Of major consequence is the usually microscopic nematode, or eelworm as it is often called. Certain species of these nematodes are plant-parasitic and do great economic damage to crop plants; others are nonparasitic and live on organic matter; a third group is parasitic on animals of the soil.

Plants Many types of microscopic plants are present in the soil in great quantities. These include bacteria, fungi, actinomycetes, algae, and the roots of higher plants. Bacteria are single-celled plants which occur in vast numbers—often estimated to exceed 1 billion per gram of soil. Conservative estimates put the fresh weight of bacteria at 1100 kg per hectare-furrow-slice. Having fantastic reproduction rates, bacterial populations which have been decimated by adverse climatic conditions, such as drought, can rebuild numbers very rapidly. Bacteria of the soil are classified according to their source of energy, the two types being heterotrophic and autotrophic. *Heterotrophic* bacteria obtain energy from organic matter and thereby function in its decay. *Autotrophic* bacteria obtain energy by oxidizing inorganic substances; examples are those bacteria which oxidize ammonium ions to nitrite ions and those oxidizing nitrite ions to nitrate ions. These reactions are discussed further in the chapter on nutrition. Bacteria in the soil are also classified according to their requirements for oxygen from the air. Those requiring oxygen are *aerobic;* those not needing oxygen are *anaerobic;* and those growing with or without oxygen are *facultative.* The anaerobic and facultative bacteria actually require oxygen but obtain it from oxygen-containing compounds or ions such as sugars, NO_3 or SO_4. The importance of bacteria is difficult to overemphasize as their presence in the soil is of utmost importance. Without bacteria, neither nitrogen fixation nor nitrification would occur.

Fungi are also of major consequence in soils, as they are present in vast numbers (normally 10 to 20 million per gram of dry soil) and are very active in the breakdown of organic matter. The most important group of soil fungi are the molds, which are adapted to a wide range of soil pH and are thus abundant in acid soils where bacteria and actinomycetes are less well adapted and therefore present in lesser numbers. In addition to the beneficial effects of soil fungi on organic matter, certain species cause tremendous damage to crop plants. Major examples would include the damping-off fungi (*Pythium* sp) and the stem, crown, and root rot fungi (*Phytophthora* sp). These diseases will be discussed further in Chapter 14.

Actinomycetes are classified as being between the molds and bacteria, having some characteristics of each. They are thought to be active in the decomposition of organic matter and the subsequent release of nutrients. Actinomycetes are very seriously hindered in acid soils and decline drastically in soils of a pH of 5.0 or below. This characteristic is used to control the actinomycete which causes potato scab. Soils are acidified by adding sulfur to keep the pH at 5.3 or below.

Soil *algae* are usually present in large numbers, but since most are chlorophyll-containing cells, they tend to be concentrated in the upper few centimeters of soil where sunlight penetrates. Certain species survive on the decompositions of organic matter and can exist at greater depths. Algae populations are encouraged by high-moisture contents such as that found in rice paddies, and under conditions of intense sun, blue-green algae may fix significant quantities of nitrogen.

TABLE 10.1
USDA CLASSIFICATION OF SOIL PARTICLES

FRACTION	SOIL SEPARATE	SIZE, mm
Sand	Very coarse sand	2–1
	Coarse sand	1–0.5
	Medium sand	0.5–0.25
	Fine sand	0.25–0.10
	Very fine sand	0.10–0.05
Silt	Silt	0.05–0.002
Clay	Clay	Below 0.002

SOIL TEXTURE

One of the major physical properties of a soil, from the standpoint of crop production and classification, is its texture. The term *soil texture* refers to the percentage by weight of sand, silt, and clay-sized mineral particles. The so-called sand particles are further broken down into five smaller categories called *soil separates* (Table 10-1). The percentage of each component is determined by a procedure called a *mechanical analysis*, which actually involves two different types of procedures. The sand particles are separated by passing the dry soil through a series of sieves, each of which collects one fraction. The silt and clay fractions are separated by suspending these components in water and determining the rate at which the particles settle out. The particles settle out in proportion to their relative size, with the largest particles being the first to settle out. By pipetting a sample of the suspension periodically, one can determine the percentage of silt and clay. The graph given in Figure 10-3 will aid in the classification of the soils into one of the twelve categories. The three scales are divided into units from 0 to 100; locate the intersection of the lines representing each of the three components. Follow the line at the same angle as the number on the scale which you are using. Examples:

40 percent sand, 40 percent silt, 20 percent clay: loam
60 percent sand, 10 percent silt, 30 percent clay: sandy clay loam
10 percent sand, 60 percent silt, 30 percent clay: silty clay loam

An experienced soils specialist can learn much about a soil by its "feel." By assessing the plasticity and grittiness of a soil sample, an experienced person can rather accurately classify a soil sample into one of the categories given in Figure 10-3.

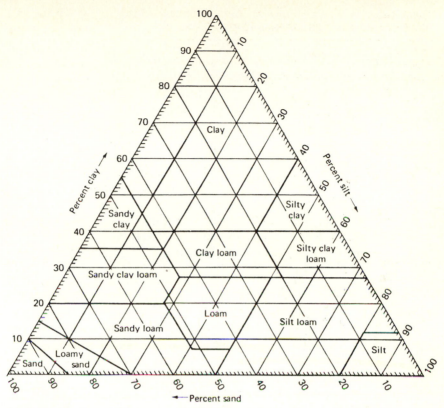

FIGURE 10-3 Guide for textural classification of soils. (From L. M. Thompson and F. R. Troeh, *Soils and Soil Fertility*, 4th ed., McGraw-Hill, New York, 1978.)

SOIL STRUCTURE

The texture of a soil is important, but by no means does it give us enough information to assess the potential productivity of a soil. In addition to the size distribution of the particles, the arrangement of the particles has major effects on the value of a soil. The manner in which the individual soil particles are grouped or arranged is called the *structure* of the soil. An example of a soil with no structure would be a sand, in which each particle is independent of all others. In soils with some smaller particles and organic matter present, at least limited amounts of clumping or grouping of individual particles occur. The grouping of particles is called *aggregation* and has profound effects on the soil. Soil aggregates may be from small to large; they may be weakly to strongly persistent, and they may occur in several distinct shapes. Whatever its character, aggregation is highly desirable, as it improves aeration, percolation, and root penetration. Since aggregates are larger than the individual particles, pore spaces among the aggregates are much larger. The porosity is more like that of a coarser-textured soil, while retaining, within the aggregates, the small pore spaces of a fine-textured soil. The major cementing agents which contribute to good soil aggregation include clay particles, partially decomposed organic matter, and certain microbes. Certain ions—such as hydrogen and calcium—promote aggregation, but sodium ions tend to cause breakup of aggregates.

Good soil structure is slow to develop, but by careless farming it can be very rapidly destroyed. A clay soil with good aggregation can be severely damaged by plowing, disking, or moving heavy equipment over it when it is very wet. The aggregates break apart and the result is a so-called *puddled soil* in which the pore spaces have been closed up and the condition of the soil very adversely affected. Upon drying, such a soil will be crusted or in lumps or clods which are very difficult to break up.

PORE SPACE

As mentioned previously, pore space is that volume of the soil which is not occupied by solid particles. Total pore space ranges between approximately 40 and 60 percent of the total soil volume. The actual percent porosity varies with both soil texture and soil structure. Coarse soils, such as sands or sandy loams, have less total pore space than a silt loam. Of major importance, however, is not only total pore space but the relative size of the individual pores. The small pores, such as are predominant in fine-textured soils, tend to fill with water and remain full of water; aeration is thus frequently limited. Good aggregation in a fine-textured soil improves aeration by providing larger pores which drain readily. Although sandy soils have less total pore space, the individual pores are larger, thus providing excellent aeration but limited water-holding capacity.

SOIL MOISTURE

Soil Moisture Tension

The moisture present in soils is held by the soil particles with a force referred to as *soil moisture tension*, which can be expressed in various units. One unit of expression used is the *atmosphere* (normal atmospheric pressure at sea level), 1.03 kg cm^{-2}. More recently the term *bar* is being used; 1 bar is equal to 1×10^6 dyn cm^{-2}. For most purposes the terms *bar* and *atmosphere* can be used interchangeably, since they are essentially the same.

The greater the amount of water present in a given soil, the lower the soil moisture tension. When a soil is saturated, the soil moisture tension is zero, but as the soil dries, soil moisture tension increases. It is important to keep in mind that, although soil moisture is often divided into categories, as a soil dries there is a gradual transition in soil moisture tension rather than any clear lines of demarcation. At a particular moisture tension, a clay soil holds more water than a loam and much more than a sandy soil. Also, at a particular moisture level, the water in a sandy soil is held much less tightly than a loam soil, but the loam soil holds water less strongly than a clay soil.

Soil Moisture Classification

From a physical standpoint, soil moisture is divided into four categories: chemically combined, hygroscopic, capillary, and gravitational. *Chemically combined* water is a component of the soil particles and as such is inactive and therefore can be essentially ignored for our purposes. *Hygroscopic* water exists as a very thin layer around soil particles and is held very tightly. At the upper limit of hygroscopic water, the soil moisture tension is 31 bar. *Capillary* water exists in films around and among soil particles at soil moisture tensions from about 1/3 to 31 bar. Since capillary water moves and is held by the phenomenon of capillarity, this portion of the soil water is greatly influenced by the size and arrangement of soil particles. Essentially all the water taken up by plant roots is capillary water. *Gravitational* water is water which is in excess of capillary water and is therefore held at soil moisture tensions of less than 1/3 bar. Gravitational water moves downward and is normally lost from the topsoil within 2 to 3 days after a rain or irrigation.

An additional set of terminology is often used in characterizing soil moisture. The *field capacity* of a soil is the amount of water which the soil can hold against the pull of gravity. After a soil has been saturated by rain or irrigation, moisture will continue to drain away for 24 to 48 h, after which the soil is at field capacity. The soil moisture tension at field capacity is 1/3 bar. Water will have left the larger pore spaces, but the smaller pores will remain full of water. Field capacity is considered to be the upper limit of *available water*, or that water which is available for use by plants. The amount of water present in a soil at field capacity is small in a coarse soil (sand) but is markedly higher in finer-textured soils (Figure 10-4).

As a soil dries, by surface evaporation or through uptake by plants, the remaining water is held by soil particles with increasing tension. If a plant is grown in a limited soil volume without adding

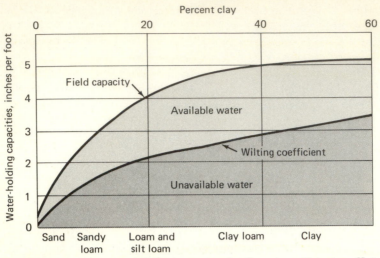

FIGURE 10-4 The water-holding capacity of soils varies with texture. Note that a clay holds much more available and unavailable water than a sandy loam. (From L. M. Thompson and F. R. Troeh, *Soils and Soil Fertility*, 3d ed., McGraw-Hill, New York, 1973.)

water periodically, the plant will eventually wilt during the day. The plant will recover turgidity each night until soil moisture reaches the *wilting coefficient*—a soil moisture level at which a plant cannot regain turgidity in a saturated atmosphere. (The wilting coefficient has also been called the *permanent wilting point*.) At the wilting coefficient, the soil moisture tension is 15 bar. Since most plants can extract little moisture at tensions above 15 bar, the wilting coefficient is considered to be the lower limit of available soil moisture (Figure 10-4). As is true at field capacity, the amount of water in a soil at its wilting coefficient is a function of soil texture.

It is apparent from Figure 10-4 that the available water which a soil can contain is markedly affected by texture. A clay loam soil holds more available and more unavailable water than a sandy soil. In fact, a clay loam holds more water at the wilting coefficient than a sandy loam does at field capacity.

The *hygroscopic coefficient* is the moisture level when the soil moisture tension is 31 bar. This moisture exists as a thin film on soil particles and is the amount which a soil can hold in equilibrium with a saturated atmosphere. Hygroscopic water is unavailable to plants.

Movement of Soil Water

Gravitational water moves downward because of the pull of gravity, which is greater than the tension with which the water is held by the soil particles. The rate of movement downward is very dependent on the size of the pores present. For example, excess water moves very rapidly through a sandy soil, but a clay loam soil may still contain gravitational water 2 to 3 days after a rain.

When a soil no longer contains gravitational water, movement is largely by *capillarity*. As can be readily seen with a small glass column dipped in water, the liquid rises in the tube. The height to which the water rises increases as the tube diameter decreases. The mechanism of capillarity is due to the attraction between water molecules and glass molecules. In soils, water movement is similar except that it moves among the innumerable shapes and sizes of soil particles.

Water movement by capillarity occurs in all directions but is most obvious in upward or horizontal directions. The process of capillarity is utilized in subsurface irrigation, in trench irrigation between raised beds (see Chapter 8), and in the watering of potted plants by adding water to the dish below. Capillary movement is relatively rapid at low soil moisture tensions, but slows dramatically as soil moisture tension rises. Although capillarity is very rapid in sandy soils, distances involved are small because of large pores. In heavy clay soils, although water may move great distances, its movement is very slow because of minute pore spaces. Therefore, capillarity is most important and useful in medium-textured soils.

In dry soils, water can move only in the vapor state. The direction and rate of movement are determined by the vapor-pressure gradient (see Chapter 8) between the areas involved.

SOIL MANAGEMENT AND CROPPING SYSTEMS

When we consider the diverse nature of horticultural crops grown and the spectrum of soils and climates involved, it soon becomes apparent that the number of *soil management systems* used is necessarily quite large. Included among the many different systems are the following examples: permanent sod, clean cultivation, both clear and black plastic mulches, a wide diversity of organic mulches, herbicide treatments, and cover crops. Each of these will be explained after the general objectives of a good soil management system are explored.

Objectives

There are five major objectives of a good soil management program: (1) to provide a favorable moisture supply, (2) to minimize soil erosion, (3) to supply necessary nutrients, (4) to minimize organic matter depletion, and (5) to maintain or improve soil structure. Some of these objectives overlap to some degree, but all are vitally important. Some are much easier to meet than others, but often the most critical are the most difficult to attain.

A favorable moisture supply is obviously necessary for optimum growth and production. Depending on the site, soil, climate, and crop, this may involve drainage and irrigation, and in many situations the selection of the best soil management system for the particular crop. Soil erosion is a major threat under many cropping systems, and the productive layer of topsoil which took thousands of years to develop can be dissipated in a matter of minutes, days, or years. With the ever-increasing pressure on our soils to produce sufficient food for the rapidly increasing population, every effort must be made to conserve this most essential natural resource. The supply of essential nutrients is obviously necessary, but this is usually the least difficult to achieve of the five objectives mentioned. The area of crop nutrition is described in Chapter 11. The importance of organic matter to soil structure and its effects on percolation rates, aeration, and plant growth have previously been described. Under some soil management systems, such as clean cultivation, organic matter levels decline. It is imperative that another system be used as a means of replenishing organic matter.

Types of Systems

The choice of the best system depends on many factors including rooting depth of the crop, slope, rainfall, and economic considerations.

Permanent Sod   One of the most widely practiced systems in apple and pear orchards is permanent sod. In the early years, some cultivation may be used; but in many cases, cultivation has been replaced by herbicides. Permanent sod is the best system for erosion control and for maintaining organic matter levels and soil structure. It is quite satisfactory overall for deep-rooted tree crops. Another advantage is reduced leaching of nutrients, since they are absorbed by the sod and are ultimately released. A permanent sod makes a very desirable surface for field work and provides a stable surface for the movement of equipment. At one time, clean cultivation was widely practiced in irrigated western orchards to eliminate transpiration by the ground cover. It became apparent that organic matter levels declined to the point where soil structure suffered, and so many such orchards are now in sod or at least cover-cropped for part of the year. It is common practice to supplement a permanent sod with band application of herbicide along the tree row (Figure 10-5). Not only does this procedure reduce competition between the tree and grass for water and nutrients, but, even more important, it eliminates the need for hand mowing beneath the trees.

FIGURE 10-5 Orchard with permanent sod in the middle of the row and herbicide-treated strips down each row.

Clean Cultivation   Probably at the opposite end of the spectrum from a permanent sod would be clean cultivation. This system entails the elimination of all competing vegetation by mechanical means. In addition to weed control and the associated moisture conservation, cultivation incorporates organic matter and may temporarily improve soil aeration. The long-term effects are rather deleterious because year-round clean cultivation leads to erosion, excess nutrient leaching, and a depletion of organic matter which in turn causes a decline in soil structure. With crops such as shallow-rooted vegetables, clean cultivation during the growing season has been a logical choice, but it is most desirable to supplement this with a cover crop or green-manure crop during the off season.

Clean Cultivation Plus Cover Crop   This system offers the advantages of clean cultivation and, to some degree, the advantages attributed to a permanent sod. By clean cultivation during the growing season, maximum productivity is attained, but the off season allows the cover crop to replenish organic matter and to reduce leaching and erosion. This is a markedly superior system to year-round clean cultivation.

Fallowing   Used mostly for wheat production, fallowing involves alternating the growing of the crop with leaving the land with no vegetation. In regions with marginal amounts of rainfall to produce a crop of winter wheat, many farmers plant one-half of their farm each year. The schedule is as follows:

First year: Plant winter wheat in fall.

Second year: Harvest wheat in midsummer to fall. After harvest, cultivate to kill weeds and to encourage stubble decomposition.

Third year: Use herbicides or cultivate occasionally to prevent weed growth. In the fall, plant wheat as in first year and repeat the cycle. The land has been fallow for 12 to 14 months.

This system of fallowing may conserve only about one-fourth of the year's rainfall, but in many cases this much additional water is reflected in considerably greater yields.

Minimum Tillage   The systems of minimum tillage consist of planting without the traditional soil preparation by plowing and disking. The concept is of recent origin but has received widespread adoption, particularly with corn. The basic idea is the complete killing of the sod, the cover crop, or other existing vegetation by the application of a contact herbicide in the spring. As compared with plowing and disking, this leaves a layer of dead plant material on the soil surface. This provides excellent erosion protection, moisture conservation, and nutrient release. Other advantages of reduced tillage are less soil compaction and slowed rates of organic matter decomposition. A slit is made in the sod, and a narrow strip is cultivated. After the seed is planted, a wheel firms the soil to ensure good germination. By saving at least one plowing or disking, minimum tillage reduces cost with yields usually equal to those obtained with conventional tillage.

Mulches   A mulch is a material applied to the surface of the soil and may consist of a wide diversity of organic and even some inorganic materials. Traditionally, mulches have been organic in nature; commonly used materials include straw, leaves, sawdust, corncobs, peanut hulls, pine bark, pine needles, wood chips, and peat moss. In many situations the choice of mulches used will depend on cost. In the corn belt, corncobs may be the least expensive, whereas in other regions pine needles, peanut hulls, or sawdust may be cheaper.

Mulches provide a great many benefits to horticultural crops. These benefits can include moisture conservation, erosion control, improved soil structure, improved water infiltration rates, reduced temperature fluctuations, increased organic matter levels, and improved nutrient availability. Moisture is conserved because an organic mulch suppresses the growth of weeds, keeps soil temperature lower than unmulched soil, and protects the soil surface from both wind and sun. Erosion control is excellent with a good layer of mulch. Part of this benefit accrues from the protection of the soil surface from the puddling effect of the beating raindrops. Perhaps equally important is that, over time, mulches improve soil structure sufficiently to dramatically increase infiltration rates and thereby reduce runoff. Temperature effects are of paramount importance in the mulching of many plants, but the desired effect of the mulch varies with both the crop involved and the time of year. As a general rule, mulches rather drastically suppress the temperature fluctuation of the soil. In summer, mulches are used to hold down the temperature of the soil, and this effect is particularly striking in the surface layers. The large quantities of heat from the sun are absorbed or reflected by the surface of the mulch, and little is conducted down to the soil because of the insulating properties of an organic mulch. The stabilizing effect of mulch on both the temperature and the moisture level in the upper soil layer makes these layers ideal for root growth. If we pull back a long-standing layer of mulch from under a perennial plant, we often see a heavy concentration of roots on or near the soil surface and often even growing up into the mulch. Because mulching tends to encourage shallow roots, it is recommended that newly transplanted shrubs and trees not be mulched until the second year so that a deeper root system is initially encouraged. Mulches are also extensively used for winter protection of roots and also for the protection of the entire plant for strawberry crops. Since mulch provides good insulation against heat loss from the soil, it tends to keep soil temperatures well above those of unmulched soil. With strawberries, the usual procedure is to wait until a few freezes have occurred before mulching with straw. The first freezes cause the plants to go dormant, and the mulch then not only reduces soil freezing but provides additional protection by preventing response to temporary warm days in winter and early spring. By delaying development in the early spring, one can reduce the danger of freeze damage somewhat. As growth does start, however, the straw is pulled to the walkways and can be reapplied to the strawberries when a freeze is imminent. More frequently today, however, freeze damage is prevented by overhead irrigation (see Chapter 7), since the labor costs for mulch application and removal are prohibitive.

Major benefits accrue from the increased levels of organic matter provided by organic mulches. These benefits have been discussed previously in this chapter. There are also major nutritional advantages to mulch usage. As an organic mulch gradually decomposes, whatever nutrients it contains are released into the soil. The significance of this effect is slight for a mulch material such as sawdust which is low in nutrients, but sizable quantities of nutrients will be added from hay or leaves. If large quantities of organic matter with a high carbon-nitrogen (C:N) ratio, such as sawdust or straw, are incorporated into the soil, a temporary deficiency of nitrogen is likely to occur. Whenever raw organic matter is turned into the soil, a tremendous growth of microorganisms which feed on organic matter occurs. Since these organisms are in intimate contact with both organic matter and soil particles, most available nitrogen is tied up in building new microbial tissues. For a while after the addition of organic material, there is a shortage of available nitrates (as well as ammonium forms) and a large liberation of CO_2. The balance gradually returns as the carbonaceous material is broken down, microbes decline, and nitrogen becomes available again for crop growth. The C:N ratio of the soil again returns to a more normal and stable level. To avoid nitrogen deficiency from this phenomenon, materials with a high C:N ratio (50:1) should not be turned under during the growing season, or else additional nitrogen should be added to meet the needs of the crop.

Organic mulches are often used for the sake of appearance as well as the beneficial effects described previously. Particularly attractive mulches are pine bark, peat moss, and pine needles. In special situations, inorganic materials such as white pebbles or rocks are sometimes used. These would be most suitable where appearance is more important than the many advantages of an organic mulch.

Certain problems are associated with the widespread use of mulch. Perhaps the most critical are cost (including the purchase price if necessary), transportation, and application. Other potential drawbacks are pest infestations, which can be devastating, and in some cases a potential fire hazard. Mulches intercept rainfall or irrigation water, reducing the amount reaching the root zone from a light shower or irrigation. Once the mulch has been saturated, however, further water penetrates readily.

The use of plastic (polyethylene) films as mulch has become widespread in recent years. The film is usually clear or black, and is most often about 0.1 mm thick. Where weed control is necessary, black plastic is used, as it completely shades out competing vegetation. In situations where the clear type is desirable, alternative weed-control methods must be used. For this purpose herbicides may be used prior to laying the plastic. In crops of very high cash value, such as strawberries in California, the soil is injected with a soil fumigant such as methyl bromide or chloropicrin and covered with thin plastic. After a few days the plastic is removed and the soil allowed to ventilate. The fumigation not only kills weeds but controls nematodes, insects, and diseases as well. After the soil is free of the fumigant vapors, clear plastic is applied, and the strawberry plants are set.

A major benefit with clear plastic and to a large degree with black plastic is the effect on soil temperatures. The clear plastic is essentially transparent to the shortwave radiation from the sun but blocks the

TABLE 10-2

SOIL TEMPERATURES AT 7.6-CM DEPTH ON JULY 9, BLACKSBURG, VIRGINIA

| TIME | AIR TEMPERATURE | SOIL TEMPERATURE, °C, UNDER FOLLOWING TREATMENTS | | |
		BARE SOIL	BLACK PLASTIC	STRAW
10:30 A.M.	23	23	26	21
2:30 P.M.	29	27	30	22
4:30 P.M.	26	27	30	23

Source: R. G. Gardner, M.S. Thesis, Department of Horticulture, Virginia Polytechnic Institute & State University, Blacksburg, Virginia, 1972.

longwave energy being radiated by the soil beneath. Thus a "greenhouse effect" occurs (see Chapter 7). Since black plastic absorbs most of the solar radiation striking it, it becomes quite warm. Some of this heat is transferred to the soil, although much is lost to the atmosphere through convection and radiation. At night the black plastic acts as a barrier to radiant heat loss, and so mulched soils do not cool off as much as bare soil. Compared with a bare soil, black plastic mulch elevates soil temperature whereas a straw mulch depresses soil temperature (Table 10-2).

The value of increasing soil temperatures can be very striking with warm-season crops (Figure 10-6). Total yields may be dramatically increased, and in almost all cases the growth and development accelerated sufficiently to allow earlier harvest. The advancement of maturity and harvest by 7 to 10 days will often mean great increases in the cash value of the crop.

The plastic can be purchased in rolls with widths of less than a meter to several meters. With vine crops, the normal procedure is to set the plants or seeds in a single row down the center of a strip of plastic 0.9 to 1.2 m in width. With many crops—such as strawberries in the hill system, sweet corn, and tomatoes—a double row is set on each strip of plastic with the rows about one-fourth of the way in from each edge.

In addition to temperature effects, plastic mulches greatly modify soil moisture relations. The plastic is applied when soil moisture is high and allows no moisture loss. The point is often raised that the plastic will shed rainfall or irrigation water and thus actually reduce available moisture. This is not normally a problem, for several reasons. Water can enter the soil around the stem and between strips and can move laterally into the root zone. Perhaps most important, however, is the drastic suppression of surface evaporation.

The use of plastic mulches is widespread with high-value crops such as pineapples in Hawaii, strawberries in California, and vegetable crops in many areas of the country. Black plastic is also very

(b)

(a)

(c)

FIGURE 10-6 Effect of plastic mulches on growth of muskmelon plants in early summer. (*a*) No mulch. (*b*) Black plastic. (*c*) Clear plastic. (Photographs taken July 22. 1976.)

frequently used beneath container-grown nursery crops. Since it provides complete weed control and a clean, no-maintenance surface, black plastic can be quite a money-saver. Although effective in weed control around fruit trees, black plastic mulches are not used near the trees, because the sheltered environment under the mulch attracts mice—one of the most serious of the orchard pests.

Many other materials have been tested as possible substitutes for plastic mulches because of the rather high cost of plastic and also the necessary removal at the end of the season. Paper mulches will break down, leaving no residue, but tend to break down too quickly, especially where covered with soil. The result is often large strips being blown around and doing no good for the crop. Other materials which have been tried are steel foil, aluminum foil, and even asphalt materials sprayed on the soil surface. The steel foil rusts and breaks down too quickly. Since aluminum foil tends to depress soil temperature, it would be advantageous only where excessive soil temperatures may be a problem. Another very different aspect of aluminum foil as a mulch is the efficient reflection of light back up into the plant canopy; the potential benefit from this reflection has been explored with various crops, but it is not economically feasible. Asphalt is efficient in transferring absorbed heat to the soil but has proven only moderately effective for weed control.

In choosing a soil management system, the horticulturist has a number of choices, each of which has certain advantages and disadvantages. The final choice may be made on the basis of the relative importance of many factors. In an arboretum, appearance will be more important than cost; in commercial crop production, however, economics will outweigh appearance and the ultimate selection would be quite different.

SOIL CHEMISTRY

The previous portions of this chapter have dealt with the physical aspects of soils, but many chemical aspects must also be considered. Among these are soil pH, cation exchange capacity, soil fertility, and fertilization.

Soil pH

The *pH* of a solution indicates the degree to which it is acidic or basic. In the case of soils, the pH is also referred to as the *soil reaction.* A pH is the logarithm of the reciprocal of the hydrogen ion (H^+) concentration. At neutrality, the pH is 7.0, which indicates an H^+ concentration of 10^{-7}

mol L^{-1}. The product of the normalities* of H^+ and OH^- is always 10^{14}, and so at a pH of 7 the concentration of OH^- is also 10^{-7} mol L^{-1}. Since pH is based on a log relationship, at a pH of 6 there are ten times more H^+ and 10 times fewer OH^- than at a pH of 7. It should be apparent that soil pH is therefore an expression of the relative proportion of H^+ and OH^- in the soil.

Most agricultural soils fall within a pH range of 4 to 9, although beyond the realm of agricultural soils, the extremes may extend from 3 to 11. The pH depends on many factors, with two of the more important being the amount of precipitation and the type of vegetation.

Humid areas or those areas with rather heavy rainfall tend to have acid soils because many of the bases are removed from the soil by leaching. As ions such as Ca^{++} and Na^+ are leached downward, they are replaced by H^+ ions, thus lowering soil pH. This effect is accelerated by the addition of acid-forming fertilizers such as those containing SO_4. The type of vegetation can also affect soil pH over a period of years. Grasses tend to cause soils to remain at somewhat higher pH than forests do, especially coniferous types.

Unless the pH is extremely low or high, direct effects on the growth of crops are minimal. In general, the most critical effects of soil pH are indirect. The availability of certain nutrients is strongly influenced by pH (Figure 10-7). For example, copper, iron, manganese, and zinc become much less available in highly alkaline soils. In certain so-called acid-loving horticultural crops, the acidifying of the soil is the most effective method of curing "iron chlorosis." Because of this, the disorder is often referred to as *lime-induced chlorosis*. Since the nitrogen-fixing bacteria associated with legumes are very seriously hindered by acid pH, the maintenance of moderate pH levels is quite critical with crops such as beans and peas.

Cation Exchange Capacity

Clay particles and humus are very important in determining the chemical properties of soils. These materials are very complex in structure and contain sites with negative charges which attract positively charged ions called *cations*. If these cations are located there, they are in contact with the soil solution; and if the chemical bonds are not too strong, they may be replaced by or exchanged with other cations. The amount of such negatively charged sites is called the *cation exchange capacity* of the soil. By rather refined laboratory procedures, the cation exchange capacity can be determined and is often used as an index of the potential fertility of the soil. The cation

---

*Normality is one method of expressing concentration. A 1 normal (N) solution contains 1 gram-equivalent weight per liter of solution. Thus a 1N acid solution has 1 g H^+ L^{-1}; a 1N base solution has 17 g OH^- L^{-1}.

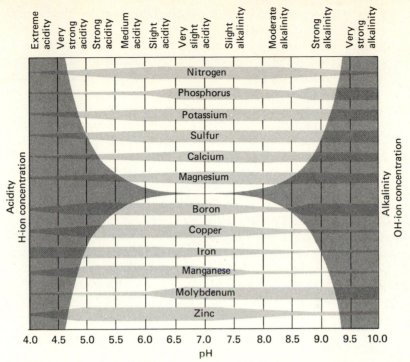

FIGURE 10-7 Effect of soil pH and associated factors on the availability of plant nutrient elements. The width of the band for each element indicates the relative favorability of this pH value and associated factors to the presence of the elements in readily available forms (the wider the band, the more favorable the influence). It does not necessarily indicate the actual amount present, since this is influenced by other factors, such as cropping and fertilization. (Reproduced from *Changing Patterns in Fertilizer Use*, 1968. Fig. 7, p. 152, by permission of the Soil Science Society of America.)

exchange capacity is typically expressed as milliequivalents* per 100 g of soil, meq g^{-1}. Depending on the type and proportion of clay present, mineral soils have cation exchange capacities ranging from 2.0 in sand to more than 50 in certain clays. Because humus has a very large cation exchange capacity, its presence greatly improves this aspect of soil chemistry—this in addition to its positive effect on soil structure and water-holding capacity, described previously.

To relate the cation exchange capacity to plant nutrition, we must also ascertain the *percent base saturation*. The latter is defined as the proportion of the cation exchange capacity occupied by basic nutrients

---

*A milliequivalent is the amount of a substance that will react with or displace 1 mg of hydrogen.

such as Ca^{++}, Mg^{++}, K^+, and Na^+. Specifically excluded are H^+ and Al^{+++}, which are acidic in reaction.

The ions in the soil solution at any given time represent only a very minute proportion of the total ions present in the soil. It is estimated that in most soils more than 99 percent of the cations are adsorbed on the various particles exhibiting cation exchange. These charged particles of clay or humus are colloidal* in nature and are called *micelles*. Various cations are adsorbed on the surface of the micelles but are in a dynamic equilibrium with cations in the soil solution. Cations are exchanged between the micelles and soil solution on the basis of two phenomena. First is the strength of adsorption, which indicates relative amounts held by a micelle if the cations are of equal concentration. From highest to lowest, the adsorption of cations is as follows: $Al > Ca > Mg > K > Na$. If H^+ is included, it falls between Al and Ca. Second, the law of mass action can be involved under certain circumstances. When a relatively large concentration of one ion such as H^+ becomes available, various other cations may be replaced on the micelles by hydrogen, because H^+ is in a much more numerous and thus more competitive position for the available exchange sites. A similar phenomenon occurs when an application of lime is made in that the Ca^{++} replaces H^+ on the micelles. The ions which are replaced go into the soil solution and thus are subject to leaching.

Adjusting Soil pH

In the humid regions of the United States, most soils are acid in nature primarily because the bases are leached downward, leaving an excess of H^+. Horticultural crops vary considerably in the soil pH to which they are best adapted. Crops requiring acid soils include azalea, blueberry, and rhododendron. Among those adapted to a mildly acid soil are strawberry, apple, cabbage, sweet corn, and tomato. Crops best-suited to a neutral to only slightly acid soil are asparagus, lettuce, cantaloupe, and lima bean. Another factor of considerable importance is the range of soil pH within which a particular crop can prosper. The blueberry will produce vigorous growth and full crops only when growing in a soil pH of 4.3 to 4.8 whereas apple trees are often equally productive in soils with pH's ranging from 5.0 to 8.0. It is recommended to keep soil pH between 6.0 and 6.5 for apple orchards, but this relates more to the maintenance of a healthy sod than to the trees themselves.

Raising Soil pH   If the soil pH is found by soil test to be lower than desired, the usual procedure is to add lime. Certain major factors

*A colloid is defined as a particle small enough to remain suspended in water without agitation. Because of its small size a colloid has a very large surface area per unit weight.

TABLE 10-3

APPROXIMATE AMOUNTS OF LIMESTONE* FOR DIFFERENT SOIL TYPES TO ATTAIN A pH OF 6.5

pH OF UNLIMED SOIL	SANDY LOAM SOILS	LOAM SOILS	SILT LOAM SOILS	CLAY LOAM SOILS
4.8	1.36 (3.00)	1.81 (4.00)	2.04 (4.50)	2.27 (5.00)
5.0	1.13 (2.50)	1.36 (3.00)	1.59 (3.50)	1.81 (4.00)
5.5	0.79 (1.75)	0.91 (2.00)	1.13 (2.50)	1.36 (3.00)
6.0	0.45 (1.00)	0.57 (1.25)	0.68 (1.50)	0.91 (2.00)

*Limestone recommended in metric tons per hectare (figures in parentheses are tons per acre).

Source: Anon., *A Handbook of Agronomy*, Extension Division, Virginia Polytechnic Institute & State University, Blacksburg, Virginia, Publ. 600, p. 60, 1974.

which must be considered before the lime is applied include soil type, how much the pH is to be raised, the fineness of lime to be applied, and the type of lime to be used. As can be readily seen in Table 10-3, the amount of lime needed to raise the pH of a sandy loam soil is much less than for a clay loam soil. This increasing lime requirement with decreasing soil particle size relates to the high cation exchange capacity of clay particles. Obviously, the more one wishes to raise the pH, the greater the lime requirement. These relative amounts are also given in Table 10-3. The smaller the particle size in ground limestone, the sooner the lime reacts and neutralizes soil acidity. The usual recommendation for ground limestone is that 95 percent should pass through a United States standard 60-mesh screen and 30 percent through a United States standard 100-mesh screen. Because of the range in particle size, this lime will react over a period of a few weeks to several years. Several types of lime are available, such as ground limestone, burned lime, and hydrated lime. Ground limestone consists of calcite or calcium carbonate, $CaCO_3$, with varying amounts of dolomite or calcium magnesium carbonate, $CaMg(CO_3)_2$. If the limestone is mostly $Ca CO_3$, it is called *calcitic.* When sizable quantities of dolomite are included, it is called *dolomitic* limestone. If limestone is heated in a kiln, CO_2 is driven off and calcium oxide and magnesium oxide are left.

$$CaCO_3 + heat \longrightarrow CaO + CO_2 \uparrow$$

$$CaMg(CO_3)_2 + heat \longrightarrow CaO + MgO + 2CO_2 \uparrow$$

This is called *burned lime* or quicklime and can be used for raising soil pH but is unpleasant to handle. *Hydrated* or *slaked lime* is produced by reacting burned lime with water as follows:

$$CaO + MgO + 2H_2O \longrightarrow Ca(OH)_2 + Mg(OH)_2$$

The product is even more disagreeable to handle than burned lime.

The primary advantage of burned and hydrated lime is that they both react much more quickly than ground limestone; but they usually are more expensive when compared on the basis of cost per hectare for an equivalent amount of neutralizing power. One exception could be where the lime must be shipped long distances, because 1 kg of calcium oxide is equivalent to about 1.8 kg of ground limestone.

The addition of lime has many direct and indirect effects on a soil. Direct chemical effects include a reduction in the concentration of H^+, an increase in the concentration of OH^-, and increased availability of Ca^{++} and Mg^{++}. As the result of these changes, there is an increase in the percent of base saturation. Indirect effects of an elevated pH are decreased solubility of aluminum, iron, and manganese (Figure 10-7). In many very acid soils, the latter effect may eliminate toxic concentrations of these ions in the soil solution. Increased availabilities of both phosphorus and molybdenum will result (Figure 10-7).

Lowering Soil pH   In order to produce an "acid-loving" crop like blueberry, azalea, or rhododendron successfully, one must occasionally acidify the soil. To reduce the problems with the scab organism, one may also find it desirable to lower soil pH in fields to be used for potato production. The substances most widely recommended to lower soil pH are sulfur, iron sulfate, or aluminum sulfate. If only a gradual reduction over a very limited pH range is needed, it is sometimes feasible to accomplish the change through the selection of the source of nitrogen. The source of nitrogen which is most effective in lowering soil pH is ammonium sulfate $(NH_4)_2SO_4$, but NH_4NO_3 and NH_3 are also acidifying. Since other nitrogen sources, such as $NaNO_3$, KNO_3, and $Ca(NO_3)_2$, are somewhat basic in reaction, they should be avoided where acidification is desired.

Alkaline Soils

In arid and semiarid regions, soils are frequently alkaline and often exhibit other related problems. Such soils are usually classified under three headings: saline, sodic, and saline-sodic soils. Although these have somewhat different characteristics, they all tend to occur in areas where there is inadequate precipitation to leach the soil. This problem is accentuated where additional bases are added to the soil by either ground water or irrigation water.

Saline Soils   These are soils which are high in soluble salts, and are usually classified as such if the soluble salt content is 2000 ppm or higher. Saline soils are relatively low in sodium; (less than 15 percent of the cation exchange capacity is occupied by sodium ions). Soil pH is usually between 7.0 and 8.5 because the sodium content is low and most of the salts present are neutral. The salts are mostly chlorides and sulfates of calcium, magnesium, and sodium. When salts are allowed to accumulate on the soil surface, a white residue is formed which has given rise to the name *white alkali soils*. Leaching is an effective means of reducing the detrimental effects of such soils on the growth of crops. To successfully accomplish leaching, one must often improve drainage. Under conditions of poor drainage, the problems with excess salts may be accentuated by the application of large quantities of water. This is especially true where irrigation water contains even moderate levels of salts. Under conditions of good drainage, however, it is often possible to rid a soil of excess salts with one or two heavy irrigations.

Saline-Sodic Soils Soils in this category not only have high soluble salts (> 2000 ppm), but are also *sodic*—a term used when more than 15 percent of the cation exchange capacity is occupied with sodium. Saline-sodic soils usually have a pH between 8.0 and 8.5. Although these soils look much like saline soils, they must be treated quite differently if they are to be successfully reclaimed. A soil amendment must be worked into the soil surface, and the two most often used are gypsum ($CaSO_4$) and sulfur. It usually takes several tons of gypsum per hectare to be successful, whereas only about 20 percent as much sulfur would be required. The addition of $CaSO_4$ leads to the following reactions:

$$\begin{matrix} Na^+ \\ Na^+ \end{matrix} \boxed{Micelle} + CaSO_4 \rightleftharpoons Ca^{++} \boxed{Micelle} + Na_2SO_4$$

$$Na_2CO_3 + CaSO_4 \rightleftharpoons CaCO_3 + Na_2SO_4$$

Since the Na_2SO_4 is soluble, it can be leached, thus removing excess Na^+ and replacing it with Ca^{++}. The second reaction indicates that Na_2CO_3 is soluble and causes the pH to be high. By reacting with $CaSO_4$, the $CaCO_3$ is taken out as a precipitate and the Na_2SO_4 can be leached out.

Sulfur added to the soil is converted to sulfuric acid, H_2SO_4. It reacts as follows:

$$2S + 3O_2 + 2H_2O \longrightarrow 2H_2SO_4$$
$$H_2SO_4 + Na_2CO_3 \longrightarrow Na_2SO_4 + H_2O + CO_2 \uparrow$$

The added H_2SO_4 can also react as follows:

$$H_2SO_4 + CaCO_3 \longrightarrow CaSO_4 + H_2O + CO_2 \uparrow$$

If sufficient gypsum or sulfur is added to take these reactions well along to completion, such soils have been converted from a saline-sodic soil to saline soil and thus can be reclaimed by leaching.

Sodic Soils   In sodic soils, more than 15 percent of the cation exchange capacity is occupied bo sodium (Na+) ions, but these soils are low in total soluble salts. The detrimental effects of sodic soil are due to high sodium and high pH (8.5 to 10). Because of high sodium, such soils tend to be deflocculated,* and their structure very poor. In severe cases, organic matter tends to be dissolved and deposited on the surface. This has been the basis for the name *black alkali soils*. Sodic soils are the most difficult of the alkali soils to reclaim, and in many situations the cost cannot be justified. Not only are large quantities of gypsum required, but the high sodium makes soil structure so poor as to make incorporation of the amendment difficult.

BIBLIOGRAPHY

Brady, N. C.: *The Nature and Properties of Soils*, 8th ed., Macmillan, New York, 1974.

Daubenmire, R. F.: *Plants and Environment*, 3d ed., Wiley, New York, 1974.

Kramer, P. J.: *Plant and Soil Water Relationships: A Modern Synthesis*, McGraw-Hill, New York, 1969.

Thompson, L. M., and F. R. Troeh: *Soils and Soil Fertility*, 4th ed., McGraw-Hill, New York, 1978.

U. S. Department of Agriculture: *Soil, the Yearbook of Agriculture, 1957*, Washington, D.C., 1957.

*Deflocculation is the breaking up or dispersal of aggregates.

CHAPTER 11
NUTRITION

The mineral nutrition of plants has held the interest of many researchers since the early nineteenth century, when it was originally indicated that the soil provided certain elements required for plant growth. One of the techniques which has been widely used in nutrition research is the chemical analysis of plant tissue. Although the methods initially used were crude by today's standards, many of the early conclusions have been confirmed by more sophisticated techniques.

ESSENTIAL ELEMENTS

The number of elements essential for normal plant growth was considered to be ten until the early twentieth century. These ten elements were carbon (C), hydrogen (H), oxygen (O), nitrogen (N), phosphorus (P), potassium (K), calcium (Ca), magnesium (Mg), sulfur (S), and iron (Fe). Many experiments to prove the requirement for additional elements have been conducted, mostly by withholding individual elements while supplying all others for which a requirement has been shown. Most often such experiments are carried out by growing plants in either washed sand or a well-aerated nutrient solution to have maximum control of the elements available to the plant. During the twentieth century the following six have been added to the list of essential elements: manganese (Mn), zinc (Zn), boron (B), copper (Cu), molybdenum (Mo), and chlorine (Cl).

Those elements required in relatively large amounts are often referred to as *macronutrients*, or major nutrients. Included in this category are nitrogen, phosphorus, potassium, calcium, magnesium, and sulfur. The *micronutrients* are those which, although also essential, are needed in much smaller quantities. These include iron, manganese, zinc, boron, copper, molybdenum, and chlorine and are sometimes called *trace* or *minor* elements. Although carbon, hydrogen, and oxygen are absolutely essential, they are not normally considered in nutritional studies because they are readily available from air and water.

PLANT ANALYSIS

In certain early research when plant tissue was analyzed, all identifiable elements were determined. Table 11-1 gives an analysis of corn plants grown in Kansas in 1920; it is interesting to note the analysis for 12 essential elements, including oxygen, carbon, hydrogen, all six macronutrients, and three of the micronutrients. Significant percentages of the corn plants were silicon and aluminum, elements which have

TABLE 11-1
ANALYSIS OF PRIDE OF SALINE CORN
PLANTS GROWN AT MANHATTAN, KANSAS

ELEMENT	PERCENTAGE OF TOTAL DRY WEIGHT
Oxygen	44.431
Carbon	43.569
Hydrogen	6.244
Nitrogen	1.459
Phosphorus	0.203
Potassium	0.921
Calcium	0.227
Magnesium	0.179
Sulphur	0.167
Iron	0.083
Silicon	1.172
Aluminum	0.107
Chlorine	0.143
Manganese	0.035
Undetermined elements	0.933

Source: E. C. Miller, *Plant Physiology*, 2d ed., McGraw-Hill, New York, 1938, p. 284.

not been proven to be essential. The "undetermined elements" presumably included the other four micronutrients as well as other elements such as sodium.

In most plant analysis research today, the tissue to be analyzed is dried and ground to a coarse powder. For the determination of most nutrients the sample is "ashed" in a furnace at 500 to 600°C to burn off the organic matter, leaving only the minerals. Because of partial vaporization during high-temperature ashing, this procedure is not suitable for nitrogen; therefore, nitrogen is determined after a wet-ashing procedure in which no nitrogen is allowed to escape. Special precautions are also necessary to avoid the loss of sulfur during dry ashing of plant samples.

As the result of many years of controlled sand or solution experiments, field studies, and nutritional surveys, certain procedures and general guidelines have evolved. For example, in apple nutrition surveys, midterminal leaves are sampled in midsummer because this procedure has given the least variation in nutrient content. Although "critical levels" have been proposed for several elements, the nutritional survey is not an exact science as yet. Many factors have an influence on nutrient levels, and these must be considered before recommendations can be made. Among the more important variables with apples are cultivar, rootstock, rainfall, previous fertilization, crop load, and the spray program (for micronutrient analysis). Certain states offer a leaf analysis service for producers to aid them in their fertilization program. It is frequently possible to confirm suspected deficiencies and excesses, but the usefulness of this in making routine recommendations is open to some debate. In many instances, a well-trained orchardist can tell more by careful observation of trees and fruit than by a leaf analysis.

NITROGEN

Although there are sixteen essential elements, nitrogen probably deserves more attention than any other. With the exception of carbon, hydrogen, and oxygen, nitrogen is present in most plants in greater concentrations than any of the other nutrients. Nitrogen is also the element to which any plant is most likely to respond. Because of its importance, nitrogen will be covered first and in greater detail than the other elements.

Atmospheric Nitrogen

It is a paradox that although nitrogen is the element most frequently deficient for crop production, about 78 percent of the atmosphere is nitrogen. As an atmospheric component, nitrogen is an odorless, tasteless, inert gas; it is utterly of no use to plants until it is "fixed." The major organisms capable of fixing nitrogen are a small group of bacteria and algae. (Nitrogen fixation is covered later in this section.)

Functions in Plants

It is difficult to overemphasize the importance of nitrogen in the normal metabolism of plants. Nitrogen is a vital component of the protoplasm, chlorophyll molecules, and amino acids from which proteins are made, as well as the nucleic acids. Without nitrogen in reasonably adequate quantities, the growth of crops is drastically suppressed.

Deficiency Symptoms

The most frequently observed symptoms of nitrogen deficiency include stunted growth and pale-green to yellow leaves which are usually smaller than normal. The older leaves are most affected because nitrogen is a relatively mobile element and is withdrawn from the old leaves and translocated to the young foliage. Older leaves may abscise early, and nitrogen-deficient perennials are characterized by early falling of leaves in the autumn.

Excess Nitrogen

Under certain conditions (for example, too much nitrogen fertilizer has been applied), symptoms of excess nitrogen can occur. The symptoms will vary with the plant but normally include very dark green foliage, weak tissues, and succulent, vegetative growth. Closely associated symptoms are delayed or scanty flowering and fruiting. Excessive application of nitrogen to apple trees can delay flowering and fruit set by one or more years. In many perennial plants, high nitrogen levels in the late summer can predispose plants to injury from freezing in the fall and early winter (see Chapter 7). The optimum nitrogen level will depend on the crop of concern and also upon the market for which it is being produced. Since rapid succulent growth is desirable with leafy vegetables, these vegetables are heavily fertilized with nitrogen. With tree fruits, however, nitrogen applications are aimed at maximizing the production of quality fruit rather than vegetative growth. For 'Golden Delicious' being grown for the fresh market, the trees are kept on the "hungry" side for optimum fruit quality, (indicated by yellow color and firm flesh). As can be seen in Figure 11-1, at leaf nitrogen levels of 1.9 percent the percent of Extra Fancy (yellow) fruit was 95 percent, but at 2.5 percent, only about 75 percent were Extra Fancy. 'Golden Delicious' being grown for processing outlets are given heavier nitrogen applica-

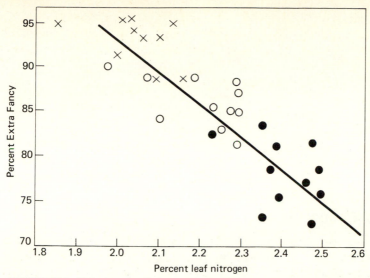

FIGURE 11-1 Relationship between leaf nitrogen and the percentage of Extra Fancy (yellow) 'Golden Delicious' apples. Different symbols indicate varying nitrogen fertilizer levels. (From M. W. Williams and H. D. Billingsley, *J. Amer. Soc. Hort. Sci.*, **99**:144–145, 1974.)

tions to increase size and yield of fruit since appearance is of lesser consequence.

Sources of Nitrogen

The gradual decomposition of soil organic matter is one source of nitrogen for plants. Although certain forms of organic nitrogen can be directly utilized by plants, most nitrogen is absorbed as inorganic ions. The two most widely available ions are nitrate (NO_3-) and ammonium (NH_4+). Although there may be sizable quantities of nitrogen in the soil, the great majority is in organic forms, either as a component of microorganisms or as a part of undecomposed organic matter.

The Nitrogen Cycle

Nitrogen is in a constant state of change which is referred to as the *nitrogen cycle* (Figure 11-2). As organic matter is decomposed, nitrogen is released as NH_4+ in the process of *mineralization*. The rate of mineralization depends to a large degree on the carbon-nitrogen (C:N) ratio. The lower the C:N ratio, the more rapidly the organic matter is broken down. The NH_4+ which is released can be utilized in different

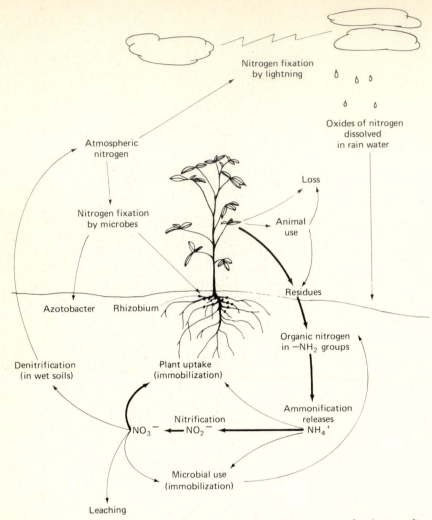

FIGURE 11-2 Nitrogen cycle. The darker lines indicate the main cycle of mineralization and immobilization. (From L. M. Thompson and F. R. Troeh, *Soils and Soil Fertility*, 4th ed., McGraw-Hill, New York, 1978.)

ways. It may be taken up by soil microorganisms, used by higher plants, held by the soil particles, or converted to NO_3-. The conversion of NH_4+ to NO_3- is called *nitrification* and results from the activities of two different types of soil bacteria as follows:

$$\underset{\text{Ammonium}}{NH_4^+} \xrightarrow{\text{Nitrosomonas}} \underset{\text{Nitrite}}{NO_2^-} \xrightarrow{\text{Nitrobacter}} \underset{\text{Nitrate}}{NO_3^-}$$

This process usually occurs rapidly enough so that there is much more NO_3- in the soil than NH_4+. Although certain plants can survive and some can thrive on NH_4+ as a nitrogen source, most nitrogen is absorbed in the form of NO_3-. The uptake of nitrogen by higher plants or microorganisms is called *immobilization*, since the nitrogen is at least temporarily unavailable. If a crop is removed, some nitrogen is removed from the cycle, but the other nitrogen is eventually returned as the microorganism or crop residue is broken down.

One of the very important processes in any discussion of soil fertility is *nitrogen fixation*, which involves the incorporation of gaseous N_2 from the atmosphere in a usable form. The microorganisms which are capable of nitrogen fixation are in three groups: (1) some free-living bacteria, such as *Azotobacter* and *Clostridium*, (2) some blue-green algae, and (3) the bacteria *Rhizobium* in a symbiotic (mutually beneficial) relationship with legumes. Some nonlegumes have also been shown to be nodulated and fix nitrogen in a symbiotic relationship with certain microorganisms. The most widely studied of the three types is the fixation of nitrogen by legumes whose roots are infected with *Rhizobium*. The specificity of the bacteria is such that a particular species will infect only certain legumes. To ensure good nitrogen fixation, legume seeds are usually inoculated with the appropriate bacteria before being planted. The process by which the nodules (Figure 11-3) are formed not only is quite complex but is influenced by a number of environmental factors, such as the pH, calcium level, and the amount of nitrate in the soil. Legumes are reported to fix fairly wide ranges of nitrogen, but average amounts appear to be from 50 to 150 kg of nitrogen per hectare per year. Although nitrogen fixation by the free-living bacteria and the blue-green algae is probably of importance, their activities are not yet well understood. It seems, however, that as the cost of nitrogen fertilizer continues at its high level, these alternative sources of nitrogen will be thoroughly explored.

Limited quantities of nitrogen are fixed by the electrical discharges from thunderstorms, but this type of fixation probably accounts for only 2 to 3 kg of nitrogen per hectare per year.

Nitrogen Losses

Through several processes, the loss of nitrogen can be sizable. Leaching is a major problem in humid regions. Because of its negative charge, the NO_3- ion is not held by negatively charged soil particles and so remains in the soil solution. Although the NH_4+ ion is held as a cation, it is eventually converted to NO_3- and can then more readily leach away. Soil erosion can account for sizable nitrogen losses. Crop removal is probably the greatest source of nitrogen loss where good soil management procedures are followed. Depending on the crop and

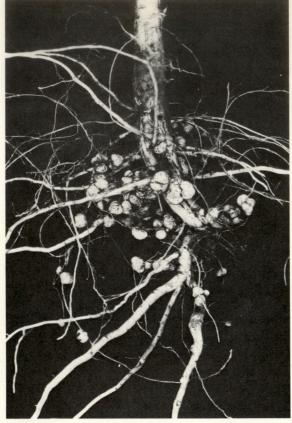

FIGURE 11-3 Nodulation of roots on a typical legume. (Photograph, U.S. Department of Agriculture.)

what part of the plant is harvested, this can account for 50 to 75 kg ha^{-1} for crops such as potatoes and tomatoes. For hay crops, this figure can be more than 150 kg ha^{-1}, since the entire top of the plant is removed repeatedly during the growing season.

The process by which nitrate nitrogen is reduced to gaseous forms is called *denitrification*. The causal microorganisms are thought to be facultative anearobes, but the exact mechanisms by which they reduce nitrate is not well understood. The volatile nitrogen gases are mainly nitrous oxide (N_2O), elemental nitrogen (N_2) and nitric oxide (NO). Certain chemical reactions in the soil can also lead to the volatilization of nitrogen. Nitrogen losses by denitrification are difficult to measure but are often estimated to be in the range of 10 percent of the nitrogen added per year. The losses are most where aeration is poor or where nitrogen levels are excessive.

Nitrogen Fertilizers

The many fertilizers which supply nitrogen are classified as *mineral* or *organic*, although some people prefer the names *chemical* and *organic*. Obviously the oldest forms are those which are organic, such as animal manure, which has been used for thousands of years. The American Indian used a fish under each hill of corn as an organic fertilizer. Other forms available today include bone meal, dried blood, and composted leaves. The release of nitrogen by organic sources tends to be slow and is by the same basic processes described previously for the organic matter of soil.

Chemical fertilizers are by far the major source of nitrogen used today. Major advantages of inorganic nitrogen fertilizers include rapid availability of applied nitrogen, ease of shipment and application, and lower cost per unit of applied nitrogen. Under certain conditions slowly available nitrogen from organic sources is advantageous, but more commonly we need high nitrogen availability followed by a declining nitrogen level.

Major nitrogen carriers are listed in Table 11-2, along with the percentage N. All these materials can now be manufactured from NH_3, which is produced from N_2 and H_2 under a combination of high temperature and pressure. The NH_3 can be stored under pressure and injected into the soil as a liquid, mixed with water and sprayed on the soil surface, or used in the manufacture of dry nitrogen carriers.

In general, the major factor in selecting a nitrogen source is cost, and therefore anhydrous ammonia is very widely used. For many horticultural crops, however, dry fertilizers are preferable. If there is no particular reason to prefer either NH_4 or NO_3, ammonium nitrate may be the best choice, as it is usually lowest in cost per unit of N. If immediate availability of nitrogen is wanted, an NO_3 source such as

TABLE 11-2
CHARACTERISTICS OF SEVERAL MINERAL NITROGEN SOURCES

SOURCE	FORMULA	% N	COMMENTS
Anhydrous ammonia	NH_3 (liquid, under pressure)	82	Cheapest source of N
Ammonium nitrate	NH_4NO_3	33	Widely used, often cheapest "dry" source of N
Ammonium phosphate	$NH_4H_2PO_4$	11	Source of N,P, used in mixed fertilizers
Ammonium sulfate	$(NH_4)_2 SO_4$	21	Used on acid-loving plants
Calcium nitrate	$Ca(NO_3)_2$	17	Imported, source of Ca + N
Sodium nitrate	$NaNO_3$	16	Originally imported from Chile, now produced synthetically
Urea	$(NH_2)_2CO$	45	Used for both soil and foliar applications

$NaNO_3$ or $Ca(NO_3)_2$ is probably best, the choice depending on the desirability of either Ca^{++} or Na^+. Urea is widely used for foliar fertilization because it is soluble in water and quickly absorbed by leaves, and therefore the response is very fast. Another advantage of a urea spray is that the effect is quite temporary, so that rather precise control of N levels is possible. Since the optimum amount of applications of nitrogen varies with the crop, soil management, soil type, organic matter level, rainfall and other factors, generalizations are quite meaningless and are not offered here.

In recent years, the use of *slow-release* fertilizers has become widespread for some crops. The fertilizer particles are coated with waxes, resins, or other chemicals which dissolve slowly and at a predictable rate, thus releasing the nutrients over a period of 3 to 6 months. Such materials are particularly useful with potted plants, greenhouse bench crops, and lawns which thrive on a uniform supply of nutrients during the entire growing season. Most often these slow-release fertilizers contain phosphorus and potassium as well as nitrogen and are therefore called *complete fertilizers*.

PHOSPHORUS

The importance of phosphorus has long been known, and phosphate fertilizers were the first to be used in large quantities.

Functions of Phosphorus

Phosphorus is vitally important in many aspects of plant growth, but perhaps the most obvious value is in the storage and transfer of energy. The formation of adenosine triphosphate (ATP) containing a "high-energy" phosphate bond is of utmost importance in plant metabolism. Other compounds of which phosphorus is a part are nucleic acids, phospholipids, and the coenzymes NAD and NADP (see Chapter 6). Thus, it is obvious that phosphorus is absolutely essential even though it is normally present in much lower amounts than the other major nutrients in plant tissues.

Deficiency Symptoms

Because of its vital role in the energy transformations of the plant, a deficiency of phosphorus is apparent in altered metabolism and growth. Growth is stunted; older leaves tend to abscise because, like nitrogen, phosphorus is mobile and moves from the older to the younger leaves; leaves are dark-green and sometimes distorted.

Carbohydrates tend to accumulate, thus encouraging the formation of anthocyanins and the associated red or purple coloration of leaves and stems.

Soil Phosphorus

Unlike nitrogen, of which there are tremendous quantities in the atmosphere, the supply of phosphorus is limited to what is in the soil or added as fertilizer. Soil phosphorus exists in both inorganic and organic forms. Inorganic phosphorus compounds are divided into two main groups: those containing calcium and those containing iron or aluminum. The mineral apatite is the source of most of the native soil phosphorus and is also found in deposits from which "phosphate rock" is mined. Most phosphorus in soils is in unavailable forms, with only a very minute portion of the total phosphorus being available at any one time. This phenomenon may seriously limit available phosphorus for crops but also serves to minimize losses due to leaching. The availability of phosphorus is affected by the pH of the soil because of associated changes in the concentration of various cations. Under alkaline conditions, phosphorus is tied up in various compounds containing calcium. These range from fairly soluble calcium phosphates to insoluble forms such as apatite. In very acid soils, the phosphorus is tied up as iron and aluminum phosphates, which are insoluble and thus make the phosphorus unavailable for plant use. As a general rule, phosphorus availability is maximized by maintaining the pH of the soil in the range of 6.0 to 7.0.

The uptake of phosphorus from the soil depends on the forms present in the soil, which vary with the pH. Most of the phosphorus is taken up in the form of dihydrogen phosphate, H_2PO_4-, but the $HPO_4=$ ion becomes more prevalent at a pH above 7.2.

Phosphorus Fertilizers

When phosphorus fertilizers are applied, the effect on the availability of phosphorus in the soil varies with the material used. The most frequently used sources today are superphosphate or treble superphosphate, both of which provide sizable percentages of soluble phosphorus. Since bone meal provides mostly insoluble phosphorus, it has little immediate effect on available phosphorus. Even the soluble forms become converted in the soil to unavailable, or "fixed," forms. To minimize fixation, fertilizers containing phosphorus are applied in a band below the seed (band application), but in other cases it is broadcast and plowed or disked into the soil. Regardless of the method of application, because of fixation, well under 50 percent of the applied phosphorus is recovered by the crop. The manufacture of superphosphate involves the treatment of rock phosphate with sulfuric acid to produce phosphoric acid and gypsum.

$$Ca_3(PO_4)_2 + 3H_2SO_4 + 6H_2O \rightarrow 2H_3PO_4 + 3Ca\,SO_4 \cdot 2H_2O$$

The content of phosphorus is about 9 percent which conventionally has been expressed as 20 percent P_2O_5. For use in high-analysis fertilizers, the rock phosphate can be treated with phosphoric acid to produce treble superphosphate:

$$Ca_3(PO_4)_2 + 4H_3PO_4 \rightarrow 3Ca(H_2PO_4)_2$$

The product here is treble or triple superphosphate and contains about 20 percent P (equivalent to 46 percent P_2O_5). The higher phosphorus content is advantageous in both long-distance shipment and in the formulation of high-analysis mixed fertilizers.

Although some phosphate is applied alone, most is used in *mixed fertilizers*. The analysis is given for nitrogen, phosphorus, and potassium—in that order. The nitrogen is expressed as N; the phosphorus has traditionally been expressed on the basis of P_2O_5; and the potassium is expressed as K_2O equivalent. Thus a 10-10-10 fertilizer contained 10 percent N, 10 percent P_2O_5 equivalent, and 10 percent K_2O equivalent. The current trend is to provide the guaranteed analysis on an elemental basis for all three. To convert P_2O_5 to P, multiply by 0.44; to convert K_2O to K, multiply by 0.83. Thus the "old" designation of 10-10-10 becomes 10-4.4-8.3 on the elemental basis.

POTASSIUM

Potassium is the third of three major nutrients supplied in a complete fertilizer. The widespread use of potassium fertilizers has lagged well behind both nitrogen and phosphorus because the native supply in most soils was quite high and deficiencies were therefore slow to develop.

Functions of Potassium

Although there has been no doubt about its essentiality, the exact function of potassium in plants remains elusive. No major organic compound in plants is known to contain potassium. Potassium is essential for photosynthesis, sugar translocation, and enzyme activation, but specific roles remain obscure. A recent hypothesis is that potassium ions are "pumped" into and out of guard cells and thus regulate water potential and the resulting opening and closing of stomates.

Deficiency Symptoms

Like nitrogen and phosphorus, potassium is mobile; therefore, the early symptoms of a potassium deficiency occur on older leaves first. Early symptoms include leaf chlorosis, which is followed by necrosis at the leaf tip and margins; the most frequently cited sign of a potassium deficiency is "marginal scorch." Metabolic changes induced by inadequate potassium are the accumulation of carbohydrates and soluble nitrogen compounds as a result of a lack of protein synthesis.

Luxury Consumption

Potassium is somewhat unusual in that excessive applications are relatively harmless; because of this, potassium is sometimes wasted. Many plants take up potassium in proportion to its availability. Therefore, if a heavy application is made every third year to save application costs, the first crop utilizes excess K, whereas the following crops may not have enough. Excessive applications of potassium can sometimes induce a magnesium deficiency.

Soil Potassium

The total potassium content of most soils is quite high, usually many times higher than either phosphorus or nitrogen. Most of it, however, is in forms which are relatively unavailable to plants, much as is true for phosphorus and nitrogen. Since potassium is not a part of major organic molecules, it is readily leached from organic matter soon after the death of the organism. Thus, it is not held in unavailable organic forms, as both nitrogen and phosphorus are. The inorganic potassium in soils includes forms which are readily available, slowly available, and unavailable. The readily available potassium constitutes only 1 to 2 percent of the total potassium. Most of this potassium exists as exchangeable ions on soil particles with the remainder being in the soil solution. Slowly available potassium ranges from 1 to 10 percent and consists of potassium ions which are in nonexchangeable positions between layers of clay crystals. There is an equilibrium among the soil solution K, the exchangeable K, and the nonexchangeable K. Therefore, as potassium is added by fertilizer or removed by plants or leaching, the equilibrium shifts accordingly. For example, as K ions are removed from the soil solution by a plant root, these ions are replaced from the exchangeable fraction. The nonexchangeable K ions are gradually released until a new equilibrium is reached. With the addition of K ions from fertilizer, the ions move from the soil solution to exchange sites and into the nonexchangeable fraction. The remaining 90 to 98 percent of the soil potassium is unavailable and is mostly in minerals such as feldspars and micas. This potassium becomes available only as the minerals gradually weather.

Because of the fixation of applied potassium, the application techniques are more similar to phosphorus than nitrogen. In soils which have a high potassium fixation potential, it is common to band the application for row crops, whereas surface broadcast applications are suitable for perennial plants with well-established surface root systems. It is not unusual for a heavy organic mulch to increase potassium uptake by large trees more effectively than potassium fertilizers applied to the soil surface. Presumably the mulch encourages rooting in the surface soil where potassium may be available, whereas the potassium applied as fertilizer is fixed in surface soil and does not penetrate into the root zone.

Potassium Fertilizers

The most commonly used potassium fertilizer is potassium chloride (KCl), frequently called *muriate of potash*. This material occurs naturally in deposits laid down as briny lakes which have dried up. Such deposits are found in Europe as well as in the United States and Canada. Muriate of potash contains about 50 percent K or the equivalent of 60 percent K_2O. In situations where the chloride ion is undesirable or sulfate is preferable to chloride, potassium sulfate (K_2SO_4) is used. This source, referred to as *sulfate of potash*, contains about 43 percent K or 52 percent K_2O equivalent. A third source is potassium nitrate (KNO_3), which provides not only potassium but nitrogen as well. The analysis of KNO_3 is about 13 percent N and 38 percent K (44 percent K_2O). The major limiting factor in the use of KNO_3 is expense. An advantage would be that it is a maximum source of nutrients with no extra ions present to cause excess salt problems.

CALCIUM

The concentration of calcium in plant tissue ranges from quite low up to several percent dry weight.

Functions of Calcium

A large portion of the calcium present in plant tissue is in the middle lamella, or cementing layer, between cells. It is in the form of calcium pectate and, being immobilized, is not available for reuse elsewhere in the plant. Because of its relative immobility, calcium must be continually supplied. Other functions of calcium are less well documented, but one relates to the normal function of meristems, especially in the roots. Calcium also functions in the maintenance of membrane structure and function and in cell division. Calcium seems somehow to regulate the amount of organic acids present by neutralizing or precipitating excess acids, as exemplified by the formation of calcium oxalate crystals.

Deficiency Symptoms

The "classical" symptoms of a deficiency of calcium are stunted growth and curled and distorted leaves often showing a hook at the leaf tip. Roots are especially hard hit by a deficiency of calcium; roots deficient in calcium have a stubby, brown appearance.

Over the past 20 to 25 years, several very serious disorders of horticultural crops have been shown to result from calcium deficiencies. Examples include blossom-end rot of tomato, blackheart of celery, and bitter pit of apple. In the past each of these disorders was related to adverse water relations and was also thought to be affected by other factors such as nitrogen level. It is now known, however, that calcium deficiency is the primary cause. As we have learned more about the role of calcium in the nutrition of crop plants, certain other more important roles have been found. Calcium level in apple fruits not only influences tissue firmness but also affects the respiration rate and storage life of the fruit. Although the classic symptoms of calcium deficiency are unknown under orchard conditions, calcium levels are in many cases well below optimum, as is reflected in the occurrence of bitter pit, excessively soft fruit, and the short storage life of apples. Although foliar applications and postharvest dips have helped, it is very difficult to elevate calcium levels in fruit.

Soil Calcium

The total amount of calcium in soils is considerably less than potassium, but a much higher proportion of the calcium is usually available. Whereas only 1 to 2 percent of the total potassium is available at any one time, it is not uncommon for more than 20 percent of the total calcium to be in readily available forms. Of the readily available calcium, well over 99 percent is held as exchangeable calcium, and less than 1 percent is actually in solution. As calcium is lost from the soil solution by plant uptake or leaching, the equilibrium is reestablished by the release of exchangeable calcium, which is usually replaced by hydrogen in humid areas. As this replacement gradually occurs, soil acidity increases and can be reversed by the addition of lime (see Chapter 10).

Calcium Fertilizers

The major source of calcium added to soils has been lime used to raise the soil pH. Since superphosphate contains more calcium than phosphorus, it is also a good source of calcium. It was usually assumed in the past that if soil pH was adequately maintained, calcium would be sufficient. The realization that devastating problems such as bitter pit of apples and blossom-end rot of tomatoes are calcium-related has led to extensive research efforts on the role of calcium in nutrition. Perhaps the greatest effectiveness has been found from foliar sprays with calcium chloride or calcium nitrate. As was noted previously, the problem with calcium is not so much the shortage of calcium in the soil as either uneven uptake by the crop or inadequate distribution within the plant. Because of the poor distribution of calcium within plants, leaf analysis for calcium is very inadequate to determine the calcium content of the fruit.

It is an interesting observation that calcium had been essentially ignored in nutritional research because calcium deficiency was very rare. Now, however, because of its involvement in various major disorders, research on calcium is very much in the forefront.

SULFUR

Sulfur is required by plants in much smaller quantities than nitrogen but in the same general range as phosphorus.

Functions of Sulfur

Since sulfur is a constituent of the amino acids cysteine, cystine, and methionine, sulfur is obviously necessary for the synthesis of proteins which contain any of these three amino acids. Sulfur is also a component of thiamine and biotin, which are vitamins, and of coenzyme A, which is involved in the Krebs cycle (Chapter 4). Certain horticultural crops such as onion, garlic, and mustard owe their particular odor and flavor to sulfur-containing compounds.

Deficiency Symptoms

In many crops the symptoms of sulfur deficiency are very similar to those of nitrogen deficiency, because both deficiencies result in a shortage of proteins. One difference, however, is that the chlorosis from sulfur deficiency is generally worse on young foliage, whereas symptoms of nitrogen deficiency are most severe on the older leaves. Other less-apparent results of deficient sulfur are the accumulation of soluble nitrogen materials and a shortage of carbohydrates. The former results from depressed protein synthesis, the latter stems from reduced photosynthesis.

Soil Sulfur

Soils contain sulfur in both organic and inorganic forms. As a constituent of proteins, amino acids, and other complex molecules, the organic fraction is released slowly as the organic materials decompose. A small portion of the total sulfur in a soil exists as the sulfate ion (SO_4=) in solution, and some is adsorbed on soil particles. A so-called "sulfur cycle" exists, which is somewhat similar to the nitrogen cycle described earlier. Small amounts of sulfur dioxide, SO_2, are present in the atmosphere owing to its release from the burning of fossil fuels and the decomposition of organic matter. When present in very low concentrations in the air, SO_2 can be adsorbed by plants, at least partially meeting sulfur needs. However, even at an atmospheric concentration of 1 ppm, sulfur dioxide is very toxic to plants and is therefore normally considered an air pollutant. Since SO_2 is water-soluble, sizable quantities are deposited in rainfall, and in many locations it has been estimated that sulfur deposited in rainfall will equal the sulfur lost by crop removal. Recently, the increased strictness of laws controlling air pollution has led to decreased levels of SO_2 in the atmosphere, especially around industrial sites. Under well-aerated conditions, the major inorganic form of sulfur is the sulfate (SO_4=) ion, and it is in this form that sulfur is taken up by plants. Upon the decomposition of organic molecules, sulfur may be released in various forms such as sulfides, but the final product is sulfate. Under conditions of plentiful calcium, the formation of gypsum ($CaSO_4 \cdot 2H_2O$) occurs; and since gypsum is only slightly soluble, precipitation occurs. Under conditions of very poor aeration there is an accumulation of reduced forms of sulfur, such as iron sulfides.

Sulfur Fertilizers

The application of sulfur-containing fertilizers is becoming more frequent than it was. Part of the reason is reduced sulfur in rainfall as a result of more stringent laws controlling air pollution. Also, as we move to higher-analysis fertilizers, the amount of sulfur present in fertilizer declines. For example, superphosphate contains considerable quantities of $CaSO_4$; but for high-analysis fertilizers, formulators switch to treble superphosphate, which contains no sulfur. Similarly, as ammonium sulfate is replaced by ammonium phosphate, potassium nitrate, or anhydrous ammonia, sulfur is eliminated from the mix.

   As sulfur is needed, it can be applied in materials such as superphosphate and ammonium sulfate, or it may be more economical to apply elemental sulfur or gypsum. Quantities needed to meet the sulfur needs of crops are much smaller than those mentioned earlier for the treatment of saline sodic soils (see Chapter 10).

MAGNESIUM

Because magnesium exists as a divalent cation (Mg^{++}), it is often considered as having many characteristics in common with calcium. There are, however, some important differences in addition to the similarities.

Functions of Magnesium

Probably the most widely known fact about magnesium is its central position in the chlorophyll molecule (see Chapter 6). Since magnesium is necessary for chlorophyll formation, it is obviously vital to photosynthesis. In spite of the importance of magnesium to chlorophyll, much of the magnesium found in leaf tissues is not a part of the chlorophyll molecule. It is thought that much of this additional magnesium functions in the chloroplasts as an enzyme activator which facilitates a wide diversity of reactions, especially in the transfer of energy. Seed also tend to be quite high in magnesium content.

Deficiency Symptoms

The most obvious symptoms of a magnesium deficiency relate to its central position in the chlorophyll molecule and are exhibited as an interveinal chlorosis. Since magnesium is a relatively mobile element in the plant, it behaves much more like potassium than like calcium. The deficiency appears first on the older leaves and advances upward. As the deficiency becomes severe, progressive defoliation may start at the shoot base.

With certain horticultural crops, symptoms of deficient magnesium directly affect the economic part of the plant. In the northeastern part of the United States, magnesium deficiencies have been common in apple orchards and frequently result in very severe preharvest fruit drop. Citrus fruit, sweet potatoes, and the stone fruits are other crops likely to suffer from magnesium deficiencies.

Soil Magnesium

Magnesium exists in the soil in forms much like calcium. It is held by soil particles, and this exchangeable fraction is in equilibrium with the magnesium in the soil solution. A large percentage of the total soil magnesium is in relatively unavailable forms. The unavailable or slowly available magnesium is in minerals such as mica and dolomite and is released only as these minerals weather by the action of soil water which contains carbonic acid. Magnesium in the soil is depleted more quickly than either calcium or potassium, and sandy soils are more likely to be deficient in magnesium than finer-textured soils with greater cation exchange capacity.

Magnesium Fertilizers

The material most frequently used to add magnesium is dolomitic limestone. Whereas calcite is mostly $Ca(CO_3)_2$, dolomite is $CaCO_3 \cdot MgCO_3$. Where magnesium is likely to be deficient, it is advantageous to use dolomitic rather than calcitic limestone. Not only is excess acidity neutralized, but sizable quantities of magnesium are added at relatively low cost.

Other than dolomite, the major source of magnesium is magnesium sulfate or epsom salts ($MgSO_4$). For quick response, magnesium sulfate is dissolved in water and applied as a foliar spray; repeated sprays may be required to alleviate a severe deficiency. For more long-term effect, the magnesium sulfate can be applied to the soil.

IRON

Iron is considered a micronutrient but is required in greater quantities than any of the other six micronutrients and was the first to be proven essential.

Functions of Iron

The involvement of iron in the metabolism of plants is very broad. Iron is necessary for chlorophyll synthesis, although it is not a part of the molecule. An iron deficiency also leads to abnormal chloroplast structure. Iron is involved in many enzymatic reactions in respiration and photosynthesis.

Deficiency Symptoms

The symptom of iron deficiency is quite specific and is termed *iron chlorosis*. The young leaves show a yellowing or even a whitish coloration in the interveinal areas. Since the veins tend to stay dark-green, they are in strong contrast to the light-colored areas between. Until the deficiency is very severe, iron-deficient plants show little leaf necrosis. As severity increases, crop plants become stunted, produce poorly, and may eventually die.

Soil Iron

In sharp contrast to the situation with many nutrients, the supply of iron in the soil is seldom the limiting factor in iron availability. Most soils contain large quantities of iron, but much of it is in unavailable forms. Iron exists in soils in both ferrous (Fe^{++}) and ferric (Fe^{+++}) forms. In well-aerated soils, iron is mostly in the ferric form; in poorly drained soils, the ferrous form predominates. The availability of iron is very dependent on soil pH. In alkaline soils, most of the iron is precipitated in compounds containing iron, hydroxyl, and phosphate ions. An iron deficiency resulting from high soil pH is often called *lime-induced chlorosis*. In more acid soils, iron availability is greatly increased, and iron may actually be available in toxic quantities.

Iron Fertilizers

Iron sulfate was often used for the treatment of iron deficiency. Foliar applications of iron sulfate can be quite effective in adding small quantities. Soil applications of iron sulfate frequently were unsuccessful because the iron was rapidly converted to insoluble forms. The introduction of iron chelates has dramatically improved the success in curing iron deficiencies. Chelates are complex organic molecules which combine with cations, but do not ionize. The iron is held in such a way that it can be absorbed by plants but does not react in the soil to form insoluble precipitates. The most widely used chelating agent is ethylene diamine tetracetic acid (EDTA). When combined with iron to form the iron chelate, it is called *Fe–EDTA*. Although rather expensive, the iron chelates are widely used and often are effective for up to 2 years.

BORON

Although the crop requirement for boron is very low, boron deficiencies are common in many parts of the United States and in a wide variety of crops. A rather unusual feature of boron as a nutrient is the narrow range between deficient, sufficient, and toxic levels. It must be emphasized that although deficient boron can cause various maladies, excess boron is lethal.

Functions of Boron

The exact roles of boron in the normal metabolism of plants have not been proven, but several hypotheses have been proposed. Boron seems to function and to somehow regulate certain aspects of carbohydrate metabolism. It has also been suggested that it is involved in carbohydrate translocation, cell wall development, and ribonucleic acid (RNA) metabolism. It remains for future research, however, to unequivocally prove the precise roles of boron in plants.

Deficiency Symptoms

The spectrum of boron deficiency symptoms is very broad, but the most common characteristic of boron-deficient plants is the stunting, discoloration, and eventual death of the apical meristems of both shoots and roots. Other symptoms include thick, brittle leaves which may become dark and discolored as well as malformed. Flowering is suppressed by a lack of boron; and even if flowering does occur, fruit set is limited because pollen germination and growth of the pollen tube are inhibited by a boron deficiency.

Major disorders of horticultural crops have been shown to be the result of inadequate boron levels. These include internal cork of apples, hollow stem and browning of cauliflower, internal black spot of beets, and cracked stem of celery, as well as a myriad of others. Legumes are particularly susceptible to boron deficiency.

Soil Boron

Total boron in most soils is quite low in comparison to many of the other nutrients. Under acid soil pH, boron availability is relatively high and some leaching may occur, especially in sandy soils. Although the mechanisms are not well understood, boron tends to be fixed at high soil pH, and overliming can induce boron deficiencies. Boron uptake is affected by moisture levels in soils and tends to be depressed at very high or low soil moisture levels.

Boron Fertilizers

The most widely used source of boron has been sodium tetraborate (borax), which contains about 11 percent boron. This material is mined from deposits in California. Recently, boron trioxide has become available under the names of fertilizer Borate 46 and fertilizer Borate 65. These contain about 14 and 20 percent boron, respectively. Soil application of boron fertilizers has been most common, but under some situations, foliar sprays are recommended. A spray on apple leaves provides quick response, danger of toxicity is lessened, and one spring treatment will satisfy the boron needs of the trees for the year. Many feel that this procedure is preferable to a soil application every 3 years as formerly was standard practice.

ZINC

The level of zinc is often deficient in plants, and the effects of a zinc deficiency can be devastating.

Functions of Zinc

Zinc serves as both a component and an activator of enzymes and is also thought to be involved in protein synthesis. Zinc has a direct influence on the level of auxin—indoleacetic acid—in plants. It is hypothesized that zinc is required either for the synthesis of tryptophan or in the conversion of tryptophan to auxin. In either case, zinc deficiency is characterized by symptoms resulting from an auxin shortage.

Deficiency Symptoms

The most common symptoms of inadequate supplies of zinc are related to striking suppressions of growth. The terms *little leaf* and *rosette* are used to describe zinc-deficiency symptoms and are very descriptive. *Little leaf* refers to severe stunting of leaves, and *rosette* describes the lack of internode elongation, which causes several leaves to originate very close to one another. Both of these effects relate to the suppression of auxin. Other symptoms include frenching or mottle leaf of citrus, and white bud of corn. Fruit trees are particularly prone to show zinc deficiency, whereas vegetable and field crops are affected somewhat less often.

Soil Zinc

As is true with many other nutrients, restricted availability of zinc is often more important than low total levels of zinc in soils. The solubility of zinc compounds in soils is low, especially when soil pH is high. For this reason the areas in which zinc deficiencies were first noted were the calcareous soils in the Western United States. Certain interactions occur with other nutrients, although the mechanisms are not well understood. High phosphorus or nitrogen levels seem to depress zinc availability, perhaps through the formation of unavailable compounds. Zinc is quite readily leached from sandy soils.

Zinc Fertilizers

Since the discovery of zinc deficiency, the major source of zinc has been zinc sulfate. Applications to the soil are effective in acid soils which are not too sandy. In calcareous or sandy soils, it is more effective to spray the plant with a solution of zinc sulfate. With some crops a foliar spray is suitable. With certain perennials such as apple, however, uptake is most efficient through pruning cuts; for this reason trees are sprayed in the late dormant season.

MANGANESE

Manganese is essential in small quantities but is toxic when excessive amounts are taken up by plants.

Functions of Manganese

Manganese is involved in many different plant processes, including chlorophyll synthesis, respiration, photosynthesis, and nitrogen assimilation. Although precise roles are yet to be clearly defined, manganese seems to be associated with multiple enzyme systems, especially in the Krebs cycle.

Deficiency Symptoms

Although symptoms vary with the crop, the most frequently encountered symptom of manganese deficiency is a mottled chlorosis of leaves, with some necrotic spots developing as the deficiency worsens. Growth is poor, and both flowering and fruiting are suppressed. The foliar manifestations are worse on young leaves because, like iron, manganese is not readily translocated.

Toxicity Symptoms

The most frequently seen symptoms of manganese toxicity are crinkling and cupping of leaves, often accompanied by chlorosis of leaf margins. In apple trees a disorder called *measles* or *internal bark necrosis* can apparently result from excessive levels of manganese, but in other situations this condition seems to be the result of other causes. In many cases the symptoms of manganese toxicity are similar to those of iron deficiency, and in fact the two ions do appear to be competitive.

Soil Manganese

Most soil manganese is in relatively insoluble forms at high soil pH but becomes increasingly more soluble as the pH is lowered. The situations in which manganese is toxic usually occur in soils at a pH of 5.5 or below. The most logical cure for excess manganese is the addition of sufficient lime to raise the soil pH. Soils most likely to be deficient in manganese are those which are sandy, organic, or alkaline.

Manganese Fertilizers

Manganese sulfate applied to the soil or sprayed on the foliage is effective in eliminating manganese deficiencies. Since manganese is tied up in alkaline soils, manganese sulfate applications are ineffective. Although more expensive, chelated manganese is often the best choice where the pH of the soil is high.

COPPER

A deficiency of copper is rarer than a deficiency of either boron or zinc, but under certain conditions it is fairly widespread. Crops are most prone to exhibit a copper deficiency on soils which are sandy, organic, or have a high pH.

Function of Copper

Copper is required to activate several enzymes in plants, is needed for chlorophyll synthesis, and is involved in carbohydrate and protein metabolism.

Deficiency Symptoms

Although the specific manifestations of deficient copper vary among species, leaf symptoms are the most general. Leaf size is suppressed, the tip often becomes necrotic, and leaf color is darker than normal. If the deficiency is severe, dieback of shoots will occur; in stone fruits gum may be exuded from the bark.

Soil Copper

Copper occurs as both the Cu^{++} and Cu^+ ions, which are exchangeable cations, with the solubility of copper being highest under acid soil conditions and declining as the pH is raised. Most soil copper comes from mineral particles as they weather; therefore, deficiencies are quite frequent on organic soils such as peat and muck. Because of unavailability of copper, calcareous soils are also occasionally deficient. Sandy soils are sometimes deficient because of leaching and lack of replacement by weathering.

Copper Fertilizers

Copper sulfate is the most frequently used source of copper. It is soluble and at one time was used in combination with lime to form Bordeaux mixture, which was a widely used fungicide. If soil applications of copper sulfate are ineffective because of a high pH, a foliar spray of copper sulfate or a copper chelate spray can be applied to the soil or plant.

MOLYBDENUM

Of all the micronutrients, molybdenum is required in the smallest amounts; in most soils, it is present in very minute quantities.

Functions of Molybdenum

The two most clearly defined roles of molybdenum relate to nitrogen metabolism. First, molybdenum is required by plants for nitrate reduction and assimilation. Second, molybdenum is necessary for the symbiotic nitrogen fixation by legumes and *Rhizobium* bacteria in their root nodules. In both of these processes, molybdenum apparently functions as an enzyme component.

Deficiency Symptoms

Because molybdenum plays a vital role in nitrogen metabolism, it is not surprising that the early symptoms of molybdenum deficiency are those of nitrogen deficiency. Older leaves become chlorotic in the interveinal areas, often having a mottled appearance. The margins may wilt, roll, and cup as the deficiency becomes more severe. In cauliflower and related species, the term *whiptail* has been given to molybdenum deficiency because the malformed leaves resemble a whip when the midrib is the predominant portion of the leaf which develops. Another common deficiency symptom is *yellow spot* of citrus. Both whiptail and yellow spot have been troublesome for many years, but the fact that they are caused by deficient molybdenum is a recent discovery.

Soil Molybdenum

Most soils contain some molybdenum, but deficiencies are not uncommon. Unlike the other micronutrients, molybdenum is more available in basic soils. Because of this, liming of acid soils will often increase availability enough to eliminate a molybdenum deficiency.

Molybdenum Fertilizers

Foliar sprays of sodium or ammonium molybdate are often effective; and because of the very small requirement, only 75 to 80 g may be necessary to treat 1 ha. In some crops, dusting the seed with a molybdenum compound before planting may meet the needs of the crop. Soil application may require about 1 kg ha^{-1} to be effective, but for citrus crops this has not been as satisfactory as foliar sprays.

CHLORINE

The general acceptance of chlorine as an essential element has occurred only recently. Presumably there are two main reasons for the rather long delay: (1) Chlorine deficiency is unknown under field conditions. (2) Chlorine is present in soils, water, air, and in many fertilizer materials; it is therefore very difficult to maintain a chlorine-free environment.

Function of Chlorine

Although the exact role is not well understood, chlorine is essential for a part of the photosynthetic process in which oxygen is liberated.

Deficiency Symptoms

Plants which are deficient in chlorine tend to wilt and show a buildup of free amino acids, but the mechanisms are not understood.

Toxicity Symptoms

Toxic levels of chlorine are known to occur and can be very detrimental to a broad array of plants. Chlorides are of major concern in saline soils (see Chapter 10) as well as in areas adjacent to highways which are salted for ice removal. Symptoms of excess chlorine include chlorosis and necrosis of leaves, growth suppression, lowered yields, and death if chlorine levels become very high.

Soil Chlorine

Chlorine exists as the Cl$^-$ ion and, as such, moves freely in the soil. The chlorine present may come from fertilizers such as potassium chloride, but perhaps the major source is rainfall. Especially along coastal areas salt spray evaporates, and the chloride is redissolved in rainwater and deposited on the soil.

BIBLIOGRAPHY

Bidwell, R. G. S.: *Plant Physiology*, Macmillan, New York, 1974.

Brady, N. C.: *The Nature and Properties of Soils*, 8th ed., Macmillan, New York, 1974.

Childers, N. F. (ed.): *Nutrition of Fruit Crops*, Horticultural Publications, New Brunswick, N.J., 1966.

Epstein, E.: *Mineral Nutrition of Plants: Principles and Perspectives*, Wiley, New York, 1972.

Gauch, H. G.: *Inorganic Plant Nutrition*, Dowden, Stroudsburg, Pa., 1972.

Hewitt, E. J., and T. A. Smith: *Plant Mineral Nutrition*, Wiley, New York, 1974.

Meyer, B. S., D. B. Anderson, R. H. Bohning, and D. G. Fratianne: *Introduction to Plant Physiology*, Van Nostrand, New York, 1973.

Thompson, L. M., and F. R. Troeh, *Soils and Soil Fertility*, 4th ed., McGraw-Hill, New York, 1978.

PART 3
Horticultural Practices

CHAPTER 12
MECHANISMS OF PLANT PROPAGATION

 Biologically, all living things reproduce and thereby perpetuate the species. Plants are unique in their ability to reproduce asexually. Salamanders can regenerate legs, as can crabs and lobsters. But only in test tubes can human beings generate animals asexually—for example, make whole frogs from frog cells. In plants, then, the unique thing called *totipotency for reproduction* is achieved. *Totipotency* is the ability to generate or regenerate a whole organism from a part. All cells in all living organisms have the potential for totipotency, but only plants can use the generative process.

Plant propagation is the increase in numbers or perpetuation of the species by reproduction. The term *propagation* indicates that human beings are involved in the reproductive process. We have learned to propagate plants, not only as they naturally reproduce, but more recently in ways that plants do not reproduce in nature. Vegetative cuttings of hollies do not occur naturally, but we have found ways to induce cuttings to root. On the other hand, humans often copy nature: apple trees are stooled by the mound layerage process, which is similar to the layerage process that occurs in nature. Horticulturists use plant propagation to perpetuate, increase, and repair plants and to control growth.

To cultivate plants from the wilds necessitated a knowledge of propagation. With the developing knowledge of propagation and other cultural methods, it became unnecessary for humans to wander in search of plants in order to survive. Plants in the wild continually evolve, but humans have selected and changed the pathway. Plants were domesticated by: (1) selecting from the wild plants such as lima beans and tomatoes, species which have evolved into types that differ radically from their wild relatives; and (2) making hybrids such as pear, maize, strawberries, and prunes, with hybridization being accompanied by changes in chromosome number. There could be relatively little progress in plant breeding without plant propagation. Most cultivated plants would be lost or revert to less desirable forms unless propagated under controlled conditions. Selection is also a major part of plant improvement, especially with tree fruits (e.g., the transition from seedlings to vegetatively propagated cultivars of the past 150 to 200 years).

Plants reproduce by sexual and asexual methods. *Sexual reproduction* is the reproduction of plants through a sexual process involving meiosis. The sexual cycle results in the production of seed or spores. A seed (embryo) results from the fertilization of the egg (female gamete) by pollen (male gamete). Because of the recombination of genetic information that occurs, sexual reproduction takes advantage of natural genetic variation inherent in the two parent plants. *Asexual reproduction* is defined as the duplication of a whole plant from any cell, tissue, or organ of that plant. It takes advantage of the precise duplication of genetic information inherent in mitosis and is possible because of totipotency. Through mitotic cell division and differentiation, a clonal population can be duplicated exactly for hundreds of years, leading to the production of thousands of genetically identical plants. Plants are unique among life forms in that they have the capacity to reproduce asexually. In addition to sexual and asexual reproduction, there are two methods which combine features of both: *apomixis* and *tissue culture*.

It is generally agreed that sexual reproduction is more highly evolved than asexual reproduction. With sexual reproduction more opportunity exists for variability, adaptibility, and change. Therefore, it is not surprising that several asexual organisms have developed a form of reproduction which may resemble, approach, or actually be sexual reproduction.

In nature most plants reproduce from seed and thereby maintain variation. But many plants reproduce asexually because they have a strong characteristic which competes effectively with the environment. The procedure or method for propagating plants commercially is dependent upon the time required, the cost, and the method required by each plant.

SEXUAL PROPAGATION

The most common type of plant propagation is the sexual method, or production of plants from seed. Most plants reproduce themselves through seed. Most agronomic, vegetable, and forest crops, in addition to many annual and perennial flower crops and some woody ornamentals, are seed-propagated (Figure 12-1). Propagation by seed is usually cheaper and easier than other methods and is used whenever possible. Certain limitations exist, however. Plants (such as hydrangea and snowball viburnum) which do not produce viable seeds cannot be propagaged from seed. Many plants, such as apple, pear, plum, peach, and raspberry, display such genetic variability in seed that

FIGURE 12-1 Geranium plant production from seed. (Photograph, Ball Seed Company.)

they are not propagated sexually. Seed planted from these fruits usually result in the production of inferior plants and fruit.

Sexual reproduction results in the creation of plants with new genotypes. In some ways, seed propagation can now approach the genetic and phenotypic features of asexual propagation, provided that the plant material is genetically homozygous. Southern Corn Leaf Blight, which caused a near disaster in corn production in the early 1970s, resulted from mutations in fungi that were able to parasitize the male sterile T source of cytoplasm. Much commercially available hybrid corn seed had this T cytoplasm, because it was the cytoplasm of the female parent that prevented the formation of viable pollen, thus allowing seed companies to sell hybrid corn seed without incurring the cost of hand detasseling. This procedure is critical to the collection or loss of germ plasm, and it is increasingly done for world populations of seed-reproduced plants that have very narrow genetic backgrounds and thus very limited genotypes. However, the process is similar to incest and could lead to serious problems in such crops. Through natural selection, seed from sexual cycles result in variability and lessen the vulnerability of a plant species to disease or insects. Humans should honor that fact and work within available sources of germ plasm.

Seed Propagation

Initially, the success of seed propagation is dependent upon the production of flowers, normal meiotic behavior in the formation of the microspores and megaspores, fusion of these male and female gametes to form the zygote, and the subsequent development of the embryo and endosperm to form the seed. Finally, it is dependent upon the germination of the seed and the establishment of the seedling.

The Germination Process   Germination (Figures 12-2*a* and *b* and 12-3*a–d*) is the initiation of active growth by the embryo, resulting in the rupture of seed coverings and the emergence of a new seedling plant capable of independent existence.

Conditions   Three conditions must be met for germination to occur: (1) viable seed, (2) favorable internal conditions, and (3) favorable environmental conditions. Usually the radicle, or embryonic root, emerges first, and the plumule pushes upward to develop the primary shoot system.

Factors affecting germination   Necessary factors accompanying germination are the absorption of water by the seed, production of enzymes and hormones, hydrolysis of insoluble stored foods into

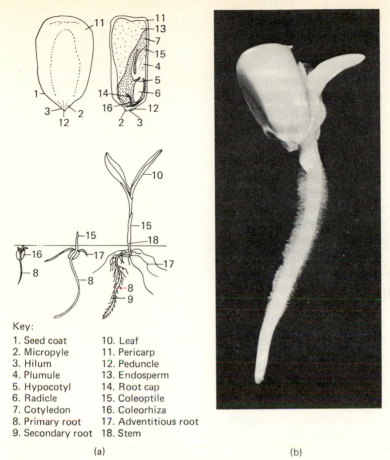

Key:

1. Seed coat
2. Micropyle
3. Hilum
4. Plumule
5. Hypocotyl
6. Radicle
7. Cotyledon
8. Primary root
9. Secondary root

10. Leaf
11. Pericarp
12. Peduncle
13. Endosperm
14. Root cap
15. Coleoptile
16. Coleorhiza
17. Adventitious root
18. Stem

(a)   (b)

FIGURE 12-2 (*a*) Development of corn seedling. (*b*) Germinating corn (*Zea*) seedling with young primary root nearly covered with root hairs. Note that no hairs are on root tip. As the corn seedling develops, its primary root grows for a time and produces secondary roots. (Carolina Biological Supply Company.)

soluble forms, and the translocation of these soluble foods to the growing points. Food reserves, water, oxygen, and temperature affect the germination process. In some cases light and physiological factors may also affect germination.

FOOD RESERVES   Food reserves nourish the seedling until photosynthesis occurs at such a rate that the seedling can become independent. The larger the food reserve, the greater the vigor of the seedling, unless another factor is limiting. Plump seed usually have more food reserves than small, shriveled seed. This is why many cultivars of sweet corn germinate more poorly than field corn cultivars.

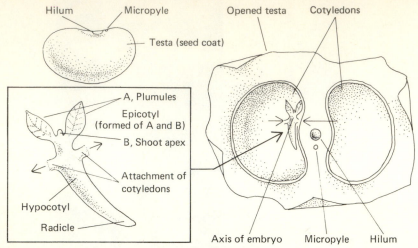

FIGURE 12-3 *(a)* Development of a bean plant: bean seed.

WATER   Water is required to soften the seed coat; soaking certain seed before planting shortens the time required for the seedling to emerge from the soil. The moisture in the seed necessary for germination varies from 43 percent by weight (for corn) to 120 percent (for sugar beets). The imbibition of water results in an increase in volume of the seed and in the ultimate breaking of the seed coat. Water is also required for activation of the enzymes which transform some of the stored food (starch, sugar, fat) into energy and chemical compounds used for the production of new cells and tissues. Water is necessary to maintain adequate moisture levels once the germination process begins, because even temporary drying can result in death of the seed or plant. Moist soil is generally all that is necessary to begin germination. However, some seed have an impervious seed coat and cannot absorb moisture or oxygen without special treatments that scarify the seed to permit gas and water exchange. *Scarification* is any process of breaking, scratching, or mechanically altering the seed covering to make it permeable to water and gases. Although some scarification probably occurs during harvesting, extraction, and cleaning, germination of most seeds with hard seed coats is improved by additional artificial treatment. Scarification can be an abrasive action against the seed coat (such as with a file), mechanical breaking by bouncing the seed over a rough surface similar to sandpaper, or a chemical action such as soaking for a short time in concentrated sulfuric acid. Acid scarification is recommended for raspberries, blackberries, and, sometimes, spinach. Mechanical breakage of the seed coat in nuts and

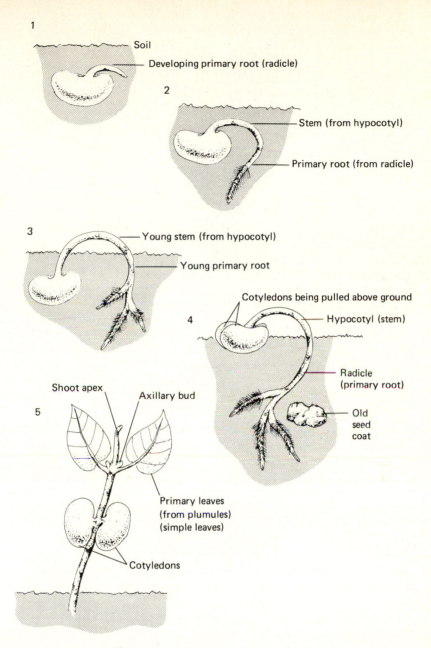

1

Soil

Developing primary root (radicle)

2

Stem (from hypocotyl)

Primary root (from radicle)

3

Young stem (from hypocotyl)

Young primary root

Cotyledons being pulled above ground

Hypocotyl (stem)

4

Radicle (primary root)

Old seed coat

Shoot apex

Axillary bud

5

Primary leaves (from plumules) (simple leaves)

Cotyledons

FIGURE 12-3 *(b)* Development of a bean plant: germinating bean seed.

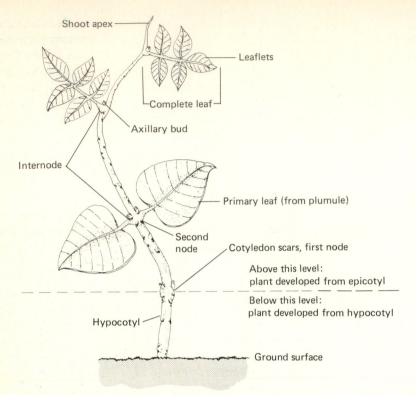

Shoot apex

Leaflets

Complete leaf

Axillary bud

Internode

Primary leaf (from plumule)

Second node

Cotyledon scars, first node

Above this level:
plant developed from epicotyl

Below this level:
plant developed from hypocotyl

Hypocotyl

Ground surface

FIGURE 12-3 *(c)* Development of a bean plant: bean plant.

FIGURE 12-3 *(d)* Closeup of a bean gemination test showing normal seedlings and ungerminated seed. Some seed that sprout make abnormal plants—that is, plants that will not survive in the field. (Photograph by Agricultural Research Service—U.S.D.A.)

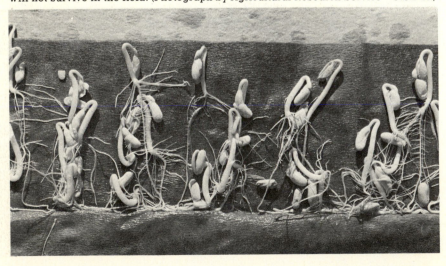

peaches is sometimes advantageous. Placing seed outdoors during the winter where they are kept moist and subjected to alternate freezing and thawing may accomplish the desired results, but organic acids and microorganisms present in the soil may also degrade the seed coat.

OXYGEN   For germination, oxygen must be present to oxidize starches, fats, and other reserves. Oxygen is also necessary in the respiration process for the oxidation of sugars. Therefore, to allow a free flow of oxygen, the germination medium should be loose, friable, and well-aerated. Seed in heavy soils may show very poor germination, especially during a wet season when the soil often lacks sufficient oxygen. Deep soil preparation usually enables the water to drain properly and improves aeration.

TEMPERATURE   Temperature affects germination percentage as well as rate of growth. Temperature influences the rate of absorption of water, the translocation of foods and hormones, the respiration rate, cell division and elongation, and the rate of many other physiological reactions. Temperatures can be either too high or too low for satisfactory germination. Fresh lettuce seed may become dormant at a temperature of 30°C or higher. Low temperature may not prevent germination, but since it greatly reduces growth rate, it may prevent emergence of the seedling from the soil. Most seed germinates well around 21°C (see Table 12-1).

TABLE 12-1
SOIL TEMPERATURE CONDITIONS FOR VEGETABLE SEED GERMINATION

CROP	MINIMUM °C	OPTIMUM RANGE, °C	MAXIMUM °C
Warm Season:			
Bean	16	16–29	35
Bean, lima	16	18–29	29
Corn	10	16–35	40
Cucumber	16	16–35	40
Squash	16	21–35	38
Tomato	10	16–29	35
Watermelon	16	21–35	40
Cool Season:			
Beet	4	10–29	35
Cabbage	4	7–35	38
Carrot	4	7–29	35
Lettuce	2	4–27	29
Onion	2	10–35	35
Pea	4	4–23	29
Radish	4	7–32	35

Source: E. P. Christopher, *Introductory Horticulture*, McGraw-Hill, New York, 1958.

LIGHT   Most seed germinate equally well whether kept in darkness or exposed to light. Some seed, like lettuce and celery, are stimulated to germinate by exposure to wavelengths of light in the red part of the spectrum (660 nm) (see Chapter 9). Others, such as garlic and onion, are inhibited by exposure to light.

PHYSICAL AND PHYSIOLOGICAL FACTORS   Viable seed that fail to germinate when environmental conditions are favorable are said to be *dormant*. Seed dormancy may be due to physical causes (e.g., hard seed coat which is impervious to water) or physiological causes such as chemical inhibitors in fruit and seed (e.g., the unsaturated lactone, Coumarin, in tomato seed, *Lycopersicon esculentum*) and immature embryos (e.g., American holly, *Ilex opaca*).

A rest or after-ripening period is sometimes required after harvest before germination will take place. The amount of rest required varies from a few days to a few years and is sometimes influenced by postharvest treatment. During this period, physiological and mechanical changes take place. Hormones are produced, enzymes are produced, water must be absorbed, and gases must be exchanged. Often the rest period is satisfied during *stratification*. Seed undergoing stratification are placed in a moist medium at temperatures usually from 4 to 7°C for 1 to 3 months. Many seed in the wild are exposed to these conditions naturally. This requirement of seed is very similar to the chilling requirement of the buds of deciduous nut and fruit trees.

Seed Production

Seed production is widespread, and its scale ranges from the home gardener to large commercial enterprises. Seed collecting from plants in their natural habitat is often done both by amateurs and commercially. Commercially, seed from ornamental shrubs and forest trees are collected in large-scale operations.

Commercial growers in this highly specialized industry devote large acreages to the production of one cultivar of vegetable or flower seed. These growers produce cereal and forage crop seed, vegetable seed, and annual, biennial, and some perennial flower seed for both commercial growers and home gardeners. Many seed companies conduct active plant-breeding programs for the major seed crops they produce. This has become especially important with the extension of plant patent laws to cover seed-propagated plants.

Environmental conditions necessary for production of high-quality seed determine the location of commercial operations. Different environmental factors affect the expression of the hereditary characteristics. The temperature, irradiance level, photoperiod, and relative humidity are considered. However, many industries producing vegetable and flower seed are located in environmental areas different from those in which the crops will later be grown, since low humidity and minimum summer rainfall are desirable conditions for drying seed for harvesting. Also, seed for many plants are produced in areas that are more likely to exhibit symptoms of the presence of a latent virus. This is especially important in asexually propagated crops, such as potatoes, strawberries, and many cut flowers. A problem encountered in areas of low humidity is premature shattering of the seed pod during harvesting. Therefore, many flower seed are produced in the Lompoc Valley in California, in Puerto Rico, and in Florida, where the moist winds from the ocean and frequent night and morning fogs aid in preventing the pods from dehiscing before or during harvest. Isolation is important in wind- and insect-pollinated crops.

Cleaning Hand labor is necessary to harvest flower and vegetable seed. This is especially true when the seed head or pod shatters easily or when seed mature over a long period of time, so that flower buds, flowers, and mature seed may be on the same plant at the same time. The chief advantage of hand picking is that the cleaning process is greatly reduced. With some crops, such as carnation, hollyhock, and sweet pea, the entire plant must be cut, placed on canvas, and dried so that the seed can be threshed out. Many crops, especially the fleshy fruits, require that the seed be milled from the fruit and then cleaned. The seed of tomato is removed after maceration of the fruit and fermentation of the pulp. However, in many cases fermentation of the seed for pulp removal may reduce germination.

Differences in size, density, and shape of seed in comparison with plant debris and other undesirable objects generally determine the cleaning practices to be followed. Screens with different sieve sizes are used to separate the larger particles from the seed. Smaller, lightweight particles can be removed by blowing an air current through the seed as they pass from one screen to another or as they pass across a porous bench or against an inclined plane. The heavier seed remain at the base while the lighter materials are blown into another plane. By using an "indent machine," desirable seed is separated from other seed or particles of the same density but of different shape. A wheel covered with indentations—the size and shape of which are determined by the particular crop being cleaned—is passed through a batch of seed, and each "indent" picks up a seed. Some seed can be separated on the basis of color (beans and peas). A single seed is picked up by suction through perforations on a hollow wheel and is then passed through a photoelectric cell; upon detection of an object of the wrong color, the vacuum is released and the seed is ejected. Seed separation and cleaning is a delicate operation. The machinery must be adjusted carefully so that the seed is not damaged or chipped. Damaged seed may fail to germinate, may show reduced viability, or may produce weak seedlings.

Storage It is usually necessary to store seed after harvest. The viability of seed after the storage period depends on: (1) the initial viability of the seed at harvest, which is a result of the production factors and handling methods, and (2) the rate of physiological deterioration inherent in the particular species and influenced by the environmental conditions under which the seed was stored. The approximate longevity of well-stored seed of some common vegetables and flowers is presented in Table 12-2.

Seed, like all living organisms, carry on respiration even when dormant. The respiration rate of seed is largely dependent on water content and temperature. Most seed have the ability to retain a certain moisture level by absorbing moisture from the air. The respiration increases with increasing temperature. The storage life of many seeds can be extended by lowering the relative humidity of the storage atmosphere and by lowering the storage temperature.

Pregermination moisture relationships can also influence germination percentages in many species. Before storage, seed is often dried to approximately 5 to 7 percent of its dry weight. Only 10 to 12 percent is needed to prevent decay. Storage temperatures of −18 to 0°C appear to be optimal for most species. A common rule for most seed is that for each 1 percent decrease in seed moisture or for each 10°C decrease in storage temperature, storage life is doubled. But the seed of some

TABLE 12-2
APPROXIMATE LONGEVITY OF WELL-STORED SEEDS OF SOME COMMON VEGETABLES AND FLOWERS

1 year	Sweet corn, onion, parsnip, okra, parsley Delphinium, candytuft (*Iberis umbellata*)
2 years	Beet, pepper, leek, chives Aster, strawflower, sweet pea
3 years	Asparagus, bean, carrot, celery, lettuce, pea, spinach, tomato Aster, phlox, *Verbena grandiflora*
4 years	Cabbage, cauliflower, Brussels sprouts, Swiss chard, kale, squash, pumpkin, radish, rutabaga, turnip
5 years or more	Cucumber, endive, muskmelon, watermelon Shasta daisy, cosmos, petunia, scabiosa, marigold, pansy, sweet alyssum, pink, hollyhock, stock, nasturtium, zinnia

Source: E. P. Christopher, *Introductory Horticulture*, McGraw-Hill, New York, 1958.

species (wild rice, silver maple, citrus, oaks, and hickory) lose viability if their moisture content is decreased. The seed of a few species, such as apples and pears, must be kept moist in order to maintain good viability.

Seed dried to a low moisture content is often stored in airtight cans or other containers. Again, the seed must be dried to 5 to 7 percent of the dry weight. Any fluctuation in the moisture content of seed during storage will usually reduce longevity.

In some cases, it is possible to modify the storage atmosphere with beneficial results. A vacuum can be created, carbon dioxide content raised, or oxygen replaced by nitrogen or other gases not affecting respiration.

Home gardeners who have been pleased with the performance of a particular plant often save the seed of the plant for the next growing season. However, they may be disappointed with the performance of the plants resulting from these seed, since the plants may have cross-pollinated and produced seed of different and undesirable genetic characteristics. A home gardener may have been planting hybrids produced from a particular cross, such as hybrid corn from inbred lines. Such seed will segregate in the F_2 generation, and the results will be undesirable.

Longevity   The seed of maple, elm, willow, and some other trees lose their viability very rapidly. Desiccation, the drying out or loss of moisture from the seed, plays a role in reducing the longevity of seed. Seed of medium life are those which normally retain their viability for 2 to 3 years and may remain viable as long as 15 years, depending on storage conditions. Certain weed seed have even germinated after hundreds of years, and seed of the Egyptian lotus that were found in pyramids have also germinated.

Treatment   Seed can be treated to protect the seed or the seedling, or both, from fungi, bacteria, and insect and animal pests. Fungi which cause "damping off" of seedlings are often found in greenhouse, hotbed, and cold-frame soils. These fungi attack a wide range of host plants by attacking the seedling stem at ground level, weakening the stem and causing the seedling to collapse.

Three types of seed treatments are used: disinfestants, disinfectants, and protectants. Disinfestants, such as calcium hypochlorite, eliminate organisms present on the surface of the seed, whereas disinfectants, such as formaldehyde, mercuric chloride, and hot water treatment (50 to 56°C for 15 to 30 min), eliminate most organisms within the seed itself. Protectants are insecticides (such as lindane, aldrin, and heptachlor) and fungicides (such as captan and chloranil), which prevent insects or fungi from attacking the seed after treatment.

Seed Testing Careful protection of the seedling and retention of viability are worthless unless the seed possesses certain desirable characteristics. Good seed is clean and free from soil, debris, and weed seed. This can be accomplished by good milling and cleaning operations. Good seed possess a high percentage of viability. Seed vitality encompasses the percentage of a lot of seed that will germinate, the speed at which they germinate, and the health and vigor of the seedlings. Viability can be tested by planting seed in sand or soil or by germinating the seed between folds of blotting paper, cotton flannel, or burlap (Figure 12-3d). Commercial seed houses employ various methods.

Seed certification programs with production standards set by the International Crop Improvement Association maintain and make available all crop seed of good seeding value and true to name. Seed is classified as breeder's seed, foundation seed, registered seed, and certified seed. *Breeder's seed* is directly controlled by the originating or sponsoring plant breeder. *Foundation seed* is seedstock handled to most nearly maintain specific genetic identity and purity under supervised or approved production methods certified by the agency. *Certified seed* is the progeny of *registered* seed stock and is the final stage in the expansion program. It is certified with a metal seal and blue tag.

Hybrid Seed A *hybrid* in its simplest form is a first-generation cross between two genetically diverse parents. Selected hybrids have the desirable characteristics of both parents combined, and vigor is sometimes intensified. A *single cross* is hybridization between two inbred lines; a *double cross* combines two single crosses. The cross between two closely related species often results in increased productivity, tolerance to adverse growing conditions, more rapid growth, or resistance to diseases. Corn, for example, may merge four inbred lines into a single plant in the production of a double-cross hybrid. Plants that are naturally cross-pollinated usually do not breed true. Much crossing also occurs in nature. By selection, a wild population may become quite similar in appearance and response. To improve plants, breeders make controlled crosses usually after self-pollinating for several generations to obtain uniform inbred lines in cross-pollinated crops. Self-pollinated plants have a high degree of *homozygosity*—having identical genes of a gene pair present—such as yellow-seeded pea plant with genes YY for yellow endosperm and yy for white-seeded plants with white endosperm. They breed true from seed in regard to endosperm color. This may be in the form of pure-breeding cultivars or (for one generation) controlled hybrids. To attain uniformity in a given line, inbreeding and selection for several generations are necessary. In many crop plants, controlled cross-pollination between inbred lines is the basis of hybrid seed production.

Seed Planting

Small areas, flats, greenhouse beds, and home gardens are usually planted by hand. Larger areas and fields are usually planted by machine. Seed drills which are used for planting may be operated by hand or may be part of the accessory equipment for a tractor. Grain drills may also be used to plant seed and are advantageous in that they place the fertilizer as a side dressing at the time of planting. A correctly adjusted drill places the seed at a uniform depth and at a uniform distance apart and firms the soil around the seed. Depth of planting depends upon the size of seed, soil structure, time of year, and moisture and oxygen content. Since small seed have less stored food, they must be planted nearer the surface to decrease the time and energy needed for the seedling to develop and start photosynthesizing.

If water is the limiting factor, as it usually is during the germination period, the seed should be planted at a greater depth to prevent desiccation of the emerging seedlings. If the soil is nearly saturated, as it is during late fall, winter, and early spring, the seed should be planted rather shallow, since oxygen is necessary for respiration. Seeding to a depth where the oxygen supply is limited may result in failure.

Temperature requirements of the crop and temperatures of the location determine the best time for planting. Warm-season crops are usually sown after the last killing freeze of spring; cool-season crops may be planted earlier. (See Chapter 7.)

Seed Tapes and Pelleting

When seed are small or irregular, even distribution is often difficult to obtain. To make the seed round and uniform in size, they are pelleted or coated with montmorillonite, a finely divided clay. Pelleting permits more uniform planting, eliminates a great deal of thinning, and permits the incorporation of fungicides onto the coat. Seed tapes or ribbons are available for many flower and vegetable cultivars. The seed is attached and spaced at the proper intervals on a plastic tape. The tapes are buried in a shallow furrow. The tapes disintegrate from moisture as the seed germinates.

ASEXUAL PROPAGATION

Asexual propagation is the duplication of a whole plant from any living cell tissue or organ of that plant. Asexual propagation is possible because normal cell division (mitosis) and cell differentiation occur during growth and regeneration. Mitotic cell division is involved, whether the form of propagation is initiation of roots and shoots or formation of callus tissue in the process of budding and grafting. A single cell can generate a new plant because each cell contains all the genetic information necessary to reproduce the entire organism. When a group of plants originate from a single individual and are propagated by vegetative means, this group is referred to as a *clone*. Clones may be maintained by humans for hundreds of years or may exist in nature by reproduction of plants by bulbs, rhizomes, runners, and tip layers. Since clonal plants are genetically identical to the parent, the desirable characteristics of the cultivar can be perpetuated. The 'Bartlett' pear (which originated in England in 1770 from seed) and the 'Delicious' apple (found almost 100 years ago) are well-known asexually propagated clones. It is quicker and more economical to propagate many plants (strawberry, blueberry, apple, sweet potato, Irish potato, narcissus, chrysanthemum, rose, and most nursery plants) by asexual means.

In addition to the perpetuation of clones, plants may be asexually propagated for other reasons. Many crops, if propagated by seed, will not reproduce true to type. The seedling progeny may not bear much resemblance to the parent and may be much less desirable. Generally, most horticultural species propagated by asexual means have heterozygous genotypes. They are phenotypically variable because of the environment, along with a heterozygous genotype when grown from seed. Thereby, valuable individuals are perpetuated through asexual means allowing for uniformity in phenotype.

Many valuable plants produce little or no seed because of the season of flowering, genetic incompatability, or sterility. Hybrids may be sterile because of a lack of homology between chromosomes of different species which results in irregular cytological behavior and reduced seed production. Polyploid plants, particularly those having odd numbers of chromosomes, as in triploidy, often show no fertility or seed viability. Many double-flowered species where flower petals replace reproductive parts, such as double petunia, carnation, and African violet, produce little or no seed. Another common problem occurs when seed is difficult to germinate. This difficulty may be due to a hard, impervious seed coat, seed dormancy, or an immature embryo. American holly and viburnum are examples.

Dwarfing effects, resistance to certain insects and diseases, better adaptability to a given soil, and hardiness of certain parts of a plant can also be obtained by combining two clones. Disease-resistant rootstocks make it possible to grow cultivars susceptible to soil-borne diseases. The European grape grown on American grape rootstocks is not attacked by root lice, nor are peaches attacked by nematodes when grown on certain nematode-resistant rootstocks. Growth-regulating rootstocks are sometimes used in fruit trees and landscape plants to control size or other characteristics.

Cuttings

Cuttings are detached vegetative plant parts which, when placed under conditions favorable for regeneration, will develop into a complete plant with characteristics identical to the parent plant. Any vegetative plant part that is capable of regenerating the missing part or parts when detached from the parent can be a cutting; hence, roots, stems, leaves, or modified stems such as tubers and rhizomes can all be used as cuttings. The type of cutting used to propagate a given plant is determined by many factors, such as ease of root or shoot formation, availability of leaves, facilities available for propagation, and season of the year.

Types of Cuttings   Cuttings are classified as stem cuttings, leaf cuttings, leaf-bud cuttings, and root cuttings.

Stem cuttings   Segments of shoots containing lateral or terminal buds are obtained for stem cuttings with the expectation that under the proper conditions adventitious roots will develop and thus produce independent plants (Figure 12-4). The type of wood, the stage of growth used in making the cuttings, and the time of year in which the cuttings are taken can be very important in rooting of the plants. There are four types of stem cuttings, divided according to the nature of the wood used in making the cuttings: softwood, herbaceous, hardwood, and semihardwood.

SOFTWOOD CUTTINGS   These cuttings are taken from soft, succulent, new spring growth of deciduous or evergreen species of woody plants. Typical examples are magnolia, weigela, spiraea, and maples. The cuttings, composed of a stem portion and a growing tip, are usually 8 to 12 cm long with the lower leaves removed. Care must be exercised to aid the wound to heal and the roots to develop rapidly. Since the tissue is immature, few carbohydrates have been stored. It is necessary, therefore, to see that the leaves are turgid for photosynthesis. Moisture will be lost from the cutting, and rot-producing organisms can invade the cutting through the unhealed wound. Immediately after the wound is made, the intercellular spaces and cells near the wound become filled with sap. Sugars in these spaces change to unsaturated fatty acids which combine with oxygen to form *suberin*, a thin, varnishlike layer which seals the moisture of the cutting inside the tissue and keeps rot-producing organisms out. Suberin is inelastic and therefore is effective for only a short period. If the cutting heals properly, a permanent layer of tissue is soon formed. These cells become meristematic and produce new cells which become filled with suberin, tannin, and other materials producing a corky layer. This layer is constantly renewed and is elastic enough to withstand the

(a)

(b)

FIGURE 12-4 Rooted stem and leaf cuttings. (*a*) Stem cuttings of *Ligustrum japonicum* ; (*b*) leaf cutting of *Crassula argentea* (jade plant) showing the initiation of shoot tissue. (Courtesy of W. S. Jordan, Clemson University.)

pressure changes due to water loss and intake. The root system arises from the pericycle, from callus, or from the epidermis of young stems and the cambium of older stems.

HERBACEOUS CUTTINGS These cuttings are made from succulent herbaceous plants such as chrysanthemums, coleus, geraniums, and carnations. Most florists' crops are propagated by herbaceous cuttings. They are cut 7 to 10 cm in length and are rooted under similar conditions to softwood cuttings. Under proper conditions, rooting is rapid and in high percentages. Herbaceous cuttings of plants that exude sticky sap, such as the geranium, cactus, or pineapple, do better if the basal ends are allowed to dry for a few hours before they are inserted in the rooting medium. This practice prevents the entrance of decay organisms.

HARDWOOD CUTTINGS Hardwood cuttings are divided into deciduous species and narrow-leaved evergreen species. Deciduous hardwood cuttings from the current season's growth do not require leaves for rooting. Examples are forsythia, wisteria, fig, grape, currant, and mulberry. In fact, they wilt when softwood is used. Hardwood cuttings made during the dormant season should be from vigorous stock plants. They are taken after the chilling requirements have been satisfied, but before midwinter or spring. Roots developing from the cut are derived from the cork cambium; roots developing from the nodes are derived from the vascular cambium. Cuttings can vary from 10 to 75 cm in length. During the winter the wound heals and roots form, and in the spring the rooted cuttings are planted in the field or in pots, where they remain a year or so before being transplanted. Hardwood cuttings, such as grapes, can also be made and stored over the winter and then planted out in the spring, when they will root. Ideally, roots form before bud break so that the transpirational needs of the new leaves can be met.

Narrow-leaved evergreen cuttings are taken between late fall and late winter. Cuttings of this type have leaves and must be rooted under moist conditions that will prevent excessive drying, since they are slow to root, taking several months to a year. In general *Chamaecyparis*, *Thuja*, the low-growing *Juniperus* species, and the yews (*Taxus* sp.) root easily; whereas the upright junipers, spruces (*Picea* sp.), hemlocks (*Tsuga* sp.), and firs (*Abies* sp.) are more difficult. The cuttings are made 10 to 20 cm long, with all the leaves removed from the lower half of the cutting. Mature terminal shoots of the previous season's growth are usually used. In certain of the narrow-leaved evergreen species, some type of basal wounding is often beneficial in inducing rooting.

SEMIHARDWOOD CUTTINGS Cuttings of this type are usually made from woody, broad-leaved evergreen species. They are usually taken during the summer from new shoots, just after a flush of growth has taken place and the wood is partially matured. Ornamental shrubs such as camellia, ligustrum, pittosporum, evergreen azaleas, and holly and fruit species (such as citrus and olive) can be propagated by semihardwood cuttings.

The cuttings are made 7 to 14 cm long, with leaves retained at the upper end. If the leaves are very large, they should be reduced in size to lower the water loss and to allow closer spacing in the cutting bed. The shoot terminals are often used in making the cuttings, but the basal parts of the stem will usually root also. The basal cut is usually just below a node.

Leaf cuttings   Leaf cuttings produce more plants per parent plant and per greenhouse space than stem cuttings. Entire leaves with or without the petioles are used in making leaf cuttings. Adventitious shoots and roots develop from the leaf blade, the petiole, or both, but rarely does the blade itself become part of the new plant. *Begonia rex, Sansevieria, Peperomia, Crassula argentea* (jade plant; Figure 12-4b), and *Saint-paulia ionantha* (African violet) are propagated by this method. Plants with thick, fleshy leaves, such as *Begonia rex*, are propagated by severing the large veins on the underneath side or cutting the leaves into sections and laying the leaf flat on the propagating medium. With high humidity a new plant will form where each vein is cut.

Leaf-bud cuttings   To make a leaf-bud cutting, a leaf blade, a petiole, and a short piece of the stem with the attached axillary bud are placed in the medium, with the bud end in the medium and covered enough to support the leaf. Care must be taken, because if the bud is injured, the plant will produce roots but not a new plant. This method is used when plants are able to initiate roots, but not shoots, from detached leaves. It is most valuable when propagating material is not readily available, because more new plants can be produced by this method than with stem cuttings. Blackberry, lemon, camellia, rhododendron, and black raspberry can all be propagated by leaf-bud cuttings.

Root cuttings   Plants which develop shoots from the root system, such as blackberry (Figure 12-5) and raspberry, are easily propagated by this method. Several techniques are applicable for making root cuttings. Root cuttings are made from root sections 5 to 15 cm long in the fall or winter. These cuttings are stored in sawdust or sand in a cool place to allow new shoots to develop from adventitious buds on the root or from the old root. This is done because true roots do not have buds and therefore must be capable of forming an adventitious bud. The plants are then transplanted in the spring. Cuttings may also be started in hotbeds or greenhouse beds in the winter and transplanted

FIGURE 12-5 Propagation of *Rubus* species (blackberry) by root cuttings set vertically. (Courtesy of W. S. Jordan, Clemson University.)

in the spring. These cuttings may be placed in the medium in either a vertical or a horizontal position. When the vertical position is used, the end originally nearest the crown of the plant should be turned upward. In other plants, such as sweet potato, the whole fleshy root is planted in a warm medium where shoots develop with their own root system. Each rooted shoot (slip) can be transplanted. Root cuttings are used as rootstocks to reproduce a cultivar which is grafted to a special rootstock. Lilacs can be dwarfed by grafting to root cuttings of privet.

Environmental Factors   Environmental factors play an important role in the ability of cuttings to heal and develop roots. Since cuttings do not have roots, the top growth must be slowed until a root system can be developed to balance it. A balance between humidity, water, temperature, and sunlight should exist for optimum growth without wilting. Removal of leaves should be minimal, since as large a turgid leaf area

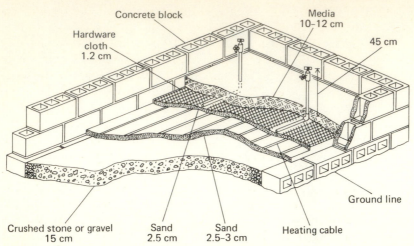

Concrete block
Media 10–12 cm
Hardware cloth 1.2 cm
45 cm
Ground line
Crushed stone or gravel 15 cm
Sand 2.5 cm
Sand 2.5–3 cm
Heating cable

FIGURE 12-6 A cutaway view showing construction of a plant bed for outside conditions. Width of bed shown is 152 cm. Nozzles are spaced 76 cm apart and 38 cm from side and end of bed. The first run of concrete blocks are laid horizontally for drainage. (Redrawn from North Carolina Agricultural Extension Bulletin 506.)

as practical should be maintained for the production of hormones and carbohydrates. A low air temperature slows respiration and the growth of the top. With high relative humidity, less water is required. Thus, carbon dioxide can enter the leaves, photosynthesis can occur, and carbohydrates and hormones will be manufactured. If the medium in which the cuttings are placed receives "bottom heat" (in which case the base of the cutting is heated to 24 to 27°C by lead-covered electric resistance wire, steam or hot water in pipes), the rate of root cell division will be increased (Figure 12-6). The high root-area temperature encourages rapid oxidation of the fatty acids to suberin, which heals the wound and aids in the development of a new root system. Oxygen is also required for the oxidation of the fatty acids to suberin and for the activities of the meristem. This requirement can be satisfied by using a rooting medium which is adequately aerated.

A relatively high light level must be maintained to enable photosynthesis to be carried out. The products from photosynthesis are consequently used for root initiation and subsequent growth. However, high irradiance also causes rapid transpiration so that wilting may occur, resulting in decreased photosynthesis and possibly desiccation of the cutting. To combat this, mist propagation (Figure 12-7) is used. Mist maintains a thin layer of water on the leaf. Because of the evaporative effect under mist, leaves are often 5 to 8°C cooler than the surrounding air temperature. The intermittent mist system syringes the cuttings with water and gives best results if applied only during the daylight hours. Continuous mist may result in severe leaching of the foliage nutrients.

Time clocks, humidistats, solar mechanisms, or electronically controlled units are used for controlling the intermittent mist. With the normally open system, the solenoid valve is opened if the electrical current stops. This characteristic is especially desirable if the power should fail; water will continue to flow through the pipes. With a normally closed solenoid system which requires the passage of an electric current to open the valve and allow the water to pass through, a power outage could result in severe damage or loss of plants.

The optimum length of a mist period varies with the type of plant material and the environmental conditions during the rooting period. Except during extremely hot, dry weather, mist for 6 sec min^{-1} from the time the cuttings are stuck until rooting begins is recommended. A gradual reduction in the frequency of misting as the roots elongate is needed, so that by the time the roots are 2 to 3 cm long the cutting is "hardened off." This reduces the severity of shock and consequently minimizes the transplanting losses. In a continuous succession of propagation where this is not possible, the newly dug cuttings should be placed in a shady area and frequently syringed until they harden off.

FIGURE 12-7 Mist propagation beds.

FIGURE 12-8 Outdoor intermittent mist propagation beds. (Photograph, John Sorozak, Monrovia Nursery Company.)

Outside propagation areas (Figure 12-8), cold frames (Figure 12-6), and greenhouses (Figure 12-9) are the structures for propagating plants. The medium, generally 10 to 12 cm deep, should give the necessary support and environmental conditions for the development of roots. The medium should be moisture-retentive while allowing good aeration and drainage. Mesh galvanized wire of 0.6 cm can be placed underneath the medium to provide better drainage and aeration. The medium should also be lightweight. Clean sand, peat moss, vermiculite, and perlite, or a combination of these materials are frequently used as a rooting medium. Generally, a combination of several media gives the best results. For most plants, a wide range of media may be used, but the plants that are more difficult to root are influenced by the medium with regard to percentages which root and the quality of the root system which develops.

FIGURE 12-9 Propagation of plants in miniature greenhouses. Note the ventilation holes in plastic.

Growth Regulators In many species the formation of root initials can be enhanced by the application of externally applied plant growth regulators. They are used to accelerate root initiation and development, to increase the rooting percentage, to increase the number and quality of roots produced per cutting, and to increase uniformity of rooting. The commonly applied plant growth regulators—3-indoleacetic acid (IAA), 3-indolebutyric acid (IBA), and 2-naphthaleneacetic acid (NAA)—can be applied separately or in combination. Other materials, such as ethylene, acetylene, and propylene, are also effective in stimulating root formation.

To be effective, growth substances must be used in a specific concentration range for an individual species. High concentrations may injure or kill the base of the cutting or cause excessive callusing, while low concentrations may be ineffective. The duration of treatment is inversely proportional to the concentration. Within given limits, a high concentration for a short period of time is comparable to a lower concentration for a longer period of time.

Auxins are applied using one of several methods. For commercial powder preparations, the powder is simply spread on wax paper or aluminum foil, and freshly cut stem tips are dipped in it. Cuttings may also be treated by soaking the basal end in a dilute solution for 24 h before placing them in the propagation medium. A third method requires that a concentrated solution of the chemical in 50 percent alcohol be prepared and the basal ends be dipped for a short time—generally 5 sec. This method is advantageous because it takes less time and equipment and results in more uniformity.

The roots produced following growth-regulator treatments generally have the same origin as natural roots. The roots that arise on a cutting may originate from any mature tissue within the cutting as a result of the action of secondary meristems.

Layering

Layering, a vegetative method of propagation, produces new individuals by producing adventitious roots before the new plant is severed from the parent plant. Conditions necessary for layering are quite similar to those necessary for rooting cuttings. The wood should be young, so that it will form adventitious roots easily. When the new plant has developed a self-sustaining root system, it can be severed from the parent plant. Species that are hard to root lend themselves to layering since the parent plant provides the food and water for as long as it takes the roots to develop. This method has two basic disadvantages. One is that only a few individuals can be started from a given stock plant at any one time; the other is that the expense of layering, which involves considerable hand labor for the small number of plants, is prohibitive for most large-scale usages. Brambles, grapes, blueberries, and other plants which naturally propagate by layering are sometimes propagated this way by growers. Various modifications of layering are used.

Tip Layering   In *tip layering* (Figure 12-10) rooting takes place near the tip of the current season's shoot, which naturally falls to the ground. The shoot tip begins to grow downward into the soil; but after a period of time the tip begins to curve upward, with roots developing from meristematic tissue at the curve. Trailing blackberries, dewberries, and black and purple raspberries are frequently propagated by tip layering. These plants root naturally by this method since the canes fall to the ground where they then root behind the tip. Limitations of new plants depend on the number of canes available.

Simple Layering   *Simple layering* (Figure 12-10), which is used for *Forsythia*, grapes, jasmine, and various broad-leaved evergreens, is the same as tip layering except that the stem behind the end of the branch is covered with soil and the tip remains above ground.

Trench Layering   Trench layering (Figure 12-10) is a method whereby a number of new plants may be obtained from a given stock plant of a certain species. Rose, muscadine grapes, spirea, and various decidu-

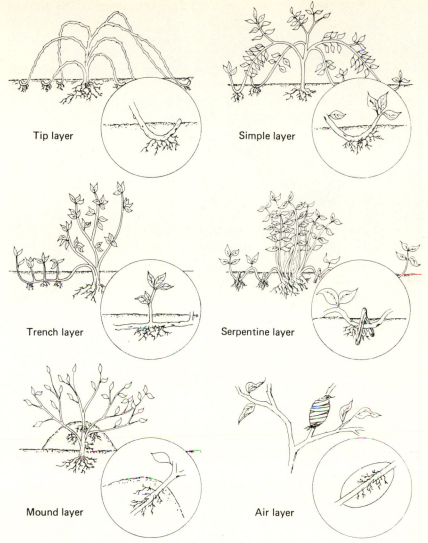

Tip layer

Simple layer

Trench layer

Serpentine layer

Mound layer

Air layer

FIGURE 12-10   Methods of layerage for propagating plants. (Adapted from Ervin L. Denisen, *Principles of Horticulture*, Macmillan, New York, © 1958. Used by permission of Macmillan Publishing Co., Inc.)

ous shrubs lend themselves to this method. The basal and middle portions of a young stem are placed in a shallow trench, injured by notching the nodes to increase the likelihood of root formation, and covered with 5 to 10 cm of moist soil. The exposed tip is left to ensure continued growth and translocation of food and water. If the tip is left exposed, the method is called *continuous layering*. When only the nodes are covered, trench layering becomes *discontinuous* or *serpentine layering* (Figure 12-10).

FIGURE 12-11 Mound layering showing new plants still attached just after soil has been removed.

Mound or Stool Layering When pruned severely by cutting back to the ground during the dormant season, some plants develop many new shoots from the base (apple rootstock, quince, and currant). In *mound* or *stool layering* (Figure 12-10) when the new shoots develop in the spring, soil is mounded around their bases excluding light and enhancing root formation. This process is repeated three to four times as the shoots elongate, leaving the upper portion uncovered. The soil is removed at the end of the dormant season, and the new plants are detached (Figure 12-11). In this method the parent plant can be used year after year.

Air Layering Air layering (Figure 12-10) can be used for foliage plants, such as rubber plant, *Ficus*; dumb cane, *Dieffenbachia*; and *Dracaena*. A ring of tissue to the xylem and about 2 cm in width from the previous years' growth is removed from the stem below the tip of the plant and surrounded with a moist medium such as peat or sphagnum moss. Plastic wrap, such as polyethylene, which is permeable to CO_2 and other gases but impermeable to water, is used to cover the rooting medium. This is tied with "twist-ems." When the roots have developed, the stem is severed from the parent plant below the root ball.

Grafting

Grafting consists of joining two separate structures, such as a root and stem or two stems, so that by tissue regeneration they form a union and grow as one plant. The upper part of the union is the *scion* (cion), and the lower part is the *stock* (rootstock). Some graft combinations contain an *interstock*, or intermediate stem piece, which is grafted between the scion and stock. This is called *double-working*. The ability of certain dwarfing clones, inserted as an interstock between a vigorous top and vigorous root to produce a dwarfed and early-bearing fruit tree, has been known for centuries.

Grafting is used to reproduce clonal cultivars, such as almond, apple, walnut, and eucalyptus, which are not easily or successfully reproduced by other methods. Special growth habits, such as those in a rose tree or weeping cherries, can be obtained only by grafting. Branches of trees can be changed to a more desirable cultivar by grafting. This is called *topworking*. Several cultivars of flowers or fruits can be grown on an individual root system, such as with pink and white dogwood. Some stocks may tolerate adverse conditions such as heavy, poorly drained soils or may be insect- and disease-resistant, thus influencing the growth of the scion. Rootstocks can shorten the juvenile period and thereby accelerate fruiting.

The formation of the graft union is an intricate process and depends on several variables and conditions for its success. The first step in the healing of the graft consists of aligning the freshly cut scion and stock tissue to fit tightly. It is necessary when grafting that the cambium layers of the scion and the stock be in contact so that callus cells from the cambium tissues may intermingle (Figure 12-12). The

FIGURE 12-12 Cross section of graft showing matching of scion and stock cambiums. (From New York State Agricultural Experiment Station Bulletin 817.)

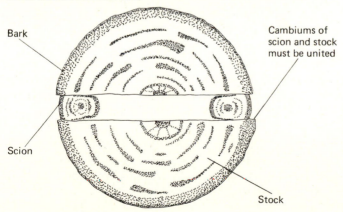

Bark

Cambiums of scion and stock must be united

Scion

Stock

outer layers of cells in the cambium of the stock and scion produce parenchyma cells which soon interlock, forming the callus tissue. Cells of the new callus tissue begin to differentiate into new cambium cells along the lines of the two intact cambiums between the stock and the scion. These cambium cells produce vascular tissue, consisting of xylem to the inside and phloem to the outside, which establishes a continuity between the cambiums of the scion and stock. Food materials and plant sap can then translocate throughout the newly grafted plant.

The success of the graft depends on several predetermined factors and environmental conditions during the healing process.

The *first* phenomenon of importance is compatibility. Problems having to do with incompatibility are often virus-related; latent virus may be present in a clone or rootstock and may cause no problem until grafted onto or used with a susceptible cultivar. Incompatibility may result in failure to form a graft union, weak unions which eventually die, nutritional disorders, stunted plants, or physiological abnormalities. Delayed symptoms of incompatability often develop many years later. The extent of this may vary with cultivars, species, and genera. Past experimentation and trials are the criteria for determining whether or not specific plants will be compatible. Generally, the closer the plants are in the botanical or taxonomic classification, the better the chance of a successful graft, provided that there are no problems having to do with viruses or incompatibility. A graft between tissues that are unrelated botanically will not generally be successful.

A *second* consideration is the kind of plant involved in the combination and the kind of graft to be used. Even though the stock and scion may be compatible, the union may be extremely difficult to initiate; but once the union is successful, the plant grows very well. Hickories, oaks, and beeches are examples of this. An additional aspect is that some plants respond to specific techniques better than to others. For example, the topworking of *Juglans hindsii* (black walnut) to *Juglans regia* (persian or English walnut) is more successful with the bark graft method than with the cleft graft method. Often, some species are so difficult to graft that approach grafting is employed. This allows both plants of the graft to remain on their own roots until a graft union is formed. *Vitis rotundifolia* (muscadine grape), *Mangifera indica* (mango), and *Camellia reticulata* are all propagated by the approach graft method.

Third, environmental conditions, especially temperature and moisture, exert a definite influence on the graft union. Temperatures that stimulate high cell activity range from 13 to 32°C. Attempts at grafting are usually done when stock tissues, especially the cambium, are in an active state and the scions are dormant. The callus tissues of thin-walled tender parenchyma cells are easily desiccated by drying air and soon die. Therefore, a high moisture content is necessary to prevent dehydration. With most plants, a wax layer over the graft is all that is necessary to maintain the natural moisture of the tissues. In addition to temperature and moisture, a sufficient supply of oxygen is necessary at the graft union to produce callus tissue. In this region, cellular activity and respiration reach a high level which requires oxygen. In some plants, waxing resists air movement, and an enclosure of water-saturated air is more beneficial.

Fourth, the success of the graft also depends on the stage of growth of the stock plant. Most successful unions in many species take place when the bark is "slipping." This is the state of rapid cell division when the vascular cambium is producing thin-walled cells on each side. This process is initiated in the spring as a result of the swelling of the buds and may be the result of a stimulus from auxin in the expanding buds. Plants which exhibit extensive bleeding or abundant sap flow when cut for grafting will not heal.

Poor grafting techniques result in failure. Delayed waxing, uneven cuts, lack of contact between cambium layers, and grafting under unfavorable conditions may result in failure. Virus infection, insects, and diseases can also cause failure. If the propagating materials are not virus-free, the success of the union may be reduced, or the vigor of the plant may be reduced. The use of pesticides to prevent disease and insect damage is encouraged.

Grafting of deciduous species is generally performed in late winter and early spring using dormant scion wood of shoots from the previous summer's growth. Scion wood with strong leaf buds is taken from one-year-old shoots of a size suitable for grafting. Older wood can be used if buds are present and some species, such as the fig, graft more successfully when using 2-year-old wood. When scion wood is obtained, care must be taken to see that it is properly stored. Scion wood should be stored in a cool place such as a cold-storage refrigerator at a temperature near freezing, but never below 0°C, since freezing temperatures may cause injury. A cool cellar or similar location has proven satisfactory for short periods. To keep the wood slightly moist, bundles of approximately 25 sticks may be wrapped in heavy, waterproof paper or polyethylene, with moist material sprinkled through the bundle, taking care not to soak the wood. More damage results from too-wet scion wood than from wood that is too dry. The bundle should release only minimal moisture when pressed. Fruits and scion wood cannot be stored together, because ethylene gas produced by ripening fruits is toxic to propagation materials. It is not necessary to store scion wood for broad-leaf evergreens, since grafting is done in the spring before active growth starts, and the scion wood may be taken as needed from basal wood with dormant buds.

Whip and Tongue Grafting   The whip graft often used for fruits (such as 1-year-old apple and pear rootstocks) and ornamentals (such as *Hibiscus syringa*) is sometimes referred to as *bench graft*, since the work is commonly done at a bench or table (Figure 12-13). The graft is made in winter or spring before bud development. With scion and rootstock of approximately 0.6 to 1.2 cm in diameter, similar cuts should be made at the top of the stock and at the bottom of the scion to ensure an almost perfect cambial fit. A long, smooth sloping cut 2.5 to 5 cm long is made in the scion. About one-third the distance from the top of this cut, a second cut is made about one-half its length and roughly parallel to it. The stock is prepared in a similar manner. The scion and stock are tightly fitted together, wrapped, and covered with wax. Whip grafts heal quickly, forming a strong union.

Approach Grafting   Approach grafting, the joining of two plants on their own rootstocks, is necessary for some plants such as conifers which are difficult to graft by other methods. A length of bark 2.5 to 5 cm long is removed from both plants, and the wounded surfaces are pressed and held tightly together. The union process is slower than in other methods but can be speeded up if the grafting takes place during the growing season. Upon successful union, the top of the stock is removed, and the root system of the stock is retained along with the new scions, usually by a gradual process as opposed to complete removal in a single step.

Cleft Grafting   Cleft grafting, performed during the dormant season when buds begin to swell but before they rupture, is used to topwork fruit trees and some large ornamentals, such as camellia. The scion should be 1-year-old wood, cut in the fall and stored or cut early in the spring. The stock is usually 5 to 7 cm in diameter and is split smoothly down the center for several centimeters (Figure 12-14). Usually two scions with the basal end cut in a long, gradually tapering wedge are inserted into the stock, which has been split. The cambium of the scion must be in contact with the cambium of the stock. The use of two scions allows selection for vigor and position, promotes speed in the healing process, and generally ensures a successful graft union. The second scion also prevents dieback of one side of the stock, which can occur if only one scion is used. When the poorer scion is removed, it is usually done gradually. The objective is to have the scion reach the diameter of the stock in a minimum number of years. Both scions should never be left because of the weak crotch which forms (Figure 12-15).

FIGURE 12-13   Whip grafting procedure. (*a*) Scion and stock cut into one-sided wedges; (*b*) scion and stock cut one-third of distance down for about 1.5 cm; (*c*) ends inserted so that cambium will match; (*d*) graft wrapped with tape. (Courtesy of John Ridley, Clemson University.)

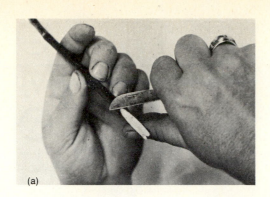

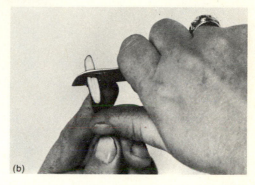

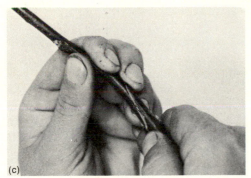

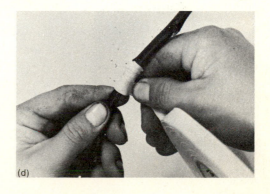

(a)

(b)

(c)

FIGURE 12-14 Procedure for making a cleft graft. (a) Scion after wedge has been cut; (b) inserting scions into stock with thick side of scion wedge toward the outside; (c) graft after waxing cut surfaces. (Courtesy of John Ridley, Clemson University.)

(a)   (b)

FIGURE 12-15 Cleft graft on apple tree. (*a*) Both scions left on stock, showing development of weak crotch; (*b*) one scion was suppressed and then eliminated, showing stock and scion the same size.

Bark Grafting Bark grafting can be adapted to stock of any size but does depend on the stock bark's readily separating from the wood. The procedure, therefore, is performed after active growth has started in the spring. The scion, shaped like a wedge, is inserted between the bark and wood of the stock. Several scions can be inserted to ensure selection. One of the most vigorous is selected, and the others are pruned to keep them subordinate until they are eventually removed. Bark grafting is rapid and easy to perform and has a high percentage of successful takes. It is used as inlay with pecan trees.

Notch Grafting Notch or saw kerf grafting may be performed after the stock initiates growth. It is generally used for topworking trees with branches of large diameters or for curly-grained wood which generally will not split evenly for cleft grafting. Usually three scions 10 to 12 cm long with two to three buds are inserted into the stock, where cuts have been made on the perimeter 1 to 2 cm into the center and 5 cm down the side. The wedge-shaped scions, with the outer side slightly wider than the inner edge, are inserted where the cambium matches. The exposed surfaces are then waxed.

Wedge Grafting In wedge grafting, which is used widely with rhododendrons and lilacs, a V-shaped notch is removed from the center of the stock (Figure 12-16*b*). The wedge-shaped scions with two to three

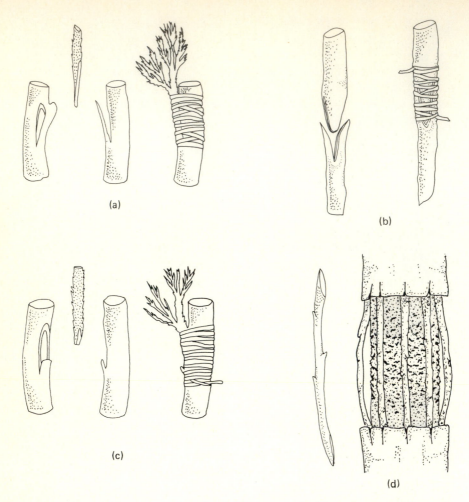

FIGURE 12-16 Other types of grafts. (*a*) Veneer graft; (*b*) wedge or saddle graft; (*c*) side graft; (*d*) bridge graft. (Redrawn from North Carolina Agricultural Bulletin 326.)

buds are inserted to match the cambium, and the wounds are waxed afterwards.

Side Grafting   Side grafting is practiced with small plants such as narrow-leaved evergreens (Figure 12-16*c*). The scion is placed inside the stock in several variations of this technique. The procedure adapts to stock that is larger than the scion.

Inarching and Bridge Grafting   Inarching and bridge grafting (Figure 12-16*d*) are used mostly for repair rather than propagation. These procedures are used when valuable trees have been damaged by implements or rodents, when the tree phloem is severely damaged, or when the tree is girdled so that food materials can not be translocated to the root system, resulting in eventual starvation.

Inarching is used to replace damaged root systems. Seedlings or rooted cuttings of the same or compatible cultivars are planted encircling the damaged tree at 15-cm intervals to provide a new root system. In the spring, when active growth begins, a 10- to 15-cm strip is taken from the seedling on the side next to the tree and a 0.8-cm notch is made on the opposite side, removing the top and giving the seedling a wedge-shaped top. A bark strip of the same size as the strip is taken from the tree trunk, and above this a flap of bark is lifted. The seedling wedge is inserted under the flap, and the raw strips of seedling and tree are nailed together and covered with wax, forming a graft union.

Bridge grafting is used to bypass damaged or incompatible bark. Bark in the damaged area is removed until healthy tissue is reached. Scions are prepared from water sprouts and are wedged at the ends. Scions are placed under bark flaps above and below the injured area, tapering slightly to ensure good cambial contact at the end and to permit the tree trunk to bend and sway under wind and other stresses. The inserted pieces are nailed, and the entire wounded area is waxed to prevent moisture loss from both the stock and scion, to allow O_2 and CO_2 exchange, and to prevent decay-producing fungi from entering the wound.

Budding Budding is a type of grafting. Budding results when a vegetative bud (scion) is placed in a stock plant (stock). Only one bud is used; this economizes the use of scion tissue.

T budding The bud is attached by placing the bud against freshly exposed cambium tissue of the stock (Figure 12-17). The stock bark,

FIGURE 12-17 Procedure for making T bud. (*a*) Scion: 1, bud; 2, petiole; 3, leaf blade. (*b*) T-shaped cut through the bark of the stock with bark raised to admit the bud. (*c*) Bud stick. (*d*) Bud in place. (*e*) Bud wrapped with raffia. (Redrawn from North Carolina Agricultural Bulletin 326.)

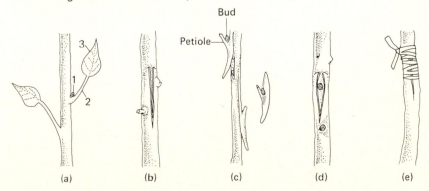

mainly all tissues external to the cambium, is split into a small T to allow the bud piece to slip under the bark. Raffia, rubber bands, soft string, or adhesive tapes are used to hold the scion and stock tightly together. Xylem and cambium tissue of the stock and scion produces the callus tissue. Budding is usually performed in June and early July in the Southeast; after the bud has "taken," the stock is cut off. The bud grows into a tree during the summer and is sold as a "June bud" tree. In the North budding is done in August and September, when the stock bark is slipping and will accept the bud. In Northern areas the stock above the bud is not removed until the next spring, so that the bud remains dormant for several months before growth. This is the propagation method most commonly used for fruit trees, as it is much quicker than whip grafting.

Patch budding In patch budding, which is similar to T budding, a rectangular bark patch is taken from the stock plant, so that the xylem is exposed (Figure 12-18). A bud is removed from the desired cultivar with bark patch of the identical size of the stock plant and positioned in place with waxed cloth or budding tape. Thick-barked trees, such as pecans and walnuts, adapt well to patch budding. This procedure is slower and more difficult to perform than T budding. It is usually done in late summer and early fall, when both stock and bud bark can be slipped easily. In *flute budding*, a variation of patch budding, a bark strip is removed from the circumference of the stock, leaving a thin strip approximately 3 mm. A strip of the same size is removed with an attached bud of the new cultivar and placed on the stock. The connecting strip serves to supply the stock top with photosynthates until a union is formed.

FIGURE 12-18 Procedure for patch budding. (*a*) Use budding tool with two parallel blades to make horizontal cuts about one-third the distance around the stock and connect the horizontal cuts at each side by vertical cuts. (*b*) Cut patch containing the bud followed by the two vertical cuts on each side of the bud. The bud is then removed by sliding the bud to one side. (*c*) Remove the bark from the stock; insert the bud and prepare it for wrapping. Wrap the union with tape or other suitable material and leave the bud exposed. (Redrawn from Hudson T. Hartmann and Dale E. Kester, *Plant Propagation, Principles and Practices*, Prentice Hall, Englewood Cliffs, N.J., © 1975. Reprinted by permission of Prentice Hall, Inc.)

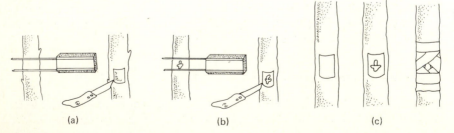

(a) (b) (c)

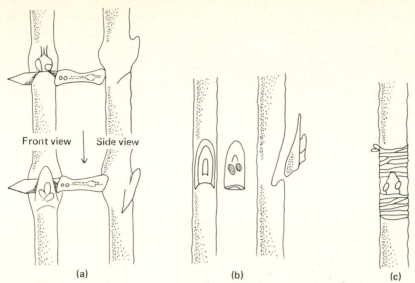

Front view   Side view

(a)   (b)   (c)

FIGURE 12-19 Procedure for chip budding. (*a*) Make a 45° angle cut about one-quarter through the stock; then about 2.5 cm above the first cut make a second cut downward and into the first cut. (*b*) Make similar cut to remove the bud, which will permit the removal of the bud piece. (*c*) Insert the bud into the stock and wrap with budding tape or suitable material. (Redrawn from Hudson T. Hartmann and Dale E. Kester, *Plant Propagation, Principles and Practices*, Prentice Hall, Englewood Cliffs, N.J., © 1975. Reprinted by permission of Prentice Hall, Inc.)

Chip budding   In chip budding, it is not necessary that the bark be slipping; thus this procedure can be performed in early spring before growth starts or in late summer when growth is halted. It is not usually done on commercial deciduous fruit trees. Chip budding procedure takes longer to complete and is more complex than T budding. On a desirable cultivar, a downward 45° angle cut is made below the bud (Figure 12-19). A second cut is made about 12 mm above the bud, inward and downward in a transverse manner, meeting the first cut. The bud chip is removed and placed near the base of the stock between two nodes, where an identical cut has been made. It is necessary that the cambial layers of stock and scion match. Wax, string, or nursery adhesive tape must be used to secure the graft. Spring grafts are topped after approximately 10 days; summer grafts are topped the following spring.

Specialized Stems and Roots

Bulbs, corms, tubers, rhizomes, and fleshy roots whose chief purpose is food storage can be used to propagate plants. Propagation by bulbs and corms utilizes the naturally detachable parts in a process known as *separation*. Monocots develop either tunicate or scaly bulbs. Tunicate bulbs, such as onion and daffodil, have continuous scales arranged concentrically around the base of the bulb protected by a papery sheath called the *tunic*. Scaly bulbs, such as the lilies, have overlapping scales and are not protected by a sheath. Meristems develop in the axil of these to produce miniature bulbs, known as *bulblets*, which when grown to full size are known as *offsets*. In various species of lilies, bulblets may form in the leaf axils either on the underground portion or on the aerial portion of the stem. The aerial bulblets are called *bulbils*; the underground parts are called *stem bulblets* (Figure 12-20). Some monocots, such as gladioli and crocus, produce enlarged stems surrounded by dry, scaly leaves, with roots extending from the bottom of the corm and a single stem developing from the top. The mother corm may also produce cormels from axillary buds which can be planted to produce new plants. Some dicots produce *tubers*, which are enlarged underground stems serving as storage organs of starch or related materials, such as inulin in

FIGURE 12-20 Bulb scales of lilies producing new plants. (Courtesy of J.P. Fulmer and E.V. Jones, Clemson University.)

Jerusalem artichoke. Tubers contain "eyes," or buds (Irish potato, caladium). The tubers may be cut into sections and placed in a rooting medium to produce plants. Monocots or dicots can produce rhizomes, stems, and nodes and internodes growing horizontally to the surface of the ground that are separated by "divisions," each segment containing a node to produce a new plant. Nodes and internodes are not produced on fleshy roots, but adventitious buds can develop to produce new shoots and plants as in the sweet potato.

APOMIXIS

In apomixis, embryos are formed without sexual cycles or products of cellular fusion; this is really a type of asexual propagation, since an embryo is developed through asexual reproduction rather than by meiosis and fertilization, but it does have some characteristics of sexual reproduction. *Obligate apomicts* produce only apomictic embryos, whereas *facultative apomicts* can produce both sexual and apomictic embryos. Apomixis ensures uniformity in seed propagation, since the apomictic cultivar is clonal in type, as in *Citrus* sp and 'Kentucky Bluegrass'. Since viral diseases are not usually transmitted by seeds, apomictic seedlings can be used to avoid some viruses in a clone which has become infected through nucellar embryony.

Sometimes, in addition to the regular embryo, an additional embryo develops outside the embryo sac, as in *Citrus*. The development of two or more embryos within a single seed is referred to as *polyembryony*. One type is a true-fusion embryo, as in *Citrus*, with up to 16 embryos from one seed, only one resulting from fertilization and the others resulting from sporophytic tissues such as the integuments or nucellus. For most plants that reproduce through apomixis, pollination takes place even though fertilization of the egg does not occur.

TISSUE CULTURE

Plant tissue culture techniques have become vitally important for pursuing a wide range of fundamental and applied problems in research and development. The techniques encompass a variety of procedures used for specific purposes. The growing of masses of unorganized cells (callus) on agar (Figure 12-21) or in liquid suspension is widely employed in biochemical and growth studies. The culture of segments of stems, roots, leaves, or callus provides systems to study differentiation, morphogenesis, and plant regeneration. Shoot apex culture methods leading to plant regeneration have been adopted for

FIGURE 12-21 Culture of *Nephrolepsis exaltata* cv. bostoniensis (Boston fern) on agar. (Courtesy of Ron C. Cooke, Flow Laboratories, Rockville, Maryland.)

plant propagation and production of virus-free stock. The culture of anthers and pollen provides new approaches to haploid plant formation. Recently the technology has been extended to include the isolation and culture of plant protoplasts which are employed in fusion and somatic cell hybridization.

Plant tissue culture is useful for rapid asexual multiplication of plants. This technique is frequently employed on such plants as dracaena, carnation, orchids, pothos, gerbera, and various ferns. Presently, herbaceous plants have best adapted to this method, but as more research on cultural requirements develop, woody types such as nandina and *Ficus* are being propagated by tissue cultures.

Another important aspect of plant tissue culture is the ability to recover disease-free material. Through research this method can also be used to improve plants genetically.

For plant tissue culture, laboratory and trained personnel are necessary. Three areas should be provided in the laboratory: (1) sections for preparing nutrient formulations, equipped with autoclave, chemicals, glassware, and balances for weighing chemicals; (2) an area equipped with a dissecting microscope and sterile bench for isolating and placing the plant part in culture; and (3) an area with controlled temperature and light regime for growing the plant cultures (Figure 12-22).

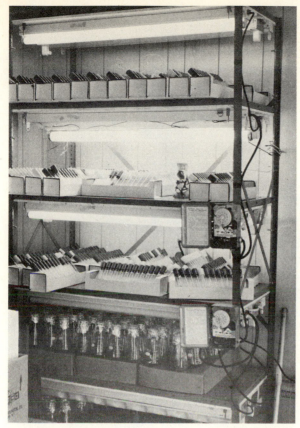

FIGURE 12-22 Photoperiod exposure of tissues cultured on nutrient agar media in test tubes. (Courtesy of A. J. Pertuit, Clemson University.)

Basic techniques are callus culture, organ culture, and cell-suspension culture. The source of the plant should be selected—shoot tip, leaf, or stem—and then dissected (to give a macroscopic piece of leaf or a microscopic shoot apical meristem embryo). The source organ is disinfected and rinsed. The plant is placed in a container prepared with liquid or semisolid medium and placed in an appropriately controlled environment (Figure 12-23). The growth and culture of the tissues are observed until plantlets reach desirable size, when they are subcultured and transferred to larger containers. To transfer plantlets to soil containers, they must be properly acclimated.

The future of plant tissue culture holds many interesting unknowns. The technique shows tremendous potential in the production of a given clone and in plants that are slow or difficult to propagate. Under proper conditions tissue culture can provide for an increase of a

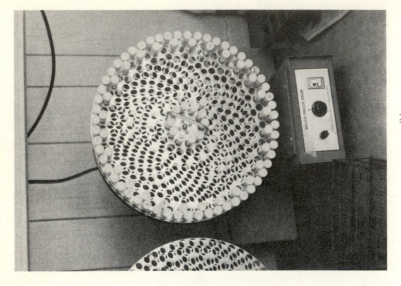

(b)

(a)

FIGURE 12-23 Controlled environment. (*a*) Multilevel roller drum used for constant agitation of tissues in liquid nutrient media in test tubes. (*b*) Closeup of roller drum used for constant agitation of test tubes. (Courtesy of A. J. Pertuit, Clemson University.)

374

million genetically uniform plants from one plant part in a year's time, enabling plant breeders to quickly develop a large volume of plants. Pathogen-free plants—such as geraniums free of xanthomonas, chrysanthemums without stunt and other viruses, and carnations free of ringspot mottle—are all advantages of tissue culture.

PLANT PATENT LAW

In 1930 an amendment to the United States Plant Patent Law became effective which enabled plant growers to obtain a patent for new cultivars. The opportunity to patent new plants has enabled horticulturists and plant breeders to protect their investment and obtain monetary rewards. This has led to many new and improved cultivars which are available today.

The statute states that "any distinct and new variety of plant, including cultivated sports, mutants, hybrids, and newly found seedlings other than a tuber-propagated plant or a plant found in an uncultivated state" is patentable. In 1970 the Plant Variety Protection Act extended the Plant Patent Law to include certain sexually propagated cultivars of cotton, alfalfa, soybeans, marigolds, and bluegrass which are considered lines or types of a given plant.

When judging whether a plant is distinct and new, we must consider such factors as growth habit; immunity to disease; resistance to cold, drought, heat, wind, or soil conditions; the color of the flower, leaf, fruit, or stem; flavor; degree of productivity; storage life, form, and method; and ease of reproduction. Only the breeder or discoverer of the new cultivar is eligible to patent the product. Any other person applying for a patent and subsequently obtaining one would find the patent declared void and face possible criminal charges if it were proven he or she had not developed or discovered the cultivar.

The patent, which is good for 17 years, allows no one but the patentee to grow, sell, and use the plant which is so patented unless a royalty is paid to the patent owner. This patent applies only in the United States and its territories. It indicates only that the plant is distinct and new, and says nothing about the quality or desirability of the plant.

BIBLIOGRAPHY

Adriance, G. W., and F. R. Brison: *Propagation of Horticultural Plants*, 2d ed., McGraw-Hill, New York, 1955.

Denisen, E. L.: *Principles of Horticulture*, Macmillan, New York, 1958.

Hartmann, H. T., and D. Kester: *Plant Propagation Principles and Practices*, 3d ed., Prentice-Hall, Englewood Cliffs, N.J., 1975.

Hill, J. B., H. W. Popp, and A. R. Grove, Jr.: *Botany*, 4th ed., McGraw-Hill, New York, 1967.

Mahlstede, J. P., and E. S. Haber: *Plant Propagation*, Wiley, New York, 1957.

CHAPTER 13
PRUNING
AND GROWTH
CONTROL

Plant growth is modified for functional and aesthetic reasons. Control of growth by removing plant parts dates back to the Egyptians and is among the oldest of horticultural practices. More recently, dwarfing by rootstocks, scion cultivars, genetic mutations, and chemicals has been used to control growth.

PHYSICAL CONTROL OF GROWTH

Pruning is the removal of plant parts, such as buds, developed shoots, and roots, to maintain a desirable form by controlling the direction and amount of growth (Figure 13-1). If the natural form of the plant is undesirable, growth can be directed to a limited degree to the desired form through pruning techniques. For example, the 'Delicious' apple tree develops an upright growth habit with narrow-angled branches which have weak crotches. Therefore, pruning a 'Delicious' apple tree to improve crotch angles would improve its fruiting ability. Constant follow-up must be practiced to maintain the desired form. Although pruning is an invigorating process, it is also a dwarfing process, because growing points are removed and the total weight of a pruned plant is usually less than that of an unpruned plant. Growth is

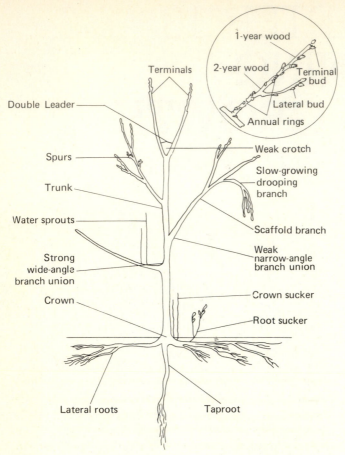

FIGURE 13-1 The major gross structure of a tree. (Redrawn from Everett P. Christopher, *The Pruning Manual,* Macmillan, New York, 1954. Copyright 1954 by Macmillan Publishing Co., Inc., and used with their permission.)

frequently quite rapid following pruning, because the top-root ratio (balance) is temporarily altered; but this does not compensate for the portions of the plant removed. Removal of foliage and branches reduces stored carbohydrates and—more importantly—reduces the leaf area available for carbohydrate production. The new leaves which develop after pruning may be larger than the older leaves, but they will be fewer in number.

Disbudding is the removal of vegetative or flower buds. Generally, the terminal bud of the shoot is dominant. Shoots grow from the apical buds more rapidly than from the lateral buds. Auxins, plant growth regulators produced by dominant buds and young leaves, inhibit the development of the lateral buds below. Removal of the terminal bud as a source of auxin production allows lateral buds to develop. In turn, the uppermost lateral buds will grow into shoots and apical dominance again will be established on each shoot.

It is not only the practice of pruning itself which governs the results obtained, but the condition of the plant. Pruning can influence the number and quality of flowers and fruits. For example, if fewer fruits are allowed to develop, the ones which do develop will be larger in size and of better quality. Excessive top pruning stimulates vegetative growth and can suppress flowering; root pruning usually increases flowering. Since pruning the top of a plant removes growing points, the remaining growing points have a greater supply of water, available nitrogen, and other elements. The vegetative stage may become dominant, and the reproductive stage may be suppressed. Root pruning results in a decrease in the amount of available nitrogen, other essential elements, and water that can be absorbed and thus utilized for growth. Cell division and enlargement, therefore, slow down, and carbohydrates are stored rather than utilized. Root pruning promotes the reproduction phase of growth. If growth initially is low, pruning may stimulate flowering and fruiting. Pruning trees deficient in nitrogen induces more fruitfulness, while pruning young, vigorous nonflowering trees delays the production of fruit. It is desirable for fruit growers to get their orchards bearing as soon as the trees are large enough to bear a commercial crop. The pruning of young, vigorous fruit trees should be light. Heavy pruning of the top will delay the formation of flower buds. Orchards that have weakened owing to age or lack of maintenance often benefit from a severe pruning to rejuvenate them. By removing the older, less productive wood, young vigorous shoots develop, rejuvenating the plant. Pruning can repair injuries. Pruning is utilized to remove diseased parts, insect-damaged parts, or wood which was killed by winter injury.

Pruning helps to develop a strong and durable framework, which is critical for fruit and nut-bearing trees and very desirable on ornamentals. No branch should be directly above another unless there is a space separating them, and unwanted branches should be removed at a young stage (Figure 13-2). A branch that is 1 m above the ground on a young tree will also be 1 m above the ground in 50 years. If two branches are growing less than 15 cm apart on a young tree, one should be removed to avoid crowding as the branches enlarge. The crotch angle should also be considered in selecting scaffold branches. The cambium divides continuously, forming new layers (Figure 13-3); therefore, branches that grow upright should be removed because if the angle is Y-shaped, the layers of cambium are not continuous and

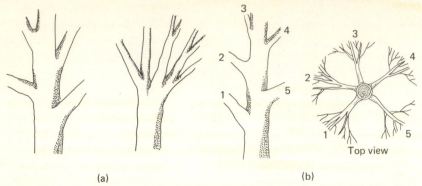

(a)

Top view

(b)

FIGURE 13-2 (*a*) Well-spaced branches have stronger attachments than those growing close together or in a cluster. Spacing of scaffold branches. (*b*) Branches with good scaffolding require proper vertical and radial spacing on the trunk. (Adapted from Ohio State University Bulletin 543.)

the branch may be easily broken. Enclosure of phloem, together with the formation of abundant wood parenchyma in the xylem of the crotch union, makes the crotch weak. Branches arising from the trunk at angles ranging from 40 to 90° are the strongest because the cambium is continuous without phloem inclusions. An ice storm can destroy trees with poor crotch angles.

FIGURE 13-3 A wide-angle crotch makes scaffold branches stronger. Note the lack of continuous cambium and phloem inclusions in the narrow angle on the right. (Adapted from Cornell University Agricultural Experiment Station.)

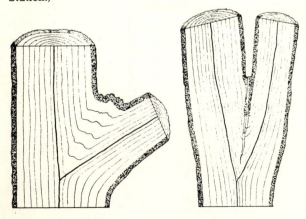

Pruning Techniques

Pruning procedures are determined by the life span, structure, and growth habits of the individual plants. Pruning may consist of as little as pinching buds to produce a more compact and bushy plant or as much as removing large branches. Herbaceous perennials generally are pruned either to remove dead or diseased parts or to thin the plants. Chrysanthemum buds are pinched to control the number of flowers and thereby increase the size of the flowers. Most frequently it is the woody, deciduous, and evergreen plants that are pruned. They can be grouped into two main categories: (1) shrubs, which have several trunks and (2) trees, which have only one or a few distinct trunks.

The two major types of pruning cuts, heading back and thinning, cause different responses in plants. Both thinning and heading-back cuts are normally used in almost every type of pruning.

In *heading back*, the terminal portion of the shoot is removed, but the basal portion is not (Figure 13-4). When heading back is used on flowering plants and foliage plants to increase branching, it is referred to as *pinching*. This method can be used effectively on all plants. Heading back can also be used to rejuvenate vegetative growth in

FIGURE 13-4 Removing one-half the growth by (*a*) heading back and (*b*) thinning gives different results in pruning. (Adapted from Everett P. Christopher, *The Pruning Manual*, Macmillan, New York, 1954. Copyright 1954 by Macmillan Publishing Co., Inc., and used with their permission.)

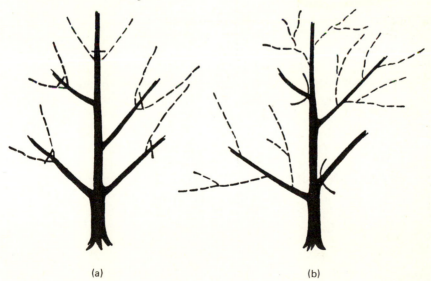

(a)

(b)

older trees and shrubs. When heading back a leader, the cut should be made near a lateral branch, leaving no stub to delay the healing process. When shrubs, small trees, old trees, and vines are headed back, it is desirable to cut above the nodes, leaving the uppermost bud or buds to develop. A typical response of many plants is three to four breaks, giving rise to the term "crows' feet." To control the direction of growth, choose a bud that points in the desired direction.

In *thinning*, entire shoots are removed; thus thinning is an extension or complement of heading back (Figure 13-4). Growth and light penetration are more evenly distributed, since thinning produces an open type of growth. Thinning can be used to rejuvenate old fruit trees. Old and weak wood is removed; this results in better-quality fruit on the remaining branches. When removing an entire branch, make the cut clean and smooth and as close to the trunk as possible.

When removing a sizable limb, make the first cut 25 to 40 cm from the trunk and one-third of the way through the limb from the bottom upward. The second cut is made a little farther out than the first, but from the top down. The last cut is made clean and flush with the trunk. Cutting in this manner eliminates the possibility of splitting or peeling the bark below the branch being removed.

After pruning cuts have been made, rapid healing is essential. Cuts less than 2 cm in diameter heal rapidly and usually do not need protection from disease or insects. If the cut is over 2 cm in diameter, a wound dressing is desirable to prevent desiccation. A good wound dressing should be waterproof and durable and should include an antiseptic to kill fungi.

Training Methods

The training systems of plants vary throughout the world. For example, trellis systems are sometimes used with apples in Europe, but in the United States the central-leader or modified central-leader system is most popular. These systems provide good light penetration, which in turn provides maximum bearing surfaces. The critical time for training is when the tree is young. The objectives of training young trees in the fruit industry are (1) to develop a well-spaced system of scaffold limbs with strong crotches and the ability to bear a heavy load of fruit and (2) to promote early production. In landscape gardening, training methods can be used to develop plants for screening, accent, space saving, and architectural purposes. Trellis and espalier training are also used. Bonsai is a training system involving root pruning and top pruning to control growth.

Central-Leader Training   In central-leader training, a single central leader is selected, and subordinate side branches are developed

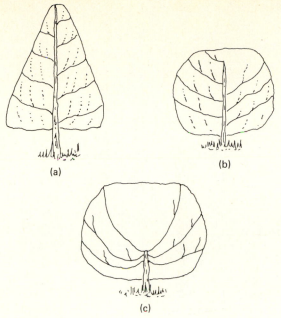

FIGURE 13-5 Plants are pruned to three basic forms: (*a*) central leader; (*b*) modified leader; (*c*) open center.

along this leader (Figure 13-5). Attention is given to selecting well-spaced lateral branches and maintaining the proper degrees of subordination among these laterals. The result is a tall, narrow tree ("Christmas-tree" shape). Some fruit types, such as apples, readily lend themselves to this system; but others, such as plums, are difficult to train.

The tree should be headed back to 0.7 to 0.9 m when it is planted (Figure 13-6). Lateral shoots should be removed below 0.5 m of trunk height. Two or three well-spaced branches with wide crotch angles and good position should be selected to remain on the main trunk, and other shoots should be eliminated. During the subsequent summer any competing branches, as well as new shoots on the bottom 0.5 m of the trunk should be removed. During the first dormant season, lateral branches are selected to form the lower scaffold limbs. From above, the tree can be visualized as the hub of a wheel with the laterals forming the spokes of the wheel. After scaffolds are selected, all other shoots should be removed. Some heading back is necessary on the leader and scaffold limbs to induce branching as well as to strengthen the limbs.

(a)

(b)

(c)

(d)

FIGURE 13-6 Training a young apple tree. (*a*) 1-year apple tree being pruned immediately after planting. (*b*) Well-grown 'Red Delicious' following its first year's growth. (*c*) This is the same tree shown in *b* after the first dormant pruning. Note the early selection of framework branches and maintenance of the central leader. (*d*) 4-year-old tree after pruning. (From U.S. Department of Agriculture Bulletin 1897.)

Modified Leader Training   The modified leader system combines the best qualities of the central-leader and open-center systems (Figure 13-5). A leader develops on the young tree until it reaches the height of 2 to 3 m. It is dominant until the tree is formed and is then restricted. Laterals are selected to ascend in a relatively spiral fashion up the central trunk, and these are cut back until the proper number and distribution of branches have been obtained. The lowest lateral is 60 cm or more above the ground. When the central leader is removed, the tree develops a top that is rounded and open. The limbs are low and well-spaced, and the fruiting wood is well distributed; this is not drastically different from the central-leader system. Many apple growers are now using the modified leader system.

Open-Center Training   In open-center training, the main stem is terminated and growth is encouraged from the upper end of the trunk, where a number of branches originate (Figure 13-5). Three to five primary lateral branches are trained to equal dominance by pruning each year. The selected branches should not originate too closely together, because any lateral might split from the trunk under adverse conditions such as a heavy crop load or snow. This training method affords maximum light penetration, which gives a more uniform distribution of fruit on the tree and thus highly colored fruit. A low-spreading tree develops; this facilitates pruning, thinning, spraying, and harvesting. The main disadvantage is that the arrangement and equality in size of the scaffold system often produce weak, crowded crotches. Limbs must sometimes be artificially propped or braced to prevent breaking. This system is widely used with peaches produced for the fresh market. It also is good for the shake-and-catch type of mechanical harvester, because the fruit tends to hit fewer branches in its fall from the tree to the catching frame.

Espalier Training   Espalier training is one of the most intense cultural practices that a horticulturist uses. Espaliered plants are much used today, as a result of the limited space characteristic of urban living. Espalier plants are trained in a pattern on a flat surface, such as a fence or wall. With proper care, plants can be trained to almost any shape. However, unless constant maintenance is planned to continue such training indefinitely, it is better not to espalier plants.

To develop a simple espalier from a 1-year whip, the whip is cut to within 25 to 30 cm above the ground the first spring after planting. From the uppermost bud, a new terminal shoot will develop. About mid-June, the new shoot should be cut to less than 1.5 cm of where it started. This technique will result in the development of at least two new shoots at the same level and of about equal strength. The two shoots are then tied to a frame or wire to keep them horizontal. After 30 to 40 cm of growth, the tips should be turned upward. The following spring and again in June, after the turned-up branches have grown sufficiently, both branches should be cut back to 30 to 40 cm above the bend and treated as the original shoot. This will produce four uprights. During the first few years other shoots will start to grow; these should be pruned back to less than 1.5 cm from the trunk or branch from which they develop. In time, flowers and fruit will develop. Ornamental plants may be trained as espaliers to serve as the focal point on the wall of a building, or more prominently displayed around borders of the private and public spaces in a landscape.

Trellis Training   Many vines and annual herbaceous plants are trained to a trellis. This is another rigid form of training. Grapevines may be trained on a trellis with canes—previous season's growth from the arms or trunk—tied to wires. Fruit trees are occasionally trained to a trellis or some other framework, and bramble fruits are also grown commercially on some type of trellis system. Some vegetable crops, such as pole beans, tomatoes, and cucumbers, are also grown on a trellis.

Bonsai Training   Bonsai is a very highly specialized training system involving root pruning and top pruning to control growth. Bonsai is pinching, bending, and tying rather than cutting back the growth after it has occurred. The objective is to keep the plant attractively healthy in a miniature form. Bonsai pruning consists of reducing the size of branches, changing their direction, and thinning dense growth that hides the line of trunk and roots.

Top and root growth are to a certain degree interdependent. If the plant is to keep the bonsai form, unwanted growth and large branches must be removed often. It is better to remove an unwanted branch in the bud today than to cut it with shears tomorrow. At first the plants are drastically pruned to shape and then headed back. Sometimes the plants are started when they are small. The thick roots are trimmed back when plants are repotted. This is done by wiring, twisting, and tying down. The shoots are then pinched back during the growing season so as to form a balanced tree.

In order for bonsai culture to be successful, it is necessary to maintain a constant balance between the top of a plant and its roots. The roots require periodic pruning to fit a particular container, and the branches are pruned to fit the form and size of tree required. The dwarfed plant which results is a product not only of root restriction but also of this constant stem-pruning process (Figure 13-7). Patience is an obvious and vital requirement of bonsai. The Japanese have excelled in this art.

FIGURE 13-7 Bonsai culture involving root pruning and top pruning to direct growth.

Physical Control of Fruit Crops

Pome Fruits After the tree is planted, the goal is to produce a vigorous tree with early fruit production. The training of nonbearing trees will determine the productivity of the tree later. The system should allow even distribution of sunlight and spray materials throughout the tree, build a strong framework able to withstand a heavy crop of fruit, and make the tree easily accessible for machinery or laborers.

Young developing trees are trained by one of two methods: (1) dormant pruning and (2) summer pruning. Generally, most pruning is done while the trees are dormant in the later part of the winter; this avoids freeze injury.

Dormant pruning affects young apple trees in several ways. First, when the terminal bud is removed, the lateral buds develop during the following growing season; the topmost side shoots have narrow crotch angles, but farther down the branch the crotch angles increase. Second, heading back will cause the branches to be rigid and stiff instead of weak and slender. Third, the removal of diseased wood sometimes aids in controlling diseases, such as fire blight.

Summer pruning is used to remove shoots growing below the first scaffold limb, to remove shoots competing for dominance of the central leader, and to reduce the vigor of the tree and thus stimulate the development of fruit buds. One of the major advantages of summer

(a)   (b)

FIGURE 13-8 (*a*) Six-year-old spur-type 'Red Delicious' on MM 106 rootstock. This tree
was trained as a central leader, but the branches were not spread. Note the crowded
branches, the lack of fruiting wood, and the small area occupied by the tree. (*b*) Tree of
the same age and cultivar as that shown in *a*. This is a central-leader tree that has
been spread since its second growing season. Note the open branches in the center of
the tree, the large amount of fruiting wood, and the large area occupied by the tree.
(From U.S. Department of Agriculture Bulletin 1897.)

pruning over dormant pruning is that much less vegetative response
occurs. Summer pruning can also help prevent "blind bud," a
condition where lateral buds fail to initiate leaf and shoot growth,
creating an unproductive limb.

Wire or wooden spreaders are used to develop a desirable tree
shape, to encourage early fruiting, to provide the best possible light
penetration, and to eliminate breakage or splitting which occurs with
narrow angles (Figure 13-8). The angles should be approximately 60°,
and the tip of the branch should be higher than its point of origin. If the
shoot apex is below the origin or other part of the branch, its growth
will stop and a strong sucker will usually originate at the uppermost
part of the branch.

Apples bear fruit terminally on spurs. Not every spur is likely to
bear fruit every year; therefore, a large number of spurs must be
present for adequate productivity. However, with good light exposure
more spurs fruit annually. When the spur terminates in a plump, oval
flower bud, it normally produces five blossoms (Figure 4-9). Usually one
to three flowers set fruit, and the others abscise. As the apple is
growing, an axillary bud breaks to form a shoot which may terminate
in a vegetative bud or flower bud. The next spring, shoot growth
occurs, adding length to the spur. This may differentiate into another
shoot bud or a flower bud. Proper pruning and thinning during the
bearing year will make annual bearing possible. Suckers and water
sprouts should be removed.

Stone Fruits Stone or drupe fruits, such as peach, apricot, almond, plum, and cherry, bear fruit on 1-year-old wood. During the first part of the season, the shoots elongate, and toward the end of the growing season they develop lateral flower buds. The following spring the flowers open and produce fruit. Generally, shoots grow 15 to 45 cm, if moderately vigorous.

Peach trees, an example of the stone fruits, usually are trained to an open-center form (Figure 13-9). The juvenile phase of the peach lasts 2 to 3 years. The first year the trees are headed back to a lateral branch or bud 0.6 to 0.9 m from the ground. For 2 to 3 years, lateral branches are allowed to grow before scaffold branches are selected. A maximum leaf area aids in rapid growth and development of the tree. Three to five lateral branches, arranged spirally around the tree, 10 to 15 cm apart, are then selected to form the scaffold branches. Any secondary lateral branches developing in the center of the tree are removed. Therefore, the leaves on the inside of outer branches are exposed to full sunlight, and the open-center habit is promoted.

A transitory period during the third and fourth years requires light pruning. Upright branches are cut back to an outside lateral branch. The objectives of pruning during this period are to maintain strong, evenly distributed scaffold limbs, to expose all parts of the tree to adequate sunlight, and to maintain a relatively low-headed tree with well-distributed fruiting wood.

FIGURE 13-9 (*Left*) unpruned and (*right*) pruned peach trees at different stages of growth: (*a*) at planting; (*b*) after one year's growth; (*c*) after two year's growth; (*d*) bearing tree.

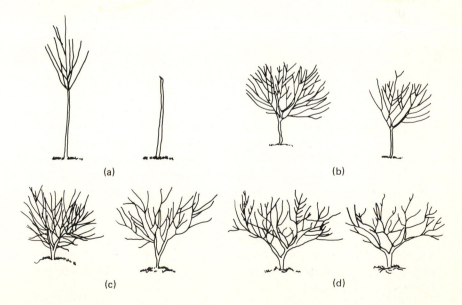

(a)   (b)

(c)   (d)

The fruiting period of peach trees generally lasts for 5 to 20 years. Since peach fruits are on 1-year-old wood only, the tree needs to be pruned every year.

Since moderately vigorous trees produce more flower buds than can grow into marketable fruit, heading back and thinning is necessary to reduce the number of flower buds. Shoots at the terminal of each branch are removed. The remainder of the previous season's growth of each branch is headed back or thinned depending on its length and vigor.

During the fruiting period of peaches, pruning is practiced to promote moderately vigorous wood, since that is the most productive. If a tree has relatively weak fruiting wood, severe pruning will invigorate the plant.

Cherries and plums are usually trained to a modified central leader. Three to four laterals and a leader are selected. Only light pruning is necessary during the first 3 years. Cherries require little pruning, since a full crop load is not too heavy for the tree to bear. Some cherries are now trained to approximately three main scaffold branches and are mechanically harvested with limb shakers. With both crops, older trees need to be pruned more severely than young trees to promote vigor and to renew the fruiting wood.

Citrus Fruits Citrus fruits, such as orange, lemon, and grapefruit, produce optimum marketable fruit on moderately vigorous stems. Pruning is directed at maintaining a balance between stem and leaf growth and flower and fruit production. The object of pruning is to maintain a moderate quantity of relatively vigorous wood. Suckers and damaged or dead branches should be removed yearly. Periodic thinning of old wood stimulates the production of new and more vigorous wood.

Nuts The pecan, walnut, and filbert are monoecious plants. Pistillate flowers arising from mixed buds are borne in terminal clusters on current season's wood. The second season's wood produces staminate flowers borne in catkins at the leaf base. Pruning should remove damaged and dead branches. Removal of other branches should be minimal, since female flowers are located terminally and severe pruning would curtail fruit production. Pecans and filberts are pruned to a central leader, walnuts to a modified leader.

Small Fruits Grapes are grown worldwide. They are trained to a trellis or support to establish the form and direction of the trunk, canes, and shoots. This regimentation is required to maintain a form which will save labor, facilitate vineyard operations, and obtain an even distribution of bearing wood over the vine for maximum high-quality crops.

Grapes can be trained to almost any form. They can be trained along fences; they can be used as a screen; they can be trained on an overhead trellis to produce·shade. In the home garden, an arbor is often used. A good pruning system is necessary, or the vine will become a tangled mass. The average person is very reluctant to prune grapevines with proper severity so that vigorous growth will result.

Pruning grapes requires an understanding of the names of certain parts of the grape plant. The *trunk* is the main unbranched stem, which starts at the ground line and terminates at the top wire; *arms* are all main branches 2 years old or older. *Canes* are the previous season's growth from the arms or trunk. *Shoots* grow from buds in the current season and are called *canes* the following season. *Spurs* are parts of the cane with one or two buds.

The fruit is borne near the base of the current season's shoots; therefore, an annual supply of 1-year-old wood is necessary to allow the fruit-bearing shoots to develop. Since some grape types cannot be grown in some systems, there are several different training systems: the *four-cane Kniffin system* (Figure 13-10), the *umbrella Kniffin*, and the *arbor system*, such as the *Geneva double-curtain system* (Figure 13-11). All these systems have two purposes: (1) to balance carbohydrate production to fruit production and to optimize sun exposure as related to carbohydrates, and (2) optimize plant coverage per hectare especially to air drainage and ability to get good spray coverage.

The *four-cane Kniffin system* consists of posts with horizontal wires—one 0.75 m from the ground, the other 1.5 m from the ground. When planted, the young vines should be cut back to one cane of two to three buds to produce a main trunk. The third year, all but four side canes should be removed. The four remaining canes should be cut to three or four buds, with one tied to each side of the upper and lower wires. Renewal spurs (one to two buds) are left near the trunk at each cane to ensure that there will be bearing canes close to the trunk.

The mature vine requires annual pruning. All wood beyond the first strong shoot on the four bearing canes should be removed. Since old wood results in bearing wood farther and farther from the trunk, renewal is required; but not all canes should be renewed in the same year.

The *umbrella Kniffin system* also results in high crop yields. Canes originating near the head of the trunk are bent over the top wire and tied to the lower wire, where side shoots will droop from the downward portion. Canes from the head are used because of their vigor. Spurs are also left to provide renewal wood.

Vigorous vines of cultivars such as 'Concord,' 'Delaware,' 'Niagara,' and 'Catawba' are more productive with the *double-curtain system*. This system allows the leaves and shoots to receive more sunlight, resulting in higher yields and better-quality fruit. A trellis is

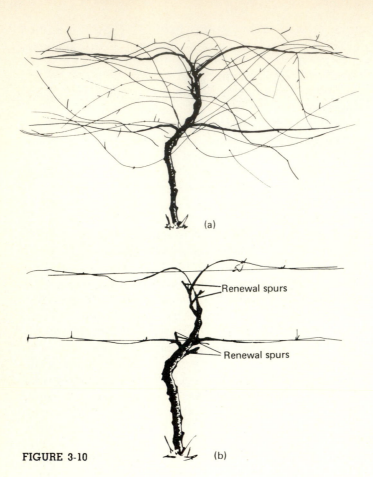

(a)

Renewal spurs

Renewal spurs

FIGURE 3-10   (b)

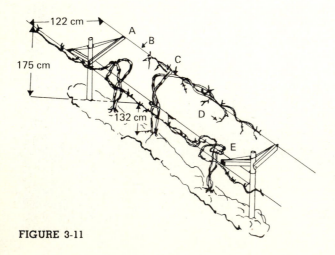

122 cm

A
B
C

175 cm

132 cm

D

E

FIGURE 3-11

constructed with two cordon wires. Vines can be spaced at 4.9 m on adjacent wires, resulting in trellis space about double that of more conventional methods (Figure 13-11). The trunk of the vine is trained to the lower wire, and generally two trunks per vine are utilized. Branches are developed from the trunk and extended along the cordon wires. The branch should be allowed to develop only five bud spurs and one bud renewal spur; otherwise, shoots from the spurs will form too dense a foliage canopy. The shoots must be positioned to grow in a vertical downward direction from the cordon wire. Machines are being developed to do this, as hand labor is too expensive. Hence a double curtain of foliage extends from each grape row (Figure 13-12). The

FIGURE 13-12 *Below:* Geneva double-curtain training system for Concord and other bunch grapes. With this training system, the shoots receive more exposure to the sun than with other systems. (Photograph by Fred Cochran, North Carolina State University.)

Geneva double-curtain system gives increased yield per shoot and per vine without delaying fruit maturation. The fruit can also be easily harvested mechanically (see Figure 16-8).

When muscadine grapes are planted, all canes should be removed except one, and this one should be cut back to two or three buds. When growth begins, the best cane should be selected and the others removed. The grape should be trellised with a stake at least 1.7 m in height with the cane tied to it. Allowing the unstaked cane to grow on the ground results in tipburn and undesirable lateral branching. When the grape reaches 1.7 m, the tip should be pinched to induce lateral branching. As lateral branches develop, two of these should be selected about 0.75 m above ground level and two at the tip for the permanent arms of the plant. The remaining lateral branches should be removed. At the end of the first year's growth a trellis can be constructed. By the end of the second year all permanent arms should be established at full length. The pruning is then limited to (1) reducing previous season's wood to two or three buds, (2) removal of dead canes, and (3) thinning of undesirable canes and spurs.

Bramble Fruits The bramble fruits (black, purple, and red raspberries; blackberries; and dewberries) develop from mixed buds borne on 1-year-old canes. The canes develop the first season, and fruit and then die the second season. Severe pruning is required for black and purple raspberries, since they tend to produce more flowers than the canes can produce into marketable quality. Black raspberry canes are supported. Canes are pinched off at 30 cm in California and 45 to 60 cm in the Northwest. Purple raspberry cultivars are pinched at 75 to 90 cm. Pinching, a form of heading back, promotes lateral branch development and prevents top-heavy canes. This should be done in spring prior to growth, since fruit develops on the lateral branches.

The red raspberry rows should be 50 cm wide. The plants should be placed 15 cm apart in the row. In the spring, the year-old canes are cut to one-fourth their length, and winter-killed or dead canes are removed (Figure 13-13).

FIGURE 13-13 Pruning of red raspberries during dormant period: (*a*) unpruned; (*b*) pruned. (From Clemson University Agricultural Bulletin 123.)

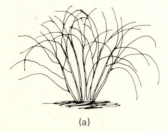
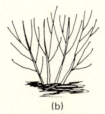

(a) (b)

Blackberry canes which have fruited should be pruned back to the ground in the fall after harvest. New canes should be pinched at 90 cm to allow new side branches in spring. In spring, these side branches should be pruned to 45-cm length.

The dewberry may be trained on trellises or stakes or allowed to run along the surface of the ground. All canes are removed after harvest. Canes that develop later are used for the next year's crop.

Currant and Gooseberry   The fruits of these plants are borne on 1-year-old wood and on 1-year-old spurs on 2- and 3-year-old wood. Since older wood is not productive, a pruning program should be established for its removal. Usually eight to ten branches per plant produce an adequate quantity of marketable fruit.

Strawberries   Strawberries need no pruning at planting time. During the first season, all blossoms should be removed to encourage runner production. For larger yields, allow four to five runners from each plant to develop a space not over 60 cm wide. The berries are harvested the second and third years. Beds are plowed under and replanted in a different location when productivity declines. In California and Florida most strawberries are grown without the growth of runners. High plant populations are set on clear or black plastic; flowering and fruiting occurs quickly, and the crop is grown as an annual.

Blueberries   Blueberries should be pruned while dormant. Old and dead wood is removed, and the fruiting shoots are cut back to three to four buds to increase fruit size. Some species, such as the rabbiteye blueberry, produce good crops with only moderate pruning.

Physical Control of Vegetable Crops

As stated earlier, vegetables which are grown as annuals need less pruning than other plant materials. Exceptions to this are tomatoes, which are intensively trained when grown under glass.

Tomatoes   Much research has been conducted during the last 20 years on modifying the growth of the tomato. The *cordon system* is used most often by home gardeners and by commercial growers. This traditional method trains the one or two main stems to a support with lateral shoots removed so that the fruit is borne on the main stem. Under glass, the main stem is twisted around a string hung from overhead wires. Outside, it is common to tie the stem to a stake. When a desirable height has been reached, the main growing point is removed, and the plant nutrients are used for the formation and development of the fruit instead of more vegetative growth. However, many tomatoes are grown on the ground with no pruning or staking.

Physical Control of Floriculture Crops

Many cut flowers and potted plants, such as chrysanthemum and roses, are pruned. The chrysanthemum will grow a single stem until a flower bud is formed. Both pinching and disbudding operations are part of the pruning process. *Pinching* is the breaking off of the terminal growing point which allows the axillary buds to start to grow. This is usually done a few weeks after the cuttings are planted, to increase the number of flowering stems on a plant. However, for the natural-season crop pinching is important for timing, spray type, and flower formation. *Disbudding* is the removal of flower buds; this is done after the flower buds are visible, to increase terminal flower size.

Roses require severe pruning one or more times while they are in the greenhouse bench to keep them from growing so tall that cutting, tying, and other operations become difficult. The two methods used are direct pruning and gradual cutback. When roses are not in great demand, the plants are direct-pruned to the desired height. The height to which the plants are cut back the first time should not be less than 70 cm for most cultivars, because the wood is generally quite hard below this point. Each time the plants are pruned, the cuts should be made about 15 cm above the previous cutback in the soft wood, which grows out more readily. Gradual cutback is often called *knife pruning*, since the stems are cut with a knife when the flowers are taken from the plant. With this method not all stems are cut off at once. No matter how rapidly the gradual cutback is practiced, it is important that not all cuts be made at the same time. Some shoots are in flower at the same time as others are breaking out following the cutback. The plants should not be cut below 70 cm in height.

Physical Control of Nursery Crops

The number of ornamental trees, shrubs, and vines is great, and each has its own individual characteristics. For instance, a small tree can be trained and pruned to a large shrub, or a large shrub can be trained and pruned to a small tree. The variations are infinite, but some basic rules do exist. One is knowing when to prune the shrub or tree. If the plant blooms before the end of June, it is generally referred to as a *spring-flowering* shrub or tree. Because the blooms are formed on the previous season's growth, pruning should take place only after flowering. When spring-flowering shrubs such as forsythia, daphne, lilac, and spirea are pruned while they are dormant, flowering will be reduced accordingly in the spring. If the plant blooms after June, it is referred to as a *summer-flowering* shrub or tree. These blooms are formed on the current season's growth; therefore, it is most beneficial to prune while the plant is dormant in fall or winter. Examples are hydrangea and althea. Nandina, Japanese quince, and pyracantha

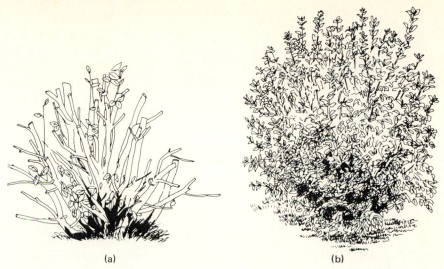

(a)   (b)

FIGURE 13-14 To rejuvenate old, overgrown, and leggy shrubs prune close to the ground in early spring. (*a*) Pruned shrub; (*b*) shrub after new growth in spring and summer.

produce flowers on 1-year-old wood; therefore, these should be pruned lightly. If they are overgrown, they can be cut back to the ground; this results in thick, bushy regrowth (Figure 13-14). A partial renewal can be accomplished each year by removing one or two old branches.

Evergreen shrubs are grown for their year-round foliage effect and are classified as either broad-leaved or narrow-leaved. The broadleaf evergreens should be pruned annually. Proper pruning controls size and shape, resulting in a more healthy, vigorous plant (Figure 13-15). Moderate pruning of plants can be done year round, but

FIGURE 13-15 In most cases shrubs should be pruned by (*a*) thinning rather than by (*b*) heading back. (Adapted from Everett P. Christopher, *The Pruning Manual*, Macmillan, New York, 1954. Copyright 1954 by Macmillan Publishing Co. Inc., and used with their permission.)

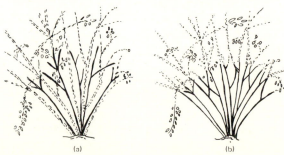

(a)   (b)

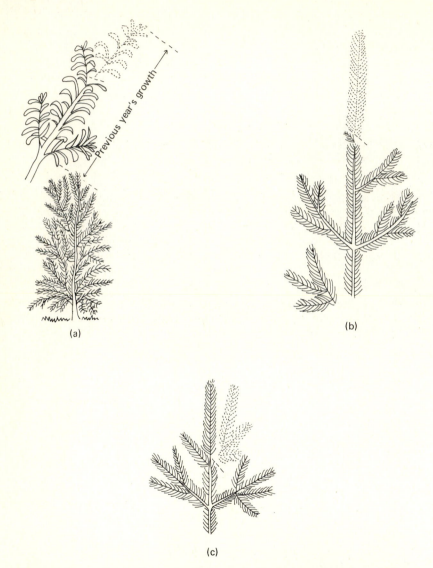

FIGURE 13-16 (*a*) Cut back one-fourth to one-half of previous year's growth on narrow uprights (note top drawing). On pyramidal plants, cut back just enough to maintain compact growth. (*b*) Cut back central leader only when it outdistances the rest of the plant. This forces a lower bud to become new leader. (*c*) Remove multiple leaders on spruce and pine, leaving the best one to become the new central leader. (From Ohio State University Bulletin 543.)

severe pruning should be done just before new growth begins in the spring. The narrow-leaved evergreens need occasional pruning to maintain their shape and size. Also, they are pruned to allow light penetration to keep the inner leaves alive. These types can be pruned back at the ends of their branches (Figure 13-16). New buds grow quickly. The best time to prune is when the new foliage is soft. Narrow-leaved evergreens do not tolerate severe pruning. Examples are yews, junipers, and pines.

Most plants require removal of a portion of the top growth when transplanted. This is done to reduce transpiration and balance the reduction in water uptake due to root damage (Figure 13-17).

Hedges are often used in home gardens, office areas, and formal settings. These hedges can be either formal or informal. In the formal types, it is important to prune correctly when the plants are planted. Tips and laterals should be pruned back on a regular basis to induce dense growth. The plants should be narrower at the top than at the bottom for good light distribution. The informal hedge requires little pruning. If a hedge of a certain height is desired, the main shoots should be cut back to a much lower level. Plants selected for hedges can be pruned at any time, but the ideal times are late spring or early summer and late summer or early fall.

FIGURE 13-17 Pruning of tree when transplanted to develop strong leader. Note that two-thirds of branches are removed to reduce transpiration loss.

Pruning Equipment

Hand Pruners   Hand pruners are of two types. In one type, sharpened blades overlap, producing a cut similar to that made by household scissors (Figure 13-18). The second type, the anvil-type pruner, has a straight-edged top blade which comes down on a soft-metal anvil. Either type is acceptable; however, a closer, cleaner cut can be made with the first type. *Lopping shears* are essentially large hand shears. They have longer handles which give more leverage, so that cuts too large for hand shears can be made. *Pole pruners* are used to make cuts in hard-to-reach areas or in trees without using a ladder. *Pruning saws* are necessary to make cuts too large for pruning shears or loppers. They have coarse, wide-set teeth to avoid sticking.

FIGURE 13-18 Pruning equipment: (*a*) pruning saws; (*b*) pole runners; (*c*) lopping shears; (*d*) hand pruning shears; (*e*) hedge shears. (From Ohio State University Bulletin 543.)

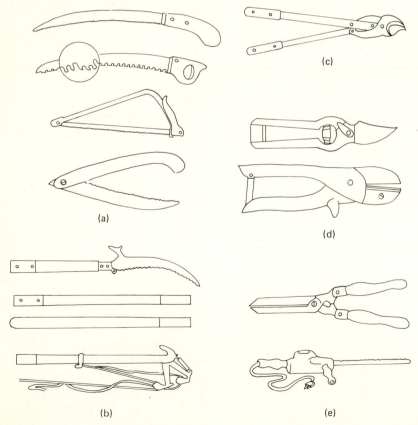

(a)   (b)   (c)   (d)   (e)

Power Pruners Electric hedge shears are commonly used today. Most of the large-scale pruning in orchards is done with power pruners, which take much of the hard work out of the job. Both power loppers and saws are available. These operate by the use of hoses and compressed air from a compressor powered by a power takeoff on a tractor.

Mechanical Pruners An innovation of recent years has been mechanized pruners. These vary widely in design, operation, and power source but are all made to prune a plant (usually a fruit tree) mechanically. They include large circular saws which rotate on arms, small overlapping circular saws, and cutter-bar type mechanisms. The large saws will cut branches up to 5 to 6 cm in diameter, whereas the cutter-bar type cuts branches up to 1 to 2 cm diameter. Some are used as "toppers" for cutting back the top of the trees, either horizontally or at an angle. Others cut the sides of the tree, either vertically or at an angle. It is possible to prune a hedgerow of citrus or apple to the shape of a box or even a Christmas tree. Much of this type of pruning is done on a custom basis.

Because of the speed and relatively low cost of mechanical pruning, it is becoming widely used. Serious problems, however, can develop. If the tree is cut back too severely, as is quite possible with some of these pruners, the tree can be thrown into a vigorous vegetative state. One partial solution to this problem is to anticipate the pruning by a year and to omit nitrogen applications to reduce the rank vegetative growth. A wise precaution is to bring trees back down to the desired size over a period of 3 to 5 years rather than all at once. Another problem is that if the tree is hedged and topped only moderately, a vast number of new shoots will develop as a result of the mass removal of apical dominance. The problem arises when each branch headed back is replaced by two to four new branches. The result can be thick growth, heavy shade, and poor fruit quality. The only answer to this problem is, after hedging, to thin the branches left by some hand pruning. Mechanical pruning can greatly aid in getting pruning done, but it does only part of the total job.

Positioners In recent years fruit growers have found it increasingly difficult to hire willing, able, and skilled workers for winter pruning. One partial solution has been the purchase and use of positioners which put the pruner up in the tree, eliminating the need for climbing the tree or using ladders. Positioners are expensive, but they not only increase the efficiency of workers but make the job more attractive. Many of the positioners are attached to a tractor or are equipped with a power source so that pneumatic pruners can also be used.

CHEMICAL CONTROL OF GROWTH

Today, the use of chemicals in the regulation of plant growth is becoming more widespread. Some chemicals have a very broad and general use, and others are very specific for individual species or cultivars.

Chemical Pinching

Hand pinching the apical portion of plant shoots is a time-consuming and expensive method of pruning. Chemicals can be used (1) to increase the number of new shoots per plant, (2) to control the shape of the plant, (3) to increase the number of flowers, and (4) to control the time of flowering. Pruning with chemicals is, therefore, a very necessary part of growing ornamental plants.

It has been found that fatty acid methylesters and alcohols of chain lengths C8 to C12 are effective in killing or inhibiting axillary bud growth. Experimentally, it has been determined that the effectiveness of the chemical varies directly with the chain length, with C10 being the most effective. The chemicals, referred to as *chemical pruning agents*, are not translocated to other parts of the plant and kill only the meristematic cells. Lower alkyl esters of fatty acids destroy terminal buds, effectively pinching the plant; but the effectiveness varies with the concentration of the chemical and woodiness of the tissue.

Chemical pruning has become adapted to azaleas (Figure 13-19) and chrysanthemums. Chemical pinching stimulates branching of azaleas and increases the number of flowers per plant. Growth of treated plants is suppressed initially, but within several weeks the growth may exceed that of an unsprayed plant. Chemicals such as methyl caproate, methyl caprylate, methyl laurate, and methyl stearate are effective chemical pinching agents.

To produce chrysanthemums commercially, many tedious hours of disbudding are required. For each flower that develops, growers remove 10 to 25 lateral flower buds. The alklapthalenes can cause death of the chrysanthemum apical meristems. The commercial product HAN of C10 to C13 is most effective in killing meristematic tissue. The methods and time of application differ with the desired purpose. For example, research has shown that when the plants are under 8h-short-day treatments for flower induction, treatment with HAN on the ninth and twelfth days kills the apical meristem but allows lateral stems of seven to eight nodes to develop flowers. Treatment on the thirteenth, fourteenth, and fifteenth short days causes the lateral buds to die, but not the terminal flower buds. The response varies greatly, with some cultivars being more susceptible to HAN treatments than others.

Another area of chemical pruning is root pruning. Often dominant

(a)

(b)

FIGURE 13-19 Chemical pinching of azalea bud. Note new shoots in *b*. (Photograph, Fred Cochran, North Carolina State University.)

taproots develop if the root is not pruned to induce strong, fibrous growth. Since hand pruning requires time, chemicals are used to treat layers in the seedbed. When the roots reach these layers, they are pinched. For example, cork oak acorns were planted 8 cm above layers of Osmocote and Perlite soaked in copper napthenate; the taproots were unable to penetrate the treated layers, thus causing a more compact fibrous root system.

Flower and Fruit Thinning

The fruit industries, specifically the apple industry, use chemicals for flower and fruit thinning. The general purpose of fruit thinning with chemical sprays is to encourage annual bearing and to improve fruit size. Each orchard owner may experience different results because of variables such as timing of application, cultivar, weather conditions, vigor of trees, pollination, and frost damage. Many cultivars of apples tend to develop a biennial bearing habit in which very heavy flowering occurs every other year with a light crop the alternate year. The only effective way to get annual flowering is to thin in the "on year," but this must be done early in the year. Flower bud initiation occurs within 30 to 60 days after bloom; to affect next year's crop, therefore, this year's crop must be thinned well before hand thinning is practical, so that chemical thinning is the only genuine alternative.

The apple tree regulates itself naturally to some degree, since less than 5 to 20 percent of the flowers develop into fruit. However, chemical thinning is still necessary. Although hand thinning is time-consuming and expensive, some hand thinning may be necessary to supplement the chemical thinning process. Growers in the West use dinitro (DN) sprays to reduce fruit set. The pollen of unfertilized blossoms may be killed if sprayed within 10 h after pollination; to be effective, then, dinitro sprays must be applied right after the bloom stage. Dinitro sprays are most satisfactory on clear days with moderate temperatures. If frost or rain follows application, the spray may greatly overthin the blossoms.

Growers in the eastern United States thin with hormone-type sprays, which are applied well after bloom. Because of the everpresent danger of spring freezes, eastern growers are hesitant to thin in bloom. Secondly, the likelihood of rain makes DN sprays very risky in the East.

If polyoxyethylene sorbitan monolaurate or Tween 20, a detergent, is added (as a wetting agent) with napthaleneacetic acid (NAA), the amount of NAA needed for thinning is reduced. The wetting agent itself has been shown to be capable of thinning. In addition, experiments have proven that frost before or after spraying increased the permeability of the leaves, allowing a greater percentage of hormone spray to be absorbed and resulting in increased thinning. Cultivars also vary considerably in their response to NAA.

Height Control of Plants

The floriculture industry is dependent upon chemical growth control. In floral crops these chemicals change plant growth so that the crop can be grown without extra labor. Growth regulators are extensively used in the production of flowering pot plants. These are frequently used to produce more salable compact plants. Chemicals used for this purpose have the desirable effect of reducing internode length without reducing the number of leaves.

In 1950, the growth retardant Amo 1618 was developed. The compound was very effective for dwarfing plants, but it was too expensive for commercial use. Later, a related material, Phosphon (2-4-dichlorobenzytributylphosphoniumchloride), appeared on the market; it is now used for retarding the height of chrysanthemums and lilies. This material is applied as a solution to the soil. Cycocel (2-chloroethyltrimethylammoniumchloride), a more recent introduction, can be used as a foliar spray or a soil drench for controlling the height of poinsettias and azaleas. Daminozide (succinic acid 2,2-dimethylhydrazide) used as a foliar spray effectively controls the height of chrysanthemum, bedding plants, poinsettia, azalea, and hydrangea. One of the newest growth retardants is A-Rest (a-cyclopropyl-a-(4-methoxyphenyl)-5-pyrimedinemethanol). This chemical controls plant height and is active in a low concentration (Figure 13-20).

FIGURE 13-20 Growth of Easter lily controlled with A-Rest:(*a*) treated; (*b*) untreated. (Photograph, Roy Larson, North Carolina State University.)

(a)　　　　　　　　(b)

GRAFTING FOR CONTROL OF GROWTH

There has been increasing acceptance among United States apple growers of the idea that the future of the industry lies with trees smaller than those produced by seedling rootstocks. Rootstock can affect the size and growth habit of a tree. Researchers at the East Malling Research Station in England have done the pioneering research on size-controlling apple rootstocks. They released the Malling (M) Series, starting about 1920, and to date have made available 27 different rootstocks which vary from dwarf to vigorous. After their release, the early members of the Malling series were crossed with the cultivar 'Northern Spy' at the John Innes Horticultural Research Institute at Merton Experiment Station in England. The logic behind these crosses was to incorporate the wooly aphid resistance of 'Northern Spy' with the size-controlling rootstocks of the Malling series. The outcome of this program is the Malling-Merton (MM) series, which range in vigor from semidwarf to vigorous. There are United States programs at both the New York Agricultural Experiment Station, Geneva, and at Michigan State University in the breeding of new apple rootstocks. The need for improved apple rootstocks for the control of tree size and productivity is apparent to all associated with apple production today. Other desirable attributes are winter hardiness, disease and insect resistance, freedom from viruses, and ease of propagation.

Increasing proportions of the trees being planted are on the Malling and Malling-Merton rootstocks, but problems have arisen in many areas. One of the more troublesome problems in warmer climates has been the death of trees just as they begin to fruit. Several different factors seem to be contributing to the deaths, and precise identification of the problem has been difficult. One widely discussed cause is collar rot caused by *Phytophthora cactorum*; others are winter injury and excess soil moisture. Other problems which have been troublesome are the breaking of trees at the graft union, signs of incompatibility, and susceptibility to fireblight, *Erwinia amylovora*. Another problem—the significance of which is not well defined as yet—is the presence of a broad spectrum of viruses. One approach taken by many growers is to plant several rootstocks so as to reduce the risk. Other growers have gone more heavily to seedling rootstocks with spur-type scion cultivars. A third alternative is the use of interstocks.

Citrus fruit is altered by the rootstock selected. For example, if sour orange is used as a rootstock, the tangerine, sweet orange, and grapefruit are all smooth, juicy, and thin-skinned. The same is true when the sweet orange and grapefruit are used as rootstocks. However, when the rough lemon is used as the rootstock, the fruits are of inferior quality, thick-skinned, and low in sugar and acid.

For many species, rootstocks are available which tolerate unfavorable conditions such as heavy or wet soils, disease organisms, and low temperatures. Rootstocks of plum, peach, apricot, and almond differ greatly in their response to adverse soil conditions. Trees with the myrobalan plum as a rootstock are more tolerant of excessive soil moisture than those on peach, apricot, or almond rootstocks. Some rootstocks are more resistant to nematodes or oak root fungus. Some stone fruits infected by bacterial canker are more resistant when grafted on certain rootstocks. Apricots or plums normally grafted on myrobalan plums are seriously affected by pathogens when they are grafted on peach roots. In citrus, the rootstock can have an influence on the cold hardiness of the scion. For example, during a severe freeze during the winter of 1950–1951 in Texas, grapefruit trees on 'Rangpus' lime roots survived better than those on rough lemon or sour orange stocks, while the trees of the 'Cleopatra' mandarin were most severely damaged.

Effects of graft combinations can also be evidenced with intermediate stock graftages or interstocks. An interstock is placed between the rootstock and scion in a process known as *double-working*. Interstocks are used to overcome incompatibility or to utilize a desirable characteristic, such as disease resistance or cold hardiness, not found in the scion or stock that will contribute to a better framework of the specimen. Interstocks also may influence the growth habit of the tree. For example, if a Malling 9 interstock is placed between a vigorous stock and a vigorous scion, flowering will be stimulated, with subsequent growth reduction. The ultimate tree size will vary with the length of interstock used. A tree with a 20-cm interstock will be smaller than one with a 10-cm stempiece.

Graft combinations can direct plant growth in the case of topworking. A fruit tree or an entire orchard may be of an undesirable cultivar. They may be extremely susceptible to the prevailing diseases and insects, or the trees may exhibit poor growth habits. In any case, the tree may be topworked by removing selected portions of branches and grafting on the desirable scion. This technique can also be used to provide adequate branches for cross-pollinating an orchard containing only one cultivar which is self-unfruitful. Also, in certain dioecious plants such as the hollies (*Ilex*) a staminate branch may be grafted onto the pistillate plant in order to ensure pollination and fruiting.

During the past 20 years the pear orchards in California, Washington, and Oregon have been devastated by a disease called *pear decline*. An intensive research program was initiated, and the results indicate that the cause is a virus or viruslike organism which is transmitted by the pear psylla, a sucking insect which attacks pear trees. It became apparent that susceptibility of the tree depended to a large degree on the rootstock. Trees on *Pyrus communis*, the French or

European rootstock, are resistant; those on the Oriental rootstocks *P. serotina*, and *P. ussuriensis* are highly susceptible. Many trees on the Oriental stocks have been killed and are being replaced by new orchards on *P. communis*.

BIBLIOGRAPHY

Childers, N. F.: *Modern Fruit Science*, 7th ed., Horticultural Publications, New Brunswick, N.J., 1976.

Christopher, E. P.: *The Pruning Manual*, Macmillan, New York, 1954.

————: *Introductory Horticulture*, McGraw-Hill, New York, 1958.

Galston, D.: *Control Mechanisms in Plant Development*, Prentice-Hall, New York, 1970.

Gourley, J. H., and F. S. Howlett: *Modern Fruit Production*, Macmillan, New York, 1941.

Grounds, R.: *The Complete Handbook of Pruning*, Macmillan, New York, 1973.

Hartmann, H. T., and D. Kester: *Plant Propagation: Principles and Practices*, 3d ed., Prentice-Hall, Englewood Cliffs, N.J., 1975.

Janick, J.: *Horticultural Science*, Freeman, San Francisco, 1972.

Leopold, A. C., and P. E. Kriedemann: *Plant Growth and Development*, 2d ed., McGraw-Hill, New York, 1975.

Mahlstede, J. P., and E. S. Haber: *Plant Propagation*, Wiley, New York, 1957.

Ware, G. W., and J. P. McCollum: *Producing Vegetable Crops*, 2d ed., Interstate, Danville, Ill., 1975.

Weaver, R. J.: *Plant Growth Substances in Agriculture*, Freeman, San Francisco, 1972.

CHAPTER 14
PEST
CONTROL

 Since their earliest days on earth, humans have been faced with a wide diversity of pests which attack them and their food crops. In this chapter we present an introduction to subjects which relate to the pests of horticultural crops and how they are controlled. The disciplines include entomology (the study of insects), plant pathology (the study of plant diseases), and weed science.

When we consider current trends in world population, it is readily apparent that we cannot tolerate crop failures like those that humanity has occasionally faced in years past. The total devastation of plants by hordes of locusts in Egypt is recorded in the Bible. Late blight, a disease of potatoes, had a tremendous impact on the history of the Irish people. In the mid-1840s late blight wiped out the Irish potato crop, upon which Irish farmers were almost totally dependent for food. Over a period of years, thousands of Irish people died of starvation and even greater numbers emigrated to North America. There are numerous other examples in recorded history, but these two examples offer striking evidence of the important battle between humans and the enemies of their food crops. Perhaps more subtle, but equally important, is the battle for survival between crop plants and the seemingly endless array of weed competitors. The damage from weeds varies from reductions in yield and quality to total crop failure.

INSECTS

Insects can be found in most parts of the world, and from fossil remains it is known that they have been on earth for at least 250 million years. The number of insect species in existence today is not known, but there are certainly well over a million and probably closer to 3 million. Current estimates indicate that insects account for approximately 75 to 85 percent of the total number of animal species. In North America (north of Mexico), entomologists estimate that there are in excess of 80,000 species, but their distribution varies with locality and climate. Although the great majority of insect species are either beneficial or neutral as far as human beings are concerned, insect damage combined with control costs runs into billions of dollars annually.

In the animal kingdom, the phylum Arthropoda includes insects, mites, and spiders as well as many other similar types of animal life. The Arthropods are further subdivided into classes. The class Insecta includes the true insects which are characterized as adults by one pair of antennae, three pairs of legs, and three body regions (head, thorax, and abdomen). The class Arachnida includes mites, ticks, and spiders, all of which have two body regions, four pairs of legs, and no antennae.

Like most other members of the animal kingdom, insects typically reproduce sexually. The fertilized eggs are laid in a wide diversity of places depending on the type of insect. Another variable feature is the number of eggs laid which may vary from one egg for some species to thousands of eggs for others. In certain types, the female can produce offspring without fertilization, a process called *parthenogenesis*. In certain aphids (*Aphis sp.*) normal sexual reproduction occurs in the fall, but during the summer months, the young are produced in the absence of males.

A newly hatched insect may or may not resemble the adult of the species. Some types go through distinct changes of form from the young to the mature insect, whereas others change little in appearance or form but merely enlarge. *Metamorphosis* is the term applied to the changes in form or appearance which may occur in the developing insect. In types showing a *complete metamorphosis*, the distinct stages are the egg, larva, pupa, and adult. An example of complete metamorphosis is found in the Japanese beetle *(Popillia japonica)* which hatches from the *egg* to become a very small white grub *(larva)* in the soil. Because the outer "skin" or exoskeleton is hard, the larva must shed its outer skin (molt) three times during its development to the adult stage. The larva changes to the *pupa*, which neither feeds nor moves about. During the pupal stage, however, the larval structure is transformed into adult characteristics. The *adult* beetle emerges from the soil and feeds for 4 to 6 weeks (Figure 14-1). It then burrows into the soil and lays eggs from which the larva develop. In the Japanese

FIGURE 14-1 Japanese beetles on peach. Note the large mass of beetles clumped together as well as the damage to the leaves. (Photograph, U.S. Department of Agriculture.)

beetle, there is only one generation per year. Other examples of insects exhibiting complete metamorphosis are butterflies, beetles, and flies. In some of these both the larval and adult stages cause damage, as with the Japanese beetle; but with others, such as the European corn borer *(Ostrinia nubilalis)* and cabbage looper *(Trichoplusia ni)*, the larval stage alone causes damage (Figure 14-2).

Those insects said to exhibit only a gradual or slight metamorphosis already resemble adults when they hatch. The major changes are in size and the growth of wings in winged species. Insects with this type of development include grasshoppers and leaf hoppers. A limited number of species have no metamorphosis because the young are essentially minute forms of the adult. An example of this type is the silverfish, but no agriculturally important insects are included in this group.

Benefits from Insects

As noted previously, many insects are beneficial in man's effort to feed the expanding world population. Commercial producers of crops such as cherries, apples, pears, cucumbers, and melons are almost totally

FIGURE 14-2 Damage to cauliflower plants from cabbage looper. (Photograph, U.S. Department of Agriculture.)

dependent on honeybees *(Apis mellifera)* for pollination. Today wild bee populations are usually too small for the task of pollinating tree fruits when, during bloom, the entire crop may be "set" during a few hours of good weather. Therefore, hives of domestic bees are moved into orchards during the bloom period. The ladybird beetle (*Hippodamia sinuata*), commonly referred to as the ladybug, is very beneficial as it feeds on aphids. There are numerous other examples of predator insects and mites. As our emphasis on biological control* of insects increases, such predators as the ladybug will likely become increasingly important.

Insect Damage

Although insects are classified in several ways, we will use a system based on mouth parts. The two major classes are those with *chewing mouth parts* and those with *sucking mouth parts*. Not only is this distinction important in the type of damage which occurs, but it is also the major consideration in selecting an effective chemical control. This aspect is covered later in this section.

*Biological control (also known as biocontrol) is the reduction in the population of harmful insects or other pests, by means of other living organisms.

Chewing Insects The damage resulting from chewing insects is particularly apparent. The total devastation which can result from large populations of voracious grasshoppers is legendary in many parts of the world. They not only denude the land but, by leaving the soil barren, make it vulnerable to the ravages of erosion by wind and water. The damage by chewing insects to commercial crops is normally much less striking but is still of major proportions. Important horticultural crop pests with chewing mouth parts include the Mexican bean beetle *(Epilachna varivestis),* Colorado potato beetle *(Leptinotarsa decemlineata)* and Japanese beetle which cause damage as adults. Other important insects are destructive in the larval stage and are exemplified by the cabbage looper, corn earworm *(Heliothis zea)* (Figure 14-3), codling moth *(Laspeyresia pomonella),* Elm leaf beetle *(Pyrrhalta luteola),* leaf miners, and the gypsy moth *(Porthetria dispar).* The injury from both of these groups of insects is above ground, but major damage can also result below ground from chewing insects. Most of this damage occurs from larval forms of a variety of insects including weevils, maggots, borers, and cutworms. Root damage is often severe before the top of the plant shows sufficient symptoms for a grower to check on the cause. Thus, control may be especially difficult, not only because the problem is underground and difficult to locate, but because the insect population may be very numerous by the time the symptoms become apparent.

Certain chewing insects are particularly damaging and also very hard to control because they feed internally. Eggs are laid on or near the surface of the plant, but the young larva quickly go inside. Examples are borers and bark beetles, of which there are many different species.

FIGURE 14-3 Corn earworm on sweet corn. (Photograph, U.S. Department of Agriculture.)

FIGURE 14-4 Tulip bulb aphid on iris bulb. (Photograph, U.S. Department of Agriculture.)

Sucking Insects Although the injury resulting from the feeding of sucking insects and mites is more subtle, the damage is still tremendous. Groups which are of major importance include the aphids (Figure 14-4), scale insects, mites (considered here although not true insects), leaf hoppers, mealybugs (Figure 14-5), and white flies. Some of these, such as aphids and mites, are very troublesome in the greenhouse as well as out of doors, whereas others, such as white flies, are worst in greenhouses.

The damage resulting from sucking insects is usually seen as discoloration, curling, and malformation of meristematic tissues. When damage becomes severe, growth is suppressed, and the overall productivity of the plant is decreased. Many of these insects excrete a sugary substance called *honeydew* which supports fungal growth and thus leads to an unattractive plant.

The damage from insects discussed above is largely the result of feeding activities. Certain insects, however, cause commercially significant damage in other ways such as occurs as a result of egg laying and, perhaps most important, in the spread of disease.

The periodical cicada *(Magicicada septendecim)* spends most of its life in the soil as a nymph and emerges as an adult only once every

FIGURE 14-5 Mealybugs on croton. (Photograph, U.S. Department of Agriculture.)

13 or 17 years, depending on the race. Since there are several broods, some may emerge at irregular intervals. The aboveground damage is caused by the egg-laying activities of the female. Each female lays groups of eggs in twigs about the size of a pencil. In laying the eggs, a slit is made sufficiently deep to lead to the breaking of the twig. With

heavy populations of cicada, injury can seriously damage trees, especially young ones.

Certain diseases are carried by insects, especially those with piercing or sucking mouth parts. During the late 1950s and 1960s "pear decline" led to the death of thousands of pear trees in the Western part of the United States. An intensive research effort was launched to rescue the pear industry. The cause of pear decline is now thought to be a virus which is transmitted by the pear psylla *(Psylla pyricola)*. Other than the use of rootstocks which have resistance to pear decline, the best method of control is to minimize the population of pear psylla. Other important insects which spread plant diseases are leaf hoppers, aphids, thrips, and white flies. Mites can also transmit diseases of crop plants. Examples of important diseases and associated insect vectors are Dutch Elm Disease *(Ceratocystis ulmi)*, a fungus disease transmitted by the elm bark beetles; and fire blight of apples and pears *(Erwinia amylovora)*, a bacterial disease transmitted by flies and bees.

Control of Insect Pests

Pest control methods can be grouped into several categories, including chemical, biological, cultural, and legal. The use of chemicals has been widely practiced in controlling pests, but increasing concern about environmental pollution has increased interest in alternative methods. Biological and chemical control of insects are discussed below. Cultural techniques often involve soil management practices, modified planting dates, isolation procedures, and the elimination of alternative hosts. Legal efforts include such techniques as quarantines imposed against the importation of potentially infected plant material.

Chemical Control Insecticides are classified into stomach poisons, contact poisons, fumigants, systemics, and microbials. Some materials may be classified in more than one of these categories.

Stomach poisons act when ingested by the organism so their effectiveness is limited to chewing insects. A major stomach poison in recent years has been carbaryl (Sevin). Because of its very low mammalian toxicity and its effectiveness against a broad spectrum of insects which it controls, its use has been very widespread. It was introduced in 1956. From 1930 through the mid-1950s, the arsenicals, such as lead arsenate, were widely used stomach poisons.

Contact poisons are absorbed by the insect through its body surfaces and thus need not be ingested. Although effective against chewing insects, contact insecticides have been most widely used for controlling sucking insects such as mites, aphids, leaf hoppers, and scale insects. Commonly used contact poisons include the organophosphates, such as malathion, which has low mammalian toxicity, and parathion, which is highly toxic to humans and animals as well as insects.

Systemics are materials which are readily translocated throughout the plant and thereby give effective control of sucking insects feeding anywhere on the plant. Some systemics are applied to the soil, become dissolved in water, and are taken up through the roots. Other systemics may be applied to the aboveground parts of the plant but are absorbed and translocated. Systemic insecticides are not as effective against chewing insects as they may not ingest enough plant sap to receive a lethal dose.

Fumigants are usually small organic molecules with vapor pressures high enough to become gases at temperatures above 5°C. Most fumigant insecticides contain one of the halogens such as chlorine, bromine, or fluorine.

Biological Control   There is tremendous interest today in the use of *microbials*, or microorganisms which cause diseases in insects without causing injury to plants or other animals. *Bacillus thuringiensis* is a bacterium which causes a fatal disease in certain insect larvae such as the cabbage looper, cutworms, armyworm, and tent caterpillars. A suspension of spores is sprayed or dusted onto the host plants and upon ingestion by the insect, the spores produce materials which cause disease and death. *B. thuringiensis* provides a very safe means of insect control and is likely the forerunner of many such safe, selective insecticides which are currently being developed.

NEMATODES

Another group of pests in the animal kingdom which causes widespread injury to crop plants is the nematode. Some nematodes are important as parasites of higher animals. Certain additional species live by feeding on decaying organic matter, and other species feed on other small animals. The discussion here will be limited to those nematode species which are parasitic to higher plants. Other names for nematodes include *threadworms, roundworms*, and—perhaps the most descriptive—*eelworms*.

Plant-parasitic nematodes are nonsegmented worms, generally about 1-mm long, although they vary both in size and shape. The eggs laid by females hatch into nymphs or juveniles which go through four molts after which they are adults. Plant-parasitic nematodes have a *stylet* or spear which is used to puncture cell walls. After the cell is punctured, a secretion is injected into the cell which presumably contains an enzyme which liquefies the cell's contents and makes them easier to ingest. The secretion may also alter the plant cell metabolism sufficiently to cause changes such as suppressed cell division, necrosis,

FIGURE 14-6 Tomato plant infested with root-knot nematode. (Photograph, U.S. Department of Agriculture.)

and cell proliferation. The macroscopic effects of these changes are often used in classifying nematodes into categories such as root knot, cyst, lesion, and stubby root nematodes.

Most plant-parasitic nematodes are not particularly specific as to the plants which they attack. There is, however, wide variation in susceptibility between and within plant species. The rootknot nematodes are worldwide in distribution, and some members of this group can attack most cultivated plants (Figure 14-6). Some plant species however, have cultivars with sufficient resistance to provide some control. The oriental peach cultivars 'Yunnan' and 'Shalil', for example, are highly resistant to all the rootknot nematode species except *Meloidogyne javanica* and therefore may be of considerable value where this particular nematode species is not prevalent. Resistance to damage caused by these species is also being bred into tomato, bean, pepper, potato, and sweet potato.

Although specific symptoms result from the feeding of particular nematodes, the general effects may be quite similar to a number of other plant ills. The typical symptoms of nematode damage to infected plants are lowered vigor, lessened resistance to adverse environmental factors, and low productivity. Certain nematodes feed on aboveground parts of plants and may cause malformed stems and leaves, necrotic spots on leaves, and leaf galls. But compared with the root feeders, leaf, stem, and bud feeders are of minor importance.

Control   The difficulty in controlling nematodes depends largely on the species, some being relatively easy to control while others are very difficult. One of the oldest methods of nematode control is heat treatment. Soil pasteurization by steam, hot water, or even heating in an oven will kill nematodes, but all of these techniques are only suitable for relatively small-scale treatment. Certain types of fleshy plant material such as bulbs, corms, tubers, and fleshy roots can be treated with hot water, but the temperature control must be exact because the lethal temperature for the plant material may be quite close to the lethal temperature for the nematodes.

On a field scale, many techniques have been tried, and their success varies. Leaving land fallow may provide control because host plants are not available. Unfortunately, it usually is impractical economically to leave land without crops long enough to provide control. Crop rotation has somewhat the same difficulty because nematode-resistant crops are seldom moneymakers and cannot therefore be grown often enough to give good control. The golden nematode of potato is found in many parts of the world but in the United States has been confined largely to Long Island, New York. The host plants include the potato, tomato, and related species. A four-year crop rotation has been of some value, but the eggs remain viable in the soil for many years. Strict quarantines have been very effective in retarding the spread of this particular species.

Chemical control has become widely adopted as the most effective method in spite of the relatively high costs involved. Soil fumigants such as DD (a mixture of dichloropropane and dichloropropene) and EDB (ethylene dibromide) have been widely used because they are especially effective against nematodes. Small quantities are injected into the soil and gradually vaporize. Unfortunately, these two materials are not very effective in killing soil fungi or weed seeds. Chloropicrin (tear gas) provides good soil pasteurization in that it kills fungi and weed seeds as well as nematodes. Methyl bromide is also a good material because it is effective against a broad range of soil pests as is chloropicrin. Since methyl bromide is a gas at atmospheric pressure, it is applied under a plastic cover to prevent its escape.

In crops of very high cash value, the soil fumigant is injected into the soil and the entire field or bed covered with a continuous cover. After a few days the cover is removed and the fumigant allowed to dissipate. The field is then thoroughly disked and allowed to ventilate before the crop is planted. This system is required when clear polyethylene is used as a mulch (see Chapter 10).

DISEASES

The yield and quality of horticultural plants are reduced by a broad spectrum of plant diseases. The plant pathologist is a scientist who studies plant diseases to gain a better understanding of them and also works to accurately diagnose and control them. The definition of a diseased plant often varies with the person who is giving the definition. To a plant pathologist, a diseased plant may be a plant whose anatomy, morphology, or physiology is sufficiently abnormal to cause visible symptoms or to have lowered growth, yield, or quality. In such a broad definition, adverse environmental conditions, nutrient deficiencies, and air pollutants often lead to "diseased plants." Horticulturists generally prefer a more restrictive definition and limit the term *diseased plant* to one which is abnormal because of an infectious pathogen. A *pathogen* under this restricted definition can be defined as a microorganism which is capable of causing a plant disease under proper environmental conditions. The three major groups of plant pathogens are the bacteria, fungi, and viruses. Some people may also include nematodes, but we have chosen to treat them separately.

Fungi

The diseases which cause the most widespread damage to horticultural crops are those caused by fungi. Not only is the financial loss usually heaviest from fungal diseases, but the number of different diseases is also greatest. Fungi are members of the plant kingdom and generally have the following characters: rigid cell walls, lack of chlorophyll, presence of nuclei and nucleoli, hyphae (collectively, the mycelium) which elongate terminally, chitin in the hyphal walls, and reproduction by spores.

Not all fungi are harmful, and in fact many are very beneficial. Examples of desirable fungi are: wine and bread yeasts, edible mushrooms, fungi used in making certain cheeses, and others used in the production of antibiotics, such as penicillin.

The fungi are divided into four groups, largely on the basis of the way in which spores are produced and the appearance of the fruiting body. These four groups are the Phycomycetes, the Ascomycetes, the Basidiomycetes, and the Deuteromycetes (Fungi Imperfecti).

The Phycomycetes are characterized as having no cross walls in the mycelium and are considered to be rather primitive. Examples of diseases caused by members of the Phycomycetes are late blight *(Phytophthora infestans)* of potato and tomato (Figure 14-7), some downy mildews (Figure 14-8), and certain soft rots of fruits.

FIGURE 14-7 Tomato leaf showing the dark, water-soaked spots
characteristic of late blight. (Photograph, U.S. Department of Agriculture.)

FIGURE 14-8 Early stages of downy mildew on cucumber leaf. The spots
are yellow-green and eventually kill the older leaves. (Photograph, U.S.
Department of Agriculture.)

FIGURE 14-9 Peaches infected with brown rot. This disease also affects plums, prunes, and cherries. (Photograph, U.S. Department of Agriculture.)

Somewhat more advanced on an evolutionary scale are the Ascomycetes, which have cross walls in their hyphae (mycelium) and also form *asci*, saclike structures in which the ascospores (sexual spores) develop. This group includes the largest number of disease-causing fungi. Examples of diseases caused by Ascomycetes are peach leaf curl *(Taphrina deformans)*, Dutch elm disease *(Ceratocystis ulmi)*, peach brown rot *(Monilinia fructicola)* (Figure 14-9), apple scab *(Venturia inaequalis)*, and black spot of rose *(Diplocarpon rosae)*.

The most highly developed group of fungi are the Basidiomycetes, characterized by sexual spores found in club-shaped structures called *basidia*. This group contains the very damaging diseases called *rusts* and *smuts*. These include white pine blister rust *(Cronartium ribicola)*, cedar-apple rust *(Gymnosporangium juniperii-virginianae)* and corn smut *(Ustilago zeae)*.

The fourth group, Fungi Imperfecti, is not a clear-cut classification because it overlaps partially with the other three groups, but its members are in this group because no known sexual production of spores is known or else it is rarely produced in nature. Representatives of this group cause anthracnose of beans *(Colletotrichum lindemu-thianum)*, *Verticillium* wilt, *Botrytis* rots of fruits and vegetables, and early blight of potato.

The spread of fungi is largely passive in that only in certain types and parts of life cycles are fungal cells mobile. Some members of the Phycomycetes form spores with flagella which can propel themselves, while some spores are discharged forcibly. Thus, the spread of fungi by either spores or fragments of mycelium is most often carried out by agents such as wind, water, animals, and insects. The numbers of spores produced are tremendous, and since they are so small, light, and easily carried in wind and water, their spread tends to be rapid.

One of the more widely studied fungal diseases is apple scab (Figure 14-10). This disease is present in most apple-growing regions and is especially severe in humid climates. The fungus attacks both leaves and fruits and can also infect young shoots. The scab lesions start as velvety spots which range from brown to green but turn black as they age. The leaves and fruits are distorted because the affected tissues fail to enlarge normally. Severe infections can defoliate trees and cause the fruit to be unmarketable.

The annual disease cycle starts with overwintering fungi on dead leaves. Ascospores (primary inoculum) are produced as temperatures rise in late winter, and they are released during early spring. The spores are forcibly ejected into the air and are carried to young leaves and fruit. The danger of an infection increases with both increasing temperatures and the number of hours that the trees remain wet. If the

FIGURE 14-10 Apple scab on Stayman apples. (Photograph, U.S. Department of Agriculture.)

temperature and wet period is sufficient, an infection can occur and is called a *primary infection*. After these primary lesions mature in a period of 10 to 17 days, the secondary inoculum (conidia) is produced which can lead to a *secondary infection*. Since these spores are spread primarily by rain, a few primary infections in the top of a tree can lead to great numbers of secondary infections throughout the tree.

The main control method for apple scab is the use of fungicides applied before the rain. These protectant fungicides prevent the germination of spores or the penetration of the mycelium. Some fungicides can eradicate a very young infection and are called *eradicants*. Eradicants are said to have "kickback" action and may be effective for a few hours or days after the infection occurs. If a thorough job in controlling the primary infection is not done, one must continually fight to control secondary infections.

Bacteria

Bacteria are one-celled, microscopic plants which vary in shape from spherical to spiral and rod-shaped. Although each bacterium cell is independent, they are usually grouped in colonies. Essentially all the plant pathogenic bacteria are rod-shaped, usually between 0.4 and 0.7μ in diameter and between 1 and 3μ in length. Bacteria differ from the fungi in that they lack a membrane-bound nucleus, but like the fungi they do not contain chlorophyll. Many bacteria have flagella. Reproduction is by cell division, and under optimum conditions can be very rapid. The general symptoms induced by bacterial diseases are blights, rots, galls, leaf spots, and vascular wilts.

Like the fungi, certain bacteria are beneficial to man. A striking example is the nitrogen-fixing bacteria which infect legumes and convert gaseous nitrogen into forms which can be used by plants (see Chapter 11).

The entrance of bacteria into plants can occur in many ways, but the most common sites for entry are natural openings and wounds. The fire-blight bacterium *(Erwinia amylovora)* enters the apple or pear tree through the open blossom, presumably through the nectaries or other natural openings. The bacterium causing blackleg of cabbage *(Phoma lingam)* is thought to enter through the hydathodes. Many of the rot-causing bacteria enter the host through mechanical injuries. Once inside the plant, movement may be in the sap stream, or motile cells may be propelled by flagella over short distances. As individual cells are killed by bacterial action, the cells walls are broken down, and the bacteria enter.

A very important disease of apple and pear, fire blight, serves as an example of a bacterial disease, its symptoms, life cycle and control.

The symptoms of fire blight occur on blossoms, leaves, shoots, spurs, and in severe cases may kill the whole tree. When a flower becomes infected, it takes on a water-soaked appearance, but within a few days the whole spur is often affected and turns brown. Infected terminals wilt, and the leaves and shoot dry and turn from brown to black. The characteristic crook in the terminal and the persistence of dead leaves are common symptoms. Depending on the susceptibility of the tree, the bacterial infection may be confined to small shoots and spurs, or the bacteria may kill scaffold limbs or entire trees. When bacteria are multiplying during wet weather, a liquid material referred to as *ooze* is often seen coming from infected tissue.

The life cycle of fire blight can be considered as starting in the overwintering cankers on branches. Ooze is formed in early spring and is spread by rain and insects, the "primary" infection occurring in the open blossoms. Secondary infections occur as the bacteria infect a blossom and then are spread by bees or other insects as well as rain. A major factor influencing the severity of fire blight is temperature, with warm temperatures encouraging rapid spread. Further infections occur through injuries such as those from insect feeding or hail. The susceptibility to infection declines after bloom both because the flowers are gone and because the vegetative tissues gradually harden and become more resistant to infection. Unless removed by pruning, the cankers provide inoculum the following spring.

The control for fire blight involves several different techniques. Cultivars of pears, such as 'Bartlett,' which are very susceptible are grown in arid to semiarid regions, such as California and Oregon, or in areas which are quite cool in the spring, such as Michigan and New York. An additional precaution with susceptible cultivars or species is to keep tree vigor quite low and thereby reduce the succulence of the vegetative growth. Present research emphasis concerns the breeding and selection of resistant cultivars. The use of antibiotic sprays has given good control when used correctly, but such treatments have been expensive. Orchard sanitation is one of the most effective procedures because, by removing infected branches, spurs, and twigs, the source of inoculum can be dramatically reduced. The infected tissues must be removed from the orchard and burned to destroy the bacteria.

Viruses

Viruses are generally considered to be infectious, submicroscopic particles composed of nucleoprotein. Virus particles lack metabolic capability, they do not reproduce by growth and division, and thus they are not considered living. Viruses are reproduced only within living host cells, and the virus redirects the metabolism of the host cell to produce virus particles rather than normal protein. The plant viruses typically contain both protein and ribonucleic acid (RNA). Since virus particles are not visible under a light microscope, their structure was unknown until the electron microscope was developed. The plant viruses are classified into three general categories: rods, spheres, and bacteroid-shaped particles.

Although various systems of nomenclature for viruses have been prepared, the one most widely used is based on the common name of the disease or the host in which it was first described. Thus common names include the following: *tobacco mosaic virus, peach yellows virus, cucumber mosaic virus,* and *aster yellows virus.* The two main groups of virus diseases are the yellows and the mosaics. The *yellows* diseases are characterized by yellowing, stunting, leaf curling, and excessive branching. The plant symptoms caused by the *mosaics* include a mottled appearance of leaves as well as flowers in some cases. Portions of the tissue become chlorotic, and these areas may be small to large or may involve the whole leaf. Some plants may also exhibit necrotic spots on leaves, fruits, and stems. Recent work has demonstrated that the yellows are caused by *mycoplasma* which are microorganisms seemingly between viruses and bacteria.

The transmission of viruses occurs in several ways. The tobacco mosaic virus can be readily transmitted by mechanical means. The sap from an infected plant will infect a healthy plant if rubbed on the surface or injected. It is therefore spread quickly by suckering (removing shoots from the lower stem) of tobacco or tomato. Most viruses are much less readily transmitted by mechanical means. For viruses affecting horticultural plants, transmission by vegetative propagation is particularly troublesome. With the exception of seeds, essentially all cells of an infected plant contain the virus particles so that propagation by cutting, grafting, and budding leads to new plants which contain the same virus or viruses as the mother plant. The mature seeds of most plants do not contain the virus which may be in the rest of the plant. There are exceptions to this, but fortunately seed propagation often avoids virus problems.

Among the insect vectors of viruses, the most important are the aphids and the leaf hoppers. The yellows group of viruses (mycoplasma) is transmitted largely by the leaf hoppers, whereas the aphids are very instrumental in transmitting the mosaic group. Some virus transmission also occurs from mites and nematodes, but these vectors are considered to be much less important.

As with other pathogens, virus damage can be controlled by one of two basic techniques: *protection* which prevents entry by the pathogen or *therapy* aimed at the pathogen after entry. The most effective choice between the two depends on the particular situation. If one is propagating plant material, it is obviously desirable to "clean up" the source of buds or cuttings before propagation. This often involves the virus "indexing" of the plant material because some species and cultivars can be virus-infected but not show symptoms (symptomless carriers). These species and cultivars are therefore said to contain *latent* viruses. Considerable progress has been made in recent years by virus indexing sources of propagation material. In *virus indexing* the material is virus-tested by propagation on specific cultivars or species which will produce symptoms of each known virus. If viruses are found to be present, a variety of treatments can be used to destroy the viruses and thereby provide "virus-tested" material. Such stock is not labeled as virus-free but can be specified as free of specified or known viruses. The use of heat has been effective in destroying many viruses in horticultural plants. The technique is based on a combination of temperature and time to inactivate the virus but at a temperature-time combination which will not do permanent damage to the host plant. Dormant plant material such as budwood and potato tubers can often withstand higher temperatures than actively growing material. Meristem culture is another widely used technique to rid a plant of viruses. Very small pieces of meristems are cultured and are often free of viruses present in the rest of the plant. Various types of chemotherapy have been tested, but results have not been promising.

WEEDS

In today's world, weeds not only are a nuisance but suppress crop yields and cost millions of dollars to control. A precise definition of a weed is hard to find, but one of the most useful is "a plant growing where it is not wanted." It is quite possible for a plant to be considered as a weed in some situations but a desirable plant in others. A corn plant in a field of green beans is a weed. Certain plants, such as pigweed, are essentially always weeds, others, such as Bermuda grass, can be a turf and pasture crop but can also be a very undesirable weed in a vegetable field.

Damage from Weeds

Weeds can be detrimental in many different ways, but the damage can be grouped into five main categories. *First,* from an agricultural standpoint the most costly and direct damage from weeds is the reduction in crop yields. Because of their great numbers and rapid growth rate, weeds effectively compete with crop plants for moisture, nutrients, and light. The effect of weeds on the crop can range from slightly reduced yields to total crop failure. The degree of damage varies with many factors but is often most devastating where moisture or nutrients are barely sufficient for the crop and thus the competition is most serious. One factor which makes many weeds particularly competitive is that many have the C_4 pathway of carbon fixation (see Chapter 6). Another feature of weeds is that they typically produce large numbers of seeds which remain viable for extensive periods.

A *second* major problem with weeds is the contamination which they cause. If weeds are harvested with food crops, their presence can markedly lower the quality of the crop. The severity of contamination is minimal in tree fruits but is particularly serious in leafy vegetables, such as spinach and kale. Not only can weeds lower the grade of the product, but they can be very difficult and expensive to remove. Contamination of crop and grass seed with weed seed can also be troublesome, and reputable seed growers make every effort to produce seed that is as free as possible of weed or other foreign seeds. A very wise expenditure of money is the extra percentage paid for "certified seed" which has minimal contamination by weed seed.

A *third* problem with weeds is that they can serve as hosts for insects and diseases and in some cases may serve as alternate hosts. For example, cedar apple rust must have both apple and red cedar plants to complete its life cycle. Before modern fungicides became available, the main control for cedar apple rust was the mass cutting of the red cedar, but in recent years very effective fungicides have largely eliminated the necessity of removing cedar trees near apple orchards. Certain weeds may be hosts for insect pests of crop plants, and weedy areas often harbor troublesome insects and encourage their reproduction.

The *fourth* major category of weed damage is that caused by the poisonous species of weeds. In horticultural crops the problem is usually most serious with poison ivy which can cause severe discomfort to many people who touch it. Being a perennial climbing vine, it is particularly troublesome in orchards, around fence rows, or in other areas which are not cultivated, tilled, or mowed closely. Adequate control of poison ivy is also important in fruit and vegetable operations which are open to the public such as pick-your-own strawberry fields.

Fifth, weeds can be very undesirable from an aesthetic standpoint. Millions of dollars are spent annually in controlling weeds in lawns, gardens, golf courses, and other areas where the presence of weeds is particularly undesirable because it detracts from the appearance of the area.

Weed Control

Attempts to control weeds can be classified into four main categories: (1) mechanical, (2) competitive, (3) biological, and (4) chemical. In any particular situation it may be most advantageous to use a combination of methods rather than a single one.

Mechanical Among the mechanical methods used are pulling by hand, hoeing, cultivation, mowing, burning (flaming), and although slow, it is still very practical in small areas such as flower beds. With many crops hoeing was standard procedure, but where possible it has been replaced by cultivation. Various types of tillers were first pulled by draft animals but more recently by tractors. The tremendous quantity of weed seed present in most soils makes hoeing and cultivation almost an endless task. Burning is useful in certain situations. In the production of low-bush blueberries, for example, competing vegetation is kept in check by periodic burning of the area. With certain crops a system has been developed in which the field is flamed to destroy the young weeds but with the flame directed so as to not harm the crop plants. Of major importance in horticulture are many techniques which are aimed at smothering weeds. These include organic mulches such as straw, sawdust, ground corn cobs, and peat moss which not only provide excellent weed control but offer other benefits such as stabilizing soil temperature and conserving moisture. Black plastic (polyethylene) mulch is also very effective in weed control and is used in vegetable production. Both the organic mulches and black plastic prevent young weed seedlings from getting light and thus effectively smother them.

Competitive The control of weeds by competitive means is possible in some circumstances and where practical can be not only effective but probably the least expensive alternative. A good example of this technique is in turf. If a good stand of grass is maintained in a lawn, weed problems are minimal because germinating weeds tend to succumb to the competition of the grass.

Biological Biological control of weeds is in its infancy but has been successful in certain instances. The prickly pear was a very serious weed problem in Australia but by introduction of the Argentine moth borer, large areas have been largely cleared of the prickly pear. As with any biological control mechanism, great care must be taken to ensure that the insect or pathogen introduced will not be harmful to desirable plants or animals or to the environment. As continued pressure is brought to bear on users of chemical pesticides, it seems very desirable for weed researchers to seek suitable biological control mechanisms.

Chemical By a wide margin, the most widely used techniques of modern weed control involve chemicals called *herbicides*. Herbicides include a wide variety of compounds that regulate the growth of plants in some fashion. Although *herbicide* literally means "weed killer," compounds which modify plant growth in any manner may be classified as herbicides. Certain chemicals, such as salt, have been used for centuries for complete killing of vegetation, but the selective action of herbicides was not discovered until the mid-1890s. Selective herbicides are those which can be applied to a mixed stand of plants and kill some while not harming others. Early work was with a variety of salts which would eliminate broadleaf weeds from grain fields.

Herbicides are classified on the basis of certain characteristics which may include (1) chemical structure, (2) selectivity, and (3) whether the herbicide acts as a contact or a translocated herbicide. Early herbicides were mostly inorganic materials and consisted largely of salts and acids. Among the salts were sodium chloride, copper sulfate, and sodium arsenite; the acids used were sulfuric and nitric. Although under ideal conditions some of these materials were useful, many factors limited their widespread application. Among these were the inconsistent results, the large quantities required, and, in the case of acids, the great difficulty in handling.

The age of modern herbicides began in 1944 with the introduction of 2,4-D [(2,4 dichlorophenoxy) acetic acid]. This was the first of the hormone type of herbicides; it is not only effective at very low dosages but is selective and readily translocated within the plant. The tremendous potential of phenoxy herbicides was soon realized and many similar compounds followed. Among these, the more important were 2,4,5-T [(2,4,5 trichlorophenoxy) acetic acid] and 2,4,5-TP [(2,4,5 trichlorophenoxy) propionic acid]. This group of herbicides is highly toxic to broad-leaved weeds, but it does not cause appreciable harm to grasses. Their use has therefore been tremendous on a worldwide scale for weed control in grasses such as turf, cereal grains, and corn. An interesting sidelight of certain of these phenoxy herbicides is that other uses have been found for them. For example, 2,4,5-TP has been widely used in apple production for preharvest fruit drop control in the fall. The rate of application is low enough to avoid typical phytotoxicity but high enough to be a very effective growth regulator.

Other groups of organic herbicides include the amides, aliphatics, benzoic acid derivatives, carbamates, and many others. A discussion of herbicide chemistry is well beyond the scope of this book, but the reader is directed to references at the end of the chapter.

Herbicides are classified as contact or translocated, depending on whether they kill only the tissues which they contact or affect the entire plant by being readily translocated. The *contact herbicides* are largely the inorganics, petroleum oils, and certain organics—such as paraquat and diquat. A major use of paraquat is to provide rapid "knockdown" of existing vegetation, especially annuals. For minimum-tillage corn production (see Chapter 10), paraquat is often used to kill the cover crop. Since it is deactivated as soon as it comes in contact with the soil, it has no effect on the germination or growth of the corn. The *translocated* type is applied to the foliage or to the soil, but in either case is absorbed and translocated. Major herbicides applied to the foliage include 2,4-D, 2,4,5-T, dalapon, dicamba, and glyphosate. Soil-applied herbicides which are absorbed by the roots include a very large number of compounds.

Soil-applied herbicides are classified as preplanting, preemergence or postemergence herbicides. Those applied before seeding are referred to as *preplanting*. They may be applied a few days or weeks before seeding and are effective against annual weeds. If sprayed onto the soil and worked in before seeding, the herbicide is called a *preplant incorporated* treatment. *Preemergence* herbicides are put on after seeding but before the seedlings emerge. *Postemergence* chemicals are applied after crops and weeds emerge and must therefore be either highly selective or applied in such a manner as to avoid contact with the crop plants.

In spite of tremendous strides made in recent decades, the control of weeds continues to be a major worldwide problem. Chemicals have provided us with the means to increase food production with decreasing labor inputs, but they have not been without their problems. One of these has been the potential for damage to human beings, animals, and the environment. As increasing pressure is exerted by various sectors of our society, such as the environmentalists, chemicals will be examined with ever-increasing scrutiny. It seems likely that with adequate precautionary measures, chemicals will continue to be a major weapon against weeds. Certainly, however, alternatives such as biological control must also be more fully explored.

PESTICIDES

Application Equipment

Because chemical pesticides are manufactured in so many formulations and applied to so many different crops, there is a wide variety of equipment for their application. The various types include sprayers, dusters, granular spreaders, fumigant injectors, foggers, and bait applicators.

Sprayers The most widely used method of pesticide application is the sprayer. Not only are most pesticides available in formulations suitable for mixing in water, but there is a sprayer available for almost any situation. These range from small, hand-operated types for spraying a rose bush up to large truck-mounted or tractor-drawn sprayers used in fields, orchards, golf courses, and nurseries. The small hand sprayer used by the average homeowner is usually operated by air pressure which may be in an aerosol can or is generated in a compressed air sprayer with volumes ranging from 4 to 20 L. These are inexpensive and are practical for small-scale treatments.

For commercial or large-scale spraying, mechanical or power-driven models are used. These may be of many types which are operated on the ground or may be attached to either helicopters (Figure 14-11) or fixed-wing aircraft. Ground sprayers are classified in various ways, one of which is the pressure at which they operate. A *high-pressure* sprayer has a pump which can produce sufficient water pressure to distribute the spray. This required pressure will range from 14 to 42 kg cm^{-2} depending on the particular situation. The

FIGURE 14-11 A fungicide is applied to an orange grove near Lake Wales, Fla. by helicopter. (Photograph, U.S. Department of Agriculture.)

FIGURE 14-12   Row crop sprayer with a 12.2-m boom which can be raised or lowered. (Photograph, Hagie Manufacturing Company, Clarion, Iowa.)

spray may be pumped through a single nozzle (gun) or through multiple nozzles (boom) (Figure 14-12). The combination of high pressure and the nozzle breaks the stream of water into small droplets.

An alternative system is the *air-blast* sprayer which has become increasingly popular. A relatively low-pressure pump is used to push the liquid through nozzles. A fan provides large volumes of high-speed air both to break up the stream into droplets and to distribute the spray onto the plant. One of the initial advantages of air-blast sprayers was that this system eliminated the need for workers to spray with the guns typically used on high-pressure sprayers. Thus one person with an air-blast sprayer could do the spraying of a three-person crew consisting of one driver, and two persons with spray guns (Figure 14-13).

Further refinements are continually being made. One of the most significant improvements has been the development of "concentrate sprayers." Both high-pressure and conventional air-blast sprayers were designed to spray plants to the point of "runoff" where the spray started to drip. This is considered as 1X or normal concentration of pesticide. *Concentrate sprayers* are designed to deliver similar amounts of pesticide per hectare but in lesser amounts of water. If, for example, 2.27 kg of a fungicide are needed per hectare of apple trees, in a conventional sprayer this would be applied in about 1500 L of spray. In a 10X concentrate sprayer, one would apply the 2.27 kg of

FIGURE 14-13 A nursery crop sprayer utilizing a high-pressure pump and a large fan to distribute the spray. (Photograph, Monrovia Nursery, Azusa, Calif.)

fungicide but do so in 150 L of spray. Efficiency of spraying increases markedly with concentrate applications as the time spent refilling the tank is reduced. By the elimination of runoff, it is also possible to reduce amounts of pesticide per hectare by 15 to 20 percent, thus providing additional savings over dilute spraying.

Herbicide sprayers are usually fairly simple, often consisting of a tank, small pump, and several nozzles mounted on a horizontal boom. Since low pressures are used to minimize drift, droplet size remains quite large. Many of these sprayers are mounted on a tractor's three-point hitch, and the pump is driven by the power takeoff.

Aircraft are being used to increasing degrees in pesticide applications. Major advantages over ground equipment include speed of application, avoidance of wet soils or other conditions which might preclude the use of ground sprayers, and elimination of crop damage caused by the movement of equipment through the field or orchard. Helicopters have the advantage of being able to land almost anywhere, whereas fixed-wing aircraft need some type of runway. Much of the aircraft spraying is done by custom applicators.

Dusters   As with sprayers, dusters are available for everything from a few small plants in a yard to sizes suitable for large fields, orchards, or nurseries. Formulations of dusts normally contain the appropriate concentration of pesticide mixed with a diluent such as talc which acts as a carrier. The dust is blown onto the plants through nozzles. Small dusters are hand-operated while commercial models are power-driven.

Granular Spreaders   Certain pesticides are prepared in granular forms. This involves coating the pesticide with some innocuous material. Examples are certain herbicides and systemic insecticides. Because dosage is critical, these materials must be applied uniformly and at the appropriate rates. Considerable progress has been made in recent years in development of tractor-mounted precision spreaders.

Fumigant Injectors   For fumigation purposes, the pesticide is vaporized in an enclosed area, such as a greenhouse, or is applied beneath the surface of the soil. When very volatile materials such as methyl bromide are applied for soil fumigation, an impervious cover such as polyethylene must be used. With other materials with lower vapor pressures, they may be injected into moist soil and vaporize slowly enough that covering is not required.

The advantages of fumigation over the other methods of pesticide application include very complete coverage and penetration and thus excellent pest control. In greenhouses, fumigation is also very rapid compared to spraying or dusting. One of the major drawbacks of fumigation is often cost, especially for soil fumigation.

Foggers   The application of pesticides by a fogging technique involves the breaking up of liquid carrier into droplets so fine that they remain suspended in the air for long periods of time. Because the fog particles float in the air so readily, they provide good penetration and coverage.

Bait Applicators   The control of certain pests is best effected by the use of baits. Examples are slugs, mice, rats, and certain other mammals. Usually an effective poison is mixed with an edible attractant or pest food in a formulation that is lethal but still palatable. Examples of baits used for orchard mouse control have included grain or pieces of apple treated with zinc phosphide. These may be applied into mouse runs by hand or in some cases a machine has been used to place the bait into an artificial mouse "run."

Pesticide Formulations

Pesticides are marketed in several forms depending both on the characteristics of the active ingredient and on the use for which it is intended. The two main categories are *liquid* and *dry* formulations.

Liquid   The liquid group includes the following types: emulsifiable concentrate (EC), solution (S), flowable (F), fumigant, and aerosols (A). *Emulsifiable concentrates* have the pesticide mixed in a liquid which will mix with water to form an emulsion. A *solution* contains the pesticide dissolved in a solvent which in turn mixes with water for application or may be applied directly. A *flowable* formulation is a thick suspension of the active ingredient, usually in water, which is diluted further for application as a dilute suspension. *Fumigants* are usually purchased under sufficient pressure to keep them in a liquid state but may be applied either as a liquid under pressure or as a vapor. *Aerosols* are usually dilute solutions of pesticides to be applied directly from pressurized containers.

Dry   Dry formulations include wettable powders (WP), soluble powders (SP), dusts (D), granules (G), and baits. *Wettable powders* are fine powders which will not dissolve in water but with continuous agitation will remain in suspension. A wetting agent is usually added to facilitate the mixing of water and powder. *Soluble powders* dissolve in water and once well mixed do not require further agitation. *Dusts* may be an undiluted product, such as finely ground sulfur, or a mixture of the active ingredient with a powdered carrier or diluent, such as clay or talc. These are applied dry without further dilution. *Granules* are larger particles containing the pesticide plus a diluent and are applied by spreaders, usually to the ground. *Baits* are mixtures of an edible component with the pesticide mixed in or applied to the surface of the pieces. The active ingredient may be applied to the bait as a liquid or dry formulation depending on the bait components.

PEST CONTROL VERSUS PEST MANAGEMENT

Recently there has been a significant change in attitude toward the goal of *pest control*. In the past, the objective in most pest control programs was to eliminate the pest, but success was seldom if ever achieved. Currently, there is increasing interest in and justification for a new approach, and this new approach is frequently called pest management. The term "management" indicates an ongoing process rather than the finality of control. As used in this context, *pest management* implies keeping the damage below some preselected level of injury which is compatible with good economics. This level has been termed the *economic injury level*.

Modern concepts of integrated pest management often involved a combination of techniques rather than attempts to "kill only." Most often, pest management programs involve appropriate types of biological control, such as predators of specific insects or the introduction of large numbers of sterile males. Ladybird beetles are very effective in keeping aphid populations under control. In some areas, large numbers of these beetles may be released, but in other situations it may be more practical to encourage an increase in native populations of the predator. A vital aspect of pest management is in the selection of insecticides, when and if their use becomes necessary. Chemicals which will kill the target insect but will not harm the predator population must be selected. To maintain a predator population, a minimal population of the target insect must survive. An equilibrium must be reached between the host and parasite where both can survive but where the pest is below economically important levels. Continuous monitoring of pest populations is a vital component of a successful pest management system.

AIR POLLUTION

A problem of increasing concern to horticulturists as well as those in the medical fields is air pollution. We are just beginning to realize that air pollution can be detrimental to plant material in much the same way as natural pests. It has been widely accepted that high levels of certain air pollutants could cause obvious injury to plants, but it is becoming increasingly evident that growth and production may be adversely affected even in the absence of obvious symptoms.

Some of the air pollutants which can harm plants are (1) photochemical oxidants such as ozone and peroxyacetyl nitrate (PAN), (2) oxides of nitrogen (NO, NO_2, etc.), (3) sulfur dioxide (SO_2), (4) hydrogen fluoride (HF), and (5) ethylene. Damage from air pollutants is often very complex and difficult to identify, partially because the effects are often very subtle. Further, there is evidence of complex interactions when more than one pollutant is present, as is very often the case.

The pollutants present in the air come from many sources. Among these are transportation vehicles, electric power generating plants, industry, heating, and waste disposal. Automobiles are responsible for much of the ozone and PAN—two of the most detrimental pollutants. Combustion of sulfur-containing fuel accounts for much of the SO_2, and the fluorides come largely from the processing of aluminum, phosphate, and glass.

Air pollution will probably become an increasingly important problem until pollution abatement laws become more and more stringent.

BIBLIOGRAPHY

Chapman, S. R., and L. P. Carter: *Crop Production Principles and Practices*, Freeman, San Francisco, California, 1976.

Christie, J. R.: *Plant Nematodes*, Agricultural Experiment Stations, University of Florida, Gainesville, Florida, 1959.

Crafts, A. S.: *Modern Weed Control*, University of California Press, Berkeley, California, 1975.

Pfadt, R. E.: *Fundamentals of Applied Entomology*, Macmillan, New York, 1971.

Strobel, G. A., and D. E. Mathre: *Outlines of Plant Pathology*, Van Nostrand, New York, 1970.

Walker, J. C.: *Plant Pathology*, 3d ed., McGraw-Hill, New York, 1969.

Ware, G. W.: *Pesticides An Auto-Tutorial Approach*, Freeman, San Francisco, California, 1975.

CHAPTER 15
MARKETING, STORAGE, AND FOOD PROCESSING

 The handling, storage, and processing of horticultural crops after harvest are a vital part of the industry today. Depending on the product, these activities may entail minimal or very extensive expenditures of time and money.

PREPARATION FOR MARKETING

Fruits and Vegetables

After harvest, many steps are necessary for the preparation of fresh fruits and vegetables for market. The number and type of treatments vary with the particular product, but with most, the following steps are involved: cleaning, sorting, sizing, and packaging (Figure 15-1). With some crops, additional treatments are typical; these may include removing field heat, trimming, treating with disinfectant or fungicide, waxing, and curing.

During the harvest operation, fruits and vegetables are put into a diversity of containers for hauling to the packing house. A few vegetables, such as head lettuce, are trimmed and packed in the field, as are many table grapes and most hand-harvested berries. Certain vegetables are packaged by a crew of workers on a moving packing

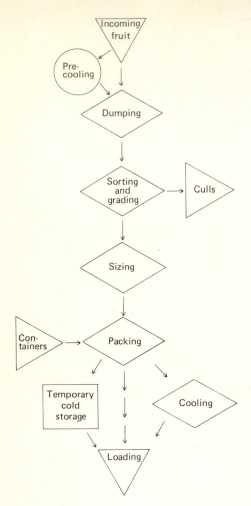

FIGURE 15-1 Flow diagram of the operations in a plum packing house which is typical for a fruit or vegetable operation. (From U.S. Department of Agriculture.)

house which travels through the field with the pickers. Examples of crops packed with this system are sweet corn and cantaloupes. The great majority of fresh fruits and vegetables are hauled to a cannery or to a central packing house where they are prepared for market. Containers used for the trip from the field to the packing house or processing plant include field crates holding from 10 to 40 kg and pallet boxes or bulk bins (Figure 15-2) which hold from 15 to 22 bushels. Some of the fruits and vegetables are loaded directly onto trailers and trucks. The type of container used varies with many factors including the type of harvest (hand or mechanical), the perishability of the item, its resistance to damage, and the equipment available. The major advantages of bulk handling are lower cost, lower risk of bruising, and much lower labor requirements.

FIGURE 15-2 Field-graded peaches are transported in bins from the orchard to the cannery for processing. (Photograph, California Canning Peach Association.)

Removal of Field Heat   Crops which are particularly perishable can have their market life extended significantly by prompt removal of field heat. The two most widely used techniques are vacuum cooling and hydrocooling. Most head lettuce is *vacuum-cooled* today. Lettuce is packed in the field into fiberboard cartons containing from 18 to 30 heads depending on head size. After packing, water is sprinkled over the lettuce, and the cartons are closed. The lettuce is hauled from the field and placed in large vacuum chambers in lots ranging from 320 cartons up to a loaded freight car (Figure 15-3). The pressure is lowered from the normal 760 mm Hg to about 4.6 mm Hg or slightly less. At a pressure of 4.6 mm Hg, water "boils" at 0°C. As the water vaporizes, it absorbs tremendous quantities of heat (heat of vaporization—see Chapter 7). The lettuce can be taken from a field temperature of 24° to 0–1°C in about 30 min. In the process, from 0.5 to 0.75 kg of water per carton may be lost through evaporation. Sprinkling water over the lettuce during packing provides some of the water; the lettuce may

FIGURE 15-3 Vacuum-cooling of lettuce at Salinas, Calif. (Photograph, courtesy of J. M. Harvey, U.S. Department of Agriculture.)

also lose 2 to 3 percent of its weight. For leafy vegetables, vacuum treatment allows rapid cooling, eliminates the need for ice, and thus allows the use of fiber cartons.

Hydrocooling has become standard commercial practice with many fruits and vegetables. The market today demands hydrocooled peaches because hydrocooling improves quality and shelf life. The advantage of hydrocooling is even more striking with sweet corn. The flavor of sweet corn is to a large degree the result of relatively high concentrations of sugars. As soon as corn is harvested, conversion of sugars to starch accelerates, and the quality therefore declines at a rate which is very dependent upon temperature. Data indicate that the loss of sugar follows van't Hoff's law. The rate doubles for each 10°C temperature rise. It is obvious therefore that rapid cooling is an absolute necessity and hydrocooling is the most effective technique. The products are slowly moved through the *hydrocooler*, which flushes cold water over the produce to withdraw the heat (Figure 15-4). Water temperature should be just above freezing and sufficient volume pumped over the product. If desirable, a fungicide or disinfectant can be included to minimize infection by disease organisms. Since sweet corn is packed in the field, hydrocooling is done before loading for

FIGURE 15-4 It is now standard for fresh-market peaches to be hydrocooled to remove field heat quickly. Cold water is pumped over the fruit as the bins move slowly through the tunnel. (Photograph, Durand-Wayland, Inc., LaGrange, Georgia.)

shipment. Peaches are hydrocooled upon receipt at the packing house and are subsequently run over the packing line.

Unloading Containers The transfer of produce from the field container to the packing line has changed markedly with the transition from the use of field crates to large containers. Produce in field crates was hand-poured onto the grading line, but bulk containers are mechanically emptied. When initially introduced, bulk bins were mechanically tipped to pour the product out, but this method resulted in considerable bruising. Many modern plants are now using *hydrohandling* systems in which the bin is mechanically submerged in a tank of water, and the fruits are floated out. A stream of water floats the apples or any other product out of the bin and onto a conveyor. The conveyor takes the fruit to the washer. This method results in a minimum of bruising.

Cleaning Before fresh fruits and vegetables are marketed, various amounts of cleaning are necessary. Cleaning typically involves the removal of soil, dust, adhering debris, insects, and spray residues. With some products, such as onions and melons, a dry brushing may suffice;

with others, various types of washers are necessary. Perhaps the most difficult crops to clean are root and tuber crops, such as sweet potatoes, Irish potatoes, and carrots, which have numerous crevices in which soil adheres rather tightly. These normally require both sprays of water and wet-brushing. Chlorine is often added to the wash water as a disinfectant. In some situations, a fungicide may also be added to aid in the control of diseases since chlorine is rapidly dissipated and therefore provides no protection after drying. Drying is often practiced to minimize growth of decay organisms. This may involve absorption of excess water by sponge-rubber rollers or, in some cases, a blast of warm air to evaporate moisture from the surface of the fruit or vegetable.

Sorting   After cleaning, the next operation required is sorting; this entails the removal of items which are unsuitable for packing. In most packing operations sorting is done by hand as the product moves along a roller or conveyor in front of the workers (Figure 15-5). Items

FIGURE 15-5   Sorting area in apple packing house. Fruits rotate as they are conveyed along in front of the workers. Fruits removed are dropped into the large tubes and move along the conveyor belt below. (Photograph, Durand-Wayland, Inc., LaGrange, Georgia.)

damaged by insect, disease, or mechanical injury, which are obviously unfit for sale, are separated from the better produce. It is also a common practice for sorters to segregate fruit into grades on the basis of color or other visible characteristics on which a grade is based. The products removed from the main packing line may be placed on another conveyor or conveyors which in turn transport them to a second area for packing or to a cull collection point for salvage or disposal. For example, in an apple packing house, three grades may be packed such as U. S. Extra Fancy, U. S. Fancy, and U. S. No. 1. Fruits not meeting these grades are often considered "table culls" and are sold to juice plants for processing into cider.

Waxing   Some fresh fruits and vegetables are waxed as a part of the packing operation. With cucumbers, waxing is done to suppress moisture loss, thereby keeping the cucumbers fresh and turgid. With apples, the waxing is usually for improved appearance only. Other products often waxed are turnips, rutabagas, and citrus. With root crops such as turnips, the wax is usually applied by dipping, and thus the coating is quite thick and hard. With cucumbers the wax used is much softer and is brushed on. Apples are coated with a thin layer of relatively hard wax which gives a higher gloss than that of waxed cucumbers. Apples to be waxed are washed in a warm detergent solution to remove natural wax so that the applied wax will adhere and give a uniform clear layer. The wax is sprayed on and the fruit run through a drying tunnel.

Sizing   In normal marketing channels, fruits and vegetables are bought on the basis of both grade and size. With most items, the larger sizes command higher prices, but regardless of this consideration, uniformity of size is necessary. Uniform-sized fruits and vegetables are demanded by retail outlets for several reasons. Most consumers prefer uniform-sized items for a particular use. Many products such as cucumbers, oranges, grapefruit, and melons are often sold by the unit; thus uniformity is necessary. As a general rule, a display of uniform-sized items is much more attractive than a display with a wide range of sizes mixed together. On many small farms, sizing is still done by hand for both fruits and vegetables. Although the operation can be mechanized, a person with a well-trained eye and quick hands can do an effective job. In most large-scale packing operations today, however, sizing is done by mechanical systems. Machine sizers operate on two basic principles: weight and diameter. Sizing on the basis of diameter is most effective with essentially spherical products such as certain apple cultivars, oranges, and tomatoes. One of the earlier types consists of a series of square-mesh chains, the size of which increases progressively from the start to the end of the packing line. The first chain is usually designed to be a size eliminator which allows items below some minimally useful size to drop through. As the fruit or vegetables move from chain section to section, increasingly large items drop through the chain. Because chain sizers tend to cause excessive bruising, more refined systems have been developed, particularly for very tender products. These operate on various principles, such as belts with perforations or rollers at increasing distances apart. Weight-sizers are widely used because they can effectively size asymmetrical items such as the elongated 'Delicious' apple and baking potatoes, but they are equally effective as diameter sizing for round items. A popular type uses cups which hold the item, and as the rows of cups pass along the line, the tripping mechanism is automatically adjusted so that it takes progressively less weight to release the fruit. Some weight sizers handle the product more gently than diameter sizers and for this reason are preferred for easily damaged items, such as tomatoes and peaches.

Presizing and Presorting Once cleaned, sorted, and sized, fruits and vegetables may be handled in several ways. Some may be packed for immediate marketing, but in the peak season some may be returned to bulk bins for storage. Formerly, putting apples and other fruits into storage immediately after harvest without any sorting or sizing was standard practice. However, with the rising cost of cold storage, growers now commonly presize and presort before storage. This has two major advantages: avoiding the expense of storing low-grade fruit and facilitating packing of fruit out of storage to meet orders as received. For example, if an order for 500 boxes of a particular size and grade is received, only that number of bushels need be removed from storage and run over the packing line; but for ungraded produce, many times that amount would have to be sorted to select those needed to fill the order. With other products, where storage costs are low, such as potatoes stored in common or air-cooled storages, presorting may not be advantageous. Another approach is to size, sort, and package before storage. The major disadvantage of this technique is the possibility that the product will have to be repacked if a storage disorder occurs or if the pack does not match what is desired by the buyer.

Packaging The degree to which a particular product is prepared for marketing depends to a large degree on the marketing channel to be used. Using apples as an example, several packing and marketing systems are employed. An apple grower, producing fruit only for processing, harvests the crop into bulk bins, usually belonging to the processor. The fruit is trucked to the processing plant with no grading, sizing, sorting, or other postharvest handling. Upon receipt of the load by the processor, samples are taken, and the grower's account is credited on the basis of weight, cultivar, size, and defects. The processor decides whether to process the fruit immediately or to put it in storage for later use. The only sorting done is in the processing plant immediately before use.

Years ago most fresh-market produce was packed loose in a so-called "jumble pack" in boxes or baskets and displayed in bulk in the retail store. Because of changes in the preferences of consumers and produce buyers, packaging is a constantly changing part of the produce business. Current packaging for fresh market ranges from putting watermelons in a bulk bin which travels all the way to the retail store, to the use of a broad spectrum of consumer packages such as polyethylene bags, shrink films, plastic boxes with overwraps, plastic net bags, and paper bags (Figure 15-6). The advantages of prepackaging by the producer include lessened risk of product damage during shipment and marketing, less contamination, and in many cases reduced moisture loss. Much of the packaging is done in the packing house, but some may also be done at the wholesale distribution center or even in the retail store.

Whether packaging is done at harvest or after storage, the process is the same. After cleaning, sorting, and sizing, the fruits or vegetables

FIGURE 15-6 Consumer-packaged produce in polyethylene bags and overwrapped trays. (Photograph courtesy of R. E. Hardenburg, U.S. Department of Agriculture.)

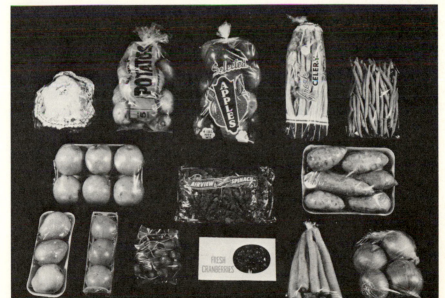

are packed in market containers. The package varies widely with the item. As examples let us consider apples and potatoes. Most apples are shipped in cardboard boxes holding from about 16 to 20 kg. The smaller-sized apples are often placed in 1.4- to 2.2-kg polyethylene bags with from 8 to 12 bags in each master box. Larger sizes are usually packed in trays with the number of apples per box varying with the fruit size. Some apples are now put into fiberboard trays, and a clear overwrap or in some cases a shrink film is used as the overwrap. A shrink film is put on and sealed by heat, and then the package is passed through a heat tunnel. The heat causes the film to shrink and thereby to hold the product snugly and prevent movement during handling. Potatoes are often packaged in heavy paper bags, usually with a mesh window through which the potatoes can be seen. Some are currently being sold in perforated polyethylene bags, but disease can be a problem in these bags because relative humidities are much higher than in paper bags. Many potatoes are also packed in cardboard crates in a loose pack for bulk display and sales.

A fresh produce grower may have a packing house in which the fruit is cleaned, sorted, sized, and packaged or may belong to a cooperative packing house which handles crops for many growers. In some areas packing is offered by custom packing houses which charge a fee.

Cut Flowers

The preparation of cut flowers for marketing is less involved than most fruits and vegetables. The grower must cut, grade, and package the cut flowers. A major goal should be cutting flowers at the proper stage of development to ensure optimum quality upon receipt and resale by the retail florist. The flowers must be graded, usually on the basis of flower size, stem length, and freedom from defects. After grading, the flowers are packed "dry" in rigid cardboard cartons. Very expensive and perishable flowers, such as orchids, may be put into vials containing water as a preservative. For maximum retention of quality, cut flowers should be refrigerated. A grower may refrigerate flowers before shipment especially if a load is shipped only every 2 or 3 days. For this holding period the stems would be put in water. Upon receipt by the wholesaler, flowers are refrigerated. Long-distance shipping is usually by air freight; short hauls are typically by truck.

Pot Plants, Nursery Stock, Foliage Plants

These items usually require minimal preparation for marketing with the exception of field-grown nursery stock which must be dug, balled, and burlapped. Since the nursery business is tending more and more toward the use of container stock, these tasks are gradually being eliminated. Most of these items are now loaded on trucks for shipment. Since these plant materials are not nearly as perishable as cut flowers, considerably more flexibility in handling is possible.

STORAGE

Types

Several types of storages are used for holding horticultural plants and products: common, cold, and controlled-atmosphere (CA) storages. A *common storage* is unrefrigerated storage but utilizes cooler outside air in fall and winter to hold temperatures at acceptable levels. The underground "root cellars" of earlier days would be an example. *Cold storages* are mechanically refrigerated to provide both rapid product cooling during loading and accurate temperature control during the holding period. Normally a room is filled gradually to prevent over-loading the cooling equipment and to lower the temperature of the product as soon as possible. Particularly perishable crops are pre-cooled by hydrocooling, vacuum cooling, or the use of ice before being placed in storage. Most fruits and vegetables suitable for long-term holding are cooled in the storage room. Stacking containers to provide good air circulation is an absolute necessity if adequate cooling and temperature control is to be maintained. An example of the effect of prompt cooling on water loss is given in Table 8-5. *Controlled-atmosphere storages* utilize not only cold temperature but altered levels of CO_2 and O_2 to maximize storage life. By lowering the oxygen from the normal atmospheric level of 21 percent to 2 to 3 percent and elevating CO_2 from the normal concentration of 0.03 percent to 2 to 5 percent, the storage life of certain products can be extended, largely by lowering the rate of respiration. The atmosphere may be modified by sealing the fruit in a completely airtight room and allowing respiration by the fruit to utilize O_2 and generate CO_2. To provide the desired atmosphere more quickly and to reduce the necessity for completely airtight rooms, many storages now use gas generators to produce the desired atmosphere. One such system burns propane gas, and the burner is vented into the storage. In any case, frequent gas analysis is mandatory to prevent injury from excessively low O_2 or high CO_2. Controlled-atmosphere storage is widely used with apples and pears and seems to hold promise with other horticultural products.

The most recent development in storage technology is the principle of *hypobaric storage.* (Other terms are subatmospheric and low-pressure storage.) The product is held at a stable, subatmospheric pressure in a refrigerated vacuum chamber which is ventilated with fresh humid air. The principle behind hypobaric storage is that at low pressure the concentration of oxygen is reduced so that respiration is suppressed, and internal ethylene concentrations are dramatically reduced. Ethylene is known to be an extremely active ripening hormone, and by holding the product at low pressure, the ethylene concentration is lowered in the fruit and its diffusion rate is increased. Early results with this technique are indeed exciting, not only for fruits and vegetables, but for flowers, cuttings, and potted plants as well.

Storage of Nursery Stock

For short-term holding, as in a garden center, little extra care is necessary other than providing adequate moisture. With balled and burlapped stock, the root ball is often wrapped in polyethylene, or a group of plants are held upright and the root balls covered with sawdust. Long-term storage becomes necessary for several reasons. Because they are handled bare-root, such products as fruit trees, strawberry plants, and asparagus roots must be shipped in late winter and early spring to enable planting as soon as the soil can be worked. To meet their shipping schedule, most northern nurseries dig their stock in late fall and early winter and thus must store it for a few months. Similar conditions exist in the handling of deciduous ornamentals which are also dug and shipped bare-root. Evergreens must be handled quite differently as they remain in full leaf. If field-grown, they are balled and burlapped; but increasing proportions are being produced and sold in containers. If field-grown, some may be dug in the fall and held in storage, like deciduous stock; but container stock is usually overwintered outdoors or in minimal-expenditure structures. In Northern parts of the country, much of the nursery stock is held in common storage. In warmer regions, such storages may be underground with dirt floors. High-value crops which require minimal space are usually held in refrigerated storages to maintain optimum quality.

Container stock is subject to winter injury (see Chapter 7). The roots are in a restricted soil volume, and if the container is exposed to the air, the entire soil mass will freeze rather quickly. Freezing of the soil leads to desiccation of evergreens exposed to warm winds on sunny days (see Chapter 7). Secondly, freezing injury to roots is a problem with container stock. Inexpensive plastic structures are often used to provide some protection against both the nightly minimum and the drying winds. A technique to provide protection to small deciduous container stock is to cover the containers and plants with mulch as insulation (Figure 15-7). An unusually severe winter may cause injury to nursery stock, but one must balance the cost of protection against the frequency with which severe winters can be expected.

FIGURE 15-7 Container stock being protected from cold by covering with straw during the winter.

Storage of Cut Flowers

The optimum temperature for holding cut flowers ranges from 0° to as high as 15°C. If the volume justifies the expense, a wholesaler will often have two storage rooms, one at 0 to 2°C for holding the majority of cut flowers and one at 7 to 10°C for cold-sensitive flowers, such as orchids. If only one room is available, the best compromise for short-term holding is about 4 to 5°C. The relative humidity should be kept high in the storage room, and flowers are often wrapped loosely to maintain a higher moisture content of the air. High humidity is particularly important when cut flowers are stored "dry" or without the cut ends in water. Even under optimum storage conditions, the storage life of cut flowers is relatively short, ranging from 1 or 2 days for some up to 2 or 3 weeks for others. As flowers are held for longer periods, not only does appearance and salability decline, but shelf life is shortened.

The use of chemical preservatives in the water is strongly recommended as a means of lengthening the life of the flowers both in storage and after they are purchased by the consumer. If the grower, wholesaler, retailer, and consumer use preservatives in the water, flower life can be doubled or even tripled. Several types of preservatives are available. Most contain a variety of substances, such as sugar, a bactericide, and an acidic substance. The sugar serves as a substrate for respiration. The bactericide suppresses growth of micro-organisms and thereby slows the plugging of the xylem elements. By lowering the pH of the solution, water uptake is improved. Other materials may also be added; these include respiration inhibitors to slow the rate of respiration. Research results with preservatives have opened some very exciting possibilities in the storage and marketing of cut flowers.

Storage of Fruits and Vegetables

Depending on the crop, the location, and the intended length of holding, various types of storages are utilized for fruits and vegetables. Although most are held under refrigeration, potatoes, onions, and carrots are often stored in common or air-cooled storages. A modern potato storage consists of an insulated, aboveground structure with an automatic ventilation system. Air is continually circulated through the bins of potatoes to maintain relative humidity at about 90 percent, and outside air is incorporated as needed for cooling. Potatoes for fresh market are stored at 3 to 4°C and 90 percent relative humidity. Potatoes for processing are held at temperatures of about 10°C; at lower temperatures starch is converted to sugar which leads to browning when the potatoes are fried—a particularly undesirable characteristic in french fries and potato chips.

Most fruits and vegetables are stored in refrigerated storage because optimum temperatures are below those normally possible in common storage. The majority of these products have maximum life at approximately 0°C (Figure 15-8). Exceptions are tropical and subtropical fruits such as avocados, bananas, citrus, melons, pineapples, and papayas which are subject to *chilling injury* at temperatures above freezing. Vegetables subject to chilling injury include snap beans, cucumbers, eggplant, peppers, squash, and sweet potatoes. Products subject to chilling injury are usually held at temperatures of 4 to 13°C, depending on the particular item. The temperature in cold storage should be held within 1 to 2°C of optimum as even a few degrees above will shorten storage life, and freezing or chilling injury can result if temperatures get too low. A problem with widely fluctuating temperatures at the high relative humidities used in storages is condensation of water on the product, which can lead to decay.

Apples and pears are widely held in CA storage because it can double the life of some cultivars, such as 'McIntosh' apples, and because it adds from 2 to 4 months to the storage life of most other cultivars. An added advantage is the improved shelf life of the product after storage. The additional expense over cold storage has generally been a sound investment.

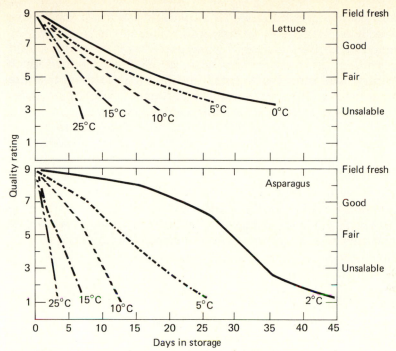

FIGURE 15-8 Quality ratings of asparagus and lettuce stored at five temperatures for various numbers of days. (Data from U.S. Department of Agriculture.)

MARKETING

Marketing of Fruits and Vegetables

Types Large chains of food stores handle the biggest part of the fresh fruit and vegetable business today. Even the independent grocers are often served by a large cooperative *distribution center*, serving 50 to 200 stores in a particular region. Tremendous buying power is in the hands of the produce buyer, who can buy in railcar and tractor-trailer loads rather than small quantities. Such centers assemble products from producers all over the country and the world and then distribute them to the local stores. Perhaps the major drawback of this system is that most of the decisions concerning the produce to be sold is in the hands of the distribution center's buyer; the local store manager, therefore, may have little choice. This disadvantage is often overcome by the buying power of a large distribution center.

Farmers' markets are regaining some of their lost popularity as means of marketing fresh produce. These markets are designed so that individual producers can market their produce directly to consumers, often directly off the back of a truck. Some cities are encouraging the reactivation of farmers' markets by providing modern facilities and renting areas to individuals who wish to sell their produce. Such markets are often attractive to small and part-time farmers because they draw large numbers of potential customers with whom these producers might otherwise not be able to make contact.

Another alternative in the marketing scheme is *roadside marketing* (Figure 15-9). By selling their produce directly to the consumer, growers realize many advantages. These include minimal costs for packaging and the elimination of shipping and brokerage fees. By dealing directly with the customer, it is also possible to market a wider group of cultivars and thereby reduce the problems associated with the 3 to 4 major cultivars which are most in demand in wholesale marketing channels. By growing 8 to 10 cultivars, the harvest operation

FIGURE 15-9 Fruits and vegetables may reach the consumer through various marketing channels.

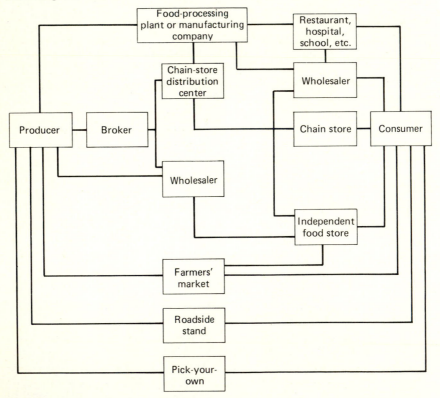

can be spread out over a longer period of time, thereby putting much less strain on labor, equipment, and cooling facilities during storage.

Diversification into several tree fruits, small fruits, and vegetables can reduce the gamble with a particular crop. For example, if a spring freeze destroys the apple crop, another crop may salvage the season's profits; but a wholesale apple grower would be much more severely affected.

Greatly increased adoption of *"pick your own"* is taking place in much of the country. In this system, since the customer picks the produce, the grower has eliminated much of the peak-harvest labor problem. Often such an operation can market crops from early summer through late fall. For example, a sequence might include strawberries, raspberries, blueberries, cherries, peaches, nectarines, apples, and pears intermixed with vegetables such as green beans, squash, tomatoes, peppers, corn, and potatoes. Pick-your-own marketing seems to appeal to today's consumers in several ways. A major feature is the recreation offered to a family by getting out in the country and visiting a farm. Many operators encourage families by providing picnic facilities and recreation areas. The educational aspect is also important. Many urban children are totally unfamiliar with how crops grow, and visits to a farm can certainly be informative in this regard.

The pick-your-own system offers one of the few remaining ways in which a young person can break into the fruit or vegetable business without great financial resources. It is quite possible to start small and expand as the business grows.

Marketing Costs for Fruits and Vegetables With the previously mentioned diversity of product preparation and marketing systems employed, it is certainly not surprising that the consumer's dollar is distributed in different ways. Taking lettuce, canned tomatoes, and frozen orange juice as examples, we can see considerable variation (Figure 15-10). The lettuce and tomato producer gets only 12 to 14 percent of the consumer's dollar, whereas, an orange grower gets 28 percent. Assembly costs vary little but processing ranges from 16 percent on lettuce to 48 percent on canned tomatoes. Wholesaling and transportation take 29 percent of the lettuce dollar, but only about half as much on the other two products. Retailing takes a very large proportion of the lettuce dollar because of perishability, required refrigeration, and wrapping.

In the marketing costs for orange juice and canned tomatoes, about 25 percent is for labor, 20 to 30 percent is for packaging, and about 10 percent is for transportation. The remainder of the costs are split among many functions such as advertising, taxes, and interest.

What the food dollar pays for:

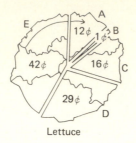

Lettuce

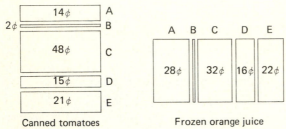

Canned tomatoes   Frozen orange juice

FIGURE 15-10 The food dollar is divided quite differently with different products. *Key:* A, production; B, assembly; C, processing; D, wholesaling and transportation; E,
retailing. The figures are estimates based on prices,
costs, and margins in 1975. (Data from U.S. Department of
Agriculture.)

Marketing of Nursery Stock

Wholesale Nursery The marketing of nursery stock involves several
different types of operations. Many of the larger nurseries sell strictly
on a wholesale basis and make deliveries by tractor-trailer loads
(Figure 15-11). Their operations are geared to mass production techniques, mechanization, and rapid turnover of plant material. Much of
the plant material sold by garden centers, used by landscape nursery
growers, or sold by large discount department stores comes from large
wholesale nurseries. Some nurseries specialize in a limited range of
related species (e.g., fruit trees) for a particular market.

Retail Nursery A vital link in the marketing of ornamentals is the
retail nursery which caters to the general public. Such nurseries often
buy much of their plant material from wholesale nurseries—frequently
more cheaply than they could grow it. The modern retail nursery
usually sells a broad spectrum of other items through a garden shop.
Items for sale frequently include fertilizers, pesticides, mulches, tools,
seed, bedding plants, and many other associated items. As a part of

FIGURE 15-11 Shipping area and loading dock for a large wholesale nursery. (Photograph, Monrovia Nursery, Azusa, Calif.)

the operation, many such retail operations provide landscape planning, installation, and maintenance. Like the roadside marketing mentioned earlier, the retail nursery offers the chance for a hardworking young person to start a horticultural business without a tremendous capital investment.

Marketing of Cut Flowers, Foliage Plants, and Pot Plants

The market for flowers, foliage plants, and pot plants has increased markedly in recent years. Much of this growth has been in the foliage plant business which has increased tremendously. Most floral sales at one time were by retail florist shops which sold services as well as products. In recent years, there is a trend toward more mass marketing, usually with much less service provided. Traditional flower sales are very seasonal and peak at Valentine's Day, Mother's Day, Easter, and Christmas. Many florists are attempting to induce more impulse buying without the traditional service associated with design arrangement, delivery, or credit. Another aspect which has added greatly to the retail sales of florists are the foliage plants which are so popular and are not tied to specific days or seasons. As with businesses everywhere, the retail florist must keep abreast of trends and outlooks and adapt to changes in the marketing outlook.

FOOD PROCESSING

The commerical preservation of food is becoming increasingly important in providing food for today's consumers. As Americans have
moved from rural to metropolitan areas, they have become dependent
on the purchase of both fresh and processed foods from commercial
grocery stores rather than producing and processing their own. In
recent years there has been a rapid acceleration in the utilization of
"convenience" foods. Most convenience foods involve processing and
prepackaging to minimize the time and effort required for final
preparation; examples are frozen french-fried potatoes, dried instant
mashed potatoes, frozen fruit pies, and dehydrated fruits.

There are several principles involved in the preservation of fruit
and vegetable products. These include canning, freezing, drying,
fermenting, and using a variety of preservatives. Each of these
processes is aimed at preventing the growth of microorganisms so that
long-term holding of the product is safe and practical.

Canning

The preservation of foods by canning utilizes heat to sterilize the
product and sealed containers to prevent reinfection thereafter. The
term *sterilize* is used, although in many cases the product is not truly
"sterile" from a microbiological standpoint. Any microorganisms
which are pathogenic or which can cause spoilage of the product must
be destroyed. There are, however, some heat-resistant organisms
(thermophiles) which may survive normal processing procedures.
Thermophiles thrive at high temperatures and so may occasionally
cause spoilage of canned foods stored at high temperatures.

In addition to the destruction of microorganisms, canning modifies
the product in a variety of ways. Temperatures sufficient to preserve
the product inactivate enzymes, thus preventing further metabolic
activity. The heat applied cooks the food so that the texture, color, and
flavor are altered. Normally, both the maximum temperature and time
of cooking are kept at adequate levels for safety, but excessive heating
is avoided so that alterations in the quality of the product are
minimized. Excessive cooking can lower the quality of the product as
determined by texture, color, flavor, or vitamin content.

In the canning of foods, the two major features are the temperature
used and the length of time involved. Sterilization is about 100 times as
fast at 121°C as at 100°C. Both of these factors must be considered in
any canning operation. As a general rule, the quality of the product is
preserved better by heating at high temperatures for shorter times
than at lower temperatures for longer times. The former also makes
more efficient use of equipment in the canning plant. The typical
procedure is to cook adequately and then cool as quickly as possible.

TABLE 15-1
EFFECT OF ALTITUDE ON THE
BOILING POINT OF WATER

ALTITUDE		BOILING POINT OF WATER	
FT	M	°F	°C (APPROX.)
0	0	212	100
2,063	629	208	98
4,169	1271	204	96
6,304	1922	200	93
8,481	2586	196	91

Source: W. V. Cruess, *Commercial Fruit and Vegetable Products*, McGraw-Hill, New York, 1958, p. 122.

Such a procedure tends to keep the quality of the product at its best and most consistent level. Cooling is normally accomplished by submerging cans in cold water or spraying them with cold water. The latter utilizes the vaporization of water from the surface of the containers as a cooling mechanism.

Depending on the product as well as the overall system used, canning may be done at temperatures of 70 to 120°C. The highest temperature obtainable in an open, boiling-water bath is 100°C, and this is only possible at sea level. As elevation increases, the atmospheric pressure decreases, and the boiling point of water therefore decreases (Table 15-1). The same principle is utilized with high-pressure cookers, called *retorts*, in which steam pressure is used to elevate the boiling point of water (Table 15-2). At a pressure of 1.05 kg cm^{-2}, the

TABLE 15-2
RELATION OF STEAM PRESSURE TO
TEMPERATURE OF CANNING RETORT

PRESSURE		TEMPERATURE	
LB IN^{-2}	KG CM^{-2}	°F	°C APPROX.
1	.07	215.2	102
3	.21	221.3	105
5	.35	226.9	108
7	.49	231.9	111
9	.63	236.6	114
11	.77	241.0	116
13	.91	245.3	119
15	1.05	249.1	121

Source: W. V. Cruess, *Commercial Fruit and Vegetable Products*, McGraw-Hill, New York, 1958, p. 122.

boiling point of water is 121°C. By using high temperatures, the time required for sterilization is dramatically reduced.

Foods are classified into *acid* and *nonacid* types. This distinction is of major importance in determining the processing times and temperatures required. The nonacid foods are those with a pH above 4.5, whereas those products with a pH below 4.5 are classified as acid. With the exception of figs and ripe olives, canned fruits and fruit juices are of the acid type. Other than tomatoes and the fermented and pickled products, most vegetables are of the nonacid type. Nonacid foods are canned under pressure and thus at temperatures in excess of 100°C, whereas the acid products are processed at 100°C or below. Because of the acidity of acid foods, the microorganisms present in the product are readily killed by relatively low temperatures. Many of the fruits and fruit products which are acid and can therefore be canned at temperatures of 100°C or less would become quite low in quality if cooked above 100°C. This is particularly true of such products as canned cherries, plums, and many of the fruit juices. Quality control is a very important part of food processing (Figure 15-12).

The major concern in canned nonacid foods is the possible presence of the bacterium *Clostridium botulinum*, which produces an extremely toxic poison. The potential presence of this organism deter-

FIGURE 15-12 Quality checks are a constant procedure with the processing plants. Both U.S. Department of Agriculture and plant quality-control personnel continuously inspect the product as it comes off the line. (Photograph, California Canning Peach Association.)

mines the relatively high temperatures and long times used for canning nonacid foods. It is the toxin from this organism which is responsible for the rare, but not unknown, outbreaks of deadly food poisoning. Perhaps the most frequent product involved is home-canned green beans which have been processed without a pressure cooker.

Freezing

The preservation of fruit and vegetable products by freezing is a very important aspect of the food processing industry. The freezing of food preserves by low-temperature inhibition of microbial growth. Some organisms may be killed by the cold, but many are merely held in check. Because of the potential for the presence of harmful microorganisms, it is highly desirable that food products be frozen as soon as possible after harvest and utilized as soon as thawed.

The additional cost of a frozen item over its canned counterpart is to a large degree caused by the need for continuously freezing temperatures from the processing plant through the marketing channels. Whether or not the consumer is willing to pay the extra price is dependent on many factors. Perhaps the major criterion is the relative quality of the canned and frozen product. Many fruits and vegetables are so greatly modified by canning that they only vaguely resemble the fresh product. Examples would be asparagus, spinach, raspberries, and strawberries. These particular foods are all much closer to the fresh product when frozen. Certain other foods, such as tomatoes, do not freeze well and therefore are primarily canned. Many others are widely canned and frozen and are excellent after either type of processing in spite of color, texture, and flavor differences. These would include snap beans, peas, corn, blueberries, apples, and peaches. The choice between the frozen or canned product will often be based on personal preference, cost considerations, or intended use. For example, canned blueberries are suitable for use in pies; but for use in a fruit salad or in blueberry muffins, frozen berries are much more desirable. Most products are packaged before being frozen, and the final product is therefore in a solid block as with fruits in syrup, spinach, broccoli, and asparagus. Some foods are individually quick-frozen before packaging. Such products are loose in the package and are exemplified by cherries, strawberries, English peas, diced carrots, blueberries without syrup, and green beans. Because the particles are loose, part of the contents of a package can be easily removed. This aspect is of importance when large containers, such as large drums, are sold to institutional buyers.

In addition to minimizing microbial growth, freezing must preserve the quality of the product as much as possible. Enzymatic changes can continue to occur during preparation prior to freezing and can lead to deterioration of quality, as with the conversion of sugar to starch in corn. To eliminate enzymatic changes, many fruits and vegetables are "blanched" by heating in boiling water or steam for a period of 1 to 10 min depending on the product. The blanching time is kept as short as possible, and the product is cooled immediately. Minimal blanching and quick cooling not only avoid overcooking but minimize the growth of microorganisms.

Containers for frozen foods are designed to protect the contents, exclude oxygen, minimize moisture loss, avoid contamination, and in many cases to appeal to the customer. The most widely used retail market package is the folding paperboard carton which is coated with wax or plastic. These can be used for a broad spectrum of products and the waxed overwrap changed with each item frozen. Polyethylene bags are also widely used with free-flowing products such as peas. Kraft paper which has been coated with polyethylene is used for bulk packages of such items as french-fried potatoes, peas, or carrots for restaurants or institutions.

Storage temperatures have a major effect on the length of time during which frozen foods will maintain optimum quality. In general, the lower the storage temperature, the longer the product will keep. The standard procedure is to hold frozen foods at $-18°C$ or lower. This temperature not only minimizes growth of microorganisms but keeps chemical changes low. Because of low temperatures in freezers, containers must be sealed tightly to avoid loss of moisture from the product. The atmosphere surrounding the frozen food in the package is at 100 percent relative humidity and has a higher vapor pressure than that of the storage-room air which loses moisture by condensation on cold refrigeration coils. The loss of moisture from the product leads to a desiccation known as "freezer burn" which leaves the product whitish, dried, and unsatisfactory for use.

Drying

One of the most ancient methods of preserving foods, drying, is still very important. The moisture content is lowered to a point where microorganisms cannot grow and enzymatic changes no longer occur. In addition to being an effective means of preserving fruits and vegetables, drying offers certain very real advantages over either freezing or canning. Among these are lower costs of packaging, storing, and shipping because of the dramatically lowered weight and volume of the product.

One technique used today is *sun drying* of certain fruits such as raisins, figs, and apricots. This system has been used in sunny, dry climates since ancient times but in the United States has been widely utilized only in California. In sun drying, the fruit is spread in a thin layer on racks in a "drying yard." Grapes may be spread on plastic or paper strips between the rows (Figure 16-6). Artificial drying techniques (dehydration) are becoming more widely used. By means of mechanical equipment, artificial heat, and controlled conditions, drying can be done much more rapidly and to lower, more consistent moisture levels. Fruits are usually dried to 5 percent moisture or less. Vegetables are dried slightly further because of the lower sugar content.

The treatment of fruits and vegetables with sulfur dioxide (SO_2) fumes before drying is widely practiced. The sulfur preserves the color of the product by retarding the normal enzymatic browning reactions and also slows the destruction of carotene and ascorbic acid. The fruit is put in a tight chamber or building and sulfur is burned for a few hours. After the "sulfuring" treatment, the fruit is dried either in the sun or in one of several types of dehydrators. An example of sulfur-treated fruit is white raisins.

Fruits are more commonly dried than vegetables are. Some of the commonly available fruits are raisins (dried grapes, mostly 'Thompson Seedless'), prunes (dried plums), dates, figs, apricots, peaches, and apples. Of the dried vegetables, onion, garlic, and parsley are well known. Irish potatoes, in a variety of forms, have become a leading dehydrated vegetable. Instant mashed potatoes have been well accepted by consumers. Dehydrated potatoes are being used in tremendous quantities in producing potato chips from reconstituted potato granules.

Considerable impetus for the development of dehydrated foods has come from the military services. The advantages of light weight, long-storage potential, and no cold-storage requirement are obvious. There is also considerable interest in dehydrated foods by camping and, particularly, backpacking enthusiasts.

A relatively recent and still quite limited process is "freeze-drying." The item is frozen and then put in a vacuum chamber. By drastically lowering the atmospheric pressure, the water is transformed from ice to water vapor by sublimation (see Chapter 7). The water vapor condenses on refrigerator coils. Since the entire process is carried out at low temperatures, few chemical or biological changes occur and losses of volatile aromatics are minimized. Freeze-dried products are very light in weight and can be stored for long periods with no refrigeration. The only major requirement is that such products be sealed in airtight containers; this is necessary because many tend to be hygroscopic.

Freeze-dried products available in supermarkets today include instant coffee, sliced mushrooms, green asparagus, parsley, and paprika. For specialty markets such as campers and backpackers, many other products are available, such as eggs, meats, and various vegetables and fruits. The cost is high enough to discourage everyday use of these products.

Pickling and Fermentation

A diversity of fruit and vegetable products are produced by pickling and fermentation. Pickles are products which have been preserved and flavored in a brine containing a mixture of salt and an acid, typically acetic acid (vinegar). Pickles are most often made from cucumbers; but they are also made from green tomatoes, watermelon rind, peppers, cauliflower, and onions. Important pickled-fruit products include olives, peaches, figs, and pears. Sauerkraut is a very important product of the pickling industry.

For the production of cucumber pickles, a typical procedure is as follows: Cucumbers are put into vats containing a salt brine for the initial fermentation to occur. The requirements are anaerobic conditions, suitable salt concentration, presence of appropriate lactic acid bacteria, and proper temperature. The pickling vats have wooden covers which are held down and thereby keep the cucumbers submerged and under anaerobic conditions. The salt concentration is started and held at about 10 percent by adding additional salt periodically because the brine withdraws water from the cucumbers and is thereby diluted. This level of salt greatly suppresses growth of unwanted organisms which could cause spoilage but has little effect on the desired lactic acid bacteria. Since a multitude of microorganisms is present on the incoming cucumbers, no innoculation is necessary.

Fermentation normally lasts for several weeks. After fermentation is complete, the pickles are ready for processing. The excess salt from the brine treatment is removed by soaking in water. They are then packed in plain vinegar or sweetened, spiced vinegar, depending on the type of pickle.

The procedure varies widely, not only with the various fruits and vegetables, but even with cucumbers. The many choices of pickles available to consumers include: dill, sweet, sour, whole, strips, sliced, and a broad spectrum of relishes.

Wine

In recent years, wine consumption in the United States has increased sizably. This increase has been reflected not only in expanded production of grapes in traditional areas such as California and New York, but in the establishment of commercial acreages in other areas which have not been in the wine-grape business.

Wine making is a very ancient art, and the diversity of wines is seemingly endless. Very expensive wines are made from very specific cultivars, grown in particular regions, and aged for many years.

Although there is a definite market for such wines, much of the increase in demand has been for less expensive wines which are usually much less specific as to cultivar, locality grown, and age. Many wines are also available today which are made from less expensive juices and flavored with fruit juices such as blackberry, strawberry, or cherry.

The production of wine is an anaerobic fermentation process in which the sugars present in the fruit are converted to ethyl alcohol by the action of microorganisms. The fruit is usually washed and crushed, and the pulp, seeds, and skins (on red cultivars) are often left in the juice to ferment for a few days. A specific fermenting yeast is inoculated into the crushed fruit, often after treatment to inactivate the native yeasts. Although great numbers of yeasts are present naturally, the introduction of particularly desirable strains causes more consistency in the final product. After the yeast has worked for a few days, the skins, seeds, and pulp are removed, and the juice is allowed to ferment. Initially the fermentation is very rapid with the production of large volumes of CO_2 which keeps O_2 levels at a minimum. As the fermentation slows, some method must be used to prevent contact with O_2. This may be done in several ways such as covering the fermentation vats. The reason why the fermentation process must be kept anaerobic is to prevent the conversion of the ethanol to acetic acid as occurs in the making of vinegar.

The making of fine wine involves not merely the conversion of sugars to ethanol but many, many more subtle changes which occur. Many of these changes are not completely understood, but the use of modern equipment such as gas chromatographs has allowed the identification of many of the flavor components of wine. In spite of the improved equipment and technical expertise, the making of fine wines is still largely an art.

The alcohol content of wine varies according to how it is made and whether or not it has been fortified. Even the most resistant wine yeasts are killed when the alcohol content reaches about 14 to 16 percent. Thus a wine with more than this much alcohol is *fortified* by the addition of ethyl alcohol, usually made from fermentation of grain. The ethyl alcohol is distilled to produce a colorless liquid with little flavor but is usually 95 percent alcohol. The so-called "pop wines" are mass-produced by fermenting a fruit juice, such as apple, and subsequently adding enough unfermented juice to flavor it. The alcohol content is usually much lower and will reflect not only the fermentation but subsequent dilution.

Sparkling wines and champagnes are bottled before the fermentation process is complete. By careful checking of the sugar and alcohol concentrations, the winemaker knows how much further fermentation will continue. At the proper time, the wine is bottled in thick glass bottles able to withstand the pressure generated by the fermentation. It is the CO_2 produced by the remaining fermentation which gives the "pop" when a champagne bottle is opened as well as the bubbly character of such a wine. Cheaper wines may be made into "sparkling" wine by addition of CO_2 gas, much as with carbonated soft drinks.

Jams, Jellies, and Preserves

Another sizable part of the fruit processing industry is the making of jams, jellies, and preserves. The principle involved in this type of preservation is the use of sugar levels high enough to prevent the growth of microorganisms. Jellies are made from fruit juice which is concentrated to a certain consistency so that it will gel when cooled. To produce a clear jelly of good consistency it may be necessary to add sugar and acid, pectin, or both. Jams are made by cooking the fruit with sugar to form a thick product. Depending on the fruit involved, the seeds may or may not be left in the product. Other products include fruit butters (particularly apple butter), marmalades (jelly with pieces of fruit mixed in), and candied fruits.

BIBLIOGRAPHY

Ashby, B. H.: *Protecting Perishable Foods during Transport by Motortruck*, U. S. Department of Agriculture Handbook 105, 1970.

Cruess, W. V.: *Commercial Fruit and Vegetable Products*, 4th ed., McGraw-Hill, New York, 1958.

Ginder, R. G., and H. H. Hoecker: *Management of Pick-Your-Own Marketing Operations*, Coop. Extension Service, University of Delaware, Newark, Delaware, 1975.

Haard, N. F., and D. K. Salunkhe: *Symposium: Postharvest Biology and Handling of Fruits and Vegetables*, Avi, Westport, Conn., 1975.

Luh, B. S., and J. G. Woodroof: *Commercial Vegetable Processing*, Avi, Westport, Conn., 1975.

Lutz, J. M., and R. E. Hardenburg: *The Commercial Storage of Fruits, Vegetables, and Florist and Nursery Stocks*, U. S. Department of Agriculture Handbook 66, 1968.

Pantastico, E. R. B. (ed.): *Postharvest Physiology, Handling and Utilization of Tropical and Subtropical Fruits and Vegetables*, Avi, Westport, Conn., 1975.

Pfahl, P. B.: *The Retail Florist Business*, 3d ed., Interstate, Danville, Ill., 1977.

Ryall, A. L., and W. J. Lipton: *Handling, Transportation, and Storage of Fruits and Vegetables.* vol. I, *Vegetables and Melons,* 1972, vol. II, *Fruits and Tree Nuts,* 1974. Avi, Westport, Conn.

Smith, O.: *Potatoes: Production, Storing, Processing,* Avi, Westport, Conn., 1968.

United States Department of Agriculture; *Marketing, Yearbook of Agriculture,* 1954.

Woodroof, J. G., and B. S. Luh: *Commercial Fruit Processing,* Avi, Westport, Conn., 1975.

PART 4
Branches of Horticulture

CHAPTER 16
POMOLOGY

 Since the earliest times, humans have enjoyed eating fresh fruits, as is indicated by the frequent mention of fruits throughout recorded history. In today's world, fruits play a vital role in our diet, and fruit growing is a major portion of horticulture. Rapid shipment, modern storage techniques, and large-scale food processing provide us with a year-round supply of a wide variety of fruits and fruit products (Figure 16-1). The early colonists had fresh fruits in season, but for several months of the year depended on dried fruit, fruit preserves, or juice which had been fermented to produce "applejack" or similar alcoholic fruit drinks. Today it is quite common to visit a grocery store in the spring and find the following fresh fruits: strawberries from California, oranges from Florida, peaches from Chile, melons from Mexico, pineapples and papayas from Hawaii, grapes from Ecuador, bananas from Costa Rica, pears from Oregon, apples from Washington, and avocados from Florida. Obviously, some of these fruits have been in storage for several months; but some, such as the strawberries, have only recently been picked. It is indeed a tremendous credit to horitcultural science and technology for such an array of fresh fruits to be available to the American consumer on any given day. When one considers the tremendous choice of canned, dried, frozen, or otherwise preserved fruits and fruit products, the list becomes even more impressive.

Although fruits can be classified in many ways, perhaps the first

FIGURE 16-1 Attractively displayed fresh fruit is appealing. (Photograph, Keyes Fibre Company, Montvale, N.J.)

distinction should be on the basis of climatic requirements, the three major divisions being *temperate, subtropical,* and *tropical.* Temperate fruits are those which require a cool period and are deciduous. Subtropical fruits are intermediate between the temperate and tropical types; some are deciduous, and others are evergreen, like the tropicals. The subtropicals can ordinarily withstand temperatures slightly below freezing but are seriously injured by temperatures of −4 to −5°C. The tropical species are evergreen, can withstand no temperatures below freezing, and some—such as banana—suffer chilling injury at temperatures above freezing.

TEMPERATE FRUITS

The temperate fruits are subdivided into the *pome* fruits (apple, pear, quince), the *stone* fruits (peach, nectarine, cherry, plum, apricot), and the *small* fruits (strawberry, blueberry, cranberry, red raspberry, black raspberry, blackberry and—most important of the temperate fruits—the grape).

Pome Fruits

The only two pome fruits of commercial concern are the apple *(Malus domestica)* and pear *(Pyrus communis)*. Although the quince is a pome fruit, its production is minimal because it is used almost exclusively for jelly and preserves. Apples are by far the most widely grown temperate tree fruit, surpassing the production of pears and the stone fruits combined (Table 16-1). Apples and pears are grown at similar latitudes as their climatic requirements are not too different.

Apples Major apple-producing countries are France, the United States, Italy, with the European continent providing over one-half of the world's crop. In the United States, apple production is concentrated in Washington, New York, Michigan, California, and in the Appalachian region from Pennsylvania to North Carolina (Figure 16-2). These regions combined produce in excess of 80 percent of the apple crop, but actually commercial apple production is reported in a total of 35 states. The northern limit for commercial apple and pear production is determined by the coldest nights of winter because it is usually impractical to grow either of these species where minimum winter temperatures are below −30 to −40°C.

The southern extremity for apples is determined by two considerations, both of which relate to temperature. Apples need relatively cool nights during the ripening season for the development of good flavor, color, and quality. In the Southeastern states, apples mature during mid- to late summer when temperatures are very high. For optimum fruit quality, orchards should probably be confined to the higher elevations where temperatures are considerably cooler than at the lower elevations. However, the availability of very highly colored strains of red cultivars has encouraged the setting of orchards in very warm regions. Since these plantings are quite recent, their long-term economic stability is subject to debate. The other temperature consideration is that the warm winters may not provide adequate chilling. Most commercial apple cultivars require from 1200 to 1500 h below 7°C to satisfy the chilling requirement (see Chapter 7).

Although apple cultivars number in the thousands, the number grown commercially not only is quite small but is declining annually. In 1976, 42 percent of the total United States apple production consisted of the 'Delicious' cultivar with another 18 percent being 'Golden Delicious'. Other cultivars of major importance were 'McIntosh' (8 percent), 'Rome Beauty' (7 percent), 'Jonathan' (5 percent), and 'York' (3 percent). These six cultivars combined accounted for 83 percent of the total United States production. The state of Washington produces 50 to 60 percent of the total United States production of 'Delicious' and 'Golden Delicious', and these two cultivars account for about 90 percent of that state's total crop, essentially all of which is grown for the

TABLE 16-1
WORLD FRUIT, NUT, AND BEVERAGE CROP PRODUCTION, ANNUAL AVERAGE FOR 1973–1975, IN THOUSANDS OF METRIC TONS

	U.S.A.	CANADA	MEXICO	SOUTH AMERICA	ASIA	EUROPE	AFRICA	OCEANIA	U.S.S.R.	WORLD
Pome:										
Apple	3,074	409	206	798	3,688	13,400	323	506	—	22,408
Pear	675	36	40	176	1,965	4,040	147	169	—	7,248
Stone:										
Peach and nectarine	1,351	52	222	508	627	2,746	201	125	—	5,831
Plums	624	9	84	113	358	2,748	33	26	—	3,995
Apricot	127	3	10	24	369	642	121	30	—	1,326
Cherry	240	17	18	6	141	1,148	—	10	—	1,581
Small:										
Grape—total	3,847	70	188	4,800	5,892	38,507	2,643	670	4,730	61,349
Wine	1,544	54	15	3,232	171	23,697	1,457	340	2,820	33,330
Raisin	212	—	1	4	413	184	9	59	—	992
Strawberry	235	14	94	7	198	693	—	7	—	1,249
Raspberry	11	6	—	—	—	112	—	3	—	132
Currants	—	—	—	—	—	345	—	1	—	346
Gooseberry	—	—	—	—	—	147	—	3	—	150
Cranberry	94	5	—	—	—	—	—	—	—	99
Blueberry	2	12	—	—	—	10	—	—	—	25
Nuts:										
Almond	146	—	—	—	68	395	42	1	18	670
Cashew	—	—	—	39	237	—	371	—	—	647
Chestnut	—	—	—	8	110	231	—	—	18	367
Filbert	9	—	—	—	299	119	—	—	13	441
Walnut	160	—	6	8	138	285	4	—	155	757

	U.S.A.	CANADA	MEXICO	SOUTH AMERICA	ASIA	EUROPE	AFRICA	OCEANIA	U.S.S.R.	WORLD
Citrus:										
Orange	8,881	—	1,755	7,548	5,089	4,230	2,999	358	109	31,518
Tangerine	571	—	—	650	4,089	1,003	438	33	—	6,806
Lemon and Lime	837	—	467	560	1,060	1,307	222	41	—	4,577
Grapefruit	2,378	—	36	249	746	15	224	23	—	3,784
Other:										
Banana	3	—	1,070	13,213	10,213	455	4,749	1,035	—	36,400
Mango	—	—	342	894	9,841	—	532	3	—	12,154
Olive	59	—	—	126	865	6,254	1,175	3	—	8,494
Pineapple	745	—	250	846	1,861	2	703	124	—	4,763
Date	21	—	3	1	1,247	17	905	—	—	2,194
Figs	37	—	12	35	254	579	160	—	—	1,077
Avocado	86	—	232	337	34	—	28	1	—	1,005
Beverage crops:										
Coffee (green)	1	—	217	1,992	353	—	1,233	39	—	4,447
Tea	—	—	—	38	1,285	—	153	4	80	1,560
Cacao	—	—	36	355	21	—	988	33	—	1,489

Source: 1975 FAO Production Yearbook, vol. 29, Food and Agriculture Organization of the United Nations, Rome, Italy.

FIGURE 16-2 Apples ready for harvest. Each apple is produced on a short "spur."

fresh market. Other important apple states not only have a broader base of cultivars but also have sizable portions of the crop produced for processing outlets. The processing cultivars vary with the region. For example, the following are major processing cultivars: California, 'Gravenstein'; New York, 'Rhode Island Greening'; and in the mid-Atlantic states, 'York' and 'Golden Delicious'. For the United States as a whole, about 50 percent of the apple crop is processed. In Washington, this percentage is much lower, whereas in the mid-Atlantic area, more than 60 percent are processed.

During recent years there has been a strong trend toward intensification of apple production systems. The orchards of large, individual trees set 12 m apart and requiring 6-m ladders for harvest are being replaced. The "new" orchards being planted vary widely in spacing, design, and concept. Traditionally, in the United States, the desired scion cultivars have been propagated on vigorous seedling rootstocks.

It is apparent, however, that we must replace this type of tree with one which is much smaller. The inherent advantages of smaller trees include easier pruning, spraying, and harvesting; greater production per unit area of orchard floor; and better fruit size and color. The greatest asset of the smaller tree, however, is that a new orchard can be in production 2 to 3 years earlier than an orchard of standard trees. Several techniques are being tested to control tree size. The most widely known is the use of size-controlling rootstocks. In the early 1900s English pomologists began to collect, catalog, test, and name available size-controlling apple rootstocks from all over the world. Once the various types were named and classified, they began to make crosses and to improve on the existing types. The research was done at the East Malling Research Station and as the program developed, rootstocks were named and released for testing. The Malling series now includes 27 releases known as Malling 1 through Malling 27. These range in size from that of a vigorous seedling down to some which are only one-fourth as large as seedlings. These rootstocks must be propagated vegetatively, and the usual technique is to use a stoolbed (see Chapter 12). A major drawback of the Malling series has been susceptibility to wooly aphids which attack the roots and cause serious stunting. In order to eliminate this problem, another group of English researchers started a breeding program at the Merton Research Station in cooperation with the researchers at the East Malling Station. Crosses were made between the Malling series and the cultivar 'Northern Spy', which was known to have a high degree of resistance to wooly aphids. The result of this program has been the release of the Malling-Merton series of rootstocks which is numbered starting with MM 101. The size of trees in the MM series is relatively large with none being less than about two-thirds of seedling size, but all have at least some degree of resistance to wooly aphids.

In the Malling series, the most widely tested in the United States have been M 7, M 9, and M 26, which produce trees of approximately 60, 30, and 45 percent the size of seedlings, respectively. Of the MM series, the most widely planted are MM 104, MM 106, and MM 111, which are all in the range of 75 to 90 percent the size of seedlings.

Other than control of tree size, several characteristics of rootstocks are of major importance. These include the stability of the tree (M 9 requires support), suckering (M 7 tends to sucker badly), winter hardiness (MM 106 tends to harden off slowly), precociousness (M 9 induces very early fruiting), and a tolerance to "wet feet" (MM 106 tends to suffer more than others).

Other means of controlling tree size can be of considerable help in modern orchards. Different scion cultivars vary in vigor and tree size as exemplified by the very large 'McIntosh' tree and the much smaller 'Rome Beauty'. Within many cultivars there are *spur-type* strains which are inherently compact in growth and grow to only 70 to 80 percent of the size of the standard strains. A wide range in tree size can be obtained by the various combinations of standard and spur-type scion cultivars on the many available rootstocks.

There is widespread interest in the possibility of growth control through chemicals. Although many chemicals have been tested, none has been found which will give effective control of tree size over extended periods of time. Several growth regulators have been shown to induce flower bud formation in young trees. Because of the competition between fruit production and vegetative growth, annual cropping can be helpful in holding vegetative growth in check.

Growth regulators are widely used on apple trees for several other purposes; these include chemical thinning, suppressing vegetative growth, increasing fruit firmness, increasing red color, accelerating fruit maturity, and delaying preharvest drop.

Pears World pear production is about one-third of that for apples and is also concentrated in Europe, especially Italy and France (Table 16-1). Pear production in the United States is much more centralized than apple production. California, Oregon, and Washington produce about 95 percent of the total United States pear crop; the remainder come from New York, Michigan, and a few other states. In the West Coast states, 'Bartlett' accounts for about 70 percent of the pears produced; of those, about three-fourths are processed. The rest of the pears produced in that area are for the fresh market and consist of 'D' Anjou', 'Bosc', and 'Comice'. In the other pear states, 'Bartlett' is also the most widely grown with limited quantities of other fresh market pears grown mostly for local sales.

Processing accounted for 59 percent of the total pear crop in 1976. The major products were pear halves, fruit cocktail, and canned mixed fruit. Limited quantities of pears are also dried. The major reason for the very limited pear production east of the Rocky Mountains is the devastation of pear trees by fire blight, a disease caused by the bacterium *Erwinia amylovora*. A severe infection can kill a pear orchard in a matter of a few days. The only chemical control available has been the use of multiple sprays of the antibiotic streptomycin during the blossoming period. In addition to being expensive, this treatment has not been consistently effective. Although cultivars resistant to fire blight have been introduced in the past 20 years, none has received widespread acceptance. Although devastating in humid climates, even in the semiarid pear regions of California, Oregon, and Washington, occasional fire blight outbreaks occur. The inoculum is spread by rain, insects, and pruning tools. The most dangerous period for infection is during bloom because the open flowers are very susceptible as are the tender succulent shoots. Cultural practices are aimed at keeping the vigor of the tree at moderate levels to avoid the excessively tender growth associated with very high vigor. The lack of rain is the major asset of the West Coast states; the pear regions of New York and Michigan are close to the Great Lakes which tend to have low spring temperatures and thereby reduce the hazard of infection.

Throughout the United States most of the early homesteads had a pear tree or two for home use. Most often these were of the 'Kieffer' cultivar which is a hybrid between the European or common pear, *Pyrus communis* and *Pyrus serotina*, the Japanese pear. Although resistant enough to fire blight to survive under adverse conditions, 'Kieffer' has never had commercial importance because of poor quality. Its low quality is caused primarily by a large number of stone cells in the flesh of the fruit.

Stone Fruits

Peaches   The peach *(Prunus persica)* is the most widely grown stone or drupe fruit in the United States and the world. The leading countries in peach production are the United States, Italy, and France. Peaches are characterized as freestone or clingstone on the basis of whether or not the flesh separates readily from the pit when the fruit is ripe. The texture of the flesh of freestones is somewhat "melting," whereas the flesh of cling peaches is very firm. Some early maturing fresh-market cultivars are classified as semicling peaches, but the flesh is that of the melting type. Canned clings hold their firmness and shape well, whereas freestones tend to become soft and lose their shape when canned.

All peaches produced for the fresh market are freestone except for some of the very early cultivars. Freestone peach production in the Eastern United States ranges from Florida to Michigan and New York, although the major states are South Carolina, Georgia, New Jersey, Pennsylvania, and Michigan. In the Western part of the United States commercial peaches are grown in several states, but none of these states is of major importance except for California which produces most of the United States clingstone crop and from 40 to 50 percent of the freestone crop—65 to 70 percent of the total United States peach crop. Of California's total production, about 75 percent are nonmelting clings.

The southern limit for peach production is set by the number of hours of chilling received during a typical winter. Until rather recently the southern extremity was Georgia on the East Coast, because the chilling requirement of commercial cultivars ranged from 650 to over 1000 h (see Chapter 7). Recently introduced cultivars with chilling requirements of less than 250 h have encouraged the extension of peach plantings into Florida. Some are as far south as central Florida now.

Peach cultivars are more numerous than apples. Some, such as 'Redhaven', are very broad in their adaptation, but others are regional. Although there are both white- and yellow-fleshed freestone cultivars, white-fleshed ones are grown only for local sales as they are not firm enough to ship well and are not as well accepted as they once were. Cultivar turnover is quite rapid with peaches compared to the other tree fruits because many new cultivars become available, peach orchard life is shorter, and it takes fewer years to bring a new peach orchard into full production.

Peach trees have historically been grown on peach seedling rootstocks. At one time, seeds from wild peach trees in Tennessee and the Carolinas were used by Eastern nurseries, but more recently, pits have been collected from commercial cultivars, such as 'Elberta'. California nursery growers have traditionally used pits collected from cannery waste. Certain problems which have recently become increasingly severe in the East appear to be related to rootstock. Among the difficulties encountered are: winter injury, nematodes, diseases, and perhaps the two most devastating—"peach tree short life" and "stem pitting." Rootstocks with improved winter hardiness include 'Siberian C' (about 15 percent dwarfing) and 'Harrow Blood', an introduction from the Harrow Research Station in Ontario. Peach rootstocks which are nematode-resistant include 'Shalil', 'Yunnan', and the most widely accepted, 'Nemagard'.

Although "peach tree short life" has been a problem for many decades in the southeast, it has gotten progressively worse. The average life of a peach orchard has declined to eight years in some areas of the southeast. The most widely accepted prevention procedures currently recommended include: (1) avoid replanting old peach sites or at least avoid replanting in the same spot, (2) fumigate the soil, (3) use trees from fumigated nurseries, (4) delay pruning until late winter, and (5) pay close attention to soil management and fertilization. "Stem pitting" has become a major problem for peach growers in the mid-Atlantic area in the last 10 years. The overall tree symptoms are similar to those caused by various other problems and involve stunting of the tree, curling of leaves, insipid flavor of fruits, and eventual death of trees in severe cases. The diagnosis can be readily made by examining the trunk for pits and grooves in the wood, ridges in the bark, and necrotic areas in the cambium. The casual organism has recently been shown to be a strain of the tomato ringspot virus and is presumably spread by nematodes and propagation with infected buds; control is through careful nursery practice, not replanting infected sites, and prompt roguing of infected trees from orchards.

Peach harvesting has been by hand until recent years when

FIGURE 16-3 Although mechanical harvest of cling peaches is becoming more widespread each year, most of the crop is still harvested by hand. (Photograph, California Canning Peach Association, San Francisco, Calif.)

mechanical harvesting became commercially feasible for processing peaches. A significant percentage of the cling peaches for processing in California is harvested by the shake-and-catch method but most are still hand-harvested (Figure 16-3). As with so many crops today, the grower must balance the cost, availability, and dependability of hand labor as opposed to the capital investment of mechanical harvesters.

Fresh-market peaches are still harvested completely by hand, but experimental harvesters have also been developed for fresh fruit. Future developments with this equipment as well as the labor situation will determine its practicality in peach operations. With fresh-market peaches, not only must the producer be concerned with the balance between labor and machine expenses, but the quality of the fruit must also be given adequate consideration.

Nectarines Nectarines are merely peaches without fuzz. They have increased in popularity in recent years. The close relationship between peaches and nectarines is indicated by the fact that peach seeds may produce nectarine trees, nectarine buds may mutate to produce a branch which grows peaches, and vice versa. California produces essentially the entire nectarine crop with only spotty production in other parts of the United States. With the availability of new and improved cultivars as well as better chemicals for controlling insects and disease, nectarine production in the East is making moderate advances. The major disease problem is brown rot—to which nectarines are much more susceptible than are peaches. Essentially all nectarines are sold on the fresh market.

Cherries World cherry production is about one-fourth of that for peaches, and the leading countries are Italy, the United States, Germany, and France. Total United States cherry crops have averaged 230,000 metric tons over the past 10 years, with the average crop being about equally split between sweet and tart cultivars. Sweet cherries are predominantly grown in the three Pacific Coast states and Michigan, with these four states accounting for about 90 percent of the sweet cherries grown (Figure 16-4). Distribution of the tart (sour) cherry

FIGURE 16-4 Sweet cherries are borne on short spurs formed the previous year.

FIGURE 16-5 Cherry harvester is positioned beneath the tree. It has two hydraulic shakers and a catching frame. (Photograph, U.S. Department of Agriculture.)

industry is quite different as most are grown around the Great Lakes, especially in Michigan with much smaller plantings in New York, Pennsylvania, and Wisconsin. About 98 percent of the tart cherries go into processing channels with the most popular products being frozen, canned, and brined cherries as well as juice and wine. The main tart cherry cultivar is 'Montmorency', with some 'English Morello' and 'Early Richmond' grown. The tart cherry is *Prunus cerasus*.

Sweet cherries are split about equally between fresh-market and processing outlets. The majority of the sweet cherries processed are brined for maraschinos, some are canned, and limited quantities are frozen. Many cultivars of sweet cherries are available, but the most widely grown are 'Bing', 'Lambert', 'Napoleon' ('Royal Anne'), 'Black Tartarian', and 'Windsor'. The sweet cherry is *Prunus avium*.

The availability of mechanical harvesters has made it practical to continue to grow cherries for processing outlets. Using hand labor to harvest cherries was not economical since from one-half to two-thirds of the total grower cost was in the harvest operation. Essentially all processing cherries are now harvested by the shake-and-catch method (Figure 16-5) and it is well accepted that this technique has saved the industry from financial disaster.

Plums Among stone fruits, plums rank second to peaches in world crops. Major producing countries are Yugoslavia, Romania, Germany, and the United States. About 90 percent of the United States plums are grown in California with small but significant quantities coming from Washington, Oregon, and Michigan. In California about 80 percent of the plums are dried to be sold as prunes; most of the others are sold fresh although small quantities are canned. For the other states about 50 percent are sold fresh, 30 percent are canned, 15 percent are dried, and a few are frozen.

Plum cultivars are very numerous and are selected on the basis of the geographical region and, in particular, the market to be served. Fresh-market cultivars in California include 'Tragedy', 'Santa Rosa', 'Wickson', 'Kelsey', and 'President'. The major cultivar for drying is the 'French Prune.'

Most commercially important plums belong to either *Prunus domestica* (European plums) or *Prunus salicina* (Japanese plums). The prunes belong to *P. domestica* and are often considered as a distinct group because the fruit can be dried without removing the pit. Many of the fresh California plum cultivars belong to *P. salicina*, although some are also European-type plums. Several other species of plums are used for home plantings or as rootstocks.

Apricots Apricot production is slightly less than that for cherries, and is centered in Spain, Turkey, Italy, and the United States. Apricots have been grown in the Unites States since the early 1700s, but commercial production is limited to the West Coast states where California accounts for 95 to 98 percent of the total crop, with the remainder coming from Washington and Utah. About 10 percent of the apricots are sold fresh, 65 percent are canned, 20 percent are dried, and about 5 percent are frozen. Important California cultivars include 'Blenheim', 'Royal', and 'Tilton'. Apricots are grown on either peach or apricot rootstocks. In recent years attempts have been made to breed apricot cultivars adapted to Eastern conditions. These cultivars may find a place in home gardens and for roadside and local sale, but commercial apricot production will no doubt remain on the West Coast.

Small Fruits

The small fruits include a diverse group of fruits which are relatively small in size and are produced on bushes, vines, or other low-growing forms. Many are true berries, but others are not—even though they are called berries (See Chapter 2).

Grapes The grape is classified horticulturally either as a small fruit or sometimes in a category by itself, but it is a berry from a botanical

standpoint. Grapes rank first among the temperate fruits in tonnage produced in the United States although apples are a close second (Table 16-1). In world production, however, grape production exceeds that for apples by a factor of more than 2:1. France and the United States are the world's leading countries in the production of grapes.

Grapes grown in the United States consist of four distinct types: the European grape, *Vitis vinifera*, the American bunch grape, *V. labrusca*, the French hybrids (*V. vinifera* x wild American species), and the Muscadine grape, *V. rotundifolia*. The *vinifera* grapes have largely been confined to California because most of this species is best adapted to climates with warm, dry summers and cool, wet winters. The *vinifera* grape has firm pulp to which the skin adheres, and the flesh is sweet throughout. Three cultivars most frequently seen in supermarkets are the 'Tokay' (red with seeds), 'Thompson Seedless' (green-yellow, seedless), and 'Emperor' (deep rose to purplish with seeds). The 'Thompson Seedless' (Sultanina, Figure 16-6) is also the grape used in very large quantities for raisins; about 25 percent of the

FIGURE 16-6 'Thompson Seedless' grapes are harvested and are then placed on paper "trays" between the rows where they dry to form raisins. (Photograph, California Raisin Advisory Board, Fresno, Calif.)

total grape crop or about 1 million tons are dried annually. Many of the *vinifera* cultivars are grown exclusively for wine; about one-half of the total grape crop in the United States is crushed for wine. Notable California wine cultivars are 'Zinfandel' and 'Carignane'. With the marked increase in wine consumption in the 1960s and early 1970s, increased interest in wine grapes is apparent in many parts of the country. Some plantings of *vinifera* cultivars have been made in the Eastern United States, but success has been very limited owing largely to diseases, winter injury, or poor adaptation to the climate.

The majority of the grapes produced in the Eastern United States are *V. labrusca*, or the American bunch grape. The skin is rather loosely attached to the pulp, which tends to be somewhat soft and relatively acid around the seeds. Major cultivars of American grapes in the United States include 'Niagara' (white), 'Delaware' (red), 'Concord' (blue), and 'Catawba' (red). In the Great Lakes area, more than 85 percent of the grapes produced are 'Concord' because of its wide adaptability, good productivity, and broad acceptance for juice, preserves, frozen concentrate, wine, and local fresh-market sales (Figure 16-7).

Since the regions around the Great Lakes offer the best climate in the Eastern United States for American grapes, major production areas lie adjacent to the lakes. A strip of land from 1.5 to 8 km wide along the south shore of Lake Erie contains most of the grape acreage of Ohio and Pennsylvania and much of that for New York which also has grape areas around the Finger Lakes. In the spring the cold lakes suppress temperatures, delay plant development, and thereby reduce the likelihood of freeze damage. In the fall the lakes cool slowly and thus delay the occurrence of freezing temperatures. Most cultivars of *V. labrusca* require a freeze-free growing season of 160 to 170 days for proper vine and fruit maturity. Other areas of *labrusca* production are Washington and Michigan.

The so-called French hybrid grapes include many cultivars developed by breeders in France from crosses of *V. vinifera* with the wild American species such as *V. rupestris* and *V. lincecumii*. The objective of developing these hybrids was to incorporate some of the inherent disease resistance and hardiness of the wild American species into the *vinifera*-type grape. The increased demand for wine has resulted in a renewed interest in the French hybrids in the Eastern part of the United States. Some of the cultivars being recommended for trial include 'Aurora', 'Baco Noir', 'Chelois', and several numbered Seibel selections. The degree to which the current interest in French hybrid production continues will likely vary with the region, the cultivars planted, the problems encountered, and the market situation. Since these types are limited mostly to wine production, many growers may prefer flexibility of dual- or triple-purpose cultivars.

FIGURE 16-7 Grape flowers are at the first few nodes of the current year's shoot growth.

Most of the muscadine grapes belong to *V. rotundifolia*, but there are two other closely related species. Muscadines grow wild in areas of the Southeastern United States and are adapted to areas where the temperatures remain above −12 to −15°C. Limited quantities of muscadine grapes are sold fresh, but most are utilized for wine, preserves, and blending with other fruit juices. Older muscadine cultivars are generally harvested as individual fruits rather than in bunches, but newer cultivars remain in a loose bunch. The 'Scuppernong' cultivar was selected from the wild prior to 1760; and 'Thomas' was selected in 1850. Many cultivars from breeding programs have been introduced in the last 30 to 40 years. Although muscadine grapes are of minor importance nationally, they are widely used in home gardens in the Southeast.

Mechanical harvesting of grapes for juice, wine, and other processed products is becoming standard procedure. In one system, an

FIGURE 16-8 An over-the-row grape harvester operating at Niagara Falls, Ontario. The grapes are knocked from the vines, collected below, and conveyed to a portable receptacle. (Photograph, Chisholm-Ryder Company, Inc., Niagara Falls, N.Y.)

over-the-row harvester knocks the grapes off the vine, catches them below the vine, and conveys them to a truck or trailer (Figure 16-8). A desired characteristic for mechanical harvesting is for the individual grapes to separate readily from the bunch and to have scars which do not burst easily.

Strawberries Among the small fruits in the United States and the world, strawberries rank second to grapes in total production. The commercial strawberry originated as a cross between *Fragaria chiloensis* and *F. virginiana* (Figure 16-9). In the Eastern United States commercial strawberry plantings range from Maine to Florida and as far west as Michigan and Louisiana. Heavy concentrations of strawberries are found in California, Oregon, and Washington. Total strawberry production in the United States in 1975 was 246,000 metric tons, and the percentages produced by the major states were as follows: California, 70; Oregon, 8; Washington, 4; Florida, 4; and Michigan, 3. Thus, the five top states accounted for 89 percent of the crop, while about 20 other states had commercial acreage. Winter strawberries come from Florida from December to mid-April. The spring crop comes from both the east and west, and the harvest moves northward with the Northern states harvesting from June through mid-July.

FIGURE 16-9 Ripe strawberries. Note mulch of pine needles to keep fruit clean.

The trends in strawberry acreage and production since the 1930s are truly amazing. In 1930, strawberry acreage was close to 81,000 ha, with an average yield of about 2800 kg ha^{-1}; in 1975 the acreage was about 16,000 ha, with an average yield of 15,650 kg ha^{-1}. Acreage was reduced by 5.1 times, but yield was increased by 5.6 times, thus giving a slight increase in total production. Much of this increase in productivity is based on the tremendous yields obtained from California, which averages about 9 metric tons ha^{-1}. Florida averages 3.4 tons and the other states produce about 1.2 metric tons ha^{-1}. The success story with strawberries in California is the result of many improvements including: more productive cultivars, utilization of temperature-photoperiod interactions, better control of pests through insect and disease resistance (especially virus diseases), and improved cultural techniques, such as soil fumigation and the use of clear plastic mulches.

Although strawberry production east of the Rocky Mountains has declined drastically, there is a resurgence of interest in strawberries as a pick-your-own crop, and this trend will apparently continue.

Blueberries There is a diversity of blueberry species, and groups which are classified as lowbush, highbush, and rabbiteye types. All blueberries are in the genus *Vaccinium* and have small, soft seeds as distinguished from true huckleberries (*Gaylusacia*) which have 10 large, hard seeds. Blueberry production is limited to Europe and North America.

Lowbush blueberries Several species of *Vaccinium* are included here, but the most important are *V. angustifolium* and *V. lamarckii.* The lowbush blueberry is harvested from the wild. Plants are not propagated and set out, but rather its ecological niche between the field and forest is artificially maintained, primarily by burning. By burning the fields in the spring when the ground is either frozen or wet, much of the competing vegetation is killed and the blueberry, which spreads by rhizomes, is "pruned" but not killed. A typical program is to burn one-third of the acreage each year so that two-thirds of the acreage is harvested annually.

The lowbush industry in the United States is confined largely to the northeastern states, with the major areas being northeastern Maine, northern New Hampshire, and Michigan. Limited acreages exist in Massachusetts, Wisconsin, Minnesota, and as far south as West Virginia. Eastern Canada also provides sizable quantities of lowbush blueberries, many of which are shipped into the United States.

Most lowbush blueberries are harvested with "rakes" and are sold to canning and freezing plants. Limited quantities are sold fresh. Because the terrain on which they thrive is often very rough, a mechanized harvest is more difficult than with the highbush type, but machines which simulate the hand-raking process are widely used.

Highbush blueberries Like lowbush blueberries, highbush blueberries consist of several species; the two major ones are *V. corymbosum* and *V. australe*. Commercial production areas are limited because of the rather specific requirements for temperature, soil, and water. Temperatures needed are not drastically different from those for a peach cultivar with a chilling requirement of about 1000 h below 7°C. The top of the plants may be killed if winter temperatures go below −30°C, but a snow cover can provide protection. The southern limit is set by the need for adequate chilling in winter, and they are thereby not widely grown south of northern Georgia. Soil requirements include a soil pH of 4.3 to 4.8 which in many areas requires the addition of sulfur to acidify the soil. The blueberry is shallow-rooted, needs ample water, but cannot withstand submersion during the growing season.

Major production areas include southeastern North Carolina, New Jersey, and southwestern Michigan, with lesser acreages in Washington, Oregon, Massachusetts, New York, and Indiana (Figure 16-10). In Canada, commercial high-bush blueberries are grown in both British Columbia and Nova Scotia.

As the result of breeding programs, many cultivars are available,

FIGURE 16-10 Clusters of blueberry flowers. Note that some flowers are past full bloom, some are in full bloom, and some are yet to open. Blueberries ripen over an extended period of time.

many of which bear fruits 3 or 4 times as large as wild berries. Leading cultivars in order of ripening include: early, 'Earliblue'; midseason, 'Bluecrop' and 'Berkeley'; and late, 'Jersey', 'Herbert', and 'Coville'.

Blueberries ripen over a period of several weeks, and must be harvested repeatedly. Although traditionally a hand-harvested crop, various mechanical innovations are now being used commercially. Hand-held vibrators are used to shake the ripe berries from the bush and onto a canvas catching frame which is moved from bush to bush. This system has increased output by 5 to 8 times over hand picking. Self-propelled harvesters straddle the row, shake the berries, and collect them beneath the bush.

Rabbiteye blueberries The rabbiteye blueberry *(V. ashei)* is important because it grows successfully in areas where the highbush blueberry will not. Since the rabbiteye is characterized by a short chilling requirement, it is suited as far south as northern Florida; its northern limit is central Alabama, Mississippi, and coastal North Carolina. Major advantages of the rabbiteye are excellent vigor, adaptation to wide variation in environment, and tolerance to heat and drought.

Cranberries A plant native to North America, the cranberry, *Vaccinium macrocarpon,* is adapted to acid, swampy, or marshy areas, and the term *cranberry bogs* has often been used. Commercial cranberry production is centered in the Northeastern United States, with over one-half of the total acreage in Massachusetts. Other commercially important states include Wisconsin, New Jersey, Washington, Oregon, and some areas of Canada.

Cranberry bogs or marshes are very expensive to develop because of the required land preparations, which include clearing, leveling, ditching, applying a layer of sand, and providing for both irrigation and drainage. Once established, bogs last for 60 years or more and actually considered to be "permanent" if properly cared for.

Although once picked or "raked" by hand, most cranberries are harvested by machine today (Figure 16-11). Some machines utilize teeth to pull the berries from the vines; others operate in flooded bogs by knocking the berries free from the vine, floating them to a corner of the bog, and picking them up by machine.

Sizable quantities of cranberries are sold fresh as they are widely used during Thanksgiving and Christmas holidays. Major processing outlets are for cranberry sauce and cranberry juice. The most important cultivars are 'Early Black,' 'Howes,' 'Searles Jumbo,' and 'McFarlin,' although at one time sizable quantities of wild or native berries were harvested.

Brambles The bramble fruits are in the genus *Rubus* and include the red, purple, and black raspberries, the blackberry, and other berries of lesser importance. Brambles bear fruit on biennial canes which grow one year, produce fruit the second, and then die back to the ground. The fruit of the raspberry and blackberry is an aggregate fruit consisting of many drupelets. The core remains in the blackberry and is edible, whereas in the raspberry the core remains on the bush.

Compared with most other fruits, bramble fruit production is very limited, but it is important in certain areas of the United States, Canada, and several European countries. Much of the production in the United States is for processing into frozen fruit, jams, jellies, and preserves. Since raspberries and blackberries are a high-priced item for local sales, they are being planted increasingly for pick-your-own operations.

FIGURE 16-11 Dry-harvesting cranberries in Massachusetts with Darlington pick-ing machines. Filled boxes are in the foreground. (Photograph, Ocean Spray Cranberries, Inc., Hanson, Mass.)

Major areas of production for raspberries include Michigan, Oregon, Washington, and New York. Blackberries are adapted to conditions farther south and are found as far south as Texas. A major improvement in blackberries has been the release of thornless culti-vars such as 'Black Satin,' 'Smoothstem,' and 'Thornfree.'

For several reasons—including viruses, poor yields, and especial-ly high labor requirements—the brambles were declining in acreage. But because of better cultivars, virus-free plants, mechanization, and pick-your-own operations, it appears that brambles may regain some earlier prominence. Like strawberries, blueberries, and grapes, the bramble fruits are well adapted to the backyard garden.

SUBTROPICAL AND TROPICAL FRUITS

Although subtropical and tropical fruits can be grown commercially only in very limited parts of the United States, they account for a large part of the total United States fruit crop. Major production areas are confined to Florida, California, Hawaii, and southern Texas.

FIGURE 16-12 Florida citrus groves. (Photograph, Florida Department of Citrus, Lakeland, Fla.)

Citrus

Of all the tropical and subtropical fruits, the subtropical citrus fruits are by far the most important in the United States (Figure 16-12). On a worldwide basis, citrus production slightly exceeds that of bananas (Table 16-1). Although there is a broad variety of citrus fruits, the sweet orange, *Citrus sinensis*, accounts for about 9 million metric tons, or 70 percent of the total United States citrus crop. Oranges are categorized as common or navel, seedless or seeded, and according to the season of ripening. In Florida all major cultivars are common and include 'Valencia' (late, seedless) (Figure 16-13), 'Pineapple' (midseason, seedy), 'Hamlin' (early, seedless), and 'Parson Brown' (early, seedy). Naval orange production in Florida accounts for less than 5 percent of the crop primarily because of inconsistent productivity. In California, slightly more than 50 percent of the orange crop are navels, mostly the 'Washington' cultivar, with the remainder being 'Valencia.' Utilization of the orange crop is quite different in Florida and California. In Florida about 90 percent of the orange crop is processed, but in California the proportion is quite small.

The second major citrus fruit in the United States is the grapefruit (*C. paradisi*), of which about 70 percent of the crop comes from Florida, 15 percent from Texas, 10 percent from California, and 4 percent from

FIGURE 16-13 'Valencia' orange. (Photograph, Florida Department of Citrus, Lakeland, Fla.)

Arizona. Total production in the United States is about 2.4 million metric tons which is about 65 percent of the total world crop. Grapefruit are classified as: seeded or seedless and white-, pink-, or red-fleshed. Major cultivars are 'Marsh' (white, seedless), 'Ruby' (red, seedless), 'Duncan' (white, seeded), and 'Foster' (pink, seeded). Fresh-fruit markets utilize 35 to 40 percent of the grapefruit crop, about 30 percent is used for canned juice, 5 percent is used for canned sections, and the remaining 25 to 30 percent is used for other products such as frozen concentrate. For the fresh market, grapefruit should have symmetrical shape, smooth skin, and uniform yellow skin color, and should be seedless and sweet.

The lemon, *C. limon*, is grown mostly in California and Arizona, with limited quantities in Florida. Average United States crops are somewhat less than 1 million metric tons. Consumption is about one-half as fresh fruit, and the remainder as juice concentrate. Tangerine, *C. tangerina*, is a specialty type of citrus which has a loose skin, peels easily, and has attractive orange-red flesh (Figure 16-14). A synonym is *mandarin*. Because tangerines are less productive than

FIGURE 16-14 'Dancy' tangerine. (Photograph, Florida Department of Citrus, Lakeland, Fla.)

oranges and do not hold quality as well after reaching maturity, they have never been widely grown in the United States. Tangerines are very popular in Japan and the rest of Asia.

Other citrus of lesser importance are limes, *C. aurantifolia*, tangelos (tangerine x grapefruit cross), and temples (mandarin x orange). There are many other hybrids within the *Citrus* genus, and many intergeneric hybrids have been tested.

Avocado

The avocado *(Persea americana)* is an increasingly popular fruit in the United States, although it is not widely grown outside the Western Hemisphere. The fruit is a fleshy berry, contains one seed, and varies in shape from round to pyriform (Figure 16-15). The flesh is green to yellow, buttery in texture, and rich in oils (10 to 30 percent). The skin

FIGURE 16-15 Lugs of freshly harvested avocados. (Photograph, California Avocado Advisory Board, Newport Beach, Calif.)

varies from quite smooth and thin to somewhat thick and woody; skin color varies from green to a purplish green. Commercial production in the United States is limited to southern Florida and southern California, which together produce annual crops of somewhat less than 100,000 metric tons.

Olive

The olive *(Olea europaea)* has been a major crop for centuries around the Mediterranean Sea, where it has been used for oil and in recent years has been canned whole for table use. California produces the entire United States crop of about 60,000 metric tons, of which 80 percent are canned, 5 percent are crushed for oil, and the remainder are utilized in miscellaneous ways. The olive tree is subtropical and cannot withstand temperatures of $-10°C$, but optimum fruiting is enhanced by winter temperatures of 6 to 8°C. The fruit is a drupe which is harvested at an immature stage for "green olives" but is allowed to mature on the tree for "ripe olives" (which are black), or for oil production.

Date

The date palm *(Phoenix dactylifera)* is a crop which thrives best in arid climates and is thought to be native to northern Africa. Both high temperatures and low humidity are requirements for normal fruit ripening. The date palm is an evergreen monocot and is dioecious. Hand pollination is often carried out to ensure adequate fruit set. Most dates are dried. California production averages about 20,000 metric tons, or about 1 percent of the world crop.

Fig

The fig *(Ficus carica)* is an ancient crop which grows as a small deciduous shrub or tree (Figure 16-16). In California, figs are usually trained as trees whereas in the Southeast, multiple stems are encouraged because severe winters can kill the fig plant. Commercial fig production in the United States averages about 35,000 metric tons per year, or 3 percent of the world crop. The fruiting habit of the fig is

FIGURE 16-16 Mature Calimyrna figs. If figs are not picked at this stage of ripeness for the fresh-fig market, they are allowed to fully ripen and dry on the tree. (Photograph, Dried Fig Advisory Board, Fresno, Calif.)

unique. The fruit, called a *syconium*, is a round or pear-shaped receptacle lined internally with tiny inflorescences. Major cultivars of figs are 'Mission,' 'Kadota,' and 'Calimyrna.' Some figs are sold fresh and some are canned, but the majority are dried.

Pineapple

The pineapple *(Ananas comosus)* is a tropical plant and a member of the Bromeliaceae family (Figure 16-17). It has many features which make it resistant to desiccation such as sunken stomates, narrow leaves, and funnel-shaped leaf bases which hold water. World pineapple production is in tropical and subtropical areas such as South America, Asia, Africa, and Central America. The United States crop of

FIGURE 16-17 Ripe pineapple ready for harvest. Note the "slip" arising below the fruit. For the fresh market the crown would be left on the fruit; for the cannery the crown is left in the field.

750,000 metric tons is mostly from Hawaii where the pineapple crop is second in commercial importance to sugarcane.

Since the pineapple is seedless, it is propagated vegetatively by *slips* which arise beneath the fruit, *suckers* (shoots originating lower down the stem) or most commonly the *crown* (the shoot arising from the top of the fruit). During the harvest of pineapples for processing plants, workers wearing heavy chaps and gloves walk between the rows, harvest the fruit, and break the crown off the pineapple. The fruit is placed on a long conveyor which carries it to a truck or trailer, and the crown is tossed on top of the plants to dry. One to two weeks later the crowns are collected from the field and are planted, often in fumigated soil using black plastic mulch. From newly set crowns, it normally takes 18 to 24 months for the plants to flower, a process sometimes accelerated and commonly synchronized by the use of selected growth regulators such as ethephon. The first crop consists of a single, large fruit on each plant; the second crop (called the *first ratoon crop*) consists of two smaller fruits per plant and is mature about 12 months after the first crop. The third crop, consisting of four smaller fruits (called the *second ratoon crop*), is harvested about 12 months after the second crop. After this crop is harvested, the field is prepared for replanting and the three-crop cycle repeated.

Banana

The banana (*Musa* spp) is a tropical fruit. Although it is productive between temperature extremes of 10 and 40°C, it is best suited to a narrower range of 15 and 35°C. The cooking banana or plantain *(M. paradisiaca)* is widely utilized in the tropics, but the dessert cultivars in world trade are *M. sapientum*. The only state in the United States which produces bananas commercially is Hawaii; annual crops are about 3000 metric tons. Most of the bananas sold in American markets are imported from Central and South America. Between 1972 and 1975 annual banana imports were about 2 million metric tons per year. Leading countries in exporting bananas to the United States are Costa Rica, Ecuador, Honduras, and Guatemala.

The banana is a fast-growing herbaceous plant which develops suckers (pseudostems) from underground rhizomes and reaches a height of 1.5 to 8 m. The leaves are large—up to 3.5 m long—and dark-green. The false stem is composed of compressed leaf sheaths. After growing for several months, the flower stalk emerges from the top and usually bends downward (see Figure 2-8). Female flowers with abortive stamens produce the hands of bananas, thus, the fruit develops parthenocarpically. Later developing flowers abort, and the flowers at the end of stem are male and abortive. Since a sucker fruits only once, it is cut back to the ground after the fruit is harvested.

Bananas are harvested when mature but green; they develop excellent quality when ripened. Ripening is done at the distribution center by exposing the fruit to approximately 0.1 percent ethylene for 24 h. The ethylene treatment not only accelerates ripening but makes a shipment of bananas ripen more uniformly.

Papaya

The papaya *(Carica papaya)* is a tropical herbaceous plant which grows to a height of 8 m or more. The normally single-stemmed plant grows rapidly, and fruits are borne in the axils of the leaves, which are progressively shed from the base upward (Figure 16-18). Although the papaya plant continues to produce fruit, the planting is usually cut down and replanted when the fruit can no longer be harvested from the ground.

The papaya fruit is a fleshy berry which varies in shape from

FIGURE 16-18 Papaya plant with all stages of fruit development from blossoms to mature fruit.

round to pyriform and contains many small, round, blackish seeds. The skin changes from green to yellow or orange as the papaya ripens. The texture, flavor, and color of the flesh resemble that of a cantaloupe. The papaya fruit is widely used for breakfast, in salads, and as a dessert as well as in juices, ice cream, and jams. In 1976 Hawaii produced 24,000 metric tons of papayas, many of which were of the 'Solo' cultivar. 'Solo' fruit weigh about 0.5 kg and are pear-shaped.

Papaya plants are staminate (produce staminate flowers only), pistillate (produce pistillate flowers only), or hermaphroditic (produce staminate and hermaphroditic flowers). The 'Solo' cultivar produces hermaphroditic and pistillate plants in a 2:1 ratio. To establish a new planting, from three to five seeds are set in each location as the sex of the plants cannot be ascertained until flowering is initiated. At that time, the pistillate plants are removed, leaving the preferred hermaphroditic plants to fruit.

In addition to being a most delicious fruit, the papaya produces the enzyme papain, which is gathered from the latex exuded from scratches on immature papaya fruit. Papain is used as a meat tenderizer.

Tree Nuts

The total United States production of edible tree nuts averaged 430,000 tons during the period from 1974–1976. Of this tonnage, 41 percent were almonds, 38 percent walnuts, 17 percent pecans, 2 percent macadamia nuts, and 2 percent filberts. In the United States, the almond *(Prunus amygdalus)* is grown only in California because of climatic requirements (Figure 16-19); European production is concentrated in Spain and Italy. The major American cultivar is 'Nonpareil,' but since the major almond cultivars are self-sterile, three cultivars are usually planted together to ensure adequate pollination. Most almonds are mechanically shaken from the trees, allowed to dry a few days, and picked up by machine.

The Persian or English walnut *(Juglans regia)* is grown largely in California, with some production in Oregon. There are numerous other species, but they are not widely grown commercially; examples are *J. nigra* (eastern American black walnut) and *J. cinerea* (the butternut). Major cultivars of English walnuts grown in California and Oregon are 'Hartley' and 'Franquette.' Most walnuts are mechanically shaken from the trees and picked up much as almonds are. Walnuts are sold in the shell or as shelled kernels; the latter have become increasingly popular in recent years. World walnut production is largest in the Soviet Union, Europe, and Asia.

The pecan *(Carya illinoensis)* is one of about 20 species of *Carya* which are native in an area extending from North Carolina to Mexico. With the exception of the pecan, all *Carya* are classified as hickories

FIGURE 16-19 'Peerless' almonds, some as shelled meats, others still in shell. (Photograph, California Almond Growers Exchange, Sacramento, Calif.)

and have much harder, more roughened shells. Commercial pecan production occurs in several states which rank as follows: Georgia, Texas, Louisiana, Alabama, and Oklahoma. Pecan production is about equally divided between seedling and wild trees and improved cultivars grafted on seedling rootstocks. The improved cultivars have thinner shells and are often referred to as "papershells." Many cultivars are grown, and the selection varies markedly with the locality and climate.

The filbert *(Corylus avellana)* is but one of several species in this genus which are edible. Two other species are native to eastern North America and are often called *hazelnuts*. Commercial filbert production is mostly in Washington and Oregon, and averages about 9,000 metric tons per year. About 70 percent are sold in the shell, often as part of bagged mixed nuts, while much of the remainder is shelled and used in canned mixed nuts. Italy and Turkey are large producers of filberts.

The chestnut *(Castenea dentata)* has been an important nut in Europe and Asia and before the chestnut blight was of some consequence in the United States from the Appalachians north to Canada. The European or Spanish chestnut *(Castanea sativa)* is widely grown in southern Europe and is sometimes grown in America. Although widely consumed, the European chestnut is considered of only moderate quality.

Macadamia nuts *(Macadamia ternifolia)* have become a major

crop in Hawaii, and production has not been able to keep pace with the demand. The macadamia nut grows within a fleshy husk and has a very hard shell. Since they mature and fall from the trees over an extended period, macadamia nuts must be "harvested" several times, often by collecting from nets hung under the trees. The nuts are dried, sized, and cracked between rotating drums. They are vacuum-sealed in small containers and command a high price because of their very high quality and excellent flavor.

Other important nuts in the world trade are the cashew (*Anacardium occidentale*), pistachio (*Pistacia vera*), and Brazil nut (*Bertholletia excelsa*) which are largely imported into the United States from India, Turkey, and a region along the Amazon River, respectively.

Beverage Crops

Beverage crops are widely utilized around the world. The main nonalcoholic beverages of world trade are coffee, tea, and cocoa—all of which are grown in the tropics. In the United States we drink much more coffee than tea; but in the world as a whole, the reverse is true. Both coffee and tea contain caffeine, an alkaloid which serves as a stimulant. Cocoa contains no stimulants but is used as a pleasant, nutritious drink.

Coffee

Coffee is made from the ground seeds of the genus *Coffea* which contains about 25 species. The commercial coffee industry is based upon three species: *C. arabica* (Arabian coffee), *C. canephora* (Robusta coffee) and *C. liberica* (Liberian coffee) with Arabian coffee accounting for about 90 percent of the world crop. The plant of *C. arabica* is a small tree from 4.5 to 9.0 m in height when unpruned, but it is often held at 2 m to facilitate harvest. The fruits are two-seeded drupes and are often called *cherries*. The berries are fleshy and change from green to yellow to red. Harvesting of the mature berries has traditionally been by hand but attempts to mechanize the harvest operation are under way (Figure 16-20). The two greenish-gray seeds are surrounded by several tissue layers which must be removed by a combination of machine separation, fermentation, curing, milling, and polishing. The coffee beans are shipped in large burlap bags and are roasted at a high temperature to develop the characteristic aroma, flavor, and color of coffee. Many commercial coffees are blends of different types and sources to provide a particular desired characteristic. The final step is the grinding of the beans into the product used daily by millions of coffee lovers.

Coffee production is limited to tropical and subtropical regions and is heavily concentrated in South and Central America and Africa.

FIGURE 16-20 Hand harvest of coffee "beans." (Photograph, Colombia Information Service, New York.)

Brazil is the leading producer. Coffee is best adapted to an area with temperatures of 15 to 25°C. The annual rainfall must be in the range of 200 to 300 cm but a dry season of 2 to 3 months during which flowers are initiated is also important. World coffee production is from 3 to 4 million metric tons. The only state in the United States that grows coffee commercially is Hawaii; annual crops are about 1000 metric tons.

Tea

The most widely used caffeine-containing beverage is tea, made from the dried leaves of *Thea sinensis.* Allowed to grow to its full size, the tea plant is a small tree, but commercially it is kept pruned to a shrub about 1 m tall to make harvesting easy. The harvest of tea consists of the "plucking" of young tender shoots which consist of 2 to 4 leaves and the terminal bud. This has been a hand operation for centuries, but mechanical harvesters are under test in the 1970s. The number of pickings each season varies from three to four in areas where growth stops in the winter to 25 to 30 in warmer areas.

The harvested leaves are wilted, rolled under pressure, and then further dried. At this stage of preparation, the leaves are completely dried but still green and can be used as "green tea." For "black tea," which is the tea of world trade, the tea leaves are rolled once, allowed to ferment, and rolled repeatedly during fermentation. During this process, the green color is lost and the flavor changes, the end result varying with the type of leaf used and the length of time fermentation is allowed to proceed. The grade of the final product depends on which part of the shoot is used, with "orange pekoe" being the smallest leaf and "pekoe" the second leaf.

Tea production centers largely in India, China, Japan, and parts of Africa. Best growth is obtained in a tropical to subtropical climate, with production being at elevations of 1000 m in the tropics and as low as sea level in cooler regions. Tea has been propagated from seeds but, more recently, cuttings and budding have been used to increase uniformity.

Cacao

The cacao plant *(Theobroma cacao)* is native to tropical America, where it is widely grown today. Parts of Africa have also provided an excellent climate for cacao. Commercial production is limited to a zone of 20°N to 20°S latitude at low elevations with 150 to 200 cm of evenly distributed rainfall per year. Leading countries are Ghana, Nigeria, and Brazil. The cacao tree is quite small, usually 5 to 8 m in height; it thrives best when partially shaded by larger trees. Flowering and fruiting occur continuously. The fruits are large "pods," or capsules, which are football-shaped, and range in size from 15 to 23 cm long and 7 to 10 cm in diameter. The fruits are harvested by hand when fully ripe, usually at weekly intervals, and split open. The contents are removed and fermented to eliminate the pulp surrounding the seeds. After drying, the seeds or beans are dried and bagged for shipment.

The beans are cleaned, sorted, and roasted before grinding into an oily paste called *bitter chocolate*. This may be hardened and sold, or it may be converted to sweet chocolate by adding sugar and spices or to milk chocolate by adding milk as well. Cocoa is made by extracting most of the fat (cocoa butter), leaving a residue which is ground into a powder.

BIBLIOGRAPHY

Bianchini, F., and F. Corbetta: *The Complete Book of Fruits and Vegetables*, Crown, New York, 1976.

Chandler, W. H.: *Evergreen Orchards*, Lea and Febiger, Philadelphia, 1950.

————:*Deciduous Orchards*, 3d ed., Lea and Febiger, Philadelphia, 1957.

Childers, N. F. (ed.): *The Peach*, 3d ed., Horticultural Publications, New Brunswick, N.J. 1975.

————:*Modern Fruit Science*, 7th ed., Horticultural Publications, New Brunswick, N.J., 1976.

Collins, J. L.: *The Pineapple: Botany, Cultivation, and Utilization*, Leonard Hill, London, 1968.

Eck, P., and N. F. Childers (eds.): *Blueberry Culture*, Rutgers University Press, New Brunswick, N.J., 1966.

Fogle, H. W., H. L. Keil, W. L. Smith, S. M. Mircetich, L. C. Cochran, and H. Baker: *Peach Production*, U.S. Department of Agriculture Handbook 463, 1974.

————J. C. Snyder, H. Baker, H. R. Cameron, L. C. Cochran, H. A. Schomer, and H. Y. Yang: *Sweet Cherries: Production, Marketing, and Processing*, U.S. Department of Agriculture Handbook 442, 1973.

Food and Agriculture Organization of the United Nations: *Production Yearbook 1975*, Rome, Italy, 1976.

Heinicke, D. R.: *High-Density Apple Orchards—Planning, Training, and Pruning*, U.S. Department of Agriculture Handbook 458, 1975.

Hill, A.F.: *Economic Botany*, 2d ed., McGraw-Hill, New York, 1952.

Mortensen, E., and E. T. Bullard: *Handbook of Tropical and Sub-tropical Horticulture*, Dept. of State, Agency for International Development, Washington, 1964.

Schery, R. W.: *Plants for Man*, 2d ed., Prentice-Hall, Englewood Cliffs, N.J., 1972.

Shoemaker, J. S.: *Small Fruit Culture*, 4th ed., Avi, Westport, Conn., 1975.

Swales, J. E.: *Commercial Apple Growing in British Columbia*, Horticultural Branch, British Columbia Dept. of Agr., Victoria, B.C., 1971.

Teskey, B. J. E., and J. S. Shoemaker: *Tree Fruit Production*, 2d ed., Avi, Westport, Conn., 1972.

Weaver, R. J.: *Grape Growing*, Wiley, New York, 1976.

Winkler, A. J., J. A. Cook, W. M. Kliewer, and L. A. Lider: *General Viticulture*, University of California Press, Berkeley, California, 1974.

Ziegler, L. W., and H. S. Wolfe: *Citrus Growing in Florida*, rev. ed., The University Presses of Florida, Gainesville, Florida, 1975.

CHAPTER 17
OLERICULTURE

 The vegetable industry is a very large and complex part of horticulture today. Despite the current trend toward increasing numbers of home vegetable gardens, the vast majority of the vegetables consumed are produced commercially. The commercial vegetable industry is a fast-moving, intensive, competitive business and in many respects differs significantly from the fruit business. With tree fruits, and to a somewhat lesser degree with small fruits, a long-term investment and commitment are made when the operation is started. Since most of the major vegetables are grown as annuals, some vegetable growers can adjust the relative proportion of different crops to meet changing market situations and—within the limits of equipment, soil type, and climate—can sometimes switch crops completely.

As with fruits, the early American colonists grew their own vegetables and stored as many as possible for winter use. Root cellars were common in many areas for the storage of crops such as potatoes, carrots, beets, cabbage, and turnips. A *root cellar* was an underground storage area which provided a moderate and steady temperature. Home canning, drying, or pickling were widely used for preservation of more perishable vegetables. The advent of home freezers has greatly facilitated the preservation of home-grown vegetables and has also improved quality and flavor.

As cities grew and fewer people were able to produce their own vegetables, commercial vegetable growing was started. Early vegetable growers grew a wide variety of vegetables and often "peddled" them in an adjacent city or town. They were called *market gardeners*, and their production was quite diversified. As transportation methods improved, the development of *truck farming* began. In the late 1800s larger vegetable farms developed in areas particularly well suited to specific crops along the newly developing transport lines, such as railroads, rivers, and canals. Thus truck farming involved larger plantings of fewer crops for shipment to distant markets.

The advent of refrigerated transportation for vegetables has had still further influence on where produce is grown. The railroads introduced the use of ice in refrigerated cars, and now mechanical refrigeration of railroad cars and also trucks is standard. Lettuce and other perishable produce can be harvested in the Salinas Valley of California and displayed on a supermarket shelf in Boston a few days later. This capability has led to very heavy concentration of production in areas most ideally suited for a particular crop. Table 17-1 gives the five leading states in fresh-market vegetable production on the basis of acreage, production, and value. It is readily apparent that the leading vegetable-growing states are determined more by climate than by proximity to market.

Vegetable production for processing is an increasingly important part of the total vegetable industry. The production of vegetables for the fresh market has been fairly stable since 1960, but processing vegetables have shown a gradual upward trend. Per capita consumption of vegetables is on a slight upward trend, with most of the rise attributed to processed vegetables. In addition to knowing about trends in consumption of fresh and processed vegetables, a vegetable producer must keep abreast of changes of particular crops. For example,

TABLE 17-1
LEADING FRESH MARKET VEGETABLE STATES IN 1976

RANK	HARVESTED ACREAGES STATE	% OF U.S. TOTAL	PRODUCTION STATE	% OF U.S. TOTAL	VALUE STATE	% OF U.S. TOTAL
1	Calif.	31.4	Calif.	41.3	Calif.	44.1
2	Fla.	16.8	Fla.	16.9	Fla.	19.7
3	Tex.	11.0	Tex.	9.1	Tex.	7.5
4	Mich.	4.1	Ariz.	4.3	Ariz.	4.3
5	N.Y.	3.6	N.Y.	3.8	N.Y.	3.7
Totals		66.9		75.4		79.3

Source: U.S. Department of Agriculture Crop Reporting Board, 1976, Vg 2-2(76) Vegetables—Fresh Market 1976 Annual Summary, Acreage, Yield, Production and Value.

TABLE 17-2
LEADING PROCESSING VEGETABLE STATES IN 1976*

| RANK | ACREAGE HARVESTED | | PRODUCTION | | VALUE | |
	STATE	% OF U.S. TOTAL	STATE	% OF U.S. TOTAL	STATE	% OF U.S. TOTAL
1	Wisc.	21.2	Calif.	48.0	Calif.	39.7
2	Calif.	17.2	Wisc.	9.3	Wisc.	10.0
3	Minn.	12.0	Ohio	5.6	Oreg.	6.0
4	Oreg.	6.7	Minn.	5.4	Ohio	5.8
5	Ill.	6.4	Oreg.	5.0	Minn.	5.0
Totals		63.5		73.3		66.5

*Includes lima and snap beans, beets, cabbage, corn, cucumbers, peas, spinach, and tomatoes.
Source: U.S.D.A. Crop Reporting Board, 1976, Vg 1-2(76) Vegetables—Processing 1976 Annual Summary Acreage, Yield, Production, and Value.

per capita consumption of fresh corn, cabbage, and tomatoes is down, but onion and lettuce are up—lettuce particularly. Even more striking are patterns in processed vegetables. Among canned products showing large gains from 1960 to 1975 were pickles, tomatoes, snap beans, and corn; the consumption of peas has declined. Among frozen vegetables, sweet corn has shown a marked increase, and broccoli and snap beans are well ahead of earlier levels. Lima beans have declined in popularity. As with fresh-market vegetables, five states account for over 70 percent of the processed vegetable crop (Table 17-2).

In many areas of the country, an increasingly important aspect of the vegetable business is the pick-your-own and roadside-stand marketing systems. Although very small when considered on the scale of conventional vegetable farms, direct-marketing outlets have real advantages for both the grower and the consumer. In relatively high-population areas, the trend is definitely upward. This marketing system is described more fully in Chapter 15.

Although very difficult to monitor, the production of vegetables in millions of home gardens is a major factor in the vegetable situation. This has been reflected in the tremendous interest in the small-package market for vegetable seeds. The production of vegetable seed is big business in certain areas of the country.

The average production of 22 major vegetables, 3 types of melons, and potatoes is presented in Table 17-3. The data include the commercial tonnage for both fresh-marketing and processing outlets. It is of interest to note that certain crops such as lettuce, onions, eggplant, and melons are listed for the fresh market only; others, such as green peas, beets, and lima beans, are given for processing only. For 11 of the crops both fresh-market production and processing production are large enough to warrant the double listing.

TABLE 17-3
PRODUCTION OF THE 22 MAJOR VEGETABLES, 3 MELONS, AND POTATOES FOR FRESH AND PROCESSING USE IN THE UNITED STATES (AVERAGE OF 1974, 1975, and 1976)

CROP	FRESH PRODUCTION (1000 METRIC TONS)	PROCESSING PRODUCTION (1000 METRIC TONS)	TOTAL PRODUCTION (1000 METRIC TONS)
Vegetables:			
Tomatoes	944	6650	7594
Sweet corn	612	2005	2617
Lettuce	2393	—	2393
Onions	1493	—	1493
Cabbage	872	228	1100
Carrots	590	339	929
Cucumbers	220	576	796
Snap beans	138	605	743
Celery	741	—	741
Green peas	—	489	489
Green peppers	234	—	234
Beets	—	190	190
Broccoli	87	93	180
Spinach	29	150	179
Cauliflower	81	59	140
Asparagus	40	67	107
Lima beans	—	73	73
Garlic	52	—	52
Escarole	50	—	50
Artichoke	33	—	33
Eggplant	30	—	30
Brussel sprouts	28	—	28
Total 22 vegetables	8667	11,524	20,191
Melons:			
Muskmelons	444	—	444
Honey Dew Melons	102	—	102
Watermelons	1113	—	1113
Total vegetables & melons	10,326	11,524	21.850
Potatoes:			
White potatoes (Irish)	5478	7261	12,739
Sweet potatoes	673	—	673
Total Potatoes	6151	7261	13,412
Grand Total	16,477	18,785	35,262

Sources: U.S.D.A. Crop Reporting Board, 1976, *Vegetables—Fresh Market, 1976 Annual Summary Vg 2-2(76) Acreage, Yield, Production and Value, and Vegetables—Processing 1976 Annual Summary Vg 1-2(76) Acreage, Yield, Production and Value.* Vegetable Situation, February 1977, TVS 203.

Several systems have been used to classify vegetables, but because of the wide diversity, a generally acceptable system is hard to devise. In this book we are using a combination of criteria—some genetic, some morphological, and some based on the plant part utilized.

POTATOES

White Potato

The potato *(Solanum tuberosum)* is a major food crop throughout the world and ranks with wheat and rice in world production. As a source of carbohydrates (mostly starch) the potato has become a vital dietary component since it was "discovered" in South America by early explorers who introduced it to Europe. Heavy dependence on the potato as a food staple is tragically exemplified in the famine that ravaged Ireland during the mid-1840s. The potato famine was brought about by the devastation of the potato crop by a disease called *late blight.* It is probably because of the Irish immigrants to the United States, who encouraged potato production here, that the white potato is called the *Irish potato*—in spite of its New World origin.

World potato production during the period 1974–1976 was about twice the total of all other major vegetables, excluding sweet potatoes (Table 17-4). Western Europe and the Soviet Union are heavy producers of potatoes, both for human consumption and as livestock feed. Since potatoes thrive under cool conditions, they are ideally suited to more northern latitudes, whereas in more tropical regions the potential sources of food energy are much more varied.

The early history of the potato is provided by Heiser, as follows:

> The white or Irish potato, *Solanum tuberosum*, which rivals wheat in volume produced and value, had a long way to go before it became an acceptable food plant in Europe, but few plants have figured more prominently in Western history than has the potato. The story begins in South America. Wild potatoes are fairly widespread in the Americas, particularly in the Andes. They were probably found to be a valuable food when man first entered this area, and at some undetermined time, probably more than 4000 years ago, they came to be intentionally cultivated. With selection by man there was an increase in size, and the potato became the most important food plant in the high Andes, for it thrives at an elevation where few other cultivated plants will grow. Maize will not grow at elevations much higher than 11,000 feet and is not a particularly productive plant at that altitude, but potatoes do well at 15,000 feet.

TABLE 17-4
WORLD VEGETABLE PRODUCTION: ANNUAL AVERAGE FOR THE PERIOD 1974–1976 (DATA IN THOUSANDS OF METRIC TONS)

	U.S.A.	CANADA	MEXICO	SOUTH AMERICA	ASIA	EUROPE	AFRICA	OCEANIA	U.S.S.R.	WORLD
Cabbage (includes Brussels sprouts & broccoli)	1314	87	—	152	11,208	7553	491	116	—	20,921
Artichoke	34	—	—	118	16	1106	87	—	—	1361
Tomato	7595	16	1045	2162	9518	12,566	3824	236	3700	40,662
Cauliflower	140	17	—	51	1463	2131	105	98	—	4005
Pumpkin, squash, gourd	—	—	—	694	1955	2330	825	89	—	5893
Cucumber	796	70	—	42	4151	2234	251	14	—	7558
Eggplant	30	—	30	3	2697	543	342	—	—	3645
Pepper, green	234	—	408	150	1879	1983	811	—	—	5465
Onions, dry	1494	106	—	1086	7433	3622	1288	150	—	15,179
Garlic	53	—	30	128	673	413	166	—	—	1463
Green pea (in shell)	1225	75	58	175	558	2345	100	162	182	4880
Carrot	930	202	—	212	1961	3040	300	140	—	6788
Watermelon	1113	—	316	1044	11,567	2816	1933	37	3132	21,958
Melons, including muskmelon	547	1	191	283	2455	1512	490	—	—	5479
Subtotals	15,505	574	2078	6300	57,534	44,194	11,013	1042	7014	145,257
Potato	15,355	2639	679	9002	56,178	116,977	3701	923	84,942	290,620
Sweet potato	623	—	113	2652	125,584	114	4936	558	—	135,110
Total vegetables and potatoes	31,483	3213	2870	17,954	239,296	161,285	19,650	2523	91,956	570,987

Source: 1976 FAO Production Yearbook, vol. 30, Food and Agriculture Organization of the United Nations, Rome, Italy.

Frost, common in parts of the Andes, is not conducive to keeping potatoes, but the Indians found how to make it an ally. Potatoes were allowed to freeze at night and the next day the Indians would stamp on the thawing potatoes. This process, repeated for several days, removes the water, resulting in a desiccated potato, called *chuno*, which may be kept almost indefinitely and used as required. Thus was born one of man's original "freeze-dried" foods, in a sense the forerunner of instant mashed potatoes but with a somewhat different taste. Although the flavor of *chuno* is not pleasing to all foreign visitors in the Andes, it is still the staff of life to many Indians in highland Peru and Bolivia. The potato was cultivated throughout the length of the Andes in prehistoric time, but it did not make its way to Central America until introduced by the Spanish.*

United States Industry   The potato industry in the United States is of commercial importance in more than 40 states. The potato is a cool-season crop and is most productive at a temperature of 18 to 21°C. By varying cultivars and planting dates, commercial production ranges from Maine to Florida to California and Washington. Heaviest acreages and production are in Idaho, Washington, and Maine (fall crop); California, Virginia, Texas, and New Jersey (summer crop); California, Florida, and North Carolina (spring crop); and California and Florida (winter crop). Potato acreage has declined from about 3 million in the 1930s to about 1,375,000 in 1976. During this same period, total production has increased from slightly over 9 million metric tons in 1939 to more than 16 million metric tons in 1976. It is obvious that per hectare yields have increased dramatically. Much of this increase can be credited to irrigation as well as to better fertilization, pest control, and cultivars.

The Plant   The potato is an annual; it produces several upright stems which grow to 0.5 to 1 m in height. Terminal flowers may occur and result in fruit 1 to 3 cm in diameter which contain large numbers of seeds. The fruits (berries) are inedible, and the seeds are used only in potato breeding work. The fibrous root system is shallow, and multiple rhizomes develop and terminate in the tubers which we know as potatoes. The tubers are fairly shallow, and when cultivation is required, soil is pulled up toward the plant both to hold the stems more erect and to keep the tubers well covered. Since the tuber is a modified stem, it readily forms chlorophyll when exposed to light—either before or after harvest.

*C. B. Heiser, Jr., *Seed to Civilization. The Story of Man's Food*, ©Freeman, San Francisco, Calif., 1973, p. 136. (Used with permission).

Propagation   All potatoes are vegetatively propagated by planting either small whole tubers or pieces of tuber. The so-called "seed potatoes" are grown specifically for this purpose and should be "certified" to show that they are the indicated cultivar and free of major diseases. A potato tuber has several "eyes"—groups of buds that correspond to the nodes of a stem. Because of apical dominance, the terminal "eye" will tend to inhibit other "eyes"; but when cut into seed pieces, each bud will grow and develop.

Harvesting and Storage   Potatoes are harvested by machines which remove them from the soil and convey them to a truck or trailer (Figure 17-1). Early potato tubers, harvested before the plants die, are immature and very easily damaged and so must be handled carefully and shipped quickly. Most production areas that market immature potatoes use chlorine-treated wash water and prompt refrigeration to provide reasonable shelf life in the supermarket. The main, or fall, crop is harvested after the tops of the plants have died or have been killed by mechanical or chemical treatment to "set" the skin on the tubers, rendering them much less susceptible to damage. This fall crop is usually stored and sold over a period of several months. After curing for 2 or 3 weeks at 13 to 15°C to suberize injuries, the tubers are held at 3

FIGURE 17-1 Irish potato harvester in operation. The potatoes are dug, separated from the vines, and conveyed into a truck traveling beside the harvester. (Photograph, Food Machinery Corporation, Jonesboro, Ark.)

to 4°C to minimize respiration and to inhibit sprouting. For storage at higher temperatures certain growth regulators are used to suppress sprouting. If potatoes are held at temperatures lower than 3 or 4°C, their starch is converted to sugar; this not only gives a sweet taste but causes browning problems in the making of potato chips. If the conversion of starch to sugar occurs in storage, the process can sometimes be reversed by holding them at 15 to 20°C for 2 or 3 weeks.

Marketing   The marketing of potatoes has changed greatly in recent years, with greater and greater proportions of the crop going to processing outlets. Per capita consumption of fresh potatoes has declined from 38.1 kg in 1960 to 26.5 kg in 1970 and 24.8 kg in 1975. During the same period, the per capita consumption of processed potatoes has gone from 11.2 kg in 1960 to 26.9 kg in 1970 and 30.5 kg in 1975. Trends from 1967 to 1975 show a marked increase in frozen potatoes, moderate increases in dehydrated potatoes, and stable levels for chips, shoestrings, and canned potatoes. The continuing trends toward more meals being eaten away from home by Americans, especially at the so-called fast-food establishments, presumably mean an ever-increasing utilization of potatoes in the form of frozen french fries (which account for about 90 percent of the potatoes frozen).

The dependence on processing outlets has led to more contract growing of potatoes and somewhat less price fluctuation. Many of the fresh-market potatoes are now packed and sold by either cooperatives or central packing sheds to provide continuing supplies of stable quality.

Sweet Potato

The sweet potato *(Ipomoea batatas)* belongs to the morning glory family, Convolvulaceae, and originated in tropical America. The sweet potato is grown throughout the tropics and is a major food source. It was introduced into Europe by early explorers but was known in Polynesia before the voyages of early European explorers.

World production of sweet potatoes is less than one-half that of Irish potatoes. It is heavily concentrated in Asia (Table 17-4), with China estimated to account for more than 80 percent of the world's crop. Other major areas include Africa and South America. Because of the relatively high temperatures and the 4- to 5-month growing season required, commercial production is limited to regions between 40°N latitude and 40°S latitude.

There is some confusion associated with the terminology used in describing sweet potatoes. Some cultivars are called the "dry-flesh" type; they have a dry, firm, mealy type of texture after cooking, and the flesh is varying shades of yellow. Most dry-flesh cultivars have skin that is light-yellow to a yellow-tan color. The "moist-flesh" type has soft, moist flesh after cooking, and the flesh ranges from a deep-yellow to an orange-red. Skin color is from light-tan to bright-red. The moist-flesh cultivars are sometimes marketed as "yams," but this term is botanically incorrect because true yams are monocots and belong to the genus *Dioscorea*.

United States Industry The major area of sweet potato production is in the Southeast, from New Jersey to Texas, with significant quantities also coming from California. During the period 1974–1976, the leading states and their percentage of American production were as follows: North Carolina, 32 percent; Louisiana, 22 percent; and Virginia, Texas, Mississippi, Georgia, and California, 6 to 8 percent each. The importance of the sweet potato in the Southeast is indicated by the fact that the Southerner considers the word *potato* to refer to the sweet potato, whereas to a Northerner it denotes the white or Irish potato.

Acreage planted with sweet potatoes has declined markedly from the 1940s, when there were more than 280,000 ha, to the mid-1970s, when total plantings amounted to about 49,000 ha. During this period yields have approximately doubled, but total production is still down considerably. Per capita consumption has declined from 5.9 kg during the period 1947–1949 to 2.5 kg during the period 1974–1975.

The Plant The sweet potato is a perennial herb but is grown as an annual in the United States and other areas where freezes occur. Its stem is a trailing vine which ranges from 1 to 5 m in length and contains latex (as do the roots). The root system is fibrous, and the sweet potato of commerce is an enlarged root which develops in the upper 18 cm of soil. Most sweet potato plants will flower if the photoperiod is 12 h or less; others flower freely in a photoperiod of 14 to 15 h. The flowers are similar to the morning glory. Since the seed has a very hard seed coat, it must be scarified for good germination. As with the Irish potato, seed is used only in breeding work.

Propagation The sweet potato is propagated by either sprouts or cuttings, or by a combination of both. To produce slips or sprouts that are disease-free, true-to-name roots are bedded in sand or sandy loam soil. In very warm areas, this may be done in the open; heated or unheated cold frames are used in cooler areas (Figure 17-2). A temperature of 20 to 26°C is warm enough to encourage good growth of the sprouts without causing them to be excessively succulent. When the sprouts are 20 to 30 cm long, they should be well rooted and are pulled from the parent root and transplanted to the field. From one to four pullings may be made at 7- to 10-day intervals if the delay in time of setting is not detrimental. Another method of propagation is the use

FIGURE 17-2 Bed for propagation of sweet potato by sprouts. The roots are planted in sandy soil and sprouts are pulled when 20–30 cm long. Clear, plastic covers can be unrolled at night and rolled up during warm days. (Photograph courtesy of Boyett Graves.)

of cuttings. After the first slips are planted and the vines reach a sufficient length, cuttings of 25 to 40 cm long are taken and immediately set with one or more nodes in the soil. Such plants will mature later than those from slips, but in Southern areas, sufficient time is available. Because of the high labor cost of making vine cuttings, this practice has declined sharply in the past 20 years.

Harvesting and Storage Sweet potatoes can be dug as soon as the roots reach marketable size. The grower usually makes trial diggings to determine the relative amounts of marketable sizes in order to decide when to harvest the crop. For many years the standard method of harvest was to plow the roots out and to have crews pick them up by hand. Careful handling is essential because the skin is very easily injured, and the flesh is also quite tender. Harvest damage contributes to decay losses in storage and may also lower the grade if it is too extensive. With increasing costs as well as declining availability of field labor, the trend is strongly toward mechanical harvest. Several types have been tried, including both tractor-drawn and self-propelled units. Some harvesters are designed strictly to dig the roots and put them in containers, while others provide platforms for field grading by hand laborers.

Roots to be stored for extended periods should be free from disease, excessive mechanical injury, and the injury associated with cold or wet soils. The normal procedure after harvest is to cure sweet potatoes at about 30°C and 85 percent relative humidity for 4 to 8 days. These conditions encourage wounds to heal over and thereby greatly reduce losses due to rot and moisture loss during storage. Once the curing process is complete, the temperature should be lowered to 14 to 16°C with a relative humidity of about 85 percent. Temperatures should not go below 10 or 12°C because of chilling injury which results in increased decay, internal breakdown, and lowered quality.

Marketing Most sweet potatoes go to market in wax-impregnated cardboard containers of various sizes; bags have not been used widely because of increased damage to the roots. Retail sales are most commonly from bulk displays, although there is some trend toward prepackaging with overwraps. Polyethylene bags are not used, because the high relative humidity and 22°C supermarket temperatures are ideal for the development of *Rhizopus*, or soft rot, the leading cause of postharvest sweet potato losses.

The major demand for fresh sweet potatoes is for medium-sized (5 to 8 cm), well-shaped roots. Prices for jumbo sizes and small sizes are low, and thus many of these go to processing outlets. The major processed sweet potato product has been canned, either in a heavy syrup or vacuum-packed. Various other products, such as dehydrated sweet potato flakes, have been developed but are not widely marketed. It appears that new sweet potato products must be developed if the decline in per capita consumption is to be reversed.

SOLANACEOUS FRUITS

Three other major vegetables belonging to the Solanaceae family are tomato, pepper, and eggplant. Each is grown as an annual, is tender, and is grown for its fruit. The family, often referred to as the *nightshade family*, also includes the Irish potato, tobacco, and certain ornamentals such as petunia.

Tomato

The tomato *(Lycopersicon esculentum)* ranks high among the important vegetables of the world and is grown in large quantities in most regions (Table 17-4). The tomato apparently originated in South America but may have been first cultivated in Mexico. Spanish explorers took the tomato back to Europe by the middle of the sixteenth century, but it was not widely utilized for many years. Even though it was introduced into the United States in the eighteenth century, it was not widely accepted as an edible fruit for another hundred years. The reputation of the tomato varied from being considered poisonous to being associated with love, as indicated by the French name *pomme d'amour*, or "love apple."

United States Industry Tomato production in the United States is exceeded only by the Irish potato. In considering the tomato, we must distinguish between the fresh-market and processing tomato industries, because they are quite distinct. In spite of moderately declining processing tomato acreages since the early 1950s, total production has almost doubled. In 1952, California produced about one-half of the processing tomatoes; in the period 1974–1976, California grew 83 percent of the processing tomato crop. Other important states include Ohio, New Jersey, and Indiana, and there is commercial production in 26 or more states. About 88 percent of the tomato crop went to processing outlets in the 1974–1976 period (Table 17-3). The fresh tomato industry accounts for 12 percent of the crop and is quite different, because since various locations ship at different seasons, fresh tomatoes are available on a year-round basis. Spring tomatoes come from Florida, other Southeastern states, and California; summer production ranges from Georgia to Michigan to California; fall crops come from California, Florida, and Texas; winter tomatoes come from Florida or are imported from Mexico. Of the fresh-market crop, California and Florida each produce about one-third of the total crop.

Certain parts of the United States, particularly in Ohio, have developed sizable greenhouse ranges for vegetable production during cold weather. The tomato is the most important vegetable produced in greenhouses. Although this method of production was profitable at one time because the fresh produce commanded a high price, heating costs in the mid-1970s have made it very questionable from an economic standpoint.

Between 1965 and 1976, per capita consumption of fresh tomatoes remained stable at about 5.5 kg, while per capita consumption of canned tomatoes increased from 21 to 28 kg.

The Plant In the tropics the tomato is an herbaceous perennial, but in northern latitudes it is grown as an annual. The growth habit of different cultivars varies widely, but the most common types grow to a stem length of 0.7 to 2 m and develop multiple stems which originate in the leaf axils. In so-called ground production, plants are allowed to develop naturally, without support or pruning. This technique is standard for processing tomatoes and is used in most fresh-market production. In some fresh-market production, however, all but one or two stems are eliminated as they occur. These are trained to a short stake or to a string hung from above and tied loosely around the base of each stem ("string-weave" method). Reasons for incurring the additional expense of pruning and support include better disease control and cleaner fruit and continued production over a longer period of time. Another approach has been the use of "cages," cylinders made of concrete reinforcing mesh that are 45 to 60 m in diameter. These cages are made to stand from 1 to 1.5 m tall, depending on the width of the mesh used, and are placed around the young plants. This method is better than conventional trellis methods because of greatly reduced labor requirements. Yields have been excellent in early trials.

Tomatoes are borne in clusters which usually contain from four to eight fruit (see Figure 2-9). The fruit is a fleshy berry with many small seeds. Although the major commercial tomato is red and has a diameter of 5 to 8 cm, many other types are available. The cherry tomato (1½ to 3 cm in diameter) is increasing in popularity for salads because it can be used whole. Other types are yellow-fruited cultivars and the plum tomato (Figure 17-3), used for puree because of its very high solids content.

FIGURE 17-3 Processing tomato of the plum type, cultivar 'Roma' V.F. Because of its meatiness, this type of tomato is widely used in tomato paste and sauce. (Photograph, Asgro Seed Company, Kalamazoo, Mich.)

Propagation   Tomatoes are propagated from seed, which will germinate as soon as they have been removed from the flesh, cleaned, and dried. For early markets, transplants are grown in greenhouses or in southern areas and shipped in. In warm areas, they are now often direct-seeded in the field, particularly for processing.

Harvesting and Storage   Processing tomatoes in California are now almost entirely mechanically harvested, and the trend is toward mechanical harvest throughout the industry (Figure 17-4). The transition from hand harvest to machines has required the development of new cultivars specifically designed for once-over mechanical harvest. The necessary features include uniform fruit set, adequate but not excessive vine growth, uniform ripening, jointless fruit stems, and firm fruit which can withstand rolling, short drops, and the pressure in bulk containers.

Fresh-market tomatoes are largely hand-harvested, but efforts have been under way for the past several years to perfect machine harvest. It seems likely that fresh-market tomatoes will continue to be hand-harvested in many parts of the country, but mechanical harvesting may become a reality in areas of very large production.

The stage of maturity at harvest varies primarily with the market being served. Processing tomatoes are picked when fully ripened on

FIGURE 17-4   Most processing tomatoes are now harvested mechanically. Workers sort tomatoes before their delivery to the adjacent truck which hauls them to the cannery. (Photograph, Blackwelder Manufacturing Company, Rio Vista, Calif.)

the vine—i.e., at the *red-ripe stage*—since they are processed as quickly as possible after harvest. Proper harvest maturity for the fresh market depends upon marketing procedures and distance to the market. For long-distance shipment, tomatoes are picked at the *mature-green* stage. Although difficult to determine exactly, it is best described as the stage at which the dark-green color has changed to a whitish-green. These are the "greenwraps" of the trade; they account for the majority of fresh tomatoes available today. When a tomato has reached the mature-green stage, its seeds will not be cut by a sharp knife, because the jellylike texture of the tissue allows the seeds to move. This characteristic is often used in the field to aid in the training of picking crews. A slightly more advanced stage of maturity is reached when the first sign of pink appears. Fruits at this stage are called "breakers." The more mature the fruit is, the more carefully it must be handled, and the shorter time it can be held. Unfortunately, the flavor of a vine-ripened tomato from a backyard garden cannot be duplicated commercially. It is probable that the sometimes inferior quality and high price of fresh tomatoes account for the stability of fresh-market consumption and the rise in consumption of canned products.

Under most conditions, tomatoes are not intentionally stored except as required within normal marketing channels. To delay the rate of ripening, temperatures of 12 to 15°C are suitable for short periods. If harvested at the mature-green stage, tomatoes require from 1 to 3 weeks to ripen at 20 or 21°C. Usually, tomatoes are ripened in controlled-temperature rooms, often with ethylene added to accelerate the ripening process.

Marketing   Tomatoes that are shipped at the mature-green stage (as most are) must be ripened and usually repacked before retail sale. Most of this is done at the terminal market. When the tomatoes are received, some will have been partially ripened, but the majority will still be green, or breakers. They are put in ripening rooms at 19 to 21°C and 85 percent relative humidity for a sufficient number of days to induce maximum color. Upon removal, the tomatoes are packed into various containers, from tray packs holding four or five tomatoes up to cartons holding from 4.5 to 9.0 kg.

The tomatoes reach the supermarket shelf in tray packs prepared at the terminal market, in overwraps or other packages prepared at the distribution warehouse or individual supermarket, or in a bulk display.

Pepper

Both the green or sweet pepper and the hot pepper which are grown in the United States are *Capsicum annuum*. This species is native to

New World tropics. Columbus apparently took pepper seeds back to Europe, and the pepper was much more quickly accepted than the tomato, both in Europe and eventually in the United States. This species includes a very diverse group of peppers which vary from 1 to 30 cm long, from green to yellow when immature, and from red to yellow when mature. The only type not included is the Tabasco pepper, *C. frutescens*.

United States Industry   The only pepper listed among the major vegetables is the bell pepper. This pepper is sweet, rather than pungent like many of the other peppers grown. Sweet peppers are widely grown on small local farms and in home gardens, but most commercial production is in Florida, California, North Carolina, and New Jersey. In total production the pepper ranks about fourteenth among the major vegetables in the United States (Table 17-3). Per capita consumption has increased from about 1 kg from 1957 to 1959 to about 1.5 kg from 1975 to 1976.

Some peppers are processed, mostly as pickled peppers. The pimiento pepper is canned in sizable quantities. It is used to prepare pimiento cheese and to provide the red stuffing for green olives. The cultivars belonging to *C. frutescens* are very pungent and are dried and ground into powder or used in the preparation of "hot sauce." Fresh sweet peppers are a common part of tossed salads and relish trays; they are particularly tasty when stuffed with any of a vast number of mixtures of meat and vegetables.

The Plant   The plant of *C. annuum* is an annual which grows to a height of 0.5 to 1.5 m. The fruit are borne singly at nodes and are many-seeded berries. The plants are erect, do not require support, and produce multiple stems.

Propagation   Peppers, like tomatoes, are grown from seed. Plants are started in greenhouses, hotbeds, or open beds. When the plants are 10 to 15 cm tall, they are transplanted to the field, but only after soil temperatures are fairly high and the danger of freezing temperatures is past.

Harvesting and Storage   Green peppers are harvested when they reach full size but before the appearance of the red or yellow color associated with ripeness. Harvest has traditionally been by hand into various kinds of containers.

In an effort to increase the efficiency of pickers, some green peppers are now harvested by pickers walking behind (riding on) slowly moving conveyors; this method has become commonplace with many vegetable crops. Once-over mechanical harvesters have been tested and will undoubtedly be adopted by the larger growers as soon as suitable cultivars are available which can (1) produce a heavy crop of fruit at one time and (2) bear peppers able to withstand the somewhat rougher handling. Mechanical harvest of peppers for processing will probably be perfected before it is practiced on fresh-market crops.

Peppers are subject to chilling injury below 7°C, and temperatures above 10°C cause excessive rates of ripening and rots. Therefore, optimum storage conditions are 7 to 10°C, with a relative humidity of about 90 percent. Under ideal conditions, peppers will hold well for about 3 weeks.

Marketing   Peppers are shipped in various types of rigid-walled containers to minimize damage. They may be sold from bulk displays in the supermarket or in various types of consumer packages from small polyethylene bags to trays with overwraps.

Eggplant

The eggplant *(Solanum melongena L.)* is a native of India but gradually spread throughout the tropics. The type grown in the United States is *S. melongena* var *esculentum*. The eggplant is not a major world crop except in Asia, where its production exceeds that of many other vegetables (Table 17-4). Many types of eggplant are grown around the world, and the fruit vary not only in size and shape but in skin color, which ranges from white and yellow to the familiar purple and black.

United States Industry   A look at figures for commercial production of eggplant would lead you to rank it as a relatively minor crop (Table 17-3), but actually eggplant is widely grown for local markets and by home gardeners. Both total production and per capita consumption have shown moderate upward trends since the 1950s. The only states with major commercial production are Florida and New Jersey, with Florida accounting for more than 70 percent of the total crop. Although essentially the entire crop is sold fresh, some marketing of frozen slices has been tried. Major uses include fried eggplant slices, eggplant casseroles with cheese and tomato, and stuffed eggplant.

The Plant   The eggplant which is grown in the United States is a bushy plant grown as an annual—even though it is a perennial. The usual size is between 60 and 90 cm in height. The base of the stem becomes woody. The fruit is a fleshy berry containing many seeds which are eaten (Figure 17-5). The green calyx on the proximal end of the fruit is often spiny.

Propagation   Like the tomato and pepper, eggplant is seed-propagated. Seedlings are often transplanted from flats to small peat

FIGURE 17-5 When the eggplant fruit is about two-thirds full-grown, its quality is best. By cutting with a knife or clippers, the calyx is left on the fruit. (Photograph, Harris Seeds, Rochester, N.Y.)

pots or veneer bands, so that when they are transplanted to the field there is soil around the roots. Bare-root eggplants are less able to withstand the shock of transplanting than bare-root tomatoes or peppers. Since the eggplant is a very warm-season crop and is more readily injured than tomato by low but above-freezing temperatures, transplanting to the field is delayed even longer than with tomatoes. Ideal temperature ranges are from 26 to 32°C during the day and 21 to 26°C at night.

Harvesting and Storage Eggplant fruits are harvested when they reach one-half to two-thirds of full size because both the total yield and the quality tend to decline if the fruits are left until they reach full size. The most noticeable changes associated with leaving the fruit on the plant too long are a toughening of the flesh and a woodiness of the seeds. The fruits are harvested with a sharp knife or clippers to cut the somewhat woody stem, leaving the calyx attached to the fruit. Because of the rather tender skin, careful handling is necessary, both during harvest and subsequently.

Eggplant should not be stored for more than 10 days, and optimum conditions are about 10°C and 90 percent relative humidity. At lower temperatures, chilling injury is likely to occur.

Marketing Eggplant is shipped in various types of containers, the most common being the bushel basket. Individual fruits are sometimes wrapped before being placed in the shipping crate or basket. Traditionally, eggplants have been sold from bulk displays in supermarkets, but increasing quantities are being prepackaged in polyethylene bags or overwraps.

SALAD CROPS

The salad crops include lettuce, celery, escarole, and a variety of miscellaneous green vegetables such as chicory and parsley. We will consider only the first three, which are ranked among the more important vegetables.

Lettuce

Lettuce *(Lactuca sativa* L.) is in the sunflower or Compositae family. It is widely accepted that *L. sativa* developed from *L. serriola* L., the wild prickly form of lettuce native to Asia, Europe, and northern Africa. Lettuce was known and eaten by the early Greeks and Romans as well as by the Chinese in even earlier times.

There are four major types of lettuce, which are all classified as *L. sativa* L. They are readily distinguishable as follows: (1) crisphead, (2) butterhead (Figure 17-6), (3) cos or romaine (Figure 17-7) and (4) leaf or bunching. In China, asparagus or stem lettuce is widely grown; the fleshy stem is eaten fresh or cooked, and the leaves are discarded.

United States Industry Most of the commercial lettuce crop in this country consists of the crisphead type. Cultivars of crisphead lettuce are characterized by their firm heads and the brittle texture of the leaves. The heads are usually at least 15 cm in diameter and are solid enough to withstand the rigors of harvest, long-distance shipment, and final marketing.

Lettuce is available throughout the year, and California grows more lettuce than any other state during all four seasons. Of the total commercial lettuce tonnage, California shipped 72 percent in the 1974–1976 period. Arizona was second with 15 percent, and other states with seasonal shipments of lettuce include Florida, Texas, New Mexico, New Jersey, Colorado, and New York.

For short-distance shipment the butterhead and romaine types of lettuce are of some importance in the East and Midwest. These are of high quality but because of poor shipping characteristics are not produced in the lettuce areas of the Southwest. The leaf lettuce cultivars are produced mostly in home gardens and in limited quantities in Midwestern greenhouses during the winter.

FIGURE 17-6 Bibb lettuce is rather loose-headed, its leaves are rather buttery, and its quality is excellent. (Photograph, W. Atlee Burpee Company, Doylestown, Pa.)

Per capita consumption of lettuce rose rapidly between the 1920s and the late 1940s and has been showing a gradual increase in the past 30 years. Current consumption is about 10 kg per person per year.

In this country lettuce is used almost exclusively as a fresh vegetable, although in some parts of the world it is cooked like spinach. Lettuce is particularly attractive as the main constituent of salad because of its mild flavor, crisp texture, low caloric content, and attractive color.

The Plant Lettuce is an annual plant which forms a rosette of leaves at the base and subsequently a tall flower stem from 30 to 100 cm high. Commercially, lettuce is harvested well before the formation of the flower stalk. Some cultivars form a definite head, whereas others are merely a loosely formed rosette of leaves. Lettuce is shallow-rooted and therefore requires excellent weed control. Cultivation should be shallow to avoid excessive root damage.

Climate is a major concern in commercial lettuce production, because the range of temperature adaptation of lettuce is more restricted than that of many other vegetable crops. The ideal conditions for head lettuce are present in certain parts of California and Arizona, and the location varies with the time of year. Cool temperatures, low humidity, and adequate moisture from irrigation combine to make the ideal environment for lettuce. Mean temperatures should be between

FIGURE 17-7 Cos or romaine lettuce is of high quality, has an attractive appearance, and is quite widely grown. (Photograph, W. Atlee Burpee Company, Doylestown, Pa.)

13 and 18°C. If they are much above this range, lettuce tends to "bolt" (send up a seed stalk) rather than forming a head, bitter flavors tend to develop, and a disorder called *tipburn* becomes prevalent.

Propagation Lettuce is propagated by seed, which are very small, averaging over 100 per gram. For early spring planting, seedlings may be started in greenhouses, cold frames, or other protected areas and later transplanted to the field. In the major lettuce areas, however, direct seeding in the field is standard practice. Most of the areas are irrigated, and the lettuce is mechanically seeded in beds which are raised as the seed is planted. One of the problems in direct seeding is planting enough seed to get a uniform stand, because if the plants are set too thickly, the job of eliminating extra plants becomes very expensive. Within 2 weeks after seeding, the plants are "blocked"—a process in which the seedlings are thinned to clumps spaced at 25 to 40 cm. A few days later each clump is "thinned" to a single plant. These two operations add greatly to production costs. Many techniques are under investigation to enable growers to plant single seeds spaced widely enough to either make mechanical thinning practical or completely eliminate the thinning operation (precision planting).

Harvest and Storage A great deal of hand labor is involved in the commercial lettuce harvest. Crews of workers walk through the lettuce fields and use knives to cut heads whose size and firmness indicate that they are mature. The loose basal leaves are removed, and any necessary trimming is done. Before vacuum cooling became standard, most lettuce was taken from the field and hauled to a central packing shed for sorting, packing, and icing in wooden crates.

The advent of vacuum cooling dramatically changed the lettuce business. Lettuce is now packed in the field, by one of several different systems, into fiberboard boxes which hold from 18 to 30 heads depending on the size of the heads. After packing, water is sprinkled over the lettuce, the top is stapled, and the cartons hauled to the vacuum coolers (see Figure 15-3).

For many years, intensive research efforts have been aimed at the development of a selective mechanical harvester for head lettuce. Prototypes have been tested which can move through the field and "feel" the size and firmness of each lettuce head. When the size and firmness of a head are at the levels specified by the operator, a switch activates a knife which harvests that head. It seems likely that such machines will soon be developed sufficiently to receive commercial acceptance.

If lettuce is harvested, handled, and cooled properly, it will keep at 0°C for about 3 weeks. Since humidity must be high to maintain freshness, lettuce to be stored is often wrapped individually, or a perforated plastic liner is used inside the carton.

Marketing Lettuce is normally available every day of the year to the American consumer. It may be in loose displays or prepackaged in shrink film or polyethylene bags by the shipper, or the local store may wrap it before display. Regardless of the packaging, lettuce must be kept moist and cold in order to maintain good quality.

Celery

Celery (*Apium graveolens* var *dulce*) is the second leading salad crop; United States production is about one-third that of lettuce (Table 17-3). Celery is in the Umbelliferae family, as are carrots and parsley. It apparently developed from a wild marsh plant occurring throughout temperate Asia and Europe. In early times celery was used as a medicinal herb, and later it was used as a flavoring; but it was not accepted as a food until the seventeenth century. Since early forms of celery were not particularly tender or tasty, they were used mainly in soups. After mild, tender cultivars were developed, celery became popular as a fresh vegetable.

United States Industry Like lettuce, celery is available year round to the American consumer. Of the United States crop of 741,000 metric tons in 1974 to 1976, 66 percent was grown in California and 25 percent in Florida, and the remainder came mostly from Michigan, New York, Ohio, and Washington. Since the late 1940s there has been a slight decline in acreage, but increased yields have resulted in a moderate increase in total production. Overall production has been quite stable since the early 1960s. Per capita consumption has been between 3.0 and 3.5 kg for many years. Much celery is used as an appetizer, often stuffed with various cheeses or as a plain component of a relish tray. Celery is common as a component of salads and soups, and as a cooked vegetable.

The Plant Celery is normally a biennial plant, but as the result of certain stimuli it may "bolt," or produce a seed stalk, during the first year. Since bolting ruins the plant for sale, every effort is made to minimize its occurrence. In addition to requiring a long growing season and uniform moisture, celery has rather strict temperature requirements. Ideal temperatures are 15 to 18°C; temperatures above 21 to 24°C tend to slow its growth and lower its quality; temperatures below 12 to 14°C for more than a few days may induce bolting.

The edible portions of the celery plant are the thickened petioles or leaf stalks, which arise from a very compact stem (Figure 17-8). Celery is classified as an herb and grows to about 1 m. The dried fruits or seeds are used for flavoring because of the presence of aromatic oils, which are also present in leaves and petioles. The fruits are often called seeds, although the seeds are very minute and are contained within the fruits.

In the early 1900s most celery was "blanched" by covering the petioles with paper, boards, or soil so that chlorophyll would break down, leaving the celery yellowish. The so-called "self-blanching" cultivars tend to be pale by nature. Little commercial celery is blanched today, as the American consumer has learned that green celery is not only more attractive but high in quality. High-quality celery has minimal stringiness resulting from either the vascular bundles or strands of collenchyma.

Propagation Celery is seed-propagated, but only with some difficulty because the seed are very small and germinate slowly. In California considerable direct seeding is done, but in other areas most celery is started in greenhouses, cold frames, or protected beds outside. Growing transplants ready for the field may take about 2 months, since they should be 10-15 cm tall. Direct seeding requires irrigation to keep the soil surface moist during both germination and early seedling development.

Harvest and Storage Celery was always hand-harvested until recent years, when mechanization has become increasingly utilized. In the hand operation, pickers would walk through the field, cut the celery at

FIGURE 17-8 Celery which has been partially trimmed for shipment. (Photograph, W. Atlee Burpee Company, Doylestown, Pa.)

ground level, strip off the outer leaves, and place the celery on conveyors. The conveyor took the celery to a mobile packing facility, where it was washed, graded, and packed in crates. Mechanical harvesters may be used in conjunction with a mobile packing unit or may deposit the harvested celery into trucks or trailers for transport to a packing shed.

The celery must not only be stripped of outer leaves but must be trimmed back from the top and thoroughly washed. After packing, the celery must be cooled quickly by hydrocooling, icing, or vacuum cooling. If temperatures are not brought down rapidly, deterioration occurs.

Celery can be stored for up to a month if temperatures are at 0°C and the relative humidity is high. Packaging is important because one of the quality criteria for celery is crispness.

Marketing Celery is shipped in several types and sizes of containers. Increasing use of polyethylene sleeves and shrink film wraps is being made. These not only provide mechanical protection but reduce water loss and provide space for the grower or shipper to place a label. Large amounts of celery are still sold from bulk displays in supermarkets, but this requires frequent sprinkling to maintain crispness and freshness.

Escarole or Endive

Escarole, *Cichorium endivia*, is in the Compositae or sunflower family, and is also called *endive*. Endive has been eaten since the days of the Egyptians and is thought to have originated in the region of East India.

The two types of this vegetable are often lumped together in reporting such factors as production and consumption. The very curly-leaved or fringed-leaf form is called *endive*, and the broad or straight-leaved form is called *escarole*.

The production of these vegetables is limited to three states; of the 1974–1976 crop of about 50,000 metric tons, Florida grew 73 percent, New Jersey 18 percent, and Ohio 9 percent (Table 17-3). Escarole is used mainly as a salad ingredient, although it can also be cooked as greens.

The escarole plant is a moderately hardy annual or biennial which produces clusters of leaves—the margins of which vary from lobed to very cut and curled. The formation of a seed stalk may occur in response to low temperatures and is avoided if possible because it ruins the crop. Since climatic requirements are similar to lettuce, escarole is grown as a winter or spring crop in the South and as a summer or fall crop in more Northern states.

Harvest, storage, and marketing are much as for lettuce.

CORN

Grain

World corn production is given not in terms of sweet corn but rather in terms of grain. As such, corn ranks with the other major food crops, wheat and rice. It is important to note the many uses of corn, including corn oil (used in margarine), cooking oils, cornstarch, cornmeal, and grain and ensilage for livestock. Since this chapter is concerned with vegetables, however, we will consider in detail only that portion of the corn crop known as sweet corn.

Sweet Corn

Sweet corn, *Zea mays*, is a member of the Gramineae or grass family which has a large number of types. Although sweet corn, as we know it today, apparently originated only in the nineteenth century, maize or Indian corn has been known and utilized since very early times in the Americas. Corn was unknown in Europe until the return of Columbus but has since become a major food all over the world.

United States Industry It is most logical to consider the production of sweet corn for processing and for the fresh market as two somewhat distinct industries. These two industries account for 77 and 23 percent of the crop, respectively (Table 17-3). Fresh-market corn is available all year, but supplies are much greater in summer. Winter corn comes from Florida; the spring crop from Florida, California, Texas, and Alabama; and the fall crop from Florida and California. Summer corn comes from about 20 states, with New York, Ohio, Pennsylvania, and Michigan as leading producers. The summer corn crop represents over one-half of the annual production reaching the fresh market. Florida grows about 40 percent of the total fresh-market corn. Processing corn is grown in the more Northern states, with Wisconsin and Minnesota together accounting for about 45 percent of the total crop harvested during 1974 to 1976. Other important states include Oregon, Illinois, Washington, and Idaho. Of the total corn processed during 1974 to 1976, almost 70 percent was canned; the other 30 percent was frozen.

Climatic requirements for sweet corn include adequate moisture and warm temperatures, although its wide adaptability is indicated by its production in every state as well as in Mexico and Canada. Corn develops more rapidly as temperatures increase from 5 to 35°C. This accelerated growth rate probably relates to its C_4 pathway of carbon fixation (see Chapter 6). The reason why the great majority of processing corn is grown in Northern areas is the cooler temperatures during the harvest season. Since corn matures more slowly at cooler temperatures, it can be harvested at optimum quality over a longer period of time, and quality deteriorates more slowly after harvest in cooler regions.

Trends in per capita consumption since the mid-1960s vary: there has been a slight downward trend for fresh corn, a slight upward trend for canned corn, and a moderate increase for frozen corn. Most fresh corn is eaten as the American delicacy "corn on the cob." It is most unfortunate that many consumers have never tasted the flavor of *truly fresh* sweet corn because of the rapid conversion of sugar to starch. Processed corn is widely used as a hot vegetable as well as in soups, casseroles, and relishes.

The Plant The corn plant is in some ways similar to other grasses, but it differs significantly in other aspects. As with most grasses, the leaves are long, with parallel venation. Corn has a fibrous root system and a stem typical of the monocots. The corn plant usually forms aerial prop roots above the ground which grow into the soil and provide support.

FIGURE 17-9 Freshly picked ears of modern sweet corn cultivars are a most attractive and delicious vegetable. (Photograph, Harris Seed Company, Rochester, N.Y.)

The corn plant has flowers which are very different from those of other grasses: most grass flowers are perfect in that they contain both pistillate and staminate parts, but corn is monoecious, and its flowers are imperfect. The staminate (male) flowers are borne on the tassel or panicle at the apex of the plant. The pistillate (female) flower is the ear, with the "silks" being long styles. Pollination occurs primarily by wind and gravity, so that cross-pollination is common in corn. Each kernel of corn is an ovary (Figure 17-9). The cob is surrounded by the husk, which consists of leafy bracts.

The corn plant is an annual; it grows to a height of 1.5 to 3 m. It is tender and is not planted until about the average last freeze date, not only to avoid freeze injury but to ensure that soil temperatures are warm enough for germination to occur.

A most interesting feature of corn is xenia, an effect of the pollen parent on the phenotype of the female parent. If yellow corn is planted beside white corn, yellow kernels will appear on the white ears. This occurs because the pericarp or hull is derived from the mother plant, but the endosperm arises from the secondary fertilization.

Propagation Corn is grown from seed planted directly in the field. Not only does corn transplant very poorly, but the high plant populations required make transplanting impractical. The harvest season is varied by selecting cultivars which mature in different lengths of time or by planting the same cultivar at intervals during the spring. With the latter approach it must be remembered that the mean temperatures increase as the season advances, and so the number of days from planting to maturity will decline. For example, a cultivar planted at 10-day intervals in May will mature at less than 10-day intervals in August. This is the reason for the use of degree-days, described in Chapter 7.

Most of the sweet corn grown today consists of yellow hybrids. Hybrids are used because of the vigor, high yields, and uniformity of maturity associated with heterosis (hybrid vigor). Most hybrids are the result of crossing two inbred lines.

Harvesting and Storage Most types of corn are harvested in a mature stage (popcorn is one example), but sweet corn is harvested when the kernels are immature, as indicated by the milky consistency of the juice. Before the milky stage, the kernels are small and the juice is watery. The stage after the milky stage is called the *dough stage*; at this stage most of the sugar has been converted to starch and the flavor is dramatically poorer. Increasing proportions of the sweet corn crop are being mechanically harvested. With improved cultivars, once-over harvest is practical, and thus multiple hand pickings are eliminated.

The quality of corn begins to decline immediately after harvest, and the rate of decline is very temperature-dependent. For optimum quality the corn should be used as soon as possible either for processing or for fresh consumption. For long-distance shipment of corn, immediate cooling and constant refrigeration are absolutely necessary. Hydrocooling is standard practice today. A recent introduction to help in getting quality corn to the consumer is cultivars which have much slower rates of sugar-to-starch conversion.

Sweet corn should be one of the major vegetables in a roadside market operation. Probably with no other vegetable is there such a striking difference between the quality and flavor of the fresh-picked vegetable and what is available in commercial sources.

Sweet corn should not be stored except as required for transit and should then be held at 0°C. Since corn has a high rate of respiration, large amounts of heat are generated, and considerable heating will occur in piles with poor ventilation. By harvesting early in the day the amount of "field heat" in the ears is reduced, and the problems of initial cooling are greatly lowered.

Marketing Much processing corn is grown on a contract basis. One of the major advantages of this system is to give the processors control of cultivars and planting dates, and thus obtain an orderly supply at harvest. Since even short-term storage is detrimental, corn should be processed immediately. A well-coordinated schedule between grower and processor is mandatory. This is the job of the processor's representatives in the field.

Fresh corn is available all year, but the period May through September usually accounts for about three-fourths of the total for the year. Corn is shipped in various containers, with the wire-bound wooden crate being most popular. In the supermarket, corn is displayed in bulk in the husk, overwrapped in the husk, or husked. Regardless of the display and sales container, refrigeration is an absolute necessity. Occasional sprinkling of unwrapped corn is also helpful in maintaining color and general appearance.

BULB CROPS

The major bulb crop is the onion, but garlic, leek, shallot, and chive are also included in this category. All the bulb crops are hardy and belong to the genus *Allium* in the Amaryllidaceae or amaryllis family. Since most considerations such as climatic requirements for the group are similar, only the onion will be discussed in any detail.

Onion

The onion is a major vegetable crop in much of the world (Table 17-4). It apparently originated in Asia, and it has been used since the earliest records of history. Since it was the early explorers and settlers who introduced the onion to the New World, its use in the Americas is relatively recent.

Onions, as well as the other types of bulb crops, are used in many ways, depending not only upon the particular types but also upon the stage of maturity (Figure 17-10). Mature onion bulbs are widely used as a cooked vegetable; in soups, stews, and casseroles; and as flavoring in many additional dishes. The introduction of dehydrated onions as a readily available product has reduced the need for peeling and dicing small quantities of raw onions—a somewhat unpleasant task. Sweet onions of the Bermuda type are used in salads, sandwiches, and other dishes where a mild onion flavor is desired. Green onions for salads, etc., are either immature bulbing onions or the bunching or nonbulbing type. These are harvested green and handled like a salad crop.

United States Industry The United States onion crop ranks quite high (Table 17-3). Spring onions come from Texas, California, and Arizona, and account for about 20 percent of the total crop. The remainder is called the *summer crop* and is produced in many states, the most important being California, New York, Oregon, Idaho, Michigan, and Colorado.

The onion is a cool-season crop which is grown in the South in

FIGURE 17-10 Onions offer many choices. Green onions on toast with hollandaise sauce are delicious. (Photograph, United Fresh Fruit and Vegetable Association.)

winter and in many areas during the summer. The winter and spring crops are sold soon after harvest, whereas much of the summer crop is stored for year-round sales. The onion is shallow-rooted and therefore needs a good supply of moisture, but a dry period during the harvest season is desirable to facilitate curing of mature bulbs.

Per capita consumption of onions has been quite stable at about 5 kg for the past 50 years. The type and use have changed toward more convenience products, such as dehydrated onions, frozen onion rings, and frozen diced onions.

The Plant   The onion plant is very responsive to both temperature and photoperiod. The critical minimum daylength for bulbing varies among cultivars but is normally between 12 and 15 h. Even if the photoperiod is adequate, minimum temperatures must be met, or bulbing is further delayed. The ideal climate would be cool weather early in the season with increasing temperatures as maturity approaches.

Some onions do not form bulbs and are utilized strictly as green onions. Others, called *multiplier onions*, form multiple small bulbs rather than one large one.

Propagation   Onions are propagated from seed, sets, and transplants. Most of the summer onion crop for storage is grown from seed sown directly in the field. A large part of this crop comes from muck soils, which are ideally suited to onion production. For early crops, either transplants or sets are used. Onion sets are small onions, ranging in size from 13 to 19 mm diameter, which have been grown very close together. When seeded late in the season, onions grown for sets reach only minimum size before harvest. The sets are dried, stored, and shipped just before planting.

Harvest and Storage   Onions are harvested as the tops begin to die and fall over. Traditionally, onions have been lifted with a shallow plow and then pulled and windrowed by hand to dry. Today, increasing amounts of onions are lifted, the tops cut, and the onions placed in containers by machine. Storage life varies widely with both type and cultivar. For example the Bermuda onion is known as a very poor keeper and is not stored for longer than a month. For long-term storage, onions are cured, which is largely a drying process. When first harvested, the remains of the tops and roots are damp and provide an area for growth of microorganisms that can cause decay. These damp parts must be dried either in the field or under forced ventilation in storage. After curing, onions are held at about 0°C and a relative humidity of not more than 65 to 70 percent—higher humidities lead to both rotting and root growth. If temperatures are too high or the holding period is too long, sprouting can become troublesome. A spray of maleic hydrazide applied just before harvest is a widely used, effective means of controlling sprouting.

Marketing   Onions are shipped in containers including 23-kg mesh bags, 22-kg cartons, and various-sized consumer mesh bags for holding from less than 1 kg to about 4.5 kg. More recently onions are also being retailed in polyethylene bags, but these must be well perforated to avoid excess moisture and the associated rotting. Bulk displays of onion are also widely used because of the range of amounts purchased. Sweet onions are usually sold from bulk displays or as one to three prepackaged onions.

ROOT CROPS

Some important vegetables are classified as root crops; these include carrots and beets as well as sweet potatoes, which were discussed earlier in this chapter. Other root crops include radish, turnip, rutabaga, parsnip, horseradish, and several other very minor crops.

Carrot

The carrot, *Daucus carota* L., is an important vegetable in the world (Table 17-4) as well as in the United States (Table 17-3). The carrot is in the Umbelliferae family and is derived from wild forms native to Europe, Asia, and Africa. Its widespread utilization as food apparently dates back only to the sixteenth century. Before that period, early forms of the carrot were used for medicinal purposes. It is thought that early settlers introduced the carrot to Virginia in the early 1600s.

United States Industry Of the total carrot crop during the period 1974–1976, about 63 percent were grown for the fresh market. Of these, California grew 60 percent, while Texas and Michigan produced 18 percent and 8 percent, respectively. Of those grown for processing, California led with 36 percent, followed by Wisconsin, Washington, and Texas. Many other states have commercial carrot production but of relatively small consequence. Carrots are also widely grown in home gardens and for local markets. Temperature is particularly important for carrot production. A range of 15 to 21°C has been found to be optimum for both shape and color of roots.

Per capita consumption of fresh carrots showed a marked increase from about 1 kg in the early 1920s to about 5 kg in the mid-1940s but has now declined to about 3 kg in the 1970s. Consumption of canned carrots has been stable at about 0.25 kg. Frozen carrots were of very minor importance in the 1940s but now account for about 0.5 kg. Carrots are widely used as both a fresh and a cooked vegetable. For fresh use, carrots can be peeled or left unpeeled and just washed, and then cut into strips or sliced thin. They add both flavor and color to salads. Carrots may be cooked alone or in combination with meats or other vegetables in stews and soups.

The Plant   The carrot plant is a biennial. During the first year a rosette of feathery leaves and the storage root form. After a rest period, a short stem bearing flowers is formed during the second growing season. The edible part of the carrot is a taproot which varies in length and diameter but ideally is 13 (or more) cm in length and from 1.9 to 3.8 cm in diameter. When the carrot is cut in cross section, two distinct zones are apparent. The outer zone is mostly secondary phloem, and the inner core is mostly secondary xylem and pith. The most desirable carrots are those with a high proportion of outer core, as the xylem can become woody and unpalatable. Large numbers of secondary roots are also present and serve as absorbing organs.

Propagation   Carrots are direct-seeded in the field. Because of the small size and slow germination of seed it is difficult to get an adequate stand of plants without excellent seedbed preparation. Great care is therefore taken to make sure that the soil is fine and free of both clods and surface crust. Thinning of seedlings was once standard practice, but high labor costs have now made it impractical. Precision seeding, therefore, is mandatory.

Harvest and Storage   Carrots are harvested in many ways, but the trend is toward more and more mechanization. Bunch carrots, or those with tops intact, were lifted with a type of plow and then pulled and bunched by hand. The great majority today are harvested without tops by machine and hauled to the packing house or processing plant in trucks or bulk bins. Such fresh-market carrots are washed, sized, hydrocooled, and packed in polyethylene bags holding from 0.5 to 1 kg. Carrots will store quite well at 0°C and high relative humidity. Since bunch carrots lose moisture by transpiration, they cannot be stored for more than about 10 days. Although carrots can be stored for up to several months at 0°C and 90 to 95 percent relative humidity, carrots are harvested in certain areas almost year round. Long-term storage, therefore, is hardly justified.

Marketing   Almost all carrots are retailed in polyethylene bags, since bunch carrots are largely a thing of the past. Refrigeration is necessary, and the combination of polyethylene bags and cold temperatures enables the consumer to buy top-quality carrots year round.

Beet

The garden or table beet, *Beta vulgaris*, is in the Chenopodiaceae family, belonging to the same genus and species as the sugar beet and Swiss chard. Although grown in about one-fifth the quantity of the carrot, the beet is an important vegetable. The beet is believed to be native to Europe and North Africa, but its origin—as we know it—is quite recent.

United States Industry Of beet production for processing during the period 1974–1976, New York and Wisconsin each accounted for slightly more than one-third of the total. Texas accounted for about 7 percent, and the remainder was divided among several minor states. Production for the fresh market is so limited that it is no longer summarized. Climatic requirements are similar to those for carrots and other root crops. Although beets will grow in warm weather, cool temperatures improve sugar content, internal color, and texture of the roots.

Since the 1940s there has been a very marked decline in the consumption of fresh beets. The 1947–1949 per capita consumption was 0.6 kg, but per capita consumption has dropped so low since the late 1960s that it is no longer tabulated by the U.S.D.A. Consumption of canned beets has been stable for the past 20 years at about 0.6 kg. Beets are, however, widely grown in home gardens and for local markets. Beets are used as a cooked vegetable, either whole or sliced; whole beets are usually those of smaller sizes. Sliced beets may be bought as pickled beets or may be pickled at home. The tops or leaves, called *beet greens*, are excellent when prepared like spinach, but unfortunately, they are unknown to many American consumers because so many beets are now bought only as the canned product.

The Plant Like most of the other root crops, the beet is a biennial. If temperatures are particularly cool for extended periods, however, some cultivars tend to produce seed stalks even before the roots reach a marketable size. The edible root is made up of alternating layers of vascular and storage tissues. Ideally, the color is uniform, but the conducting tissues tend to be light and the storage tissues dark-red.

Propagation The seed of the beet are not individual seed but small fruit which contain from two to six seed. The presence of multiple seed makes precision seeding difficult. In home gardens, the seedlings can be thinned and used for beet greens, but on a commercial scale, thinning is not economically feasible.

Harvest and Storage Beets for processing are mechanically topped, dug, and conveyed to trucks or other large containers for hauling to the processing plant. For the limited fresh-market trade, beets may be handled much like carrots. Like other root crops, beets can be stored for extended periods if kept cold and under a fairly high relative humidity.

Marketing Limited quantities of beets are sold in bunches; in this form they must be refrigerated and sprinkled frequently. Topped beets in polyethylene bags are easier to market, since moisture loss is not a problem.

Miscellaneous Root Crops

Radishes, *Raphanus sativus*, are of limited tonnage but are available year round to the American shopper. Since the major use of radishes is in salads, per capita consumption is quite low. Radishes are available in various sizes, shapes, and colors, but are all rather pungent. The radish is an annual or biennial plant. It grows from seed to market size in a very short time, usually in about 3 to 5 weeks. Essentially all radishes are mechanically harvested, topped, and packed in small polyethylene bags in which they are marketed. All the steps are mechanized so that the individual radishes are almost untouched by human hands.

Two other relatively minor root crops are turnip, *Brassica rapa*, and rutabaga, *Brassica napobrassica* which, although closely related, do have somewhat different characters. The turnip is a cool-season crop utilized for both its root as well as its leaves which are cooked as a green. The rutabaga has somewhat thicker leaves, and the root is more dense. Commercially, rutabaga roots are yellow-fleshed, while turnips are white-fleshed. Both are utilized mainly as a boiled vegetable or as an ingredient of soups or boiled dinners.

COLE CROPS

The cole crops of major importance in the United States are cabbage, broccoli, cauliflower, and brussels sprouts. Others of minor importance are kohlrabi and Chinese cabbage. All are members of *Brassica oleracea* in the Cruciferae or mustard family.

Cabbage

Cabbage, *Brassica oleracea* var *capitata*, is by far the most important member of this group of vegetables, both in world production (Table 17-4) and in the United States (Table 17-3). Early forms of cabbage apparently originated in Europe and parts of Asia and have been eaten since prehistoric times. The wild cabbage of antiquity was presumably a nonheading type, but the "hard-heading" types are described in writings of the thirteenth century. Many types have evolved through mutation, selection, and breeding.

United States Industry Of the United States crop in 1974–1976, 79 percent was sold fresh with the other 21 percent being processed, mostly as sauerkraut. Most processing cabbage is produced under contract rather than in an open market. In 1974–1976 New York, Wisconsin, and Ohio accounted for 37, 29, and 9 percent, respectively, of the processing cabbage. The fresh cabbage crop is grown in many states, with Florida and Texas dominant in the winter and spring crops. The summer and fall crops are widely grown with 15 to 17 states reporting commercial acreages.

Cabbage is definitely a cool-season crop and thrives on plenty of moisture. In the South, cabbage is grown in the fall, winter, and early spring, whereas it is grown as a summer crop in the North. Long-term exposure to low temperatures can induce premature seeding, especially if the plants are large when exposed to the cold.

Per capita consumption of fresh cabbage has declined considerably from more than 7 kg in the late 1940s to about 4 kg in the mid-1970s. Sauerkraut dropped slightly during the same period, but only from 0.8 to 0.6 kg.

Cabbage is used as a fresh vegetable in cole slaw or cooked with meats, especially corned beef. Red cabbage adds both color and flavor to salads. Sauerkraut is cabbage which has been cut up, salted, and allowed to ferment.

The Plant The cabbage plant is a biennial, but under some conditions it will send up a seed stalk the first year. The head of cabbage is a large terminal bud, as one can readily see by cutting a head longitudinally. The outer leaves are loose; and as the head matures, the inner leaves become progressively more tightly packed. Most cabbage has smooth leaves, but some cultivars have crinkled leaves, and these are called *savoy* cabbage. Cabbage heads may be flattened, round, or rather pointed. As maturity approaches, stable moisture levels are desirable because a drought followed by rain can lead to splitting of the heads.

Propagation Cabbage is seed-propagated. It can be direct-seeded in the field, or the seedlings can be started in greenhouses and transplanted (Figure 17-11). In recent years increasing numbers of seedlings are being field-grown in the South for shipment to Northern growers. The obvious advantage of this system is greatly reduced cost in comparison with greenhouse plants.

Harvest and Storage Cabbage has been hand-harvested by having workers cut mature heads and toss them onto a truck or by having one person cut the heads and another person pick them and pack

FIGURE 17-11 Mechanical transplanters are widely used to set out young plants which have been imported from Southern areas or started in greenhouses. Some models apply water or a fertilizer solution to the young transplants. (Photograph, Mechanical Transplanter Company, Holland, Mich.)

them in crates. Mechanical harvesters are being tested, but variations in the size and maturity of heads have been troublesome.

Much less cabbage is stored than formerly, as it is available fresh year round. Fresh cabbage is green and storage cabbage pale; consumer preference is strongly in favor of the green, freshly harvested cabbage.

Marketing   Cabbage is shipped either in large mesh bags or in crates and is sold either as bulk-displayed, unwrapped heads or as individual heads wrapped in a polyethylene or some other film.

Broccoli

Broccoli, *Brassica oleracea* var *italica*, is also called *sprouting broccoli* to differentiate it from *heading broccoli*, which is similar to cauliflower.

The early history of broccoli is very vague; it seems to have been widely confused with cauliflower.

Broccoli production is centered in California, which grows about 95 percent of the total United States crop. The only other states with major commercial production are Oregon, Texas, and Arizona. Climatic requirements are similar to those for cabbage. Total broccoli consumption has risen from about 0.5 kg in 1947–1949 to about 0.9 kg in 1976. Fresh consumption per person dropped from 0.4 kg in 1947–1949 to less than 0.2 in the 1960s, but it had risen to 0.5 kg in 1976.

The broccoli plant forms a central head which should be 15 cm in diameter or larger. The head consists of masses of flower buds borne on thick fleshy stems (see Figure 4-10). The entire head is cut, including from 15 to 20 cm of stalk. Once the main head is cut, multiple small heads called *side shoots* develop from the axils of the leaves, and so multiple harvests are possible. Since currently available cultivars vary considerably in time of heading, fields are usually harvested several times to ensure optimum maturity. As soon as the flowers begin to appear, the head becomes loose, yellow, and useless. Broccoli is a nutritious green vegetable which is most often boiled or steamed and may be served alone, in combination with a cheese sauce, or in a casserole with meat. It also makes an attractive, tasty, raw, green vegetable, especially with various types of dips. Because of its color, flavor, and nutritional value, consumption of broccoli will probably increase.

Cauliflower

The cauliflower, *Brassica oleracea* var *botrytis*, is another one of the cole crops closely related to cabbage. Genetically and historically there is great confusion about the development of and distinction between broccoli and cauliflower. Horticulturally, cauliflower differs from broccoli in that the white curd or head consists of malformed or hypertrophied flowers which form a dense head (Figure 17-12).

Cauliflower production is largely in California. For the 1974–1976 period, California grew 75 percent of the fresh-market crop and 84 percent of the processing cauliflower. Other cauliflower producing states include New York and Michigan. Climatic requirements are even more strict than for other cole crops requiring cool temperatures, plenty of water, and high fertilization. Fresh consumption of cauliflower has declined from more than 0.6 kg per person in the mid-1950s to less than 0.5 kg in 1974–1976. During the same period per capita consumption of frozen cauliflower has risen from about zero to almost 0.2 kg.

Because cauliflower is not only white but tender and subject to

FIGURE 17-12 A creamy head of cauliflower ready for harvest is a unique sight. (Photograph, W. Atlee Burpee Company, Doylestown, Pa.)

discoloration, the heads are often covered by tying the leaves up around the head as soon as the head enlarges enough to spread the leaves and thus become exposed. The time from tying to harvest is normally from 5 days to 3 weeks. The "self-blanching" types do not require tying. At maturity, the cauliflower head is about 15 cm in diameter. Cauliflower is cut largely by hand because of uneven maturity in a given field. Most fresh cauliflower is trimmed, washed, and wrapped individually with a clear film over the head. Cauliflower is served much like broccoli.

Brussels Sprouts

Brussels sprouts, *Brassica oleracea* var *gemmifera*, are grown in about one-half the same quantity as cauliflower. Brussels sprouts are grown on a commercial scale only in California. Essentially the entire crop is frozen, although small-scale local production may occasionally be marketed fresh. Fresh consumption is so minimal that it is no longer listed by the U.S.D.A.; frozen consumption has increased from 0.04 kg in 1947–1949 to 0.1 kg in 1974–1976.

FIGURE 17-13 Brussels sprouts are small heads which form in the axil of each leaf up the stem of the plant. They resemble minute heads of cabbage. (Photograph, Harris Seed Company, Rochester, N.Y.)

Brussels sprouts form multiple small heads in the axils of leaves on a growing plant which is rather tall compared with that of other vegetables in the cabbage group (Figure 17-13). The miniature, cabbagelike heads form continuously and have traditionally been hand-harvested repeatedly during the season. The cost of labor has led to the development of once-over mechanical harvesters which cut the plant and then strip the brussels sprouts from the stem. This procedure obviously reduces yields, but savings in labor costs will presumably make it economical.

Brussels sprouts are cooked, usually by boiling or steaming, and are served in many ways including alone with butter or cheese sauce, or mixed with other vegetables in a variety of dishes.

CUCURBITS OR VINE CROPS

The vine crops include cucumber, muskmelon, watermelon, pumpkin, and squash—all of which are in the family Cucurbitaceae. Each is a tender annual crop and is grown for its fruit; all have similar cultural requirements.

Cucumber

The cucumber, *Cucumis sativus*, is a very popular vegetable in the United States (Table 17-3) as well as worldwide (Table 17-4). It has been cultivated since very early times, presumably well over 3000 years. Its origin was probably in India, it was known by the Romans, and it was brought to the New World by Columbus.

United States Industry Of the total crop in 1974–1976, about 28 percent was sold fresh and the remainder processed as pickles. Fresh-market cucumbers come largely from Florida (39 percent) and California (15 percent) with several other states supplying 6 to 8 percent each. From December through April a major proportion of the cucumbers available are imported from Mexico. Many states are involved in growing pickling cucumbers, with the following as leaders in 1974–1976: Michigan (18 percent), North Carolina (12 percent), California (11 percent), Ohio (10 percent), and Wisconsin (7 percent).

Per capita consumption of fresh cucumbers was about 1.2 kg in the late 1940s and gradually increased to about 1.5 kg by 1975–1976. Pickle consumption has increased even more rapidly, going from 1.5 kg in 1947–1949 to 3.5 kg in 1975–1976. Pickles consumed per person increased 48 percent between 1960 and 1974—greater than any other canned vegetable.

Cucumbers, like squash, thrive under somewhat cooler temperatures than muskmelons or watermelons, but germination will not occur unless temperatures are as warm as 13°C, and germination is much more rapid at 30°C. Since the cucumber reaches harvest maturity in a relatively short growing season and will grow at relatively cool temperatures, it can be grown almost anywhere in the United States, although for success in the far North it may need to be transplanted. Since it can withstand no freezing temperatures, winter production, even in Florida, is a risky venture. Greenhouse production of cucumbers was a sizable business in Ohio, but high fuel costs have raised serious questions about its feasibility in the late 1970s. In Europe, greenhouse cucumber production has been a large industry, especially in the Northern areas. The English and Dutch developed cultivars which are large (often 0.6 m long) and usually seedless for this type of production. Production of these European types is gaining in the southern United States.

Cucumbers are used fresh primarily as a salad ingredient, for vegetable trays, or with any number of dressings, sauces, or marinades. Pickles are widely used in many ways.

The Plant   The cucumber plant is an annual trailing vine with a spiny or hairy stem. Cucumbers have traditionally been monoecious (bearing both female and male flowers). Recently introduced hybrids are gynoecious and produce female flowers only. For commercial plantings, a small percentage of a monoecious type is included to provide pollen for pollination. Gynoecious cultivars are used because of their earliness, concentrated ripening, and increased yields. For mechanical harvesting these recently introduced cultivars are particularly beneficial.

Commercial cucumber production is totally dependent on pollination by insects, and honeybees are the primary pollinating agents. Research has shown that cucumber flowers must be visited several times by bees for normal fruit set, and that fruit weight increases with increasing numbers of visits up to 40 or 50. The number of honeybee colonies needed for adequate pollination varies with the type of cucumber being grown. With monoecious cultivars, one colony per hectare might be adequate; with high plant populations of gynoecious types for mechanical harvest, two or three times as many bees may be beneficial.

The great majority of cucumbers fall into one of two categories. Slicing cultivars are those grown for the fresh market. These are grown to a fairly large size and retain their dark-green color. The other major type is the pickling cultivars, which are smaller, somewhat lighter in color, and more productive (Figure 17-14).

Propagation   Cucumbers are usually direct-seeded in the field after the danger of damage from freezing weather is past, but market gardeners may transplant small acreages. Seeding rates vary with the cultivar and, in particular, with the method of harvest. For once-over or "destructive" harvest, very high plant populations are used. With multiple harvests—the normal procedure for the fresh market—lower populations are used.

Harvest and Storage   For the fresh market, cucumbers should be in the range of 12 to 20 cm in length with a diameter of 4 to 5 cm and dark-green in color. To supply top-quality cucumbers, the fields are picked every 2 to 4 days. Cucumbers which are too large or too ripe are removed and discarded because their presence will decrease subse-

FIGURE 17-14 Modern cucumber cultivars are gynoecious and thus produce heavy, concentrated yields. This is a pickling-type cultivar. (Photograph, Harris Seed Company, Rochester, N.Y.)

quent yields of young cucumbers. Hand-harvested cucumbers are placed in containers or onto conveyors. Modern machines transport the pickers in a prone or sitting position, and the picked cucumbers are placed on a conveyor in front of the pickers.

For the pickling industry more and more cucumbers are being harvested by the once-over, or destructive, harvester. For this method to be economically practical, cultivars had to be developed which produced high tonnage of cucumbers of similar maturity at one time. Planting distances and cultural practices had to be modified to encourage maximum production.

Marketing Fresh-market cucumbers are brushed to remove the soil and spines, and are graded. Almost all are now waxed to improve appearance and suppress water loss to maintain the firmness and crispness so necessary in a cucumber. They should be cooled rapidly and held at 7 to 10°C and 90 to 95 percent relative humidity during shipment. Under such conditions, cucumbers will hold for 7 to 10 days. Chilling injury will occur if temperatures remain below about 5°C for any extended period of time.

Muskmelon

The muskmelon, *Cucumis melo* var *reticulatus*, is referred to as the *cantaloupe* in the United States. Muskmelons are thought to have originated in Asia, eventually spreading throughout Europe and the New World. Although the exact type varies, muskmelons are widely grown around the world (Table 17-4). The muskmelon is distinct from some other melons because of its netted skin, and also because of its much better keeping quality.

United States Industry The muskmelon is available to the United States consumer primarily in the summer, with smaller quantities in the market during the spring and still less in the fall. California provided 62 percent of the muskmelons in the 1974–1976 period; Arizona and Texas grew 11 and 10 percent respectively.

Per capita consumption of muskmelons declined slightly from 3.7 to 3.2 kg between the late 1950s and 1975–1976. Muskmelons are grown and marketed as a vegetable but are utilized as a fruit. They are often served as a fresh breakfast fruit, but they are equally delectable in a fruit salad or with a scoop of vanilla ice cream or strawberries in the center as a refreshing dessert. The muskmelon fruit is from 11 to 14 cm in diameter; it has thick orange flesh and multiple seeds in the internal cavity.

The Plant The muskmelon is a tender, trailing vine which needs warm temperatures and plenty of sunshine. Low humidity and low rainfall make the control of fungus diseases much easier; fungus diseases are particularly severe on muskmelons in humid climates. Improved disease resistance is being incorporated into new cultivars. Most of our crop of muskmelons comes from irrigated regions in the Southwest, and furrow irrigation is commonly used to avoid wetting the foliage. The plants produce both staminate flowers and perfect or hermaphroditic flowers. Pollination is by insects, especially by honeybees, which are often brought into the fields for this purpose. Excess moisture just before maturity results in poor quality, due to low sugars.

Propagation For large plantings, muskmelons are direct-seeded; but for small areas or where the growing season is too short, seed may be started in greenhouses and transplanted. In many areas muskmelons are grown on black plastic mulch because they are particularly responsive (see Figure 10-6).

Harvest and Storage Maturity is very critical if top-quality muskmelons are to be offered to the consumer. Because a muskmelon does not increase in sugar level after harvest, immature fruits are not only hard

FIGURE 17-15 When a muskmelon is mature, the stem will separate completely from the fruit; this is called the *full-slip* stage. At this stage, the quality is excellent.

but insipid as well. The most widely used index of maturity is the ease with which the fruit can be separated from the vine. A "full slip" melon breaks from the vine cleanly, leaving no stem attached (Figure 17-15). This is considered to provide a melon of adequate quality if handled properly, shipped under refrigeration, and marketed promptly. "Half slip" melons have some stem remaining with the fruit and are normally considered to be of inferior quality even after adequate softening.

Although many machines have been tested to harvest muskmelons mechanically, the majority are still picked by workers who carry a large bag on their back. Muskmelons should be hydrocooled or iced down as soon as possible and held at 4 to 7°C during shipment. They should not be stored for more than a few days.

Marketing Muskmelons are sold from bulk displays with no prepackaging. If the melons are ripe, those not on display should be refrigerated; most melons are not refrigerated while on display because of the normally rapid turnover.

Other types Several other melons are closely related to muskmelons but are classified as *C. melo* var *inodorous*. The most important of these is the honeydew, a somewhat larger, smooth-skinned melon with a very pleasant, sweet, greenish-white flesh. Honeydews are grown strictly in California and Arizona. The others are Crenshaw, casaba, and Persian. Some of these are shipped, but many are consumed locally in the West.

Watermelon

The watermelon, *Citrullus vulgaris*, is usually considered to have an African origin but was also being grown by Indians in the Mississippi Valley when early explorers arrived. Today, watermelons are grown worldwide (Table 17-4).

In the United States, watermelons are grown in more than 15 states, but production is concentrated in Florida (33 percent), Texas (19 percent), Georgia (10 percent), and South Carolina (6 percent). A long, warm growing season is required for watermelons, but since leaf diseases are less troublesome than on muskmelon, high humidity is less detrimental. The per capita consumption has declined from about 8 kg in 1947–1949 to about 6 kg in 1975–1976. Watermelons are used mostly as a cold, juicy dessert. They can be sliced or scooped out to make melon balls, or the flesh can be scooped out and cubed or balled and the shell filled with a mixed fruit salad of watermelon, muskmelon, and honeydew cubes or balls mixed with other fresh fruits such as strawberries, blueberries, and raspberries.

The watermelon plant is a tender vine with deeply cut leaves, is most often monoecious, and is pollinated by bees. The fruits are round to oblong, ranging in weight from 4 kg to more than 25 kg. The small-fruited cultivars have been developed for the short growing season in the North; most of the commercial melons for the South are the larger cultivars. The commercial cultivars have red flesh, although yellow-fleshed types are available. Until recently only seeded types were on the market, but recently seedless cultivars have been introduced.

For optimum quality, watermelons should be mature but not overripe at harvest. For the inexperienced, it is most difficult to determine maturity as there is little if any change in the skin color. One of the best indices is the change in color from white to creamy yellow of the portion of the melon which is in contact with the ground. Another index is thumping the melon with a finger: a metallic ring indicates immaturity while a dull thud or hollow sound is a fairly good indication of maturity. The degree of desiccation of the tendril immediately proximal or distal to the point of fruit attachment to the vine is also used as an index.

Watermelons are harvested by hand and loaded onto trucks or trailors. Rail shipment was once common, but trucks are more commonly used today. Formerly, the melons were stacked three to five

FIGURE 17-16 There are many types of squash which can be prepared in a wide variety of dishes. (Photograph, United Fresh Fruit and Vegetable Association.)

high on straw, but the modern method uses bulk bins which may travel from the field to the supermarket.

Pumpkin and Squash

These are major crops for home gardeners, pick-your-own farms, roadside marketing, and, to a lesser degree, commercial production (Figure 17-16). This group is not listed among the 25 most important vegetables in the United States (Table 17-3) but is important in various regions of the world (Table 17-4).

The classification and nomenclature of the pumpkin and squash are very confused. All belong to the genus *Cucurbita* and are considered by many to be composed of five species, *C. pepo, C. moschata, C. maxima, C. mixta*, and *C. ficifolia*. Rather than continue the debate of botanical classification, it seems more appropriate to use the horticultural system spelled out by Whitaker and Bohn. Their descriptions were as follows:

Hence, current usage suggests that the term "pumpkin" should be defined as the edible fruit of any species of *Cucurbita* utilized when ripe as forage, as a table vegetable or in pies; flesh somewhat coarse and/or strongly flavored; hence not generally served as a baked vegetable . . . Current usage suggests that the term "summer squash" should be defined as the edible fruit of any species of *Cucurbita*, commonly *C. pepo*, utilized when immature as a table vegetable. Similarly, "winter squash" should be defined as the edible fruit of any species of *Cucurbita* utilized when ripe as feed for livestock, as a table vegetable or in pies; flesh usually fine-grained and of mild flavor, hence suitable for baking.*

BEANS AND PEAS

The two major types of beans are the snap bean and the lima bean, which are both legumes; they are classified as *Phaseolus vulgaris* and *P. lunatus*, respectively. *P. vulgaris* is native to North and South America and was taken to Europe by early explorers.

Snap Beans

This group includes snap beans and beans used in a mature stage and referred to as *dried beans*. Some cultivars can be utilized in the immature stage as a snap bean or allowed to mature and used as a shell bean. Although most snap beans are green, certain cultivars are yellow; these are called *wax beans*.

United States Industry As can be seen in Table 17-3, snap beans rank among the top 10 or 11 of our vegetables. During the period 1974–1976, more than 80 percent of the commercial crop was processed and the remainder marketed fresh. Of the processed tonnage, about 20 percent was frozen and 80 percent canned. Wisconsin and Oregon produced about 23 percent of the processing tonnage each, while New York contributed 14 percent. The fresh-market crop was heavily concentrated in Florida (41 percent), with many other states producing commercial quantities. Climatic conditions best for snap beans are mild temperatures and adequate, but not excessive, moisture. Since snap beans require a relatively short growing season, multiple crops can be grown to provide continuous harvest in areas where temperatures are not excessive.

Per capita consumption varies widely with the type of bean product. Fresh consumption has fallen from 1.9 kg in 1947–1949 to 0.7 kg in 1974–1976; in the same period, canned consumption has risen from 1.3 to 2.7 kg, and consumption of frozen beans has risen from 0.1 to 0.5

*T. W. Whitaker and G. W. Bohn, The Taxonomy, Genetics, Production and Uses of the Cultivated Species of Cucurbita, *Economic Botany*, 4:52–81, 1950.

kg. Snap beans are used as a cooked vegetable, with or without sauces; as a cold salad vegetable; or marinated.

The Plant Since the bean plant is a tender annual, it is planted after the danger of freezing has past and after soil temperatures are at least 16°C to ensure germination. Beans are characterized as either bush or pole beans. Bush beans grow on a low, erect plant which does not need support and seldom is taller than 50 cm. Pole beans grow as a vining plant which must be supported, as it grows to a length of about 2 m. Because of higher yields per hectare and distinctive flavor, pole beans were grown in some areas commercially as well as in home gardens. The development of mechanical harvest of bush beans has, however, relegated pole beans to small-scale production only.

Propagation Beans are direct-seeded in the field, and successive plantings are made to allow continuous harvest. From seeding to harvest takes from 50 to 60 days for bush beans and from 60 to 70 days for pole beans.

Harvest and Storage Complete mechanization of snap bean harvest is available for bush types (Figure 17-17). Multiple harvests provide greater yields, but economics dictate the use of cultivars giving concentrated fruit set to maximize tonnage in a once-over mechanical

FIGURE 17-17 Mechanical harvest of most types of beans is now standard commercial practice. The bin on the back of the harvester can be tipped to empty into a large trailer for transport. (Photograph, Chisholm-Ryder, Niagara Falls, N.Y.)

operation. Pole beans are more productive over a longer period of time, but since they require hand harvest, they have declined for commercial plantings.

Maturity of the beans at harvest is critical. Optimum quality is obtained when the pods are almost fully grown but the beans are only about 25 percent of mature size. If allowed to mature further, the pods tend to develop toughness and fibers or "strings"—a characteristic that plant breeders have essentially eliminated if beans are harvested at the proper stage.

Marketing   For the fresh market, snap beans are hydrocooled and shipped in refrigerated trucks. Retailing may be from bulk displays or in consumer packages, most often prepared at the retail store.

Lima Beans

Included in the category of lima beans are two distinct types: the small-seeded or baby lima and the large-seeded or standard lima. Although there is not complete agreement, most include both types in *P. lunatus*. Within each type there are both vining and bush forms.

Lima beans are consumed in the United States in quantities about 10 percent as great as snap beans (Table 17-3). In the 1974–1976 period, California produced about one-half of the lima beans for processing with Delaware, Wisconsin, and Maryland producing from 4 to 12 percent each. Of those processed, 45 percent were frozen baby limas, 27 percent frozen large seed or Fordhook types, and about 28 percent canned baby limas. Fresh consumption is too small to be tabulated any longer by the U.S.D.A. Large quantities are, however, consumed fresh from home gardens, roadside stands, and pick-your-own operations. Per capita consumption from commercial channels has been stable for 20 years at 0.15 kg of canned limas. Frozen limas accounted for about 0.2 kg in 1947–1949, rose to about 0.3 kg in the 1960s, and had declined to about 0.2 kg in 1976.

The lima bean plant is much like the snap bean in growth habit. Lima beans require a somewhat longer growing season than snap beans because they are grown for the seed rather than the pods. Nearly all commercially grown limas are of the bush type. All processing limas are mechanically harvested by machines which mow off the vines and shell the beans mechanically, and so only the seeds are hauled from the field.

Peas

The pea *(Pisum sativum)* is also called the *garden pea* or, in the South, the *English pea*. Like beans, peas belong to the legume family. Peas apparently originated in Europe or were at least well known to ancient

FIGURE 17-18 English peas are a legume, and the immature seeds are a nutritious green vegetable. (Photograph, courtesy of Harris Seed Company, Rochester, N.Y.)

Greeks and Romans. As can be seen in Table 17-4, peas are widely grown in Europe today. Most of the United States cultivars are grown strictly for the seeds after shelling (Figure 17-18), but Europeans and Asians utilize many cultivars of sugar, or edible-podded, peas.

United States Industry Almost all the commercial peas are processed today, and the U.S.D.A. no longer reports on fresh-market peas. The processing crop is produced in the Northern United States, and the following states were leaders during the 1974–1976 period: Wisconsin, 28 percent; Washington, 22 percent; Minnesota, 15 percent; and Oregon, 9 percent. Of the total processed, 62 percent were canned and 38 percent frozen. Commercial pea production is centered in the North because a cool growing season is required; as maturity is reached, the cooler the temperature, the more leeway is available for harvesting with optimum quality. Per capita consumption for fresh peas has dropped from less than 0.4 kg in 1947–1949 to essentially zero in 1974–1976; in the same period, consumption of canned peas dropped from 2.6 to 1.6 kg, and consumption of frozen peas rose from 0.4 to 0.8 kg. Peas are used as a cooked vegetable and in soups, casseroles, and stews.

The Plant Commercial pea cultivars are of the bush type, although there are also pole or climbing types available which are confined to

FIGURE 17-19 Precision seeding is a necessity in modern vegetable production. This seed drill plants twelve rows at a time. (Photograph, Stanhay, Inc., Dunn, N.C.)

home gardens. The pea is one of the hardy vegetables and is direct-seeded 2 or 3 weeks before the average last freeze (Figure 17-19). Early planting is desirable to ensure maturity before the heat of midsummer, which reduces both yield and quality.

Harvest Peas for processing are all mechanically harvested by mowing and are shelled by machine requiring minimal labor. For the fresh market, peas are picked by hand—a slow, expensive operation which no doubt contributed to the decline in fresh peas on the supermarket shelf. Pea quality depends on both sugar content and texture. For maximum tonnage per hectare, one should wait well beyond the stage of top quality. Since peas should be young, green, tender, and sweet, the grower must balance yield and quality for maximum returns.

GREENS

Among the vegetables used as greens or potherbs are spinach, Swiss chard, kale, mustard, collards, and turnip greens. Since spinach is the only green of major commercial consequence, it will be described in some detail; the others are omitted, except as covered elsewhere.

Spinach

Spinach, *Spinacia oleracea*, is thought to have originated in Asia and spread through Europe only relatively recently. Its widespread use in the United States has occurred only in the past 60 years.

United States Industry   Of the United States crop, about 84 percent is processed; the processed spinach was split almost evenly between canning and freezing in 1974–1976. California produced about 57 percent of the processing spinach, with the remainder coming from many states. The fresh spinach crop is available from California year round (47 percent), from Texas in the fall and winter (26 percent), from the mid-Atlantic states in spring, and from Colorado in summer. Per capita consumption of fresh spinach has declined from 0.9 kg in 1947–1949 to about 0.2 kg in 1974–1976; in the same period, canned consumption declined from 0.5 to 0.3 kg, and consumption of frozen spinach increased from 0.1 to 0.3 kg in the same period.

Spinach is very vitamin-rich and is considered to be one of the most nutritious vegetables. It has traditionally been boiled as a so-called potherb, but it is being increasingly utilized as a tasty addition to tossed salads or as the main salad green.

The Plant   Spinach is a hardy, low-growing, short-stemmed plant which produces a rosette of leaves. The leaves and petioles are the edible portion, and cultivars are available with either smooth or savoyed leaves. The smooth-leaved cultivars are used for processing; savoyed-type cultivars are preferred in the fresh trade. One of the most serious problems with spinach is that it tends to bolt or send up a seed stalk which destroys the value of the plant. Bolting is encouraged by long days and is also strongly influenced by cultivar. There are also complex photoperiod-temperature interactions.

Propagation   Spinach is direct-seeded, usually well before the last freeze, as it is one of the most hardy vegetables and also a cool-season crop. If planting is delayed, the onset of warm weather will induce early bolting. In Southern areas spinach is grown during the winter. Most spinach is planted in rows, but some is broadcast.

Harvest and Storage   Essentially all spinach is now mechanically harvested. For the fresh market it is cut just at the soil line, so that the leaves remain in a rosette. For processing, the plant is cut about 2.5 cm above the soil; only the loose leaves, therefore, are removed, and the growing point is left intact. The regrowth can be subsequently harvested. Since fresh spinach is very perishable, it is put into cellophane bags and shipped immediately. Although spinach can be effectively vacuum-cooled, most is iced down when packed and loaded.

Marketing   Because of its perishability, spinach is placed in consumer packages at the shipping point. It must be refrigerated throughout the marketing process as decay can develop rapidly and make the product unsalable.

Asparagus

Asparagus, *Asparagus officinalis*, is a member of the Liliaceae or lily family and is a perennial crop. As a native of Europe and Asia, asparagus has been eaten for well over 2000 years, and it was brought to the New World by the earliest settlers.

United States Industry   In the period 1974–1976, 63 percent of the commercially produced asparagus was processed. California and Washington each produced over one-third of this crop; Michigan provided about 13 percent. Of the fresh asparagus crop, California grew 76 percent and Washington grew 13 percent. It is apparent from these figures that asparagus is concentrated in a few states on the West Coast and in the Great Lakes region. Asparagus thrives best in cool areas with adequate moisture.

Per capita consumption of fresh asparagus has declined from 0.4 to 0.2 kg between 1947–1949 and 1974–1976, but consumption of canned and frozen asparagus has held quite stable: 0.3 and 0.1 kg, respectively. Asparagus is certainly one of the most delicate, wholesome, and appetizing vegetables available to the American consumer, but it is also one of the most expensive. Cost has no doubt limited the consumption of this most delectable vegetable, which can be served hot as a green vegetable, cold in salads, or hot or cold with a wide variety of sauces (Figure 17-20).

The Plant   Asparagus is a perennial plant which should remain productive for 15 to 20 years, depending on care. The harvested crop is spears 15 to 25 cm in length, which are usually cut at least slightly below the soil line. Although seed-propagated, the plants are grown for at least 1 year before field setting. The underground portion consists of rhizomes, fleshy roots, and fibrous roots. The rhizomes and fleshy roots store the materials necessary for the production of multiple spears or shoots the following spring. Normally asparagus is cut every 1 to 5 days, depending on temperature, and cutting is terminated after 6 to 10 weeks, depending on age of the planting and climate.

Propagation   Some asparagus growers produce their own crowns from seed, but many are purchased from vegetable nurseries. The asparagus plant is dioecious, with separate plants producing male and female flowers. The seeds are planted and grown for 1 year. The

FIGURE 17-20 Although not a major vegetable, asparagus is a tasty, attractive addition to a meal. (Photograph, United Fresh Fruit and Vegetable Association.)

crowns are then dug for field planting. Spacing of crowns varies considerably; but often crowns are spaced 30 to 45 cm apart in rows which are 1.5 to 1.8 m apart.

Harvest and Storage   The harvest of asparagus is a very expensive operation if done by hand. Mechanized harvesting has progressed considerably in recent years, and it is likely that such developments will save the asparagus business as an economically viable industry. Once harvested, asparagus is extremely perishable, as quality declines very rapidly. Hydrocooling is necessary, and it must be held as close to 0°C as possible during marketing.

Marketing   Fresh asparagus is sold in consumer-sized bundles of spears cut to a uniform length.

Artichokes

The artichoke, *Cynara scolymus*, is also called the *globe artichoke*, presumably to distinguish it from the Jerusalem artichoke. The artichoke is a thistlelike herbaceous perennial with deeply cut leaves; it grows to 1 m or more, with a spread of up to 2 m. The artichoke is a native of the Mediterranean region. Its use has been sporadic during history, but records indicate that it was considered a delicacy over 2000 years ago. Artichokes are widely grown in Europe today (Table 17-4).

United States Industry   Commercial artichoke production is confined to California (Table 17-3), where artichokes are grown in low-lying coastal areas between San Francisco and Los Angeles. Other areas of minor importance are along the south Atlantic and Gulf coasts. The artichoke must be grown in a frost-free region; it thrives in coastal areas with foggy, cool summers.

Per capita consumption of fresh artichokes doubled from 0.1 to 0.2 kg between the late 1940s and the mid-1970s. The processed market is much smaller, but artichokes are sold as frozen hearts or as hearts canned in oil, water, or marinade. Sizable quantities of canned artichoke hearts are also imported from Italy, France, and Spain. Artichokes are eaten in various ways: as an appetizer, in salads, or as a hot or cold vegetable.

The Plant   The artichoke is a perennial which is normally grown for 3 to 5 years before being replaced. After a harvest season is completed, the plant is cut back to the ground. As a new top develops, a crop of flower buds develops over the next several months.

Propagation   Although the artichoke can be propagated from seed, the commercial grower uses offshoots or suckers arising from the base of the older plants. Often the old base or crown is divided, with each piece having a sucker attached.

Harvest and Storage   The artichoke is hand-harvested by cutting the stem 3 to 5 cm below the flower head. During cool weather, harvesting is done about every 10 days, but the interval between harvests drops to 4 days during the peak season in the spring. The harvested artichokes are taken to packing sheds, where they are sized, sorted, and packed. For maintenance of quality, the artichoke should be cooled to less than 4°C within 24 h. Being quite perishable, artichokes are not stored except as required in the marketing channels.

Marketing   Artichokes are packed in large boxes or crates and shipped under refrigeration, preferably at 0°C and 90 to 95 percent relative humidity. For retail sale, most artichokes are put on refrigerated shelves in a bulk display.

BIBLIOGRAPHY

Anon.: *Vegetables—Processing, 1976 Annual Summary; Acreage, Yield, Production, and Value.* U.S. Department of Agriculture, Crop Reporting Bd., Statistical Reporting Service, Vg. 1-2 (76), 1976.

Anon.: *Vegetables—Fresh Market, 1976 Annual Summary; Acreage, Yield, Production, and Value.* U.S. Department of Agriculture, Crop Reporting Bd., Statistical Reporting Service, Vg. 2-2 (76), 1976.

Anon.: *Vegetable Situation,* U.S. Department of Agriculture ERS (issued periodically), 1976–1977.

Anon.: *FAO Production Yearbook, 1976,* vol. 30, Food and Agricultural Organizations of the United Nations, Rome, Italy, 1977.

Bianchini, F., and F. Corbetta: *The Complete Book of Fruits and Vegetables,* (translated from the Italian), Crown, New York, 1976.

Hill, A. F.: *Economic Botany,* 2d ed., McGraw-Hill, New York, 1952.

Mortenson, E., and E. T. Bullard: *Handbook of Tropical and Subtropical Horticulture,* Department of State, Agency for International Development, Washington, D.C., 1970.

Purseglove, J. W.: *Tropical Crops, vols. I and II, Dicotyledons,* Wiley, New York, 1968.

Smith, P. G., and J. E. Welch: "Nomenclature of Vegetables and Condiment Herbs Grown in the United States," *Proc. Amer. Soc. Hort. Sci.,* **84:**535–548, 1964.

Thompson, H. C., and W. C. Kelly: *Vegetable Crops,* 5th ed., McGraw-Hill, New York, 1957.

United Fresh Fruit and Vegetable Association: *Facts and Pointers Booklets on All Major Vegetable Crops,* UFF & VA, Washington, D.C., updated and rev. periodically.

Ware, G. W., and J. P. McCollum: *Producing Vegetable Crops, Interstate,* Danville, Ill., 1975.

Whitaker, T. W., and G. W. Bohn: "The Taxonomy, Genetics, Production and Uses of Cultivated Species of Cucurbita," *Econ. Bot.* **4:**52–81, 1950.

CHAPTER 18
FLORICULTURE

 Commercial production of flowers in the United States started in the Philadelphia area in the early part of the nineteenth century. Commercial floriculture includes the commercial production and distribution of cut flowers, flowering pot plants, foliage plants, and bedding plants. Today, production of floriculture crops is a worldwide industry. In the early nineteenth century, most flowers were grown out of doors; this resulted in small, poor-quality flowers. As demand increased, greenhouses came into use to produce flowers year round. Florist shops became the center for merchandising floriculture products. Today, plants and flowers are sold in many types of retail outlets, from drugstores to supermarkets. In many European countries the street sale of flowers is popular.

Commercial floriculture must find solutions to several challenging problems of the future. Increases in the cost of energy and decreases in its availability now necessitate research studies to find more energy-efficient growing structures as well as reliable, economical substitutes for petroleum fuel. The use of solar energy for heating greenhouses is being studied extensively. Better transportation has resulted in advantages to foreign countries where labor costs and environment are more favorable for production. Sales of cut flowers have declined because of changes in customers' interests; new markets will have to be developed with other crops.

California leads all other states in floral crop production, with a total crop valued at $200 million. Florida ranks second in production, with sales around $148 million, the chief crops being gladioli and foliage plants. Greenhouse crops are produced in all areas of the country, but California, Pennsylvania, New York, Ohio, and Illinois account for one-half of the total United States production. The results of a 1976 survey by the U.S.D.A. of the flower and foliage crops in the United States are shown in Table 18-1. In 1976 the wholesale value of these crops was almost $600 million. Table 18-2 shows that mass-market retail sales have significantly increased. The value of all domestic floricultural crop sales has risen about 100 percent since 1970. Consumer expenditures or retail sales exceeded $4 billion in 1977.

In very general terms the floriculture industry consists of whole-sale growers who are production specialists (Figure 18-1), commission merchants or middlemen, and retail florists who provide a wide range of floricultural services and products to the public. There are numerous combinations of the three branches. Wholesale growers produce the merchandise for the market; their business is highly specialized, producing one or only a few crops, usually in large greenhouse ranges (usually 0.25 ha or larger). Because of heating costs, the trend is toward greenhouse production in milder climates. The temperature advantage has to be compared with the transportation disadvantage to

TABLE 18-1
VALUE OF SALES AND PRODUCTION AREA OF FLORICULTURAL CROPS DURING 1976

CROP	NUMBER OF PRODUCERS	VALUE OF SALES,* IN MILLIONS OF DOLLARS	PRODUCTION AREA, THOUSANDS OF SQUARE METERS
Carnations, standard	539	45.6	2675
Carnations, miniature	221	5.6	252
Chrysanthemums, standard	1029	29.3	2087
Chrysanthemums, pompon	1126	34.2	3479
Chrysanthemums, potted	1339	52.9	1645
Gladioli	77	17.2	3372†
Roses, hybrid tea	230	58.7	2187
Roses, sweetheart	192	15.7	487
Foliage plants	1685	235.8	10,816
Snapdragons	572	2.8	231
Poinsettias, potted	1747	35.5	2006
Geraniums, potted	2234	30.4	1378
Lilies, potted	1219	13.9	539
Hydrangeas, potted	448	5.0	275
Bedding plants, flowering	2541	62.0	2964
Bedding plants, vegetable	2343	58.0	1418

*Equivalent gross wholesale value of all crops except foliage. Foliage data based on net value of sales.
†Hectares.
Source: U. S. Department of Agriculture.

TABLE 18-2
TRENDS IN FLORICULTURE CROP DISTRIBUTION BY KIND OF BUSINESS CLASSIFICATION, UNITED STATES

CONSUMER EXPENDITURES FOR THE GOODS AND SERVICES OF FLORICULTURE BY RETAIL OUTLET

YEAR	ALL EXPENDITURES		SPECIALIZED RETAIL FLORIST		RETAIL GROWER		NONFLORIST	
	AMOUNT, THOUSANDS OF DOLLARS	PERCENTAGE OF TOTAL	AMOUNT, THOUSANDS OF DOLLARS	PERCENTAGE OF TOTAL	AMOUNT, THOUSANDS OF DOLLARS	PERCENTAGE OF TOTAL	AMOUNT, THOUSANDS OF DOLLARS	PERCENTAGE OF TOTAL
	M. T. Fossum Estimates Based on U.S. Department of Commerce							
1950	670,000	100.0	400,000	60.0	135,000	20.0	135,000	20.0
1960	985,000	100.0	690,000	70.0	50,000	5.0	245,000	25.0
1970	2,000,000	100.0	1,250,000	62.5	0	0.0	750,000	37.5
	M. T. Fossum Estimates Based on U.S. Department of Commerce							
1970	2,000,000	100.0	1,250,000	62.5	0	0.0	750,000	37.5
1971	2,200,000	100.0	1,375,000	62.5	0	0.0	825,000	37.5
1972	2,500,000	100.0	1,550,000	62.0	0	0.0	950,000	38.0
1973	2,750,000	100.0	1,685,000	61.3	0	0.0	1,065,000	38.7
1974	3,000,000	100.0	1,800,000	60.0	0	0.0	1,200,000	40.0
1975	3,300,000	100.0	1,950,000	59.0	0	0.0	1,350,000	41.0
1976	3,650,000	100.0	2,100,000	57.5	0	0.0	1,550,000	42.5
1977	4,000,000	100.0	2,250,000	56.3	0	0.0	1,750,000	43.7

Source: M. T. Fossum, *Marketing Facts for Floriculture,* Washington, D. C.

FIGURE 18-1 Wholesale production of many floricultural crops requires extensive structures for controlling the environment for plant growth and development. (Photograph by Charles A. Conover, University of Florida.)

determine the most economical growing area. Flowers are often produced in outdoor areas under cloth houses, plastic screen houses, lath houses, or frames, or in totally open areas in Florida, Georgia, and California. Large areas of foliage plant production are centered in Florida, California, and Puerto Rico. The leading states for the production of bulbs are Washington, Michigan, Illinois, and Oregon. California leads the United States in production of bedding plant seed. Bedding plant production and pot plant production are not geographically concentrated because of high shipping costs.

With the expansion of the floriculture industry, a need for the services of the middleman or wholesale commission florist developed. Retail florists purchase almost all materials—such as flowers, plants, and greenery—which they use for arrangements. From these products, they make corsages, wreaths, sprays, and table arrangements. Hence, the retail florist industry mainly encompasses the designing and selling of floral arrangements, along with such customer services as delivery, gift wrapping, and reminder service for special dates such as birthdays. The retail florist industry consists mostly of small businesses. Figures from the U.S.D.A. and the Society of American Florists indicate that in 1970 there were 22,451 retail florists in the United States whose sales were $1.8 billion annually. Two out of every three florists had businesses with annual gross sales of less than $60,000. In 1970 only seven retail florists grossed $1 million or more.

An analysis of the merchandising structure of the retail florists showed that traditional floral sales (wedding, funeral, hospital, and holiday) accounted for nearly 55 percent of gross sales. Of these, sales for weddings, funerals, and illnesses contributed significantly to the total merchandising structure; but holiday sales accounted for only a small portion of florists' annual gross sales. Nontraditional floral sales (artificial flowers, specialty arrangements, etc.) accounted for nearly 45 percent of the gross sales of retail florists in 1960. In the early 1970s this percentage declined to about 10 percent, with some of the remaining 35 percent being replaced by the tropical foliage plants. Florists in areas with populations of 50,000 or more are the most aggressive promoters of nontraditional floral merchandise. Artificial flowers and specialty arrangements were the core of the nontraditional merchandising structure. China, silver, novelty items, giftware, garden center sales, and bedding plants were also important sources of nontraditional sales of many florists.

Floriculture crops are classified by use as cut flowers, potted plants, foliage plants, and bedding plants. The manipulation of environmental factors which makes it possible to produce a marketable pot plant or cut flower out of season is called *forcing*. This requires a knowledge of many factors related to the developmental cycle of numerous plant species. Light and temperature are the most important elements in environmental control. In most instances, forcing is accomplished by utilizing naturally occurring climatic conditions in combination with artificially controlled ones. In other cases, the plants are controlled entirely by artificial means, both environmental and chemical. The exact technique used depends on the species being forced, how and when the pot plant or cut flower is to be marketed, and the climatic conditions existing in the forcing locality.

Interest in foliage plants for home and public buildings, along with a greater demand for bedding plants, has aided the recent growth of the floriculture industry. Flower sales are irregular and seasonal; since flowers are perishable, the florist bears a degree of risk. The demand for certain floricultural crops changes rapidly. All these factors make the industry highly specialized.

CUT FLOWERS

Cut flowers (Table 18-3) are crops grown for the purpose of selling flowers and their stems rather than the intact plant. Commercial cut flowers are grown both in greenhouses and outdoors. California produces 30 percent of all cut flowers; Florida produces 10 percent. Colorado, which specializes in carnation production, produces 7 percent of all cut flowers. Flower production is influenced by market demand, the

TABLE 18-3

FLORICULTURE CROP PRODUCTION, CUT FLOWER CROPS: NUMBER OF ESTABLISHMENTS, QUANTITY AND WHOLESALE VALUE OF CROP SOLD, BY SPECIFIED CROPS, U.S., 1970

CROP	ESTABLISH-MENTS, NUMBER	QUANTITY TOTAL, 1000 UNITS	UNIT	WHOLESALE VALUE TOTAL, 1000 DOLLARS	PER UNIT, DOLLARS	PERCENTAGE OF FLORICULTURE CROPS
All floriculture crops	7969	—	—	484,669	—	100.0
Cut flowers	—	—	—	229,943	—	47.4
Acacia	13	87	Stem	14	.16	—
Anemone	45	3226	Bloom	314	.10	0.1
Anthurium	146	9496	Bloom	1236	.13	0.3
Aster	260	14,998	Bloom	908	.06	0.2
Bird of paradise	66	833	Bloom	170	.20	—
Camellia	54	1182	Bloom	156	.13	—
Carnation						
Standard	1875	640,179	Bloom	49,503	.08	10.2
Miniature	438	30,398	Stem	2758	.09	0.6
Chrysanthemum						
Pompon	2598	35,454	Bunch	30,799	.87	6.4
Standard	2462	150,845	Bloom	31,447	.21	6.5
Fuji and spider	425	20,106	Bloom	2641	.13	0.5
Cornflower	65	4878	Bloom	118	.02	—
Dahlia	84	959	Bloom	74	.08	—
Daisy	161	34,886	Bloom	2241	.06	0.5
Delphinium	118	1130	Stem	165	.15	—
Eremurus	2	21	Stem	3	.18	—
Eucharis	15	107	Bloom	38	.36	—
Freesia	199	1824	Stem	190	.10	—
Gardenia	40	2593	Bloom	683	.26	0.1
Gerbera	54	560	Bloom	108	.19	—
Ginger	29	539	Bloom	62	.12	—
Gladioli	530	23,974	Dozen	20,918	.87	4.3
Gypsophila	154	8809	Stem	1223	.14	0.3
Heather	35	2255	Stem	292	.13	0.1
Hyacinth	104	447	Bloom	101	.23	—
Iris	743	20,785	Bloom	1975	.10	0.4
Larkspur	42	382	Stem	49	.13	—
Lily of the valley	20	261	Stem	34	.13	—
Lily, calla	300	1853	Bloom	314	.17	0.1
Lily	312	3669	Bell	535	.15	0.1
Narcissus	765	45,100	Bloom	1377	.03	0.3
Orchid						
Cattleya	239	4835	Bloom	4042	.84	0.8
Cymbidium	222	8920	Bloom	4788	.54	1.0
Cypripedium	69	135	Bloom	117	.86	—
Other	153	78,923	Bloom	1191	.02	0.2
Peony	131	3686	Bloom	378	.10	0.1

(Continued.)

TABLE 18-3 *(CONTINUED)*

CROP	ESTABLISH-MENTS, NUMBER	QUANTITY		WHOLESALE VALUE		
		TOTAL, 1000 UNITS	UNIT	TOTAL, 1000 DOLLARS	PER UNIT, DOLLARS	PERCENTAGE OF FLORICULTURE CROPS
Ranunculus	38	2044	Bloom	105	.05	—
Rose						
Hybrid tea	464	335,832	Bloom	46,782	.14	9.7
Sweetheart	327	133,303	Bloom	12,783	.10	2.6
Snapdragon	1108	21,628	Stem	3205	.15	0.7
Statice	106	2974	Stem	336	.11	0.1
Stephanotis	160	4655	Bloom	331	.07	0.1
Stock	202	45,872	Stem	1375	.03	0.3
Tulip	412	9502	Bloom	1180	.12	0.2
Violet	13	550	Bunch	145	.26	—
Zinnia	114	1854	Bloom	110	.06	—
Other	247	62,526	—	2611	.09	0.5

Source: Bureau of the Census and M. T. Fossum, *Marketing Facts for Floriculture,* Washington, D.C.

general economy of the country, customs, and fashions. The popularity of mums has greatly increased because they can be produced year round. The decreased use of corsages has lessened the demand for such crops as camellias, gardenias, and orchids. Economically, carnations, chrysanthemums, and roses constitute the largest portion of cut flower crops (Table 18-3).

Carnations, chrysanthemums, orchids, roses, and snapdragons are usually forced in temperature-controlled greenhouses (Figure 18-2). Irradiance level may be regulated by shade materials, and photoperiod can be decreased by the use of black cloth and increased with electric lights.

Cut flowers are grown in raised benches, ground beds, or pots. The optimum width of a bench is 1.2 m, and the width should not exceed 1.5 m. The length of the bench is dependent on the length of greenhouses and the arrangement of benches. Good drainage is critical in both benches and ground beds and is achieved by placing a layer of tile or gravel under the ground bed or placing a layer of gravel under the bench. Aluminum alloy, asbestos rock concrete, concrete slabs, steel, tile, and wood (cypress or redwood) can all be used for bench construction.

Some cut flowers are field-grown. Outdoor production takes place throughout the United States, allowing a succession of crops to be produced. Chrysanthemums and asters are often grown in polypropyl-

FIGURE 18-2 Greenhouse production of carnations, showing the regulation of temperature by the evapo-cooling method. (*a*) Fiber pads with water dripping through them. (*b*) Fans for pulling or drawing air through the pads and greenhouse. (Photograph, Lord and Burnham, Irvington, N.Y.)

(a)

(b)

ene or saran houses to keep out leaf hoppers which transmit disease. These houses decrease irradiance level, lower leaf temperature, and reduce scorching.

The life or keeping quality of cut flowers is determined by water absorption, transpiration, respiration, and plant cultivar. Since cut flowers can absorb water only through the stem, the absorption area is small in comparison with the transpiration area. Turgidity of the tissues, temperature, relative humidity, movement of the air, and absorptive area of the cut surface determine the amount of water absorption and transpiration. Lower temperature, high relative humidity, and still air will decrease transpirational losses.

Cut flowers with high sugar content usually last longer, and respiration can be decreased by lowering the temperature. Chemicals may be applied in the water to control the activity of bacteria and fungi that rot the stems, reducing the life of cut flowers. Examples are Floralife, a solution of hydrazine sulfate, manganese sulfate, and sugar; and Bloomlife, a solution of potassium aluminum sulfate, sodium hypochlorite, ferric oxide, and sugar. The sugar provides substrate for metabolic processes, and the chemicals control the growth of bacteria and fungi, which is stimulated by concentration of sugars in the water.

Chrysanthemum

As a cut flower, the chrysanthemum (Chrysanthemum morifolium), or "mum," ranks first in importance. It is native to Japan and in fact is the Japanese national flower. It was introduced to England and brought to America about 1795. The middle of the nineteenth century saw the introduction of the chrysanthemum as a greenhouse crop; before that time it was grown as a bedding plant.

A myriad of chrysanthemum cultivars allow for variation in size, flower type, and color, and contribute to its popularity. Standards (single, large flowers) or pompons (groups of small flowers), especially popular in Europe, give variety and contrast to the designer. Chrysanthemums are noted for their keeping quality. From the grower's standpoint, the ability to produce the desired grades and types at any time during the year adds to their popularity.

In addition to forcing, a large part of the mum industry is devoted to developing new cultivars, producing disease-free stock, and selling rooted cuttings. Production can be carried on year round by manipulating daylength to alter the plant's normal growth.

Stem-tip cuttings are used to propagate chrysanthemums. The plants are then grown under long-day conditions until the stem reaches the desired length. Thereafter, short days are provided for flower initiation, which occurs when the night is 9.5 h or longer. Depending on the cultivar, 9 to 15 weeks are needed for flower development. Temperatures at night should be 16 to 18°C during vegetative growth and 13 to 16°C after flower initiation. Plants can be grown single-stem or pinched to allow two, three, or four stems per plant. The grower must determine the economics of either using more cuttings or pinching and tying up bench space about three extra weeks. If the plants are pinched, three flowers per plant are allowed to develop on the outside of the bed and two on the inside. Pompons can have three on the inside and four stems on outside plants. Spacing of plants in the bench will depend on whether the crop is single-stem or pinched and on the type of cultivar, time of year, and grade desired. About 155 cm² of space per stem is required. The cost of production in single-stem standard chrysanthemums is approximately $3.75 per 1000 cm² per year. This area generally yields around 17 flowers per year. Pinched spray chrysanthemums can be produced for approximately $3.50 per 1000 cm² per year with a yield of about three and one-fourth bunches per year. After harvesting, plants are removed and a new crop is planted.

Chrysanthemums have a variety of flower types, including in-curved, spider, pompon, decorative, single, and anemone reflexed (see Figure 18-3, page 576). The reflexed type have ray florets with long petals, but the outer florets reflex downward, forming a less formal flower.

Crown bud, summer heat delay, cold delay, and neckiness are some of the physiological disorders of the chrysanthemum. The crown bud is more prevalent in outside production, since it is caused by high temperatures and high irradiance. This can sometimes be corrected by shortening the daylength, causing the plant to flower normally, or by pinching the crown bud to encourage axillary buds to develop. Heat delay, which delays development, is often caused by absorption of heat by the black cloth used to give short-day treatment. This condition can be corrected by lowering the night temperature. Temperatures below 10°C, however, can prevent development of buds.

California and Florida lead the nation in chrysanthemum production under polypropylene or saran houses. From August 1 into October, top-quality mums can be produced if protected from frost. The August–October flowering cultivars grow well under cloth. Eleven-week pompons are used in October. Production parallels that of chrysanthemums grown in greenhouses, although diseases are more of a problem.

FIGURE 18-3 Types of chrysanthemum flowers: (*a*) incurved, (*b*) spider, (*c*) pompon, (*d*) decorative, (*e*) single, (*f*) anemone. (Photograph, Yoder Brothers.)

FIGURE 18-4 Group of hybrid tea roses. (Photograph, George E. Rose, Shenandoah, Iowa.)

Roses

Roses are produced year round, but demand is greatest around holidays. Today, all commercial roses *(Rosa* sp) are hybrids. The Chinese were the first to cultivate the rose. The prevalent form today is the hybrid tea rose (Figure 18-4), but floribunda roses are also grown commercially. One large terminal flower per stem is usually formed by the hybrid tea, the lateral buds being removed. Floribundas generally keep longer than hybrid teas.

Roses are planted during the spring, with flowers produced 2 to 3 months later. Support is needed for the plants. This can be provided by an individual plant stake or by layers of grids created by wire fabric or other materials. The plants are soft-pinched in early stages of development to encourage branching and the production of leaf surface needed for eventual flower production.

Rose pruning usually involves one of two systems. In the systematic system, the plants are cut to a height of 0.6 m or more. Successive prunings are performed 15 cm above the previous year's cut to ensure the cutting of soft wood. Gradual pruning is generally done with a knife. Not all stems are removed simultaneously, and each stem is cut so that at least one five-leaflet leaf per stem remains. Roses should be cut in the early morning, the late afternoon, or both.

It is crucial that roses be grown with their precise light, temperature, and moisture requirements. Roses should be isolated in one area to maintain these exacting requirements and to control disease and insects. Night temperatures should be maintained at 15.6°C for most cultivars. On sunny days, temperatures should range between 20 and 21°C. Humidity should be raised during the day by wetting the walks and watering mulched beds. From late May through August it is desirable to lightly shade the plants to avoid burning of the flowers. Control of aphids, spider mites, mildew, and black spot must be programed.

Dutch Iris

Iris bulbs used for forcing are usually Dutch iris *(Iris tingitana)*. Bulbs are produced in Holland, Japan, and the United States Pacific Northwest, and can be forced from December to June.

Each year the iris produces a new bulb as the old bulb disintegrates. West Coast bulbs dug in July and August are kept at various carefully controlled vernalization temperatures, depending on their later use, until shipped to greenhouse growers. The bulbs are then subjected to 10°C for 6 weeks for growth to begin; this allows normal stem elongation and earlier flowering during forcing.

Bulbs require 10 to 13 cm of soil to meet moisture requirements. Flats, boxes, or benches may be used for planting bulbs 5 by 10 cm or 7.6 by 7.6 cm apart. Uniformity of temperature, good ventilation, and light are also necessary for proper development.

One advantage of the iris is that it can be cut when the buds are only beginning to show color, allowing easier shipment. Upon receipt, placing the stems in water allows flower development to continue. Bud blasting during forcing is caused by high night temperatures (over 16°C), overcrowding, or insufficient light and water. Blindness, caused by premature digging, can be prevented if bulbs are dug after August 15, precooled, and planted by October 15.

Orchids

Orchids are produced worldwide. They are long-lasting and are therefore often used in corsages and contemporary floral designs. The orchid family has 20,000 recorded species, but most are tropical or jungle exotics; few genera are grown commercially. Cattleya (Figure 18-5), Cymbidium, and Phalaenopsis are the usual commercially produced genera grown in greenhouses.

Orchids are herbaceous perennials with a sympodial or monopodial growth habit. Sympodial orchids produce a prostrate rhizome with

FIGURE 18-5 Cattleya orchids are long-lasting cut flowers excellent for corsages and contemporary floral designs.

growth terminating periodically. The Cattleyas and Cymbidiums are sympodial examples. Phalaenopsis and Vandas, which are monopodial, have upright stems which produce closely spaced leaves which periodically develop flower stalks in the leaf axil.

Orchids require 4 to 8 years to produce flowers from seed. Propagation, therefore, is frequently done by "meristeming" or shoot-tip culture. Some cultivars respond to photoperiod, and this can be used to alter production for better marketing. Cattleya flowering can be delayed by long days and promoted by short days, but Cymbidiums do not respond to photoperiod. Phalaenopsis orchids flower from November to June.

Each cultivar requires specific irradiance levels and temperature, depending on type. Cattleyas require about 16°C and about 16,000 to 18,000 lux whereas Cymbidiums require 10°C and 64,000 lux. All growing media must be well aerated, and uniform moisture must be maintained. Good growing media are fir bark, osmunda, or a mixture of sand, peat, and fir bark.

Black leaf areas, dry sepals, and flower deformities are the main physiological disorders. Black spot is caused when intense light rays kill the leaf tissue; reduction in irradiance level alleviates the problem. Dry sepals in Cattleyas are common in high-humidity areas. Air pollution abatement may be needed to correct the problem. Flower deformity can be corrected by proper temperature maintenance during development.

FIGURE 18-6 Snapdragons are one of the few raceme inflorescence flowers produced commercially. Note flower beds at different stages of development scheduled for marketing. (Photograph, Ball Seed Company.)

Snapdragons

Snapdragons *(Antirrhinum majus)* are native to the Mediterranean area. They are one of the few commercially produced flowers with a raceme inflorescence (Figure 18-6). Snapdragons are popular in flower arrangements because they lend themselves to line development in the design composition. They are somewhat difficult to ship because the flowers shatter easily, and bended tips result from geotropism. Most are grown in greenhouses, but outdoor production is possible in southern California. About two dozen inflorescences are produced annually for each 1000 cm² of bench in the greenhouse.

Long-day perennials grown as annuals, snapdragons must be grown in steam-pasteurized soil, since seedlings are very susceptible to root- and stem-rot organisms. Snapdragons require a very porous, well-aerated soil. Pinching delays the crop by 3 or 4 weeks but results in multiple breaks per plant. When pinching is done, the pinch should be made 3 weeks after benching, and four breaks per plant should be allowed to develop.

Irradiance level and temperature influence growth and flowering. Summer-grown plants will flower within 7 weeks after planting. Different cultivars require various light conditions; some require intense light and will flower only in summer, while others tolerate low light and can be grown in winter. Flowers are cut when expansion of the lower florets is complete and the tip florets are still tight. They are then graded on the basis of flower and stem length and immediately placed in water. After cutting, a 4°C air temperature is desirable. When the flowers are cut, the plants are removed.

Poor soil drainage may result in iron chlorosis of tip foliage and encourages disease problems. Winter cultivars often develop hollow stems as a result of low irradiance level. Floret skip (lack of development of individual flowers) occurs in snaps exposed to very cool temperatures during flower formation. Floret shattering is caused by genetic factors and ethylene. Hybridists have developed new cultivars to alleviate floret shattering.

Carnations

Although it is a native of southern Europe, the carnation *(Dianthus caryophyllus)* has been a major floral crop in the United States since the late nineteenth century. The U. S. Department of Agriculture estimates that 613 million blooms were produced in 1976, chiefly in Colorado, California, and New England. Production costs vary with the methods employed but average $2 per 1000 cm² of bench area annually. Bloom prices have decreased slightly over the last 20 years, while production costs have doubled. Research has led to more flowers per 1000 cm² of bench area, with an improvement in flower quality and grade.

Commercial production is sometimes programed for only a single year, so that plants are benched in the spring, flowers cut in the fall, and plants removed the following spring after the second cutting. High light and low temperatures allow a more rapid development of the plants. During the calyx development stage, nights of 16°C reduce the incidence of split calyx.

Carnations require support, which is given by several layers of wire or wire and string grids (Figure 18-2). The first grid is placed about 15 cm above the soil, and the succeeding grids are spaced at increasing intervals, the top ones about 31 cm apart.

Since the carnation lends itself to mass-market sales, production in the United States and importation from Latin America are expected to increase. Demands for shorter stems, a variety of colors, and multiple blooms should increase. Cost reduction will create a greater sales potential and more profit.

Narcissus

Narcissus, or daffodils, include an abundance of cultivars, but only the cultivars grown in the Pacific Northwest or the Netherlands are used for forcing. Bulbs shipped to growers for forcing are distributed as early as September. Temperatures around 9°C are required for cooling. Bulbs not ready for planting may be held at 13 to 16°C. Generally, bulbs are planted in flats.

Cold and forcing temperatures influence stem length. Temperatures of 10 to 13°C cause the longest stems. Forcing procedures must be carried out carefully for the production of quality flowers. Basal rot can be a serious problem during forcing; but losses can be reduced by dusting the bulbs with a good fungicide before planting.

China Aster

Asters *(Callistephus chinensis)* are grown chiefly in cloth houses during the summer. Lighting is necessary during short days for stem elongation. Night temperatures should be 10°C to produce quality asters with strong stems. Greenhouses are usually used for aster production in the spring, when high irradiance levels and low temperatures are prevalent.

As soon as they can be handled, the seedlings are benched. No pinching is required, since they are self-branching. Eight to ten flowers can be produced per plant.

Fusarium wilt borne in the soil is controlled by steam sterilization. Yellows, a viral disease, is transmitted by leaf hoppers from weeds or other host plants. Cloth houses properly built alleviate this problem by excluding the vectors.

A market is beginning to develop for potted asters. They are long-lasting and offer colors not available in mums. Dwarf cultivars of lavenders and purples are increasing in demand.

Gladiolus

The gladiolus *(Gladiolus grandiflorus)*, a native of South Africa, is usually field-grown. Florida produces the winter crops and the Northern states the summer crops. Greatest production is centered in Florida; North Carolina is second.

Technically, the gladiolus produces a corm, but this thickened underground stem is often incorrectly referred to as a bulb. It is day-neutral and has a rest period that can be broken by 4°C storage for 8 weeks before replanting. High irradiance levels are necessary during forcing.

Blindness, the major physiological problem, is caused by low irradiance levels. *Fusarium* causes a very destructive brown rot of the corm. The soil must be either chemically treated before use or retired from gladiolus production for 10 years. Fungicidal dusting or dipping of corms is helpful.

Statice

The genus *Statice* is no longer botanically recognized but has been reclassified as either *Armeria* or *Limonium*. The common name *statice* is still popular, however. Statice is used commercially in winter bouquets and wreaths to supply dry materials for these arrangements. It is also used fresh in arrangements. *Sinuata*, the annual species, is grown in Florida. The perennial types, treated as hardy perennials, are grown throughout the United States and Europe.

Sinuata is planted in late July or early August in Florida. Flowers that develop in January or February are sold to tourists or sent to Northern markets. Later sowings produce successive crops of flowers until spring. Some *Sinuata* are sown in the Midwest in late May to produce late summer flowers.

Latifolia is the most popular perennial type, and it is quite hardy. Blooms are produced the second year. It can be dried successfully, and dieing can be done after drying.

Gerbera Daisy

Gerberas *(G. Jamesoni)*, "Transvaal Daisies," are excellent cut flowers. They are usually field-grown in California and Florida in ground beds with at least 15 cm of loose, well-drained soil containing ample organic matter. To avoid crown rot, the crown must remain above soil level. Growers have also been successful in producing a superior crop in greenhouses. Plants should be set on ridges in the bench to alleviate crown rot. Gerberas are most productive with a night temperature of 16°C. They require more ventilation than most greenhouse crops.

Gerbera flowers are long-lasting and ship well if cut at the proper time—e.g., when the first outer row of staminate flowers releases pollen. Premature cutting will cause wilting. They do not require refrigeration.

Other Cut Flowers

The decline of stocks and sweet peas is due partly to the development of other, longer-lasting cut flowers and partly to the labor requirement for their production. Stocks are largely grown outdoors in California and Arizona. Some retail growers still produce sweet peas on a small scale during the Christmas season and around Valentine's Day. Freesia is produced in southern California fields. These three crops, along with crops such as peonies, dahlias, and gypsophila, are commercial crops grown on a small scale for a limited market.

FLOWERING POT PLANTS

Potted flowering plants are grown commercially to sell in flower. Often these plants are grown to coincide with a particular season of the year or for a particular holiday. However, chrysanthemums, gloxinias, African violets, and azaleas sell well year round. Poinsettias, Christmas cactus, Christmas begonias, and cyclamen are usually produced for December sales. Hydrangeas, calceolarias, and lilies are usually forced for the spring holiday sales. Since these are all short-term crops, greenhouse space can be rotated to grow a variety of crops which are in demand during other seasons. Costs can be cut, therefore, by continuous use of the greenhouse.

Benches in greenhouses designed for pot plant production should be about 76 cm in height and 150 to 180 cm wide. There is usually one main wide aisle and smaller secondary aisles 75 cm or less in width. Benches should be placed so that wheeled carts can move easily alongside. Potting areas with soil, soil additives, fertilizers, and pots should be close by. Space must also be available for soil mixing and sterilization equipment.

Since flowering pot plants are in containers, the volume of soil is very small. The growth of roots is greatly restricted, making the supply of water and nutrients critical. Frequent waterings tend to leach nutrients from the media; thus, frequent fertilization is necessary. The soil mix must be uniform for optimum moisture and air relationships and usually contains a high quantity of organic matter such as peat. If adjustments are needed in the pH of the soil or in its phosphorus or calcium content, they are best made when the potting medium is mixed. Superphosphate and ground or hydrated lime are added in the initial mixing process.

The fertilizer program for flowering potted plants should begin as soon as new root growth is visible. This often occurs within a week after potting. The most efficient method is constant fertilization in the irrigation water. The soil must be uniformly moist so that the fertilizer will not burn the roots. Frequency of irrigation is determined by temperature, light, air movement, the soil's water-holding capacity, and the kind and size of the plant. Small growers without necessary fertilizer proportioner systems may find some of the slow-release fertilizers useful in their pot plant production.

Most pot plant production is for local markets; however, improvements in shipping have made wider market ranges more feasible. Potted plants may be sold through commission houses, but most are sold by the grower directly to the retailer. The cost of marketing the plants amounts to approximately 25 percent whether the plants are sold directly to the retailer or to the commission house.

Table 18-4 presents various statistics for flowering plants.

TABLE 18-4
FLORICULTURE CROP PRODUCTION, FLOWERING POT PLANT CROPS: NUMBER OF ESTABLISHMENTS, QUANTITY, AND WHOLESALE VALUE OF CROP SOLD, BY SPECIFIED CROPS, U.S., 1970

CROP	ESTABLISH-MENTS, NUMBER	QUANTITY TOTAL, 1000 UNITS	UNIT	WHOLESALE VALUE TOTAL, 1000 DOLLARS	PER UNIT, DOLLARS	PERCENTAGE OF FLORICULTURE CROPS
All floriculture crops	7969	—	—	484,669	—	100.0
Flowering pot plants	—	—	—	125,826	—	26.0
Azalea	1745	9750	Pot	16,770	1.72	3.5
Begonia	868	2162	Pot	1204	.56	0.2
Bromeliad	114	170	Pot	307	1.80	0.1
Cacti and succulent	295	2231	Pot	1208	.54	0.2
Calceolaria	201	323	Pot	253	.78	0.1
Christmas cherry	224	142	Pot	184	1.29	—
Christmas pepper	178	230	Pot	173	.76	—
Chrysanthemum	2174	21,542	Pot	35,241	1.64	7.3
Cineraria	380	425	Pot	453	1.06	0.1
Cyclamen	935	1028	Pot	1914	1.86	0.4
Fuchsia	864	1296	Pot	901	.70	0.2
Gardenia	245	1859	Pot	1669	.90	0.3
Geranium	2466	25,508	Pot	14,776	.58	3.0
Gloxinia	917	1109	Pot	1953	1.76	0.4
Hyacinth	1083	2608	Pot	1957	.75	0.4
Hydrangea	883	2035	Pot	3795	1.86	0.8
Kalanchoe	259	372	Pot	524	1.41	0.1
Lily	1766	5359	Pot	10,005	1.87	2.1
Narcissus	445	320	Pot	343	1.07	0.1
Orchid	191	957	Pot	2634	2.75	0.5
Poinsettia	1989	8951	Pot	18,621	2.08	3.8
Primula	140	826	Pot	224	.27	—
Rose	386	615	Pot	1235	2.01	0.3
Saintpaulia	265	2657	Pot	2088	.79	0.4
Tulip	1207	2910	Pot	3651	1.25	0.8
Other	465	6439	Pot	3731	.58	0.8

Source: Bureau of the Census and M. T. Fossum: *Marketing Facts for Floriculture,* Washington, D.C.

Azaleas

All forcing azaleas are derived from either *Azalea indica* or *Azalea obtusum.* Breeding programs have produced hybrids with very different characteristics. The length of cooling period required for flower bud maturation leads to the classification of cultivars as early, midseason, or late. Azaleas bloom naturally in the spring but can be forced to flower year round by manipulation of daylength and temperature.

Commercial producers usually buy budded plants in the fall and precool them at 4 to 10°C before forcing. Plants grown in Oregon or Washington do not require precooling, since they have been precooled naturally. Kurume cultivars require about 4 weeks and Indica cultivars about 6 weeks of uniform cold storage to break the rest period before forcing.

In the production of azaleas for forcing, pinching is required for compact vegetative growth that will yield many flowers. Manual pruning is required to initially shape the plant. After this, pruning to increase branching can be done with chemical pinching agents which contain methyl esters of fatty acids. Six weeks of high temperature should have passed before chemical pinching is begun in late July.

Chlorosis of leaf tips may develop owing to iron deficiency. Leaf drop occurs when roots are damaged by soils too dry or too wet or by overapplication of fertilizer. Very cool temperatures often cause leaf bronzing. This is especially pronounced in high irradiance levels and with insufficient nitrogen supplies.

Easter Lily

The pot lily *(Lilium longiflorum)* is generally the most profitable flowering pot plant produced in a unit area basis in the greenhouse, but since it is in demand only at Easter, production can be risky (Figure 18-7). Bulb production takes place on the West Coast, in the South, and in Japan. In a state of rest when dug in the fall, the bulbs are precooled for 5 weeks or more at temperatures from 0.5 to 3°C, depending on the cultivar, to break the rest period and promote uniformity.

Lily bulbs can be forced at 16°C nights in less than 120 days. Location, exposure, irradiance level, and cultural practices influence the number of days required for forcing. Lilies are ready for sale when the first bloom opens completely; refrigeration at 4°C in the dark one day after the first bud opens enables the grower to hold the plants for 2 weeks should they come into flower too early.

Poorly rooted plants or excessive temperatures during forcing can result in bud blast. Loss or yellowing of lower foliage is often caused by insufficient nitrogen, poor aeration, and overcrowding (insufficient

FIGURE 18-7 Pot plant production of Easter lilies. (Photograph, J. W. Love, North Carolina State University.)

light). Scorching often results from too much phosphorus. Fluoride damage also occurs from superphosphate.

Poinsettia

The poinsettia *(Euphorbia pulcherrima)*, when first introduced into the United States from Mexico by J. R. Poinsett, was grown as an exotic plant in conservatories and botanical gardens. Later it was grown as a cut flower. Today, it is chiefly produced in greenhouses as a pot plant for Christmas flowering (Figure 18-8). Short days are required to produce modified leaves (the colored bracts) which are often called the flower. The true flower of the poinsettia is inconspicuous.

Poinsettias are propagated by stem tip cuttings taken from stock plants; cuttings are rooted during the summer or early fall for Christmas. They are grown single-stemmed, producing one large "flower," or pinched to produce 3 or 4 flowers per plant.

To grow poinsettia for commercial pot production, minimum night temperatures of 16°C must be maintained. Ideal day temperatures are 24 to 27°C, but plants will tolerate considerably higher temperatures.

Full sun is necessary for optimum growth. Flowering is controlled by daylength. Flowers are initiated when nights are 12 h or longer if the night temperature is between 13 and 18°C. In warmer night temperatures, a night of more than 12 h is necessary for flower

FIGURE 18-8 Production of poinsettia pot plants in the greenhouse, scheduled for Christmas sale. (Photograph, Ball Seed Company.)

initiation. The higher the temperature, the longer the night requirement. For bract development, 16°C is ideal. The plants become reproductive in early October and flower for Christmas. Poinsettias can be grown for spring holiday sales by lighting 4 h during the night and then exposing them to short days. Extremely early Christmas sales can be made by short night treatments in mid-September.

Until recently, growth retardants were used to keep stems from elongating. New naturally short-growing cultivars have reduced the need for growth regulators.

Leaf drop, blindness, and bract burn are often seen in poinsettias. If the plant is allowed to wilt, leaf drop will occur in 3 or 4 weeks because an abscission layer is formed. Night temperatures of about 18°C with limited light produce blindness. Molybdenum deficiency will cause leaf and bract burn.

Begonia

Begonias are ideal for flowering pot plants because they will last well in the poor light found in many homes. They may be tuberous *(Begonia tuberhybrida)* or fibrous-rooted *(Begonia semperflorens)*. The Christmas begonia *(Begonia socotrana)* is a semituberous begonia. The Rieger begonia, developed in Germany by Otto Rieger, is fiberous-rooted and a valuable addition to the flowering group.

The fibrous-rooted or ever-flowering wax begonias *(Begonia semperflorens)* are sold both as flowering pot plants and as spring bedding plants. These day-neutral plants are reproduced by seed or propagated from cuttings. Minimum night temperatures of 16°C result in the best flowering.

Tuberous begonias are often used for summer hanging baskets. They can be produced from seed in 6 months. Early December planting in high humidity with light shade beginning in February will result in June flowering. Night temperatures of 18°C should be maintained for the seedlings. Most florists purchase tubers which have been grown previously in lath houses in California. For best results, the tubers should be placed upright in flats of moist peat at 21 to 27°C until several leaves appear. They are then potted in 12- to 15-cm pots with a soil mix of equal parts peat, soil, and sand.

The semituberous Christmas begonias are propagated in November or December by leaf cuttings with a long petiole. These cultivars flower in late fall and for Christmas. Natural light is normally all that is necessary, since the plants will flower at the time of greatest demand (Christmas). Norwegian or Scandinavian cultivars are desirable because they are sturdy, retain flowers well, and have a long life in the home.

Rieger begonias can be made to flower year round (Figure 18-9).

FIGURE 18-9 Rieger begonias can be flowered year round. Note tubular irrigation system for efficient and economical watering. (Photograph, J. W. Love, North Carolina State University.)

Since they are slightly responsive to photoperiod, lights can be used in winter to stimulate growth; three weeks of black cloth covering in the summer to induce flowering will shorten plant height. Crops can be produced in 10 to 16 weeks.

Chrysanthemum

Chrysanthemum, one of the major flowering pot plants, is desirable because it lasts well in the home, has varied colors, and is available year round (Figure 18-10). Demand for these plants reaches its peak at Easter and on Mother's Day.

Chrysanthemums require careful temperature regulation and daylength control in order to produce vegetative and flower growth at the appropriate times. Four or five rooted stem cuttings are placed in 15-cm pots, where they are grown under long day conditions for vegetative growth. After about 10 to 14 weeks, they are pinched and placed under short-day conditions for flowering. Tall cultivars may have short days started 1 or 2 weeks before being pinched.

Nights must be 9.5 h or longer for flower bud initiation. From the start of short days it takes from 8 to 15 weeks to flower, depending on the cultivar. Cultivars are classified by the length of time from flower initiation until flowering.

FIGURE 18-10 Chrysanthemum pot plants grown and flowered in the greenhouse utilizing automatic black shade cloth treatment to provide short light periods and long dark periods. (Photograph, J. W. Love, North Carolina State University.)

Standard cultivars and spray types are used for commercial production because they produce large flowers. Pompons are undesirable for pot plants. Good branching when pinched, relatively short stems, good shaping, and the desired flower size are factors considered when selecting a cultivar for potting. When large-flowered cultivars are used, stems are disbudded to produce only one large flower per stem. Potted mums are troubled by the same disease problems as cut mums. Growth regulators are often used to help obtain the desired height, which can be controlled somewhat by daylength; however, growth retardants such as SADH (succinic acid-2,2-dimethylhydrazide), Phosphon (2-4-dichlorobenzytributylphosphoniumchloride), and A-Rest (α-cyclopropyl-α-(4-methoxyphenyl)-5-pyrimedine α methanol) are employed.

Narcissus

Narcissus can be used for pot plant production from December through April. When the bulbs are received, the grower ventilates, inspects, and stores them just as those used for cut flower production. Daffodils are potted in 15- to 20-cm pots. A desirable height is 30 to 45 cm at the bud stage. Height is affected by the cultivar, the forcing temperature, and the length of cold treatment. If the bulbs are overcooled, tallness will usually result. Forcing temperatures of 15 to 17°C should be used because lower temperatures tend to produce taller plants. Forcing must be timed precisely.

Kalanchoe

The kalanchoe *(Kalanchoe blossfeldiana)*, a long-lasting plant with showy heads of bright flowers, was introduced by Robert Blossfeld, a German hybridizer (Figure 18-11). Grown worldwide, the kalanchoe is popular at Christmas but will flower at any time if the prescribed daylength is used.

Temperatures of 16°C and full sun are necessary for keeping the stems from becoming leggy. Grown under natural conditions, the kalanchoe will flower in the spring; however, short days produced by black cloth from 5 P.M. to 7 A.M. allow the grower to produce flowering plants in proportion to demand. If the black cloth is applied from July 20 to September 20, plants will flower in late October. If flowers are desired in early December, shade should be applied from August 15 to October 1. To produce Christmas blooms, shade should be applied from September 1 to October 20.

The kalanchoe is highly susceptible to stem rot. This can be reduced by steam sterilization of both soil and pots and shallow planting. The plants will not adapt to overhead watering systems. High temperatures prevent budding. Too much water accompanied by high

FIGURE 18-11 Kalanchoe plants are produced world-wide and flower year round if daylength is controlled. (Photograph, Pan-American Seed Company.)

humidity can result in oedema—corky tissue on the leaves. If the days are too short, flower buds do not develop, and the foliage yellows and cups.

African Violet

The African violet *(Saintpaulia ionantha)*, native to tropical Africa, is a very easy-to-grow houseplant because it is tolerant of the low light, warm temperatures, and low humidity found in the average home. Optimum growth occurs at a light intensity of 1200 lux; this makes it economical to produce African violets commercially under artificial light. The violet is day-neutral and flowers year round; however, long days generally induce more flowers. Attractive crown-type growth is formed as leaves and flowers emerge from the compact stem. African violets can be produced from seed in about 10 months or—as is usually the case—they can be propagated from leaf-petiole cuttings.

Water temperature of 21 to 24°C must be used on African violets because colder water causes leaf spotting. Nights of 21 or 22°C should be maintained. Commercial growing of African violets has increased considerably because of the demand generated by the present interest in house plants.

Calceolaria

Calceolaria *(Calceolaria hybrida)*, "pocketbook plant," has pouch-shaped flowers which are quite showy in brilliant yellows, reds, and bronzes (Figure 18-12). New cultivars with better keeping qualities will increase the demand for this old favorite. Most cultivars may wilt under normal home conditions, unless careful attention is paid to watering.

Seed is planted in late summer to produce plants for spring sales. Vegetative growth is best at night temperatures of 16°C. Three months of cooler temperatures are necessary for flowering. At present calceolarias are not suitable for the South; however, improved cultivars of calceolaria are being developed which may some day make production possible there.

FIGURE 18-12 Calceolarias coming into bloom. Favorable growth at low temperatures makes this a good crop to produce to conserve energy in greenhouse production. (Photograph, Ball Seed Company.)

Cyclamen

The cyclamen *(Cyclamen persicum)* is a showy plant which blooms in midwinter. It is very popular in Europe, where 40 percent of the sales of cyclamen are cut flowers. European growers remove the first several flowers from the potted plant for cut flower sales; then, it is sold as a pot plant.

Special care must be taken in growing cyclamens. Plants usually are reproduced from seed or from corms. Most are planted from seed in September or October and sold 13 to 15 months later. Newer cultivars have been developed that can be produced in 8 or 9 months. Flowering date can be planned at roughly 5 months from potting. Forcing temperatures from 10 to 15°C are necessary for flowers to develop. Fertilization should be low until buds are evident.

Low irradiance levels will result in weak growth. Plants will not flower if temperatures are too high, too much fertilizer is applied, or too much water is applied during flower development.

Gloxinia

The gloxinia *(Sinningia speciosa)* is produced as a greenhouse potted plant in the spring and summer (Figure 18-13), although there is an

FIGURE 18-13 Gloxinias are produced in the greenhouse in the spring and summer. They have extremely long flowering periods. (Photograph, Ball Seed Company.)

increasing year-round demand. The gloxinia, a short-stemmed plant with large leaves and bell-shaped flowers, has an extremely long flowering period. The plants are day-neutral, and bud formation is unresponsive to temperature. Gloxinias can be produced quickly from the tuberous stem, but if large quantities are needed, seed production is more economical. Production from the tuberous stem requires 2 to 3 months; production from seed takes 5 to 7 months.

Gloxinias are prone to stem rot and to red spider mites. Brittle leaves develop if the plants are not grown in high humidity and 21°C nights.

Cineraria

The cineraria *(Senecio cruentus)* is a very colorful, inexpensive pot plant to produce (Figure 18-14). It is grown from seed for sale in January through April. Seed sown in June will produce flowering plants in January; seed sown from July 15 to August 1 will flower in February; and seed sown in September will be ready for sale at Easter. Cinerarias are cool-temperature plants and will not set buds at night temperatures above 16°C. It usually takes 3 to 4 weeks at temperatures of 13°C or lower to set buds. The buds will then develop and mature regardless of lower or higher temperatures. However, a night temperature of 7 to 10°C is ideal.

Fertilizer should be applied every 2 or 3 weeks during establish-

FIGURE 18-14 Cineraria plants ready for marketing. They can be produced in greenhouses with temperatures as low as 7°C. (Photograph, Ball Seed Company.)

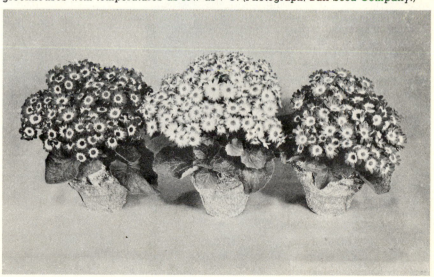

ment to obtain large specimens with dark-green leaves. Spacing is important to prevent the plants from becoming leggy. Cineraria produce a large leaf area and should be watered frequently. In the spring, when the sunlight is intense, the plants may wilt even if the soil is moist. They will need some protection from the sun. However, during the short days the shading should be removed. Since the plants grow rapidly, the cost of production is comparatively low. Several compact strains with numerous flowers are available.

Cinerarias are susceptible to *verticillium*, and all pots, flats, and soil should be sterilized. They are also susceptible to aphids, red spider, leaf rollers, white fly, and thrips.

Other Flowering Pot Plants

Other plants, such as pot roses, primula, crossandra, geraniums (Figure 18-15), hydrangea, and Christmas cactus, are produced commercially as flowering pot plants, but they are not of great economic

FIGURE 18-15 Geranium pot plant grown from seed. (Photograph, Pan-American Seed Company.)

importance. Pot roses are usually sold in the spring so that they can be replanted in the garden. Polyanthas (or baby ramblers), hybrid teas, hybrid perpetuals, and climbers are often sold as pot roses. The development of new cultivars of primula, easier and quicker to produce, has created an increased interest in them in the mass market. The crossandra, which came originally from India, is both unusual and attractive: the plants have a glossy, gardenialike foliage, and flower spikes are produced with overlapping pale-yellow florets. Geraniums are usually sold as bedding plants, but the demand for potted geraniums is increasing, especially in the South (Figure 18-15). Geraniums can be grown from seed or cuttings and flowered year round. Christmas cacti often survive very unfavorable conditions. These hardy plants grow best if they are watered only when dry and fertilized lightly in the early fall.

FOLIAGE PLANTS

Foliage plants are used indoors to create an interior landscape for a home or business. Public buildings often have areas set aside especially for indoor plantings which may include 12-m palm trees and other small and large plants. Foliage plants may provide a point of interest, serve as a screen, or soften harsh lines in contemporary design. Today, many homes include an area for a planter or indoor garden.

Since 1945 there has been a steady increase in the production of tropical foliage plants. Until 1970, foliage plants encompassed less than 10 percent of the wholesale floriculture market. In 1976, about 25 percent of the market was in tropical foliage plants.

Foliage plants generally come from the tropics or other mild climates not unlike the climate of some areas of the South and California. Greenhouses are necessary for Northern production. Florida produces about 45 percent and California about 15 percent of the foliage plants. Many foliage plants are propagated and begun outdoors or in a protective structure in the South and shipped to Northern greenhouses for further growth. In the southern parts of Florida and in California, *Sansevieria*, caladium, and palms are grown outdoors; but *Philodendron, Ficus, Dracaena*, crotons, *Dieffenbachia*, and *Epipremnum* must be grown in plastic screen houses (Figure 18-16). In central Florida, foliage plants are grown in slat sheds, plastic screen houses, and greenhouses. The stock plants of *Dieffenbachia, Maranta, Brassaia, Peperomia, Philodendron,* and *Epipremnum* are grown in heated slat sheds or plastic houses and transferred to greenhouses for

FIGURE 18-16 Production of foliage plants in California in plastic screen houses. (Photograph, John Sorozak, Monrovia Nursery Company.)

propagation and finishing. California growers are the chief suppliers of ivies, cacti, ferns (Figure 18-17), and succulents.

About 10,000 lux are necessary for most foliage plants; however, there are exceptions, as *Sansevieria* and *Peperomia* require 21,000 and *Aglaoenema modestum* needs only 7500. Daylength does not affect foliage plants. The chief problem is providing light in interior design plantings or in dish gardens. Most foliage plants need a minimum of at least 325 lux for 12 h per day. Artificial light may be necessary to meet this requirement.

Temperatures of 21 to 24°C and 75 to 80 percent humidity are necessary for ideal growth. Peperomias, *Sansevieria*, cacti, and succulents withstand dry environments. Plants used in homes and buildings

FIGURE 18-17 Fern production area in cloth houses. This photograph shows good space utilization with large specimens and hanging baskets. (Photograph, Carl E. Bell, Hines Wholesale Nurseries.)

usually require less water because of the low irradiance level and resultant slow growth.

Foliage plants thrive in highly organic soil mixes with at least one-half peat moss or leaf mold. They require a balanced fertilizer. Liquid nitrogen should be applied only after the root system is well established. Generally, foliage plants grow well in slightly acid soils with pH of 5.5 to 6.5.

Table 18-5 gives statistics on the foliage plant industry.

BEDDING PLANTS

Bedding plants are used for flower gardens, window boxes, hanging baskets, and miniature gardens. Although some biennials and perennials are grown as bedding plants, annuals generally predominate and are replanted each spring when the danger of frost has passed. Commercial growers must plan carefully so that the plants are at the right stage and size for planting. Generally, growers propagate bedding plants by seed because this is the most economical method.

Most bedding plants require full sun. Photoperiod cannot be considered, because it generally cannot be controlled outdoors. Plants

TABLE 18-5
FLORICULTURE CROP PRODUCTION, FOLIAGE OR GREEN POT PLANT CROPS: NUMBER OF ESTABLISHMENTS, QUANTITY AND WHOLESALE VALUE OF CROP SOLD, BY SPECIFIED CROPS, U.S., 1970

CROP	NUMBER OF ESTABLISH-MENTS	QUANTITY TOTAL, THOUSANDS OF UNITS	UNIT	WHOLESALE VALUE TOTAL, THOUSANDS OF DOLLARS	PER UNIT, DOLLARS	PERCENTAGE OF FLORICULTURE CROPS
All floriculture crops	7969	—	—	484,669	—	100.0
Foliage or green pot plants	—	—	—	38,375	—	7.9
Caladium	552	1737	Pot	1292	.74	0.3
Dieffenbachia	423	2096	Pot	1597	.76	0.3
Dracaena	659	2848	Pot	1732	.61	0.4
Fern	735	7580	Pot	2417	.32	0.5
Palm	448	1825	Pot	1888	1.03	0.4
Pandanus	149	389	Pot	517	1.33	0.1
Peperomia	415	3185	Pot	1103	.35	0.2
Philodendron						
Cordatum	636	16,405	Pot	3910	.24	0.8
Other	552	4265	Pot	3139	.74	0.6
Schefflera	447	1895	Pot	1387	.73	0.3
Other	638	44,156	Pot	19,387	.44	4.0

Source: Bureau of the Census and M. T. Fossum, *Marketing Facts for Floriculture,* Washington, D. C.

FIGURE 18-18 Containers for production and marketing of bedding plants. (Photograph, courtesy of E. V. Jones and J. P. Fulmer, Clemson University.)

grown in low light or tightly spaced will have weak stems and flower more slowly. Artificial light may be used for seed germination and growing young seedlings. Fluorescent light of 10,000 lux obtained by spacing lights at 1-m intervals 15 cm above the plants is effective. At least 16 h of light per day should be used until the plants are moved into full sunlight in the greenhouse. Although a warm temperature (21°C or higher) is required for germination, most bedding plants grow better under cool temperatures. Night temperatures of 13 to 16°C and day temperatures of 18 to 21°C are needed for slow growth and flowering with short, compact stems of desired diameter.

Seed is sown usually in vermiculite or peat-lite mix. Seedlings are then transplanted to artificial soil mixtures of peat moss and perlite, vermiculite, or fine sand in a variety of containers (Figure 18-18). Liquid fertilizer can be applied after germination.

Careful attention must be·given to watering, since bedding plants are grown in small containers under light and temperature conditions which quickly dry the soil (Figure 18-19). However, overwatering should be avoided since it often leads to root rot problems. Begonias being produced in the greenhouse are shown in Figure 18-20.

The most common disease problem is damping-off, produced by the pathogens *Rhizoctonia*, *Pythium*, and *Phytophthora*. Seed decay, stem rot, root rot, and—in extreme cases—rot of the upper parts of the plant may result. It may be controlled by making the environment

FIGURE 18-19 Production of bedding plants for spring sale. (Photograph, W. H. Carlson, Michigan State University.)

FIGURE 18-20 Greenhouse production of begonias as a bedding plant. (Photograph, Pan-American Seed Company.)

unfavorable for the pathogens or by the use of chemical treatments. Steam can be used to sterilize the soil, containers, benches, and equipment. Soil can also be sterilized with methyl bromide or vapam. Since damping-off pathogens thrive in constantly moist conditions, less-frequent irrigation, better-drained soil, sterile containers, and good air circulation can be used to counteract pathogen growth. A number of soil fungicide drenches give control, but prevention is best.

Bedding plants are also affected by botrytis blight, caused by *Botrytis cinera*. This often develops in the upper portions of the plant, progressing downward, The elimination of old stems, leaves, and flowers will help eliminate sources of botrytis. Irrigation early in the day, proper ventilation, and air circulation are important in controlling botrytis. Moisture on the leaves, especially at night when lower temperatures prevail, aids spore development. Humid, cloudy weather when the foliage stays moist for long periods is the time of greatest danger. Aphids, slugs, thrips, and the white fly are all pests of bedding plants and should be controlled.

Statistical information on bedding plants and their economic importance is presented in Table 18-6.

TABLE 18-6
FLORICULTURE CROP PRODUCTION, BEDDING PLANTS: NUMBER OF ESTABLISHMENTS, QUANTITY AND WHOLESALE VALUE OF CROP SOLD, BY SPECIFIED CROPS, U.S., 1970

| CROP | ESTABLISH-MENTS | QUANTITY | | WHOLESALE VALUE | | |
		TOTAL	UNIT	TOTAL	PER UNIT	PERCENTAGE OF FLORICULTURE CROPS
	—	—	—	44,824	—	9.2
Ageratum	2660	34,305	Plant	1879	.05	0.4
Alyssum	2290	30,330	Plant	1796	.06	0.4
Begonia	1882	15,146	Plant	2093	.14	0.4
Coleus	2619	25,680	Plant	1884	.07	0.4
Dusty miller	1686	10,314	Plant	1230	.12	0.3
Geranium	2085	18,808	Plant	6577	.35	1.4
Impatiens	2213	9646	Plant	1921	.20	0.4
Lantana	1184	13,662	Plant	1096	.08	0.2
Marigold	2886	45,535	Plant	3201	.07	0.7
Pansy	2101	48,313	Plant	2688	.06	0.6
Petunia	3160	156,619	Plant	10,345	.07	2.1
Salvia	2637	30,871	Plant	2211	.07	0.5
Verbena	1852	28,891	Plant	1421	.05	0.3
Zinnia	1984	33,099	Plant	1678	.05	0.3
Other	1067	74,139	Plant	4797	.06	1.0

Source: U.S. Bureau of the Census and M. T. Fossum: *Marketing Facts for Floriculture*, Washington, D.C.

BIBLIOGRAPHY

Ball, V.: *The Ball Red Book*, 13th ed., Ball, Chicago, 1975.

Laurie, A., D. C. Kiplinger, and K. S. Nelson: *Commercial Flower Forcing*, 7th ed., McGraw-Hill, New York, 1969.

———, and V. H. Ries: *Floriculture*, 2d ed., McGraw-Hill, New York, 1950.

Pfahl, P. B.: *The Retail Florist Business*, 3d ed., Interstate, Danville, Ill., 1977.

CHAPTER 19
NURSERY CULTURE

 A major area of horticulture is the *nursery industry*, which is defined by the American Association of Nurserymen as "the production and/or distribution of plant materials, including trees, shrubs, vines and other plants having a woody stem or stems, and all herbaceous annuals, biennials, or perennials generally used for outdoor planting by companies whose major activities are agricultural or horticultural." A nursery is a place where trees, shrubs, vines, and other plants are grown and maintained until they are placed in a permanent planting (Figure 19-1). The nursery may propagate the plants from seed or cuttings, purchase rooted cuttings from a nursery specializing in cuttings and grow them to a salable size, or purchase plants of the desired size and sell them directly.

GROWTH OF THE AMERICAN NURSERY INDUSTRY

The first commercial nursery in America was established by Robert Prince in 1730, in Flushing, New York. The trees and shrubs he grew from cuttings and seeds were obtained from French Huguenots who had brought them to the Western Hemisphere. By 1771, Prince's nursery was offering for sale 42 different pear cultivars. Early nurseries specialized in growing fruit trees and other food plants. Beginning in

FIGURE 19-1 Wholesale nursery producing a wide range of nursery crops. (Photograph, Monrovia Nursery Company, Azusa, Calif.)

1750, the commercial nursery industry expanded westward from the Atlantic seaboard. In 1816, the Stark Brothers' Nursery, which also specialized in fruit trees, was established. Today it is one of the largest nurseries of its kind in the world. Evergreens were first grown commercially by Robert Douglas in Illinois in 1844. The practice of using cellar storage for nursery stock during the harsh winter months was introduced in Alabama by William Heikes around 1850.

The nursery industry has grown since the start of these first nurseries to a billion-dollar industry. Technology and greater interest in the environment and aesthetics of the landscape have changed nurseries from chiefly producers of fruit trees for home and farm orchards to producers of shade trees, ground covers, shrubs, and roses as well. The technology needed for crop production has been provided through university, government, and industry research programs. New and better cultivars of plants have been introduced. Methods and materials for controlling plant pests have improved. Labor-saving devices covering all phases of a nursery operation from planting to shipping have been designed and put into use. Improved methods of transportation have made longer shipments less expensive, allowing firms to locate in areas with the best resources for production.

Garden centers became popular in the mid-1940s because they allowed the consumer to buy all supplies—including seeds, fertilizers, hand tools, and plants—at one center. Nurseries were only seasonal businesses at this time, but with improved technology and intense interest in gardening and the landscape, nurseries have become a year-round business. High-density living—more people in the same area, or even a smaller area—has made people realize how important plants are for maintaining a pleasant environment.

Wholesale nurseries are a good example of growth in the nursery industry. In 1949, there were about 640 firms in the wholesale nursery business; their combined annual sales amounted to $71 million. By 1971, there were about 5000 wholesale nursery firms with annual sales of over $25,000 each; their combined annual sales amounted to over $250 million. These increases have of course been accompanied by similar increases in the number of plants sold, and in their value (Table 19-1).

The nursery industry as a whole employs more than 110,000 people year round; during times of peaks in demand, this number more than doubles, to 235,000. Of the permanent employees, approximately 50,000 are employed by wholesale nurseries and 60,000 by retail, landscape, and other types of nurseries. In both retail and wholesale work, the seasonal demand for labor is about double the year-round demand. The annual payroll of all firms for both year-round and seasonal workers is over $9 million.

TABLE 19-1
PERCENT OF INCREASE IN NUMBER AND VALUE OF NURSERY PLANTS PRODUCED

	INCREASE IN NUMBER OF PLANTS, %		INCREASE IN VALUE OF PLANTS, %	
	1970/1959	1959/1949	1970/1959	1959/1949
Broadleaf evergreens	+20	+180	+ 81	+265
Conifers	+68	+ 65	+ 78	+122
Deciduous shade and flowering trees	+92	+107	+183	+286
Deciduous shrubs	+67	− 21	+156	+ 71
Roses	−13	− 2	+ 73	+ 44
Herbaceous plants	+24	− 17	+ 95	+ 56
Vines (woody)	+83	+125	+297	+ 46
All other landscape plants	—	—	+ 24	+379
Forest tree seedlings	−49	− 22	+ 7	+131
Deciduous fruit & nut trees and grape vines	—	—	+133	+125
Citrus and subtropical fruit trees	—	—	− 5	+627
Small fruits	—	—	+ 60	+ 92

Source: Horticultural Research Institute, Washington, D.C.

The nursery industry utilizes many products from allied agricultural fields. These include fertilizers, chemicals, and shipping materials. More than $13 million worth of fertilizers is used for the production of plants, and much more is handled by retail centers for resale. The use of fungicides, insecticides, herbicides, and other agricultural chemicals in production alone approaches $15 million annually, with more being purchased for resale. More than $50 million is spent annually for products such as containers, burlap, stakes, and mulches for growing and shipping plant materials. This figure does not include materials which were bought for resale.

The nursery industry in 1971 had a sales value of nearly $300 billion. Of this, more than $750 million came from wholesale nurseries. This was an increase of nearly 15 percent, or $100 million, over 1970. Although the largest single contribution of this income was generated through plant production, approximately 12 percent came from other activities such as operation of garden centers, landscape firms, and other types of nurseries in conjunction with wholesale nurseries.

Retail sales of nursery products in 1971 were in excess of $2 billion. This came mainly from retail garden centers and landscape firms. Like the wholesaler, retail firms seldom confine themselves to a single type of activity. Within the retail sector of the nursery industry, approximately 89 to 95 percent of all landscape nurseries and 78 to 86 percent of all garden centers engage in other retail ventures. Most landscape nurseries (up to 76 percent of all firms) engage in some garden center activities; up to 74 percent of all garden centers undertake some landscaping.

There are 545 firms engaged in turf production. These firms had a sales value of $43 million in 1970. Florida firms produced approximately 15 percent of all sod grown in the United States. However, the Eastern North Central States of Ohio, Indiana, Illinois, Michigan, and Wisconsin had approximately 30 percent of the total production of turf in the United States.

Overall, the nursery industry is growing and expanding rapidly. The rapidly increasing cost of materials, supplies, and equipment, along with the higher wages and benefits required by government legislation, combine to keep the profit margin small and hard to increase. To some extent, the use of mechanization wherever possible and chemical weed control to replace human labor has helped defray rising costs. In some nurseries improved business practices, such as inventory controls, have also been used. But the nursery has one very important factor working for it. The product it sells—"plants"—is almost the only commodity which *increases* in value after being sold. Another positive factor is that one good experience with a plant usually creates an appetite for another. Further, the demand for supplies to use with these plants also increases. These factors, along with the realization that plants are important in the creation of pleasant surroundings, are helping the nursery industry.

TYPES OF NURSERIES

Nursery businesses can be broken down into several component types including the wholesale nursery, the retail nursery or garden center, the landscape nursery, and several related types, such as the agency nursery and the mail-order nursery.

Wholesale Nursery

The *wholesale nursery* produces plants in quantities sufficient for sale to retail outlets. It is usually either a field operation or a container operation. Many wholesale nurseries propagate their plants and grow them to a salable size. Some wholesalers specialize in producing liners for sale to other wholesalers, who then grow them for the retail market. Liners are rooted cuttings which are essentially self-supporting when planted in the field or in containers.

Wholesale nurseries are usually located in rural areas where land prices and taxes are low. Rural nurseries can thus afford to hold some land in expectation of expansion without incurring too much additional expense. Important criteria for locating a nursery include a plentiful source of good water, a good labor supply, and proximity to a dependable means of transportation, such as a highway system and airport.

Retail Nursery or Garden Center

Because the *retail nursery* or *garden center* is largely dependent on homeowners for its trade, it must be located quite near a community or city. Consequently, land prices and taxes are usually higher for these operations. Zoning restrictions must be investigated before any land is purchased. Retail nursery growers must consider the traffic patterns of potential customers before deciding on a final location; they must locate where it will be easy for customers to reach them. Like the wholesaler, the retail nursery should have sufficient room to expand when necessary.

The retail garden center sells plants which it may grow itself or buy from a wholesaler. Hard goods such as fertilizers, seeds, tools, and other supplies that homeowners use in their gardens, yards, and homes are also available.

The garden center usually contains an office sales building with a showroom, a shade house or other areas where plants may be displayed, and possibly a small greenhouse. Ample parking space is also important; potential customers will not usually stop to browse or buy if parking appears jumbled and confused. Garden centers may also offer landscape services, such as planning, installation, and maintenance.

Landscape Nursery

The *landscape nursery* should be located near a population center with a high percentage of homeowners, since it is these homeowners who will be needing the landscape services. Owners of newly built homes will also need these services. The nursery may be located on the outskirts of a city or town as its clientele will be more likely to travel a little farther if necessary to obtain landscaping services. Usually, a firm conducts business within a 50-mile radius of its location.

The landscape nursery derives the major portion of its income from landscape jobs. These include drawing plans, implementing or installing the designs, and maintenance after planting. The landscape nursery may grow its own plants, or more often it may buy the plants it needs for implementing a design from wholesale nurseries or retail garden centers.

Mail-Order Nursery

The *mail-order nursery* is a specialized wholesale nursery. It depends primarily on a catalog displaying the stock it offers for sale. Customers order from the catalog and receive nursery plants through the mail or parcel service. Like wholesale nurseries, mail-order nurseries are usually located in the country, where land is relatively inexpensive, near a good source of water, sufficient labor, and a transportation system. Like retailers, they sell to individual retail customers, such as homeowners, and they may also sell hard goods.

Agency Nursery

The *agency nursery* is a nursery which sells stock through agents or sales representatives. Such nurseries are comparatively few in number, and they are a highly specialized part of the nursery industry.

PLANT PRODUCTION IN A NURSERY

Growing of nursery stock can follow two different cycles. Plants can be grown either in the field or in soil media in containers. Commercial nursery production of plants in containers on a large scale is only about 25 years old. It was brought about by the need for mechanization of the production cycle of plants as a result of higher labor costs.

Field Production

Field production consists of planting liners in nursery rows, growing these to a salable size, digging them, and selling the mature plants bare-root (Figure 19-2*a*) or balled and burlapped ("B and B"; Figure 19-2*b*). The field production cycle was once the only method of producing landscape plants, but it has been partly replaced by growing in containers. However, some growers are returning to field production of large specimen landscape plants. Also, technology has improved and streamlined the field production operation so that it is now much more efficient.

Lining Out In the field production cycle, the plants—either rooted cuttings or seedlings—are first lined out in nursery rows by hand or by one of the many kinds of transplanting machines. Liners may be left in one location until they reach a desired size, if spacing permits; or they may be shifted to beds with larger and larger spacing until they are of the desired size. Once in the field, the plants must be fertilized, pruned, and watered. Each of these cultural practices varies with the plant species being grown, the location of the nursery, and the environmental conditions, such as the soil, natural rainfall, and seasonal temperature fluctuations under which the plant is being grown.

FIGURE 19-2 Nursery plants for transplanting. (*a*) Bare-root fruit tree; (*b*) balled and burlapped shrub.

(a) (b)

FIGURE 19-3 Field production of shade trees, showing lower limbs removed for training.

Pruning Pruning is used to produce the desired plant shape. The type of pruning used depends upon the use or purpose of the plant. For example, the lower limbs are usually removed from shade trees (Figure 19-3), whereas shrubs are pruned to promote lateral growth. Pruning (see Chapter 13) should be started when the plant is a liner to obtain quality shrubs and trees. Root pruning should also be practiced in a field nursery. This forces the production of a more compact, fibrous root system, which helps reduce transplanting shock. Plants which have been root-pruned have a higher recovery and viability rate when planted permanently than plants which have not been root-pruned.

Plants such as pines, junipers, and Burfordi hollies have naturally coarse roots. To develop a better landscape plant, one that will adapt when sold and planted, it is necessary to encourage a fibrous root system. This is done by frequent transplanting beginning when the plant is young, and root pruning each time the plant is transplanted.

Large trees and shrubs are often root-pruned the year before they are transplanted. The plant then has several months to develop new fibrous roots within the root ball before being transplanted. Root pruning may be done manually or mechanically by a root-pruning machine, which is a large, hydraulically controlled knife in the shape of a semicircle that is pulled through the soil with a drawbar. The knife enters the ground on one side of the plant and moves under the plant, severing roots at a predetermined depth.

Fertilizers Fertilizers may be applied by side dressings down the rows of the plants. A soil test should be conducted on all parts of the field to determine how much of which kinds of fertilizer to use. The soil test will give recommendations about soil deficiencies and pH which should be followed in designing a fertilizer program (see Chapter 11).

Weed Control Weed control is necessary in the field nursery. Weeds compete with plants for minerals, water, and space, and also make removing the plants difficult. Herbicides for weed control are cheaper and more efficient than hand labor. But the use of herbicides is complicated by the fact that many ornamentals react differently to the same herbicide; what works well with one plant species may be phytotoxic to different genera or species, even if they are closely related. Also, an herbicide can be legally used on a plant only if the plant is listed on the label of the herbicide; these uses are tested and approved (see Chapter 14).

Irrigation Irrigation to provide the proper amount of water is important for the production of good landscape plants. Irrigation may be used to supplement natural rainfall; or it may be the main source of water, with rainfall supplementing it. Water may be applied either through overhead risers or in furrows (see Chapter 8).

Plant Removal Removing the plants from the field before selling them may be done either by hand or by machine. Plants may be dug bare-root, with no soil surrounding the roots, or "B and B," balled and burlapped, with the root ball and its surrounding soil wrapped and secured with burlap. Most evergreens are dug "B and B," while most deciduous plants are dug bare-root. Machine removal is becoming increasingly more important and more widely used because of the shortage and high cost of skilled hand labor. Machine digging is also much faster than hand digging. One type of machine digger consists of four large triangular-shaped blades, which open and encircle the plant and then close (Figure 19-4). The blades are hydraulically driven into the ground one at a time until all the blades meet under the plant. The plant is then lifted out and the root ball is wrapped with burlap. This is done by placing the plant on a square of burlap, which is then pulled tightly and pinned or tied in place. The burlap must be pulled tightly to prevent shifting of the root ball. The plant may also be placed on a burlap square within a wire-frame basket. With this method, the burlap does not need to be pulled as tightly, since the wire basket helps prevent shifting of the root ball. Neither the burlap nor the basket has to be removed when the plant is planted. The burlap will rot and the basket will rust away in a few years in the moist soil environment.

The American Association of Nurserymen (AAN) has established

FIGURE 19-4 Hydraulically powered machine for digging and balling plants in the field.

standards for "B and B" plants. These specify what size the root ball has to be for certain types of plants. For landscape jobs these specifications should be followed carefully.

Many plants, such as fruit trees, are dug while dormant by large machines moving down the row, which cut the plant's roots at a certain depth, and gently lift the plant. The plants must be in a relatively loose, sandy soil so that the roots can be easily cleaned. The plants are then handled in one of three ways. They may be "heeled in" to hold until they are ready to be shipped or planted, they may be held in storage in refrigerated storage rooms with mist to maintain high relative humidity, or they may be peat-balled—that is, the roots may be wrapped in plastic with moist peat around them for shipping. Bare-root plants are cheaper to dig, as they require less time, and the bare-root method has proven to be a satisfactory way of shipping deciduous plants.

Plant Storage Storage of plants before selling or planting them may be accomplished using several methods. Plants may be "heeled in" by covering the root balls or roots with sawdust or a sandy soil. Thus, plants can be removed any time the ground is not frozen. Also, storage houses are used where the proper environmental conditions—a low temperature (−0.6 to 1.7°C), high relative humidity (85 to 90 percent), and adequate ventilation—can be maintained. These temperatures not only minimize respiration rates but assist woody plants in satisfying their chilling requirement. High relative humidities minimize transpiration. The ventilation supplies sufficient oxygen and carries away the carbon dioxide and heat of respiration.

Container Plant Production

Growing plants in containers in commercial nurseries began in the 1950s in California and spread gradually to other production areas. Container growing, which has gained increasing importance in the last 20 years, consists of planting liners in containers and caring for the plants by watering, fertilizing, pruning, and weed control to produce healthy plants. It also involves the transplanting of plants as needed into larger and larger containers until they have reached salable size, after which they are planted permanently.

There are several important advantages to using container-grown plants in the landscape. The plants can be sold when they are the most attractive, thus requiring less handling. The planting season is extended, with less loss from transplanting. Container plants are easier and cheaper to harvest, transportation costs are lower, and there is greater production per hectare. Even though container-grown plants require precise cultural practices, they adapt more easily to new production programs than field-grown plants.

The greatest disadvantage to container production is that the plants may become pot-bound (a *pot-bound* plant has restricted root growth, and its roots grow in a circular fashion) because the containers in which they are planted are not large enough for proper root growth, and the plants become stunted. If a plant is pot-bound, it usually will not grow vigorously when it is planted in a larger container or in the landscape. Also, container-grown plants may have difficulty adapting to the landscape site and soils after being grown in artificial media. Another disadvantage is that many landscaping companies want to buy large trees and shrubs to use in their work. These large specimens have to be grown in very large containers, which are expensive to ship and hard to handle (Figure 19-5). The large containers do not fit into a standardized operation and require a nursery which is especially set up to handle them. Container-grown plants at the nursery are much more dependent upon proper and frequent irrigation than field-grown plants. Repotting is still a disadvantage in container production. Much work is currently being done to make the process more economical. Increased mechanization—e.g., potting machines—has helped in larg-

FIGURE 19-5 Handling container plants of this size is expensive and requires special equipment.

er businesses. Some nursery growers initially plant the liners in large containers to avoid repotting; but this seems to result in a lower-quality root system—if the plant is started in a smaller pot and transplanted into a larger one as necessary, a more compact root system is formed.

Containers For container growing, the perfect container does not exist. Such a container would have a neat appearance and be rustproof, lightweight, and reusable, with a relatively long life. It would be structurally strong enough to protect the root ball from physical damage and to withstand rough handling. It would be insulated to protect the roots from heat and cold, and it would not be affected by herbicides, fertilizers, other chemicals, or adverse weather conditions. It would not be brittle or have a tendency to crack. The perfect container would be stackable for ease in storage, and it would be available to growers at prices they can afford.

The first containers were used food cans. Although they rusted easily and required a large storage area, these food cans were inexpensive, strong, and available.

Eventually, a steel can with tapered sides was developed (Figure 19-6). This container was strong and thus protected the roots from being crushed. Since it had tapered sides, stacking for storage in a minimum of space was possible. The tapered sides also permitted easy removal of the plants from the cans. But, being steel, these cans readily

FIGURE 19-6 Plant container types. (Photograph courtesy of J. P. Fulmer and E. V. Jones, Clemson University.)

conducted heat and cold, and thus they offered no protection to the roots from adverse weather conditions.

Plastic containers were developed next and are still widely used. These containers stack easily, and plants may be easily removed from them. They conduct heat slowly and thus protect the plants somewhat from adverse weather conditions. They are thin, however, and may allow root damage—especially in transit—but they are relatively inexpensive.

Polyethylene bags are now being tried in California. These bags are filled, using special equipment, with a soil mix which prevents sagging. Choosing a soil mix that will prevent sagging, determining the proper amount of tamping, deciding on the number of holes to put in the bottom, and developing proper irrigation techniques are the major problems associated with the use of plastic bags. There is also a problem of acceptance by consumers.

For moderately large, fast-growing plants, treated bushel baskets have been used successfully. The continuous-stave type with no seam at the bottom is the only basket that has proven satisfactory. The chief advantage is that they are relatively inexpensive, but they deteriorate quickly and give little protection to the roots. For large-plant production, baskets of the continuous-stave type may be used. Baskets do not conduct heat readily and thus the roots are protected somewhat from heat or cold. They have little structural strength, but if treated, they will last more than one season. Drainage is good, and the plants may be transplanted in the basket as it soon rots.

Soil Mixes   Since plants in containers are not grown in field soil, a growing medium or soil mix must be supplied. This medium should

FIGURE 19-7 Soil-mixing operation at a large container production nursery. (Photograph by John Sorozak, Monrovia Nursery Company, Azusa, Calif.)

function as soil does. The soil must have certain properties: it should be well-drained but moisture-retentive, and it must have adequate aeration. The components for the mix should be easily obtainable, lightweight, easy to handle and mix, and clean. The mix must be adaptable to mechanical mixing and easily reproducible from one batch to the next. The soil mix should withstand sterilization without breaking down, and it should be capable of retaining nutrients. Only one basic mix should be used throughout the nursery, with slight modifications for different species or cultivars. Having a different mix for every variety of plant grown requires excessive storage space, excessive expense for the different components, and excessive labor time.

The components for soil mixes most commonly used are sphagnum peat moss, sand, vermiculite, perlite, bark, and sawdust (Figure 19-7). Quarry dust, rice hulls, or shavings may also be used. Soil for container production is variable in structure and nutrients, and thus any mix using soil would be less uniform and would also tend to be heavier.

The John Innes Institute in England was the first to develop a soil mix for container growing for a wide variety of plants. However, this mix used composted organic matter, which has the same disadvantages as soil. It is not uniform and not reliable for exact duplication of the mix, and it requires a great amount of storage space in the nursery.

More recently the University of California developed mixes consisting of an inorganic and an organic part. These contain sphagnum moss (the organic part) and fine sand (the inorganic part) in varying proportions, depending on the use to which the mix will be put. The most common ratio is three parts of fine sand to one part of sphagnum peat. These mixes are designed to have the advantages of a clay base soil with none of its disadvantages. Cornell University has also developed a soilless mix. This mix contains sphagnum peat and either perlite or vermiculite and is called *peat-lite* mix.

Any growing medium must be able to withstand sterilization without breaking down or having a buildup of toxic salts. This sterilization should free the soil mix of disease organisms, weed seed, bacteria, fungi, soil insects, and nematodes. The most common method of sterilization is the use of steam. Steam is cheap, easy to work with, and nontoxic; it can be used near living plants without injuring them. Steam is the most efficient method of sterilization and is effective against all but a few weed seed. It is also the quickest method, requiring only 1 h with a 1-h cool-down period after completion. Chemicals used for soil sterilization include methyl bromide and chloropicrin. Methyl bromide takes 24 to 48 h for sterilization and 24 to 48 h for aeration before the soil mix can be used. Chloropicrin takes 48 to 72 h for sterilization and 7 to 10 days for aeration. Both chemicals are dangerous to use and expensive, and they require more time than steam.

The mix may be prepared by several different means. The ingredients are mixed mechanically, such as with tractors with front-end loaders. A converted cement mixer may also be used. For large-scale operations, a specially designed soil mixer is available (Figure 19-8). The preparation of the soil mix should be scaled to the size and scope of the nursery operation using it.

FIGURE 19-8   Soil-mixing machine. (Photograph by John Sorozak, Monrovia Nursery Company, Azusa, Calif.)

Fertilization Nitrogen, phosphorus, potassium, and various trace elements must be added to container-grown plants. These essential elements may be applied either in the dry form or in solution with water.

When dry fertilizer is used, it may be applied on the surface of the soil, or it may be incorporated into the growing mix. One method of applying dry fertilizer is to use measuring spoons to apply the desired amount to each plant. A fertilizer dispenser can be used to ensure uniformity and increase the speed of the operation. Tablets of fertilizer can also be applied to the soil surface of the containers.

Liquid fertilizer can be applied by spraying or by adding it to the irrigation system. The fertilizer can be injected into the watering system with a simple pump that adds a certain unit of fertilizer per volume of water. In some of the larger nurseries a complex injector system is used. Huge tanks, prefilled with fertilizer, are brought to the nursery and attached to the watering system. Intricate electrical controls feed precise amounts of liquid fertilizer into the water at the appropriate time. In constant-feeding, 10 percent extra solution is applied; this results in some leaching at each water-feed application to prevent salt buildup. The advantages of irrigation-applied fertilization are obvious. Labor costs are reduced, and the plants are kept under optimum nutrient conditions at all times. The possibility of wasted fertilizer and the increased need for weed control in areas outside the containers are the chief disadvantages of this system.

Nitrogen, often the limiting nutrient in plant growth, is available in inorganic forms such as calcium nitrate and ammonium nitrate. Urea-formaldehyde as a nitrogen source combines the long-lasting effect of organic nitrogen with the economy and ease of mineral fertilizers. Phosphorus is available to plants in the PO_4 form. Superphosphate and treble superphosphate are the best sources; the element is always present in complete fertilizers. Potassium, available to the plant as the $K+$ ion, is furnished in potassium salts, e.g., potassium chloride. Since potassium salts are highly water-soluble, they can easily become toxic, but they can easily be leached from the soil. Potassium sulfate is not as toxic in terms of salt buildup as potassium chloride. Possibly sulfur-coated urea and KCl will soon be commercially available and useful. Researchers have recently developed glass frits of potash which are slowly available to the plant. Sequestered or chemical forms and glass frits of several trace elements, such as iron and manganese, are available commercially. In these formulations, the trace element is combined with another molecule which adheres to the soil particles, allowing the trace element to become slowly available to the plant.

Plants have different nutrient requirements at different times of the year. The soil in the containers should be tested several times a year to determine fertilizer requirements.

The trend in the nursery industry today is toward long-lasting or slow-release fertilizers. Osmocote is a slow-release fertilizer that is becoming widely used. There are three osmocote programs: 14-14-14 applied every 3 months, 18-6-12 applied every 9 months, and 18-5-11 applied once a year. In each program, each time the plant is watered, fertilizer is automatically released. The 18-5-11 program saves labor, since it requires only one application a year; it is an outstanding example of the slow-release concept. With all the programs, a quick-release fertilizer must be applied to act immediately, since osmocote takes 6 weeks to begin releasing fertilizer. Also, if a synthetic soil mix is being used, minor elements must be added, since there are none in the osmocote.

Potting and Bedding Out   Potting consists of putting the soil mix into containers and planting the liners in the containers. Again, the potting operation must be tailored to the size and scope of the nursery that is using it.

There are machines designed specifically for potting (Figure 19-9), or a machine can be built to fit the needs of the nursery, or the plants may be potted manually. The containers should be filled with soil mix.

FIGURE 19-9   Potting machine which mechanizes the canning operation. (Photograph, John Sorozak, Monrovia Nursery Company, Azusa, Calif.)

FIGURE 19-10 Bedding out containers of *Gerbera jamsonii* in the field. Note gravel base for drainage of the beds. (Photograph, Hines Nurseries, Santa Ana, Calif.)

The hole should be prepared and the liner placed in it. The mix should be watered and loaded to be taken to the growing beds (Figure 19-10).

Pruning Pruning is used to produce compact dense plants of the proper shape. With containers, root pruning is not necessary, because the container acts to restrict the root's outward development. Caution must be taken, however, that the plants do not become pot-bound. This condition can be prevented by transplanting plants to larger containers as necessary. Further, the root ball can be cut or cracked when transplanting to encourage new root formation and to discourage "pigtailing" (encircling) roots.

Irrigation Water and watering are absolutely essential to a container nursery. The source of water must be sufficiently large to meet the demands of the nursery. Water can be delivered to the plants by overhead risers or through a tubular system; either method can be operated manually or automatically through a system of clocks and solenoids. The tubular system is a series of thin, flexible pipes which feed off a main pipe to each individual pot. With this system, water goes only where it is needed and work can be done in the beds while they are being irrigated. Also, if fertilizer is injected into this irrigation system, there is less waste than with other methods. Fungal disease is held to a minimum because the foliage of the plant is not wet. However, the cost may be prohibitive for the smaller-size containers. Setting up a tubular irrigation system is expensive, and it requires a high level of maintenance.

Weed Control   Because there is less available growing space within containers, competition from weeds is even more severe than in the field. For this reason weed control is essential in container nurseries. As a precaution, the beds where the plants will sit should be fumigated to kill all undesirable weed seeds, soil insects, nematodes, and disease organisms. They should then be covered with black plastic, which will act as a physical barrier to any germinating seeds that are not killed or that were carried in before the beds are covered. Herbicides should be used to control weeds within the containers themselves, using the same precautions as with the field. The use of herbicides is one of the more inexpensive methods of weed control. It is estimated that chemical weed control costs approximately $100 per acre, whereas mechanical or hand weed control costs $3000 per acre. Generally, a combination of chemical and mechanical methods is used.

Every nursery should be equipped to apply both the granular and the liquid forms of herbicides. Since one of the most important factors in applying an herbicide is the precise calibration of the equipment, the equipment should be checked often. The sprayers for herbicide application are usually mounted on tractors and may be powered by the power-takeoff of the tractor or by a gasoline engine mounted on the sprayer. The nozzles are set low to the ground and mounted on steel booms. The person operating the sprayer should be constantly alert, since mistakes can be costly. Tractor speed, constant pressure, and nozzle height are among the most critical aspects of herbicide application.

For smaller nurseries and for situations in which a tractor-mounted sprayer cannot be used, backpack sprayers are available. These sprayers use hand pumps of various types and a hand-held nozzle. Generally, low pressures (1.4 to 2.8 kg cm^{-2}) are used in herbicide applications to keep droplet size large, thereby minimizing drift.

Herbicides may be nonselective, that is, toxic to all plants; or they may be selective, that is, toxic to some plants and nontoxic to others. Two groups of herbicides are used for nursery weed control: preemerge chemicals and contact chemicals.

Preemerge chemicals must be applied before the time the weeds germinate. Some preemerge chemicals must be incorporated into the soil before the plant is potted because they may affect root growth in some species. Other preemerge chemicals are applied to the soil surface before weed seed germination. These surface-applied chemicals kill the weed as it emerges after germinating. Most of the preemerge chemicals control grasses such as crabgrass, crowfoot, and Bermuda grass.

Contact chemicals kill only what they contact. These are used as a direct spray to kill those grasses and weeds that have already germinated. The contact chemicals kill only the top of the plant; they do not kill the root system. These chemicals are used in nurseries to keep down weeds. around beds, greenhouses, and fields. The nursery grower's control of these weeds by mowing or by spraying herbicides also controls a major source of weed seed.

Clean pots should always be used to help prevent contamination by weeds and diseases. If pots are being reused, they should be fumigated; this is accomplished by placing them in an airtight chamber and steaming them or using a chemical fumigant, e.g., methyl bromide. Plastic containers should not be steamed, as they melt at high temperatures.

Another means of controlling weeds in containers involves cutting fiberglass or rubber disks to fit around the plant stem and close to the side of the can. These disks allow fertilizer, water, and air to reach the roots in satisfactory amounts while presenting a barrier to weed seedlings.

Winter Protection   Winter, or cold, protection varies with the location of the nursery and the anticipated severity of the winters there. Winter protection is more important with container-grown plants. Where the winters are known to be severely cold—i.e., where the weather turns cold early and stays cold, as in the Northern areas of the country—unheated plastic houses may be the best and safest form of protection. Such houses are expensive but may well be worth the price because of the persistent cold. On the other hand, in areas such as the Southeast and California, where it is not consistently cold throughout the winter months (that is, the temperature may be fairly warm one day and fairly cold the next), plastic houses may not be worth the expense. In such areas, sufficient protection may be provided by using shade cloth (Figure 19-11) or by moving the containers tightly together, especially under a stand of pines, which would act as a canopy and windbreak. After the containers are placed close together, the outer edges could be ringed with sawdust, foam, or some other insulating material to provide additional protection.

NURSERY CROPS

Several hundred species of plants are produced in nurseries. They are classified into three broad groups based on growth characteristics and similarities in cultural requirements: (1) coniferous or needled ever-

FIGURE 19-11 View of saran shade house showing section of cultivars of *Hedera helix.* (Photograph, Hines Nurseries, Santa Ana, Calif.)

greens, (2) broadleaf evergreens, and (3) deciduous plants. The various species and cultivars in each group may be classified according to their habit of growth—that is, trees, shrubs, ground covers, and vines.

Coniferous Evergreens

The coniferous evergreens usually have thin, narrow foliage; the leaves are extremely slender in relation to their length. This group of coniferous evergreens includes the pines *(Pinus)*, the firs *(Abies)*, and the spruces *(Picea)*. Some coniferous evergreens have very small scalelike leaves. These leaves overlap on the stem and occur with a very high density. This group includes the Junipers *(Juniperus)* and the false cypress *(Chamaecyparis)*.

Most coniferous evergreens are produced from seed. The seed for coniferous evergreens are collected from parent plants with the desirable characteristics. Once seedlings have germinated and are established, they are either lined out in the field or transplanted into containers from which they are transplanted into larger containers until they are ready for marketing. Many coniferous evergreen seeds will not germinate immediately upon planting and require stratification, scarification, or both to break the seed's dormancy.

An exception to seed propagation of coniferous evergreens is the perpetuation of desirable "sports," or mutations, in form (pendulous or prostrate as opposed to upright), in color (blue-green or yellow instead of green), or in size (dwarf shrub versus large tree). When a sport is considered worth perpetuating, cuttings are taken from the mutated branch, and the plant is perpetuated vegetatively. Most Junipers are propagated by cuttings.

Coniferous evergreens can be produced throughout the United States, provided hardiness requirements for the particular crop are satisfied. A coniferous evergreen can be produced that will grow in almost any location.

Broadleaf Evergreens

Broadleaf evergreens, which represent a higher sales value than coniferous evergreens, have leaves that vary considerably in size, and the leaves possess some width as well as length. These variations in leaf size can be seen within the general group as well as within the individual genera which make up the group. Within the classification of broadleaf evergreens are the tiny-leaved boxwood (Buxus), some hollies (Ilex) and the cotoneasters, the medium-leaved azaleas and camellias, and the large-leaved rhododendrons and magnolias.

Most broadleaf evergreens are produced from cuttings. Once rooted, the liners are lined out in the field or are placed in containers. Most broadleaf evergreens, such as hollies and camellias, prefer an acid soil with a pH from 5.5 to 6.5. Daphne and cotoneaster, however, prefer a more alkaline soil. The Ericaceae family, which includes azaleas, rhododendrons, and heathers, prefers a soil pH of 4.5 to 5.5.

Deciduous Plants

Deciduous plants are produced throughout the United States. Many are propagated by vegetative methods such as grafting and cuttings, whereas others are produced from seed. Examples are apples *(Malus)* propagated by budding, grapes *(Vitis)* propagated by dormant cuttings, and *Viburnums* produced from seed. Each crop has specific cultural requirements for production.

BIBLIOGRAPHY

American Association of Nurserymen: *USA Standard for Nursery Stock*, Washington, D. C., 1969.

Baker, K. F.: *The U. C. System for Producing Healthy Container-Grown Plants*, University of California Division of Agricultural Sciences Manual 23, Berkeley, 1957.

Beard, J. B.: *Turfgrass Science and Culture*, Prentice-Hall, Englewood Cliffs, N. J., 1973.

Black, C. A.: *Soil-Plant Relationships*, 2d ed., Wiley, New York, 1968.

Kramer, P. J., and T. T. Kozlowski: *Physiology of Trees*, McGraw-Hill, New York, 1960.

Leopold, A. C., and P. E. Kriedemann: *Plant Growth and Development*, 2d ed., McGraw-Hill, New York, 1975.

Patterson, J. M.: *Container Growing*, American Nurseryman, Chicago, 1969.

Pinney. J. J.: *Beginning in the Nursery Business*, 2d ed., American Nurseryman, Chicago, 1967.

CHAPTER 20
LANDSCAPE
DESIGN

Landscape design functions in preserving and protecting our natural environment (Figure 20-1). The development of the environment determines our reactions to the landscape. The environment can subdue us, dominate us, overpower us, or inspire us. Through design we can manipulate and control the environment to improve our relationship with it (Figure 20-2). Within the creative process of nature and art, growth may be directed and developed to attain a better quality of life.

The renewed interest in the preservation and protection of the natural environment indicates sincere and determined efforts to heal the wounded earth around us. Thoughtless destruction and careless rebuilding have never been acceptable practices. Through the creative processes of art and science and with the judicious observance of the principles of nature, the foundation for a better environment can be built.

HISTORY OF LANDSCAPE DESIGN

Although many lessons can be learned from historical precedence and modern practice, the designer needs to consider the unique qualities of land, plants, water, and space to enhance and accomplish

FIGURE 20-1 Waterfalls developed with a few stones on a creek, showing the rewards of working with, instead of against, nature. (Photograph, Warren Uzzle, Raleigh, N.C.)

a purpose. In achieving a goal, the designer may apply engineering, architecture, horticulture, ecology, geology, and social sciences, thus joining the technology of science and the aesthetics of art to achieve success in landscape design.

To understand current design, it is necessary to understand what was done in the past and why it was done that way. Each civilization tried to fulfill the needs of its people. As time passed, needs changed; therefore, design considerations also changed.

The earliest recorded gardens were built by the Egyptians around 4000 B.C. The Egyptians began building their dwellings along the Nile River, with walls around them for defense. Inside these walls formal gardens were developed, with the house at one end. Since the Nile was a major source of transportation, every house had a boathouse on

FIGURE 20-2 Design used to manipulate and control the environment and to enhance people's relationship with it. (Photograph, Robert E. Marvin and Associates, Walterboro, S.C.)

the river. The main axis ran from the boathouse to the dwelling; there was an arbor of grapes over the path, and pools, used for irrigation, on each side of it. Flowers were planted around these pools, and double rows of date palm trees were planted around the walls. At the end of the pools on each side of the path, kiosks (which may have been used as summerhouses) were built. The serfs' quarters and grain fields were outside the walls. Everything in these formal gardens was planned for a functional purpose as well as for beauty.

The Persians made extensive use of landscaping around 1500 B.C. Most of the aristocracy built their houses in the hills and, like the Egyptians, enclosed their gardens for protection. The main dwelling unit served as a focal entry point. Since the land was hilly, gardens were bench-terraced. Wells high in the hills provided fast-flowing water for irrigation and garden fountains. The garden contained a large central pavilion where all entertaining took place. Floors were covered with colorful rugs for sitting. The garden was divided into square or rectangular compartments, each having a pool, plant materials, or mounds for viewing the surrounding countryside or polo games. The Persian garden was highly functional as well as beautiful. Trees such as peach, apricot, cherry, fig, and pomegranate provided shade, flowers, and fruit. Orange trees were planted for blossoms, fragrance, fruit, and evergreen foliage. Roses, jasmines, crocus, and cyclamens were also used.

Houses of the Greeks and Romans were very similar in design, being built around a central courtyard, or *peristyle*. Peristyles were level, with stones laid on the ground. All rooms opened out on and looked onto this area. The Greeks were not interested in plants. There may have been some plant materials in containers, but the main point of interest was statuary. The Romans, however, sloped the peristyles slightly to create interest. Trees, shrubs, and statuary were placed along the edges, and a water feature was centrally located. The gardens developed in these villas were usually formal, with many plant materials, trellises, and topiary employed.

The formality of axial symmetry was carried through into the gardens of the Renaissance Italian. Garden design during this period changed somewhat, from straight lines to curves. Plant materials were trained into very formal hedges, and trees were clipped or pleached; pathways, stairways, and ancient statuary were used to achieve the final effect. The greatest landscape work of the Renaissance period is the formal gardens at Versailles, in France.

England's terrain lent itself to the concept of landscaping regions in a natural, informal way—the first real change from formal gardens.

The first American gardens were formal, contrasting with the unsettled new land. They provided a psychological victory over nature, although an informal style rapidly developed which used native plants. The only formality that appeared was near the house.

The necessity of designing for large masses of people in modern cities and heavily populated areas has given city planning new impetus. Contemporary preferences choose both formal and informal designs. The importance of design in cities is not only aesthetic; design also has proved to have social and pscyhological effects on the quality of life.

ELEMENTS AND PRINCIPLES OF LANDSCAPE DESIGN

Landscape design is an art for people. To create landscape designs, an understanding of the basic principles (unity, focalization, balance, scale, proportion, and rhythm) common to all spatial arts is necessary. But the way these principles are ultimately applied depends on the imagination, spirit, and sensitivity of the designer. Because these principles serve as guides for design, they can help the inexperienced. Experienced designers must adhere to the principles to some extent, but they must also be flexible in situations that require variation. To fulfill a purpose, the landscape designer—like any other artist—must establish limits. The painter's canvas and the potter's clay impose limitations and, thus, form a point of reference. These limits do not indicate that the project may not be expanded; they are simply intended to give the artist a finite area with which to work. (Infinity is impossible to deal with.) The boundaries may be previously designated, such as streets or property lines, or they may be imposed by the designer. This may be done obviously (with the use of a wall, for example) or subtly (as with partial plantings). The treatment of these boundaries determines whether the project is complete within itself or will later be expanded.

Like a painter or sculptor, a landscape designer develops a composition by employing or purposefully eliminating the elements of design: line, pattern, color, light, texture, form, and space (Figure 20-3). Using these elements of design, the designer can compose a design with an overall concept.

FIGURE 20-3 Elements of design. (From A. J. Rutledge, *Anatomy of a Park*, McGraw-Hill, New York, 1971.)

FIGURE 20-4 The element of line used to create and control patterns of movement. (Photograph, Amy Mackintosh; from Robert E. Marvin and Associates, Walterboro, S.C.)

Elements of Landscape Composition

Line may be perceived as the junction of two materials, such as a border of water and land or grass and a walkway (Figure 20-4). It is used to create and control patterns of movement and attention. Straight lines denote formality and pomp, implying strong, solid structural qualities. On the other hand, curved lines are less formal because they are considered passive, soft, and pleasant, and because they encourage slower movement. Vines can soften lines, while clipped hedges can strengthen them.

Pattern provides surface interest and enrichment (Figure 20-5). It is usually the repetition of a design motif. Pattern can be directional (like a path) or static (like a pattern of leaves against the sky); both kinds of pattern are shown in Figure 20-5.

Color brings the world to life. It invokes a variety of human responses. As with line, the abstract qualities of color can create moods and emotions through the landscape. Nature can be extravagant with color, especially in the spring and fall, but one must be cautious in

FIGURE 20-5 Pattern provides surface interest and enrichment to a design. (Photograph, Robert E. Marvin and Associates, Walterboro, S.C.)

its use. Almost any color can be made to harmonize where sunshine and foliage blend; however, too much color can be visually displeasing. Many flowers, such as oriental poppies, can be used freely, since they bloom for only a short time. Most shrubs are compatible, but some (such as azaleas) need to be carefully arranged in order to have a pleasing effect. In selecting colors, one should allow for some color during every season.

Colors harmonize by contrast or analogy. Contrasting colors are opposite on the color chart, e.g., blue and orange. Adjacent colors are analogous, such as green and blue.

Colors also project coolness and warmth; for example, bright red is a warm color which may not be pleasing around a patio, while blue would give a restful feeling. Warm colors are red, orange, and yellow; cool colors are blue, green, and gray. Colors can also be used to create a feeling of depth. A bright color, e.g., red, at a distance gives the illusion that the distance is shorter. A cooler color, e.g., light blue, can be used to create a feeling of depth or distance. The colors of plants can also be used to direct the observer's attention. Color in bark and foliage should be considered as well as color in flowers.

Light is used in respect to colors and shadows. As light passes through the atmosphere, dust particles reflect the blue light waves, giving a blue hue to the sky or to mountains viewed from a distance.

This can be exploited by using blues with the proper transition in texture and size to create a feeling of depth or distance.

Light causes shadows (Figure 20-5). As the sun passes through the sky, forms created by shade continually change and move. The continual change and movement create interesting illusions, but shadows can interfere and confuse ground patterns. Artificial lighting at night for trees and shrubs casts intriguing patterns on ground or walls.

Texture is the variation of the surface of a material. Plants are classified as fine-, medium-, coarse-textured. Weeping willow, cotoneaster, and spirea have fine-textured foliage; azalea, sasanqua, and cleyera have medium-textured foliage; loquat and aucuba have coarse-textured foliage. As a general rule, fine-textured plants enrich architecture having smooth surfaces and fine lines; coarse-textured plants harmonize with large spaces and coarse building material. Differences in texture can be used for the development of depth or distance; fine texture gives the illusion of being farther away, while coarse texture gives the illusion of closeness (Figure 20-6). A sudden change in texture can also be used as an accent.

FIGURE 20-6 Plant texture used to develop depth and enrich entry space. (Photograph, Amy Mackintosh; from Robert E. Marvin and Associates, Walterboro, S.C.)

The *form*, or "architecture," of plants defines and qualifies space, giving it order (Figure 20-7). Plant masses are selected to achieve a desired visual effect. Vertical branching leads the eye upward; upright evergreens such as spruce or hemlock accent the high gables of a large house and the verticality of hilly terrain. On the other hand, horizontal branching pulls the vision to ground forms; for example, rounded shrubs echo the profile of a ranch house and the undulations of rolling terrain. Tall, upright, stiff plant forms, such as podocarpus, are useful as emphasis points, as in the repetition and regularity of formal gardens. Informal design usually calls for plants with loose form, such as rhododendron and loropetalum.

Density in growth habit also reinforces form. Boxwood and yew are heavy and compact; spirea and wax myrtle are light and open. A few compact Japanese hollies might be used to balance the loose form of a taller dogwood. Line patterning of branches varies from upright to spreading. Deciduous materials should be selected for their form during the growing season as well as for the interesting line patterns they create when they are dormant. When considering the form of a plant for a specific situation, one should consider the mature form of the plant as well as the immature form. Form as a positive element is closely related to space.

FIGURE 20-7 Tree form defines and qualifies space, giving it order. (Photograph, Amy Mackintosh; from Robert E. Marvin and Associates, Walterboro, S.C.)

FIGURE 20-8 The outdoor environment is three-dimensional. (From A. J. Rutledge, *Anatomy of a Park,* McGraw-Hill, New York, 1971.)

Space is volume defined by physical elements. We live in atmospheric space as fish live in water, looking out now beyond the surface of our global ocean of air into the black immensity of interplanetary space. This air space in which we live achieves form, volume, comprehensibility, and scale only when it is defined by tangible visible elements and only to the extent that it is so defined. Topography and trees are the great space definers in nature, (Figure 20-8), and to these we have added human construction at every scale from fence to skyscraper. Every physical element above or below the horizontal ground plane on which we are standing defines space, qualifies it, gives it height, depth, simple order, complex structure, or indefinite continuity, within the range between agoraphobia and claustrophobia.*

Space, or volume, is the negative element in design, the area in between; because of this it is usually difficult to comprehend. Space achieves form, volume, and scale when defined by tangible visual elements from the surrounding landscape. A landscape is three-dimensional: It has height, width, and length (Figure 20-8). The three-dimensionality of landscaping is an art, an art to be walked through and experienced from a variety of angles; the fact that it is to be absorbed points to the importance of the concept of space. A landscape is composed of several spaces (much like the rooms of a house): the ground, the ground fixtures, and the ceiling or sky. Each has a specific purpose, but each is also a part of a whole.

Before surfaces are applied to the earth, the natural ground forms need to be considered. A common practice of developers is to level hills and fill valleys, producing monotony. The designer should strive to accentuate the highs and lows of the terrain, e.g., building homes and planting trees on high points and creating pools in low spots (Figure

*Garrett Eckbo, *The Landscape We See*, McGraw-Hill, New York, 1969.

FIGURE 20-9 Peaceful integration of architecture and the landscape. Note the structure and plantings on high points, and the pool in the low areas. (Photograph, Amy Mackintosh; from Robert E. Marvin and Associates, Walterboro, S.C.)

20-9). Verticals determine to a great extent the emotional quality of space. Figure 20-10 shows various types of space. A vast space without verticals tends to frighten and humble an individual; yet, at the opposite extreme, too many verticals tend to excite and confuse. Although the sky is the obvious ceiling to the outside space, it becomes an element in design when viewed through a canopy of leaves or a latticed roof.

Principles of Landscape Composition

Unity means "oneness." It is the fitting together of parts that have a relationship to one another. When one of these parts is overly strong or weak, the unity is disrupted. Parts are unified by an overriding concept or idea. If successful, the unity presents one image from several angles. Unity is achieved by using similar plant textures, forms, or colors; by planning noticeable repetition and transition from one group to another; by enclosing areas which set the scheme apart; or by developing patterns.

While unity is essential, *variety* humanizes a design and saves it

(a)

(b)

(c)

(d)

FIGURE 20-10 Types of space: (*a*) static space; (*b*) linear space; (*c*) free space; (*d*) intimate space secured by the low tree canopy. (From A. J. Rutledge, *Anatomy of a Park*, McGraw-Hill, New York, 1971.)

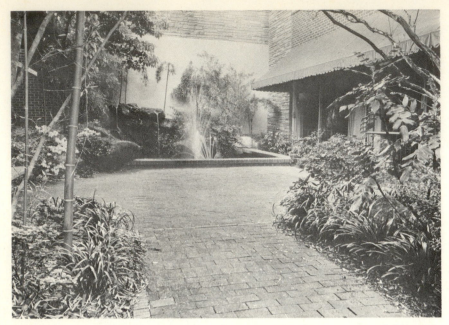

FIGURE 20-11 Brick paths and planting design pointing the way to a restful fountain in the sunlight. (Photograph, Warren Uzzle, Raleigh, N.C.)

from monotony. Too much unity can lead to sterility, whereas too much variety can lead to confusion. Variety can be tempered by dominance and subordination. To reduce monotony, different textures, shapes, heights, or colors of plants may be employed as accents. Not only must this occur on a small scale; it must also occur on a large scale. One feature should become a focal point.

Focalization is the climactic or dominant point in the design in which various parts of the design attract and hold the attention of the viewer (Figure 20-11). In the formal design, the focal point is often a terminal feature at the end of the axis, such as a statue, birdbath, sundial, arbor, fountain or pool. In some cases of formal design, the central feature of focalization is located at the crossing of two axes, called the "central motive scheme." In an informal design, the various parts of the composition lead the eye to a climactic point, usually groupings of garden furniture or plant material.

Balance (Figure 20-12) helps create unity by giving a sense of equilibrium and stability. Balance may be classified as symmetrical or asymmetrical. Symmetrical balance utilizes a central axis with an exact repetition of elements on either side. This is used in formal design and is best employed where the terrain is flat. Generally, some feature is used as a focal point. Major and minor axes are included, with lines radiating away from the focal point to create a pattern. The garden area is enclosed, and the lines in the pattern are defined with clipped plant material. This type of development is criticized for its rigidity and mechanical regularity; it is complete and finished, leaving nothing for the observer to imagine.

Contrasting with the formality of symmetrical balance is the informality of asymmetrical balance. No central axis is employed in asymmetrical design; instead, objects or areas of equal attraction are weighed against one another, with the resulting balance creating interest. Informal design adheres only to the principles of nature in its loose, open, uncontrolled growth. It invites exploration and fosters the imagination of the participant. This type of design is best suited for areas where the terrain is uneven or where there are outcroppings of rocks or informal groupings of trees. The plant material is unclipped, garden form is not geometric, and complete enclosure is not necessary.

A word of caution should be introduced here. Both formal and informal schools have suffered from academic precedent. Contemporary design is a modification of many ideas expressed in the formal and informal design styles; the overall theme is to preserve, protect, and beautify the environment.

FIGURE 20-12 Balance involving motion: the viewer's vision moves from object to object, but there is harmony in repetition of plants and patterns. (Photograph, Longue Vue Gardens, New Orleans, La.)

Turning again to composition techniques, one must remember scale and proportion. Scale refers to the overall size of the area; proportion deals with relative sizes of things within the space.

Scale is the relationship of the entire design to the environment. It gives sensations of bigness or smallness within a space. Many large-leaved plants would be as inappropriate in a small garden as a single shrub would be on a vast plain. There are many animate determinants of scale; inanimate determinants of scale derive from ground forms, trees, rocks, mountains, or plains. Obviously, when designing for humans, human scale is used. Humans in a room of gigantic scale feel dwarfed and insignificant. This is very evident in some high-rise buildings in which human scale is diminished and insensitively treated. Plants, then, are used to maintain human scale in the midst of gargantuan architecture.

Proportion is the pleasing and proper relationship of one part of a design to another. If any part seems large or ungainly in comparison with the rest, it will be psychologically displeasing. The interrelation of the size of one part or object to another should also be considered in designing space.

Rhythm is the regulated movement of similar parts. It is created by repetition and transition in which the reappearance of identifiable elements creates a feeling of motion as the eye is directed throughout. Good design technique is exemplified in good transition. Gaining a feeling for the rhythmic beat of color and form produces added pleasure in participating in a landscape.

In selecting plants for an environmental design, one must coordinate and anticipate problems. Cultural and ecological conditions must be considered. When plants are planted in an unsuitable environment, additional maintenance will be necessary. Rate of growth must be a factor in choosing a plant for a specific purpose. At times, rapidly growing plants are needed to fill a space quickly, but most can be used only temporarily or in a protected location. Therefore, the designer must be familiar with the growth habits of plants as well as design principles and techniques.

LANDSCAPE MATERIALS

Never underestimate the value of a handsome tree. Protect it, build your house and garden compositions around it, for it offers you shade, shadow, pattern against the sky, protection over your house, a ceiling over your terrace. It can also provide an enviable example of living sculpture.*

*Thomas Church, *Your Private World*, Chronicle Books, 1969.

FIGURE 20-13 Tree unifying natural and artificial elements.

Aesthetically, plants may become sculpture. One plant noteworthy for its coloring, texture, form, and pattern may serve as a focal point. Not only could this plant be a prime feature, but plants may also be used as backgrounds, to frame and highlight.

As a transition element, plants can soften the junction of two dissimilar surfaces or reduce the impact of architecture. Plants, when properly utilized, can unify natural and artificial elements (Figure 20-13).

Landscaping may be used as a microclimate control by regulating air temperature and humidity, solar radiation, and air movement. An example most familiar to designers is the use of deciduous trees to control temperature and solar radiation. Spring and summer foliage block the sun's harsh rays and shade an area or object; conversely, in the cooler months, the sun is allowed to penetrate to provide warmth. Vine-covered walls are cooler than exposed ones and, like grass, reflect less heat.

Foliage acts as an air conditioner in that it releases oxygen and moisture into the atmosphere, recaptures carbon dioxide for growth, and filters out dust. Wind force can be increased, decreased, or directed by plants. A hedge of dense foliage can actually reduce the impact of air movement, redirecting it elsewhere. Natural ventilation may be augmented or reduced by sensitive landscaping.

Engineering applications are lending more support to the expected external use of plants. Wind and water erosion can be easily controlled through planting. A tree canopy can shield rain that might erode soil beneath trees. Organic matter cushions droplets, lessening the impact of damaging water flow; and shallow fibrous root systems form blockades to further control erosion.

Another engineering application of plants is in the reduction of natural and artificial glare and reflection. The careful placing of plants, particularly those with dark foliage and heavy texture, can block dangerous glare in cities and on highways.

Last, architecturally, plants can control privacy, screen views, and articulate space. Depending on the type, age, and condition of the plant, a wall may be created. Ankle-high walls provide a floor. Knee-high walls casually direct. Traffic control and partial enclosure can be established by waist-high hedges. Chest-high walls divide spaces; walls above eye level define private enclosure.

Recalling the previous discussion of space, one might further realize that site spacing is less defined, larger, and looser than architectural spacing. The horizontal elements are of greater importance than the vertical ones. Plants can divide this vastness into meaningful spaces, spaces that are more human, and spaces that are of more perceivable scale because of the variations in their heights and densities (for example, trees can be used as canopies and ground covers as floors). They can alternately reveal and conceal focal points and direct the participant to secondary and tertiary spaces. Through the use of channeling, pooling (Figure 20-14), linking (Figure 20-15), and framing, people are quickly propelled through an area when paths are narrow and straight. Like a stream, the dynamic movement of people is enhanced by channeling. Pooling is an open space in a path, allowing a closer inspection of an attraction or providing a resting area. Plants serve as linkages between buildings and also between spaces (Figure 20-15). Plant material can be used to draw attention to a view or an important feature.

Plants serve a variety of purposes—aesthetically, architecturally, climatically, and scientifically—that give further importance to their use in developing a functional environment. Landscape designers are

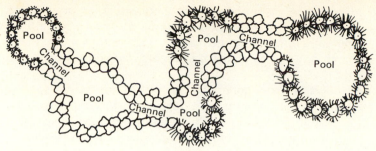

FIGURE 20-14 The channeling, pooling, and rechanneling cycle is a technique for directing movement and breaking down overly large spaces into discernible units. (From *Plants, People, and Environmental Quality*, U. S. Government Printing Office, 1972.)

primarily concerned with the development of a space. In striving to develop a space, they must use artistic concepts, economic philosophies, sociological ideals, and physical techniques. They are concerned not only with how the area is to be used but also with how much it will cost to construct and maintain and how it will actually be built.

Artistic concepts have been discussed previously. The economic variable is subject to constant change and, therefore, must be sensitively considered at the time of planning. In addition to these two factors, function must be considered. A designer must plan to fit the needs of the client or user. To prepare for this aspect, a landscaper may arrange interviews to assess particular needs and interests. For instance, in designing a residential area, the family's level of activity, expected growth, and range of interests are determinants in the

FIGURE 20-15 Linkage is the technique which joins one space with another to make a large area seem smaller and less awesome. Linkage makes it clear that a smaller space is part of a group of spaces or part of a larger space. (From *Plants, People, and Environmental Quality*, U. S. Government Printing Office, 1972.)

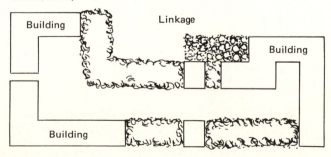

manipulation of space as an attractive composition as well as an efficient, convenient one. The design should be planned for flexibility as plants grow and family changes occur. This analysis applies to family residences but also to shopping centers, industrial parks, and apartment terraces.

Designers consider art, economics, sociology, and psychology in creating spaces. They must then consider physical techniques. The introduction or manipulation of tangible elements allows a space to have meaning. This pertains not only to natural objects—the existence or transplanting of trees, rocks, or bushes—but also to structural materials. Many varieties of structural materials require little maintenance, withstand year-round weather, and create a variety of these qualities. Wood and masonry in all their variations are the most commonly used materials. These include split, rough-sawn, and finished lumber; concrete block, brick, tile, stone, poured concrete; and stucco on a wood frame or concrete blocks. Sheet materials—such as plywood, hardboard, asbestos-cement, glass, plastics, or metals—are used; and rods and grids are also common. These structural materials are used to form a composition. Horizontal elements, such as steps, walks, drives, and terraces, plus vertical elements of walls and fences, are employed to complete the composition. Steps, walks, drives, and terraces serve as transition elements. Walls and fences are enclosing elements for spaces.

DESIGN PROCESS

Some landscape designers have an *intuitive* sense of design. This sense helps them create spaces successfully. The majority of successful designers have a disciplined sense of critical analysis and thought process to complement the intuitive feelings: this allows them to carefully sort and organize variables so as to establish a basis for design decisions. Design is a process of relating a large scope of factors into a comprehensive whole. The critical analysis can be used in any realm of activity from selecting a tomato at the supermarket to determining the growth of a computer corporation.

Critical thinking applied to design involves:

1. Research to understand all the factors.
2. Analysis to establish relationships among the factors to be considered.
3. Synthesis: discarding and retaining variables and finally establishing priorities.

Research must be objective and unprejudiced. This includes not only reading literature already written on the subject but also incorporating the results or one's own experience.

Analysis focuses on the specific problem. Answers are supplied to questions about the particular function, site, people, and materials.

From research and analysis comes *synthesis*, which involves relating relevant factors and discarding irrelevant ones. In determining a solution, preconceptions should be avoided so that the solution evolves from a compromise of ideas and is flexible enough to allow for alteration. Recognizing that there is an intuitive aspect of design that is immeasurable, landscape designers discipline themselves to employ certain elements of artistic creations to achieve a successful design.

In beginning a planting design for a site, an analysis of the existing conditions is first in order. Such factors as topography, geology, soil, surface hydrology, existing vegetation and wildlife, and climate must be analyzed and evaluted before the designer can begin to apply aesthetic considerations. Since even the most beautiful landscape design will fail if scientific requirements are not met, the designer begins with these and, through a series of overlays and eliminations, narrows down the immense possibilities that would fit a certain site.

Two considerations which designers have to deal with before they begin an actual design are the client's requirements and financial limitations. These are sometimes very restrictive factors, but it is important to know and understand the client before plans are drawn so that maintenance requirements and general tastes can be met.

A landscape designer begins a design with a visit to the site, which he or she should come to know in all conditions—morning, noon, night; summer, winter; rain, shine. The site is a changing phenomenon; a good design will take into account seasonal and yearly changes, including elements that harmonize through all these various changes. John Ormsbee Simonds quotes a Japanese designer's method of site analysis:

> If designing say, a residence, I go each day to the piece of land on which it is to be constructed. Sometimes for long hours with a mat and tea. Sometimes in the quiet of the evening when the shadows are long. Sometimes in the busy part of the day . . .
>
> And so I come to understand this bit of land, its moods, its limitations, its possibilities. Only now can I take my ink and brush in hand and start to draw my plans. But strangely, in my mind the structure by now is fully planned, planned unconsciously, but complete in every detail. It has taken its form and character from the site and the passing street and the fragment of rock and the wafting breeze and the arching sun and the sound of the falls and the distant view.*

A base layout of the site from which prints can be made on which

*John Ormsbee Simonds, *Landscape Architecture*, McGraw-Hill, New York, 1961.

to take notes is a very efficient way of beginning an analysis. Research notes, relational studies, existing problems—all parts of the analysis can be noted on these maps, and spontaneous ideas which often arise out of these studies can also be recorded. By penciling in impressions and ideas as they arise, one can compare the new and the old in a plan. This can be done in the studio if the plan is taken to the site periodically and visually projected onto the site to see if it fits. It is often helpful to make some sketches while visiting the site; the drawings will, of course, be rough, but once in the studio they can be reworked and drawn to scale in a neat, readable form.

When analyzing the site, problems should be isolated, and the nature of the problems identified. A set of requirements which will solve the problems can then be compiled. Before the requirements can be fulfilled, the designer has to synthesize all the data and research and apply previous knowledge to the site in question, and then put ideas together to form a solution.

These ideas should work from the general to the particular—i.e., the designer should first picture masses which should be placed in relation to one another and in relation to the function which they will perform. If the basic forms are good, they will fulfill specific functional requirements and will relate to each other and to the entire landscape in a meaningful and aesthetically pleasing manner.

The landscape designer develops a series of major and minor spaces, and then links them with transition elements (Figure 20-16). The development of the major space receives the most concentration. The success of the design is determined by how well all the parts are structured and related. This hierarchy of spaces, all dependent upon one another, corresponds also to the development of the landscape. As a general rule, landscape can be divided into three parts: the public receiving space (Figures 20-3 and 20-7), the service space, and the private space (Figures 20-2 and 20-4). The importance a client places upon these areas determines the major and minor spaces.

The basic design principles of line, shape, mass, space, rhythm, balance, proportion, and harmony should be evident here as well as in later stages of the design, where the principles of color, texture, emphasis, and variety join them. They should continue to apply as the masses are broken down into various plants, whose species cannot be chosen until plant form, height, color, and texture are decided upon and coordinated with cultural requirements and objective factors noted during the analysis (soil characteristics, sun and wind orientation, temperature, and precipitation, for example). Plants which meet all the requirements and possess the necessary physical properties, growth habits, and compatibility with other plants and the surrounding environment are usually few; therefore, selecting a specific species is often accomplished by the process of elimination. The same is true of

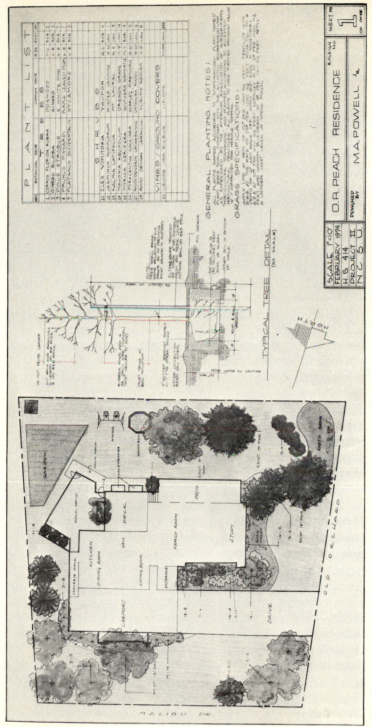

FIGURE 20-16 Residential design plan, showing relationship of spaces linked with transitional elements.

the great variety of possible configurations which are suggested during the design process; although infinite in the beginning, they are quickly narrowed down as material is synthesized, combined, and evaluated. The final plan may be drawn only after many of these possibilities have been penciled in on tracing paper and considered at the site. The possibilities are compared, design decisions are made, and the plants with maximum potential are selected to fulfill the design requirements.

CASE STUDIES

Projects presented in this section were assigned to a beginning landscape class in a horticultural design program. Each assignment was carefully selected to give students a wide range of experiences in graphics and experience in solving problems on different sites for people using the spaces. In all cases, emphasis was on the use of plant materials and structural materials to develop spaces that would be functional and aesthetically satisfying. Plans and discussions presented for each of these projects were prepared by students in this course. The reports and design plans should be helpful in gaining experience in evaluating designs and graphic techniques. The discussions of design logic exemplify the design process.

University Square Mini-Mall

Typical of the specialty-shop malls that are springing up all over the United States, the University Square Mini-Mall (Figure 20-17) is a rustic but handsome shopping area. Each of its small shops opens onto a central interior courtyard which is divided into three levels by a retaining wall and a wood deck designed for outdoor entertainment and relaxation. This central space is entered at each corner of the mall from the parking area which encircles the structure. Encircling the parking is a belt of native trees which shade cars while visitors shop. The site might be described as a series of concentric circles: trees enclosing parking area enclosing mall enclosing courtyard.

    The main approach to the mall is from a small, moderately busy highway that runs into town. Care has been taken to space the trees across the front to separate the mall from the highway and at the same time leave it open and inviting to automobile traffic. The building is on a slightly higher level than the highway, and low creeping juniper has been planted on this bank to prevent erosion and to separate the highway from the parking lot with a mass of greenery. Since shrubs would block prospective customers' vision of the "University Square Mini-Mall" identifying sign, which is tastefully located on the south side

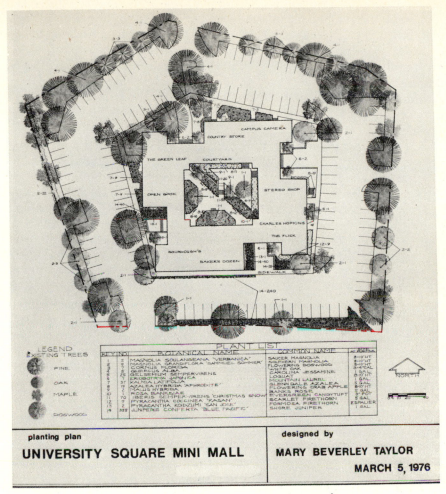

LEGEND
EXISTING TREES

PINE

OAK

MAPLE

DOGWOOD

PLANT LIST				
KEY	NO	BOTANICAL NAME	COMMON NAME	AT PLANTING
1	1	MAGNOLIA SOULANGEANA 'VERBANICA'	SAUCER MAGNOLIA	8-10'HT
2	5	MAGNOLIA GRANDIFLORA 'SAMMUEL SOMMER'	SOUTHERN MAGNOLIA	8-10'HT
3	5	CORNUS FLORIDA	FLOWERING DOGWOOD	5-6'HT
4	5	QUERCUS ALBA	WHITE OAK	5-6'CAL
5	25	GELSEMIUM SEMPERVIRENS	CAROLINA JESSAMINE	1 GAL
6		ERIOBOTRYA JAPONICA	LOQUAT	8-0'HT
7	37	KALMIA LATIFOLIA	MOUNTAIN LAUREL	1 GAL
8	37	AZALEA HYBRIDA 'APHRODITE'	GLENDALE AZALEA	5 GAL
9	2	MALUS HYBRIDA	FLOWERING CRAB APPLE	8-0'HT
10	1	ROSA BANKSIAE	BANKS ROSE	5 GAL
11	70	IBERIS SEMPERVIRENS 'CHRISTMAS SNOW'	EVERGREEN CANDYTUFT	1 GAL
12	7	PYRACANTHA COCCINEA 'KASAN'	SCARLET FIRETHORN	3 GAL
13	7	PYRACANTHA KOIDZUMI 'SAN JOSE'	FORMOSA FIRETHORN	ESPALIER
14	333	JUNIPERIS CONFERTA 'BLUE PACIFIC'	SHORE JUNIPER	1 GAL

NORTH

planting plan
UNIVERSITY SQUARE MINI MALL

designed by
MARY BEVERLEY TAYLOR
MARCH 5, 1976

FIGURE 20-17 University Square mini-mall planting design plan.

of the building and does not overhang the street, trees constitute the next canopy level. Two magnolias have been placed with the few existing trees to add a touch of elegance to the front of the mall. They are deciduous trees which bloom early, before the foliage appears, and the beautiful purple blossoms will attract the eyes of passersby, perhaps drawing them, in a very subtle manner, to come in and do some shopping.

The large Southern magnolia, which reaches to the ground, together with the Japanese magnolia, sets the landscape theme for the mall. The Southern magnolia is found on nearly every plantation in the South; selected for its aristocratic bearing and naturalness, it blends beautifully with the mall's refined yet rustic atmosphere.

These magnolias accent the main entrance to the mall, and they also function as a screen between the less-than-elegant hamburger chain to the east and the laundry to the west. By screening these undesirable neighbors with plants rather than fences, the mini-mall unquestionably identifies itself as a separate, more elite member of the community, but one which is open to everyone who wants to come in. Fences tend to isolate, while plants merely separate one site from another.

Behind the mini-mall is a residential area. In consideration of the people who live there, and who probably do not relish their proximity to a commercial—though tasteful—establishment, another subtle screen has been added by planting dogwoods and white oaks. These trees were selected because they blend well with both the existing trees around the mall and those around homes throughout the neighborhood. They camouflage the mall from these neighbors, who already know what the mall is and do not need to be told visually.

Once the mall has been identified and landscaped in its larger setting, as it relates to the surrounding development of houses, hamburger chain, laundry, and highway, more thought can be given to on-site landscaping. Moving inward, we reach the circle of landscaping which separates or unites the parking and the structure. As many existing trees as possible have been retained in this space; they help unite the building with its natural surroundings.

Proportion, one of the basic design principles, must be considered here. Small shrubs against such a massive building would seem lost and completely out of proportion, but a few large Southern magnolias added to the existing trees serve to optically decrease the immense size of the structure. Loquats are used for accent at the entrances to the courtyard; their rough texture complements the mall's cedar siding, and in turn, the architectural background enhances this tree's sturdy, interesting quality.

Once the size of the mall has been optically reduced, smaller plants to fill the space between the ground and the trees' canopy may be included. The loose form of the mountain laurel, planted along the side and back of the mall, softens the rigid line between structure and ground. The berries of the pyracantha in the sunnier spots will add fall color, and—like the blooms of the laurel—will provide a bright contrast against the structural brown background.

The interior courtyard, upon which all shops focus, demands special attention. A canopy is necessary to give this space a more intimate atmosphere, making people more comfortable and less awed by the height. The courtyard will be viewed from three levels; ground level, the first-floor deck, and the second-floor deck—all of which completely encircle this space. The canopy effect is necessary for atmosphere, but care must be taken that these trees are placed where they will not block a newcomer's view of each shop from the courtyard. Japanese magnolias and crab apples are appropriate; the crab apples will bloom later than the magnolias, at about the same time as the mounds of azaleas partially surrounding each tree.

To balance off the tree in the west corner and to cover a large hole running under the structure in the east corner of the courtyard is a Banks rose. The decks and supporting columns here provide ample room for upward growth of this yellow-flowered vine.

Beside the vine, draping over the retaining wall is evergreen candytuft, a beautiful, white-flowered ground cover which softens the wall and provides a gentle transition from the ground to the triangular deck.

This deck is not well used at present—on special occasions it functions as the stage for a band, but the casual shopper rarely sets foot in this potentially relaxing area. Picnic tables and chairs protected by a low canopy would give this area a more human scale and a comfortable place to eat sandwiches made with sourdough bread, or doughnuts from the bakery. Once drawn to this area, visitors may sit for a while and feel rested enough to do more shopping.

Hanging baskets with red geraniums would brighten up the upper stories of the mall. They must be permanently affixed, to discourage vandals from removing them during the night, however. Such embellishments would come in phase 3 of the landscaping; phase 1 includes the planting of all trees, which take years to grow and therefore need to get started; and phase 2 includes the planting of all shrubs and ground covers, which will aid immensely in softening the site. The landscaping of the mall is a subtle way of advertising; it attracts people initially, and by making them relaxed, cool, and comfortable, it encourages them to shop longer while they are there and to come again later.

Tinsley Residence

A good residential design fits a house both to its location and to the needs of the family who will be living in it. The Tinsleys are a typical American family: they have two children, a dog, and the need to feel secure, creative, and stimulated. Such needs can be filled by surrounding a residence with pleasant views and vistas which are well composed and oriented in relation to the family's movements. This should be done in a way which provides a comfortable sense of human scale in the graceful transition from occupied spaces to the larger surrounding landscape.

The Tinsley house (Figure 20-18) does this. The house is a wood

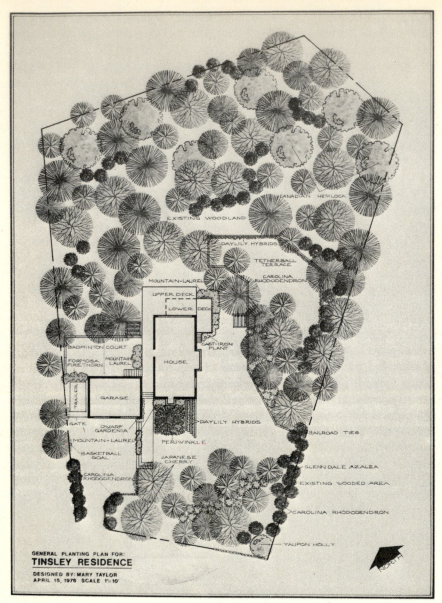

Labels within the figure:

CANADIAN HEMLOCK

EXISTING WOODLAND

DAYLILY HYBRIDS

TETHERBALL TERRACE

CAROLINA RHODODENDRON

MOUNTAIN-LAUREL

UPPER DECK

LOWER DECK

CAST-IRON PLANT

BADMINTON COURT

FORMOSA FIRETHORN

MOUNTAIN LAUREL

HOUSE

TRAILER

GARAGE

GATE

DWARF GARDENIA

DAYLILY HYBRIDS

RAILROAD TIES

MOUNTAIN-LAUREL

PERIWINKLE

BASKETBALL GOAL

JAPANESE CHERRY

GLENN DALE AZALEA

EXISTING WOODED AREA

CAROLINA RHODODENDRON

CAROLINA RHODODENDRON

YAUPON HOLLY

NORTH

GENERAL PLANTING PLAN FOR:
TINSLEY RESIDENCE
DESIGNED BY: MARY TAYLOR
APRIL 15, 1976 SCALE 1"=10'

FIGURE 20-18 Planting design for Tinsley residence.

structure which stresses verticality. Located in a residential area outside of a small town, the lot is large, steeply sloping, and heavily wooded. The Tinsleys have moved away from the unimaginative cut-and-fill suburbia to a niche of sculptured masses and interesting land forms. Form is always considered first, and they chose to begin with the sloping topography, which is generally the most interesting to develop.

Instead of flattening the foundations of the house, the house was merged with the land by angling the foundations. The diagonal roofline parallels the foundations, reflecting the slope of the land and fitting the house to the site. The vertical boards between ground and roof complement the surrounding trees, and the house reflects the landscape in a total, harmonious way. When adding ornamental plants, the landscape architect must be careful not to destroy this harmony; foundation planting should be kept low; dwarf gardenias, gumpo azalea, and hybrid daylilies will soften the transition between earth and structure without hiding these lines and destroying the harmony. The woodlands combine the elements of design naturally; when adding ornamentals, such as the rhododendron in the Tinsleys' woods, the pattern of repetition which gives rhythm must be kept irregular, as it is in nature. A tension is set up between the observer and nature, and one is stimulated to complete patterns of one's own.

Such involvement is part of an observer's urge to move into the scene and become part of what is there. This is very significant in landscape design, and the Tinsleys have satisfied this urge by building a deck uniting the indoor and outdoor living areas of their home. They can move comfortably from interior spaces to the larger surrounding landscape and experience the three-dimensional forms in a relationship which creates a sense of structure, tension, and volume. This 7-m-high deck gives space, height, and depth; topography and trees, the space definers in nature, are open to view, and awareness of these features is increased by the variation from the normal height. By placing themselves among the treetops rather than below them, the Tinsleys changed their perception of the landscape, encouraging new sensations and fulfilling a basic need for flexibility in the environment.

The landscape designer needs to assess not only the site but the particular needs of the family. The decks provide room for outdoor adult entertainment—barbecues, parties, or teas,—but the Tinsleys' two boys, ages 9 and 11, need room for active games such as soccer, tetherball, and badminton. A soccer field on a residential lot is, of course, out of place; but the side yard can be terraced and planted with grass to form an area suitable for kicking a ball around, and a natural land form in the back can be molded with railroad ties into a tetherball terrace. These terraces balance each other and provide a sense of equilibrium and stability to the steeply sloping site.

Erosion is a problem on sloping sites. Terraces solve part of this problem, and a stone retaining wall and plantings of periwinkle solve the problem in other areas of the yard. The natural stones, which Mrs.

Tinsley likes, are repeated along the driveway and in the steps between the drive and the walk. Mountain laurel has been planted between these steps to separate and screen cars from the yard. Mountain laurel is used extensively throughout the plan, as is rhododendron. These plants are native to the surrounding forests and mountains and therefore blend beautifully with this steeply sloping and wooded site, recreating a mountain atmosphere which many people travel miles to experience.

In order to preserve as much as possible of this atmosphere, all existing trees except two have been kept. The removal of the two trees allowed the surrounding, more dominant trees plenty of room to grow and develop into handsome, noteworthy trees.

A view of the house from the road is not completely blocked by these trees; it can be seen beneath their canopy. The house is too public in one area: the family dinner table, beside a large window, can be viewed from the road. An evergreen, a yaupon holly, and clumps of rhododendron have been planted to this corner to block any view of the dining room from the road. Such development of the "public" area for the benefit of the homeowner is typical of today's trend away from the traditional emphasis on creating a showpiece. The family's interest and pleasure comes first, and though care must be taken to provide an inviting and convenient entrance for guests and an attractive front area, the design does not have to revolve around passersby.

The Tinsleys' drive at one time bisected the center of the yard; a camping trailer was always parked next to the garage, making the entrance to their home very cluttered and unattractive. No one ever parked in the garage; even the family parked in the drive, blocking the steps to the walk and causing people to cut across the lawn. The driveway entrance has been moved closer to the property line, reproportioning the yard and putting it in proper relation to the whole. The entrance is articulated through the use of rhododendron and a rock wall which leads visitors to the front steps.

A wooden gate has been placed in front of the trailer and a planting of pyracantha behind it to hide this unsightly vehicle from view. The gate, which blends with the garage, opens onto the drive, so that the trailer may be moved easily and conveniently.

The Tinsleys' house is protected and secluded during the summer months, but during the winter the sight and sounds of a highway below their lot travel up the hill. An evergreen screen of Canadian hemlock and rhododendron scattered carefully through the woods will reduce the sound and make the highway less visible. The addition of such ornamental plants does not destroy the harmony of the site.

Plants in harmony with their environment require little maintenance, and since the Tinsleys are a busy family, this suits them perfectly. Combined with a natural setting and an outdoor deck, this low-maintenance landscaping has made their home a comfortable and secure and yet stimulating and interesting place in which to live.

Nocturnal House

A nocturnal house (Figure 20-19) is a zoological house designed to house animals, such as bats, raccoons, and opossums, which are active at night rather than during the day. Located in a zoological garden, the nocturnal house is an example of the landscaping of an area within an area; though a separate entity, the house is part of a whole—the zoo—which must be considered.

The house is approached from two directions along a 6-m-wide concrete walkway which runs throughout the park. Softened at intervals by islands of trees which were preserved to provide shade and visual relief from the vast expanse of concrete, the walk widens at the main entrance of the nocturnal house into an area suitable for either entering the house or standing and viewing the white rhinos, a future neighboring exhibit. The widening of this area and the introduction of a grove of trees at the entrance break the rhythm created by a continuous walk, thus encouraging the visitor to move inward rather than onward for a few minutes. Seats in this area would further encourage a break, and families who paced themselves differently could meet here.

Separating the rhino exhibit from the nocturnal house along the walkway is a ginkgo avenue. Since a zoological garden, by its very nature, stresses diversity in both plants and animals, it is not easy to obtain the continuity that repetition brings. The ginkgo avenue, recommended to extend throughout the park, would clearly define the main circulation areas and would pull together the various elements of the park, making it a unified and harmonious entity.

Ginkgo was chosen because of its interest both visually and historically. Its fan-shaped leaves turn a brilliant yellow in autumn, and, after a glorious two-week display, they all drop to the ground at once. The survival of this tree, essentially unchanged through millions of years, indicates that it is tolerant of many conditions. Considered a "living fossil," it dates back further than any other tree, to the Palaeozoic Era, and it has been grown as an ornamental tree around temples in China since ancient times.

Paralleling the ginkgos, past the sea lion exhibit toward the nocturnal house, is a screen of wax myrtle which discreetly separates the public area from the service area of the nocturnal house by forming a tall, loose hedge. Attention will be directed away from this inconspicuous plant and the service area toward the more interesting

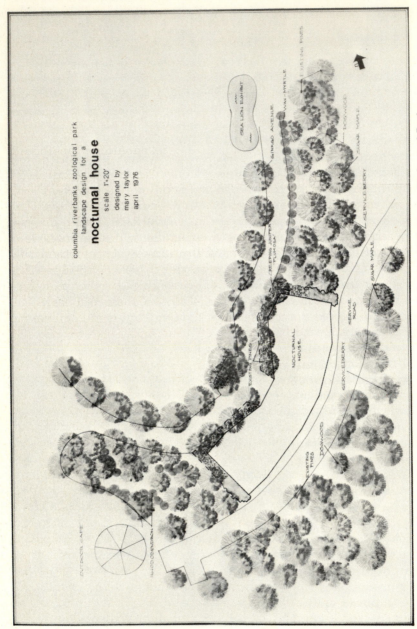

FIGURE 20-19 General planting design for nocturnal house at Columbia Riverbanks Zoological Park.

ginkgos and rhinos on the right. More trees are needed behind the myrtle to provide shade along the walk and background when viewed from the far side of the rhino exhibit. Dogwoods, serviceberry, and maples were chosen to fill in this area and the sparse existing woods behind the nocturnal house. All three are native trees and blend well in the woods, but each will add accent in spring or fall. Dogwoods are very popular for the white-layered effect they give to the woods. The serviceberry also has a beautiful white flower in spring, with the added attraction of bright red berries in the fall. The sugar maples' brilliant red fall leaves will contrast beautifully with the yellow ginkgo and native hickory leaves. The color is seasonal; in off seasons, these trees function as background trees important in providing a total setting for the house.

Rhododendrons have been scattered throughout the small woods between the nocturnal house and outdoor cafe. Also native, they provide a natural-looking screen to the presently visible service area behind the cafe, and in spring they provide the added pleasure of large pink blooms.

Since daylight is not permitted inside a nocturnal house, and windows are therefore unnecessary, berms have been added to the outside of the building. Existing pine trees in the berm area have been preserved by means of a dry well around the trunk and a gravel fill over the root area of the tree to allow necessary air flow. Drainpipes are also necessary to prevent water buildup that might cause decay.

To prevent children from running up the berm onto the roof, a 0.7-m-deep ground cover of prickly juniper has been planted on the berm. Juniper is very effective in preventing erosion, and this particular variety, 'Plumosa', was chosen for its 0.7-m-height and reddish purple color. Repetition of color will ease the transition from the outside to the inside of the nocturnal house, where infrared lights are used to illuminate the exhibits. The interior exhibits should include night-blooming plants such as datura to add interest and to draw a parallel between the plants and animals—of which some are active by day, some by night.

As can be seen, varieties of plants around the nocturnal house have been kept to a minimum. Care was taken to select a few hardy plants which will take care of themselves rather than a variety of high-maintenance species. Exotic cultivars will be stressed in the areas of the zoo specified as gardens, where attention will be focused on plants; but around the nocturnal house, plants take on more subtle functions. Here they unify, screen, direct traffic, prevent erosion, and provide background, framing, rhythm, and harmony to an area of the zoo where animals are the focal point.

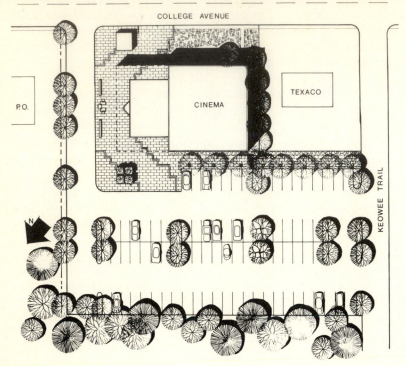

FIGURE 20-20   Site plan for Lyric Cinema I and II.

Lyric Cinema I and II

In the case of the Lyric Cinema I and II (Figure 20-20) the program constituted that of a typical cinema, and the initial concern lay in selecting the proper site for a cinema within a small university town. Since business in such towns relies heavily on college students for support, pedestrian accessibility from the university was primary among the considerations in the selection of a site. The procurable site closest to the campus was a cleared lot between a run-down service station and the local post office on College Avenue, a busy street which would allow for automobile access and advertising.

Though the site satisfied all the social requirements, it fell a little short on the physical requirements. Local codes required one parking place for every four seats in the theater, and the site as it existed was simply too small to accommodate both theater and parking. The architect was left with a choice between filling a huge gully at the rear of the site and selecting the next available site, which was a considerable distance from the campus. Deciding that the cost of filling the gully would be made up by the greater number of customers which the site would afford, he began his design. To develop harmony between the Lyric and its site, the architect and landscape architect first collaborated on the proper orientation for the building on the site, keeping in mind their aim: to have site and building intimately related. Primary among the deciding factors for orientation were:

1. *Topography*—Natural slope in the land could be used for the seating slope necessary in theaters. This would save in the construction and installation phase of development.

2. *Community pattern*—Consideration for relation between cinema and service station, post office, highway, houses and college. The neighboring service station needs screening. Consideration for neighbors in the residential area across College Avenue—these people would not like having the typical neon lights found on movie theaters flashing through their windows. Neither would they like a parking lot directly in front of their houses.

3. *Pedestrian accessibility*—Pedestrians must be attracted inward from the sidewalk to the theater.

4. *Automobile accessibility*—Drivers must have easy access and a pleasant parking area with two exits.

5. *Local codes*—Local codes require that all buildings along College Avenue be set back at least 2 m.

6. *Advertising*—Movie times and titles must be clearly visible from the road.

The best response to all these requirements was to place the building sideways to the road, with parking behind and the entry facing the post office. The natural slope of the land would then be utilized, and the service station would not be evident from the entrance.

Functional demands of the Lyric are expressed in terms of circulation and area usage. Circulation patterns originate in the parking lot and on the sidewalk, and they focus on the ticket booth and the doors on either side. Since the ticket booth faces neither sidewalk nor parking lot—as it must serve both—the complication of leading people to the atypical entry arises. The problem is solved by using glass walls, paving patterns, and a fountain as a focal point. Pedestrians approaching from either side of the Lyric are attracted by the change in surface material; the patio, which extends from the sidewalk to the parking lot, is bricked, and a diagonal inlay of white concrete which reflects the border along the top of the building draws people toward a bubbling fountain in front of the main entrance. By focusing on the fountain rather than on the building, moviegoers unconsciously find themselves beside the main entrance.

People approaching from the far corner of the parking lot are aided in finding the entrance by the glass walls. Someone rounding a corner expecting to see an entry may be confused by a solid wall, but a transparent glass wall will reveal in a glance where the doors are.

Natural elements are also used in defining circulation and area usage. A line of trees separates the parking area from the rear walk, and they provide a pleasant canopy for those walking beneath. A tall evergreen hedge screens off the service station and continues along the side wall of the cinema, softening the structure and leading the pedestrian onward.

Circulation requirements met, we return to the functional demands of area usage. The exterior patio space can function as a lobby as well as for circulation.

This area is partially enclosed by the marquee on the street side, two walls and a fountain on the drive side, and planters with trees on the parking side to balance the marquee. The walls, on either side of the fountain, separate the drive from the theater, encouraging people to focus on the fountain rather than on automobiles. They thus function as a space definer and a visual control rather than a barrier. The planter, low enough to sit on, doubles as outdoor furniture.

A self-sufficient building often limits itself, but when a continuity between building and site is established, the entire complex expands and becomes a large-scale work of art with many times the impact of a long building. By joining the interior and the exterior lobby with glass, the waiting area of the Lyric has been expanded so that it unites the building with the site. New interrelationships can be developed, and the essence of the movie can be reflected in the architecture. The concept of seeing a cinema as "night architecture" unfolds, and the architect and landscape architect can work together to recreate this essence through the play of lights against the dark night. The marquee is placed away from the building and close to the road. The marquee, 6 m high and constructed of the same materials as the main building, is lit with spotlights which cut through the night to focus on its message. The beam of light, like that of the movie projector, may be seen against the darkness of the night, and the movie atmosphere is thus created outside.

This same technique is used to light the fountain, the focal point of the indoor as well as the outdoor space. Beams of light lead the eye to the fountain, which is brighter than the dimly lit interior lobby; thus the glass walls allow an outdoor area to become an extension of the interior space, and a continuity is established. The exterior space and light become predominant even from the interior of the building. The building becomes dependent on the landscape. Building and landscape interlock in a harmonious balance, and they combine to make a positive statement regarding the nature of the building and its purpose.

Tennis Center

The A. H. Sloan Tennis Center (Figure 20-21) is part of a university which has 25 courts and a tennis house on campus. As is generally true of a tennis center, the landscape is absolutely flat, and during construction all vegetation was destroyed except some trees along the hills which border the site. The landscape designer will start with virtually nothing in an endeavor to make the site an interesting, welcoming place to play.

The atmosphere at present is hard and sterile; there is no break between courts to give relief from the hard lines and flat surface. Even the walks are made of concrete, doubling the harsh quality of the courts and increasing the temperature level by reflecting the sun's radiation rather than absorbing it as plants would.

The area around the house is not well used. The porches are always deserted; the restrooms are blocked by overgrown photinias and by a mower which is parked directly in front of the entrance; plants are walked on; and the only sloping spot on the site has become a precarious gathering place.

This latter area deserves special attention; the land, a hazardous combination of concrete and plants, slopes off at an odd angle, but it faces the only courts adjacent to the house where matches can be watched, as the varsity courts have wind screens which block the view from the house. Its popularity as a gathering place is further encouraged by a water fountain, a bulletin board, and a blackboard where players must register before playing. Though uncomfortable, players usually stand here after signing in and watch play until a court becomes available.

The ideal solution to finding a better gathering spot would be the removal of the upper half of the wind screens for viewing the varsity courts and development of the covered porch area facing these courts. Half screens on the east side of these two courts would not produce wind problems, as the winds blow from the south during the summer, from the north during the winter, and seldom from the east; but varsity players are afraid that removal of the screens would result in visual distractions and therefore prefer that they be left up.

The alternative is to develop the two courts just south of the building as an exhibition area. A wooden deck built as an extension from the house would negate the uncomfortable slope, and the raised platform will give spectators a pleasant perspective and sense of elevation while viewing the matches. Various space levels provide interest, and this interest can be carried further by the introduction of a tree between the deck and the court, its canopy creating a space which "frames" the court, makes it easier to focus on, and at the same time allows multidirectional traffic below. This is especially necessary around the water fountain, a potentially busy area.

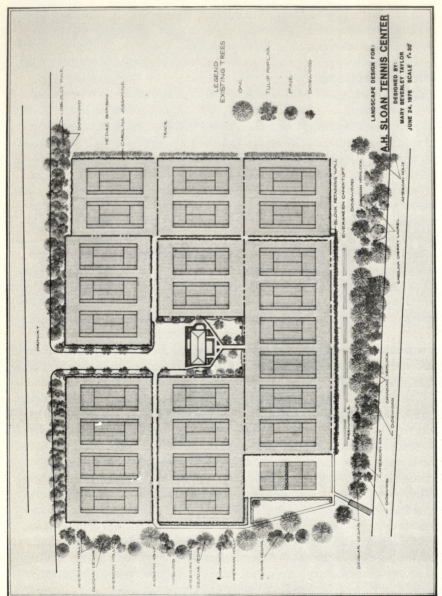

FIGURE 20-21 General plan for tennis center.

To tie the deck area and the courts more closely together, a screen is needed between the two exhibition courts and the courts behind them. There is no room for a hedge, but by drilling holes every 2 m along the edges of the walkway, vines of Carolina jessamine can be established which will grow along the fence, forming a pleasant green background for the spectators. Carolina jessamine will not grow out over the courts, and it is recommended on every fence possible in the complex because it affords a break from the stiff concrete and wire of which the area is composed. Softening these harsh lines makes the courts a much more pleasant playing area, reduces glare from adjoining courts, and provides an added wind screen and visual screen between courts. Plants define and enclose the exhibition area, creating an outdoor room open to view and achieving visual emphasis and focus on the players rather than on activities in the distance.

The line between court and yard is presently straight and hard. A curving bed of 'Rosea' India hawthorn should be planted to soften this transition and to add movement and contrast with the straight, static, court lines. Since it reaches a maximum height of only 1.2 m, the hawthorn will not block the spectators' view but will provide variety with pink flowers in the spring and red berries in the fall. Evergreen day lilies border this bed, and annuals may be added to provide year-round interest.

The yard around the tennis house is hardly an oasis at present. Deciduous trees should be planted to give protection from the summer sun; their cool canopy and shadows provide immediate relief to a player walking off the court. Here in the yard the player can relax and cool off without having to undergo the unhealthy shock of air conditioning. Dogwood and crepe myrtle will provide this shelter; they will not dwarf the already small tennis house; and being deciduous, they will allow the winter sun to shine on those occasional winter players who brave the cold. Two evergreen American hollies were added for winter accent; the dogwood and crepe myrtle both have interesting branch structures when dormant, but a spot of green is always welcome in winter.

This holly reinforces the drive to the house and serves as an invitational device by visually accenting the main access area. Since the avenue runs right beside the courts, a tree that will not drop leaves and fruit onto the courts is needed. Although the existing oaks along the west border of the tennis center will drop acorns, it would be disasterous to cut them. They are nearly 100 years old, and they provide a mandatory background for the center and a screen that could not be equalled for another hundred years. One oak has been lost, and through that gap the smokestacks—the university's most objectionable view—glare hideously. Hemlocks, holly, dogwoods, and Carolina cherry laurel will eventually fill the gaps, blocking this eyesore completely.

The border of trees between courts and campus continues along the north edge of the center. These trees receive direct sunlight and winter winds; deodar cedars should therefore be planted instead of hemlocks; but the dogwoods, holly, and cherry laurel—which take either sun or shade—are planted to provide unity throughout the center. Deodar cedars accent the rear entry to the courts in the northwest corner of the site.

A stand of existing loblolly pine forms a partial screen between the highway and the courts. This screen should be completed by continuing the line of pines and by adding dogwoods to fill in the lower canopy level. Such mixed planting will reduce sound better than the pine alone.

To define the site boundary to the south where land is too narrow for trees, hedge bamboo has been planted. As an evergreen, bamboo provides a tall windbreak which can be planted right up to the court without shedding onto the court. Hedge bamboo will not take over as some bamboo will, and the sound of its soft rustling will be very pleasant to play beside.

Gravel is hard to walk on; thus, the gravel area in front of the stands should be replaced with grass, which is easy to walk on and pleasant to sit on when the stands are full. A tanbark walk and a planting of periwinkle behind the stands will help to solve erosion problems, and periwinkle will grow well in the shade where grass would not. The grey, concrete-block retaining wall in front of the stands is very unattractive; a planting of evergreen candytuft along the wall would transform this area. Mounding along and over the wall, it would be attractive from the stands and from the walk below, softening the hard lines and minimizing the amount of concrete exposed to the eye and sun.

All the walks in the complex need to be softened; the monotony of concrete should be reduced by changing the surface material of all new walks to tanbark. The sun does not glare off tanbark, and tired legs would welcome the soft walking surface. Juniper should be planted along the existing concrete walks to soften the harsh lines and to keep people from cutting across the grass. The triangles west of the house always have cross traffic; raised planters here planted with mugo pines would solve this problem and would provide seats for weary players. Raised planters are repeated to cover the blocks of concrete which protect the drains in the yard. These may also be planted with mugo pines, which grow well in containers, or with annuals.

The foundation plants of the house vary somewhat, since one side gets sun while the other receives shade. Rhododendron at the corners gives height in a loose, informal setting. The *Skimmia japonica* is small enough to leave the "A. H. Sloan Tennis Center" sign visible once the plant reaches maturity, and the day lilies provide a unifying border which will cover the curb. Hypericum has been planted on the sunny west side, and dwarf gardenias have been planted in the protected area between the mechanical equipment and house.

The overgrown photinias around the mechanical equipment have been replaced by a board-on-board fence which includes a gate, leaving the entrances to the restrooms open for circulation. The soft-drink machine, out of place on the sidewalk, has been moved to this area, and potted plants now fill its old space next to the house.

All in all the landscaping has transformed the tennis center from a flat, dull concrete desert to an interesting, cooler recreation complex. Separated from its neighbors by groves which function as boundary delineators, wind screens, noise reducers, and visual screens, the complex provides room for both play and rest. Instead of a large, impersonal facility where people exercise and then leave immediately, there are smaller, greener, more intimate areas where people can relate socially and recreationally in a relaxed, soothing atmosphere.

BIBLIOGRAPHY

Brookes, J.: *Room Outside*, Viking, New York, 1969.

Ching, F.: *Architectural Graphics*, Van Nostrand, New York, 1975.

Church, T.: *Your Private World*, Chronicle, San Francisco, 1969.

Eckbo, G.: *The Art of Home Landscaping*, McGraw-Hill, New York, 1956.

————: *The Landscape We See*, McGraw-Hill, New York, 1969.

Haring, E.: *Color for Your Yard and Garden*, Hawthorn, New York, 1971.

Hubbard, H. V., and T. Kimball: *An Introduction to the Study of Landscape Design*, rev. ed., Macmillan, New York, 1935.

Lynch, K.: *Site Planning*, 2d ed., M. I. T., Cambridge, Mass., 1971.

McHarg, I. L: *Design with Nature*, The American Museum of Natural History, Garden City, New York, 1969.

Newton, N. T.: *Design on the Land—The Development of Landscape Architecture*, Belknap Press, Harvard University Press, Cambridge, Mass., 1971.

Robinette, G. O: *Plants/ People/ and Environmental Quality*, U. S. Department of the Interior, National Park Service, Washington, D. C., 1972.

Robinson, F. B.: *Planting Design*, McGraw-Hill, 1940.

Rutledge, A. J.: *Anatomy of a Park*, McGraw-Hill, New York, 1971.

Simonds, J. O.: *Landscape Architecture: The Shaping of Man's Natural Environment*, McGraw-Hill, New York, 1961.

Walker, T. D.: *Perception and Environmental Design*, PDA Publishers, West Lafayette, Ind., 1971.

GLOSSARY

Absorption spectrum The ability of a pigment to absorb energy at the various wavelengths of the spectrum.

Accessory buds Lateral buds occurring at the base of a terminal bud or in an axil at the right or left of the axillary bud.

Achene An indehiscent dry fruit which has only one seed which is separable from the walls of the ovary, except where it is attached to the inside of the pericarp.

Actinomorphic See *radially symmetrical.*

Actinomycetes Microorganisms which are between the molds and bacteria. They are thought to be active in organic matter decomposition.

Adiabatic cooling The cooling of air caused by its expansion as it rises. No heat is exchanged with its surroundings.

Advective freeze A freeze associated with the invasion of a large, cold air mass and usually accompanied by windy conditions and therefore no temperature inversion.

Adventitious structures Structures arising from places other than the usual; e.g., roots growing from leaves; or buds developing at locations other than in leaf axils or shoot apices.

Aerobic Requiring oxygen or occurring only in the presence of oxygen.

Aggregation Groups of soil particles which are clumped together and thus improve the structure of a soil.

Air-blast sprayer A sprayer which has a large fan to produce a high-speed, high-volume air flow to both break spray particles into small droplets and to carry the spray to the plant.

Air conditioning The use of overhead irrigation to cool a crop through the evaporation of water and the associated uptake of the heat of vaporization.

Air drainage The flow of cold air downhill. Freeze-sensitive crops are planted on hillsides so that on calm spring nights the cold air will drain down and away from the crop.

Air layering Layering procedure in which a ring of tissue to the xylem and about 2 cm in width from previous years' growth is removed from the stem below the tip of the plant and surrounded with a moist medium such as peat.

Alkali soil A soil with a pH of 8.5 or higher, or high sodium content, or both.

Alternate leaf arrangement An arrangement characterized by one leaf per node on different sides along the stem.

Anaerobic Not requiring oxygen or occurring only in the absence of oxygen.

Andromonoecious Type of sex expression where plants contain perfect as well as imperfect staminate flowers on the same plant.

Angiosperm A plant that bears enclosed seeds in fruits formed by development of the pistil of the flower.

Annuals Plants that complete their life cycle in one year or less.

Anther The upper part of the stamen in which pollen is produced.

Antitranspirant A material applied to plants to reduce the rate of transpiration. These are usually plastic or wax formulations which are sprayed and dry to form a relatively impervious film.

Apomixis Form of reproduction in which new individuals are produced without nuclear or cellular fusion. The embryo develops from an unfertilized egg, or from tissues, such as the integument, which surround the embryo sac.

Arms In a grape plant, all main branches 2 years old or older.

Artificial system Any classification system of plants devised for convenience. Artificial systems are often based on arbitrary, variable, and superficial characteristics.

Asexual reproduction The duplication of a whole plant from any cell, tissue, or organ of that plant.

Assemblage Individuals reproducing from seeds that show some genetic differences but have one or more characteristics by which the plants can be differentiated from other cultivars.

Autotrophic bacteria Bacteria which obtain energy by oxidizing inorganic substances.

Available water Soil moisture between field capacity and the wilting coefficient (one-third to 15 bars soil moisture tension).

Axial placentation Ovules arising from the axis of the compound ovary.

Axillary buds Buds borne laterally on the stem in the axils of leaves.

Bark All the tissues of a root or stem from the cambium outward.

Biennials Plants that ordinarily require 2 years, or at least part of two growing seasons with a dormant period between growth stages, to complete their life cycle.

Bilaterally symmetrical (zygomorphic) Term describing a flower that may be divided into two similar parts by division along one longitudinal plane only.

Binomial system System using names consisting of two parts. Applied to the scientific names of plants, a system in which the genus name and species name are always given together.

Biological control The reduction in the population of pest organisms by means of other living organisms. Control of aphids by ladybird beetles is an example.

Blade Usually, the flattened, green, expanded portion of the leaf.

Blanching Heating of a fruit or vegetable product in boiling water or steam for a brief period to inactivate enzymes before freezing.

Bolting Physiological process by which plants produce a flower stalk and seed and then die before the end of the season.

Bracts Small, pointed, modified leaves which subtend many flowers or inflorescences and may appear to be part of the flower.

Branch Lateral portion of the tree that originates from the trunk or from another branch and gives rise to shoots, twigs, and leaves.

Breeder's seed Seed directly controlled by the originating or sponsoring plant breeder.

Budding Type of grafting in which a vegetative bud (scion) is placed in a stock plant (stock).

Buds Undeveloped and unelongated stems composed of a very short axis of meristem cells from which arise embryonic leaves, lateral buds, flower parts, or all three.

Bud sport Mutation which occurs in the apical meristem of a bud.

Bulb A budlike structure consisting of a small stem with closely crowded fleshy, or papery, leaves or leaf bases.

Bulbils Aerial bublets.

Bulblets Miniature bulbs which develop from meristems in the axils of scaly bulbs.

Bulk bin A large container used in handling and storing fresh fruits and vegetables. Most bulk bins have a capacity of 10 to 20 bushels.

Burned lime CaO or $CaO + MgO$; also called *quicklime*. Limestone is heated to drive off CO_2, leaving the oxide form.

C_3 plant A plant in which the first product of CO_2 fixation is the 3-carbon compound phosphoglyceric acid. This group includes most crop plants except the grasses.

C_4 plant A plant in which the first product of CO_2 fixation is the 4-carbon compound oxaloacetic acid. C_4 plants include many tropical grasses, corn, sugarcane, and some weed species.

C:N ratio The ratio or relative amounts of carbonaceous materials and nitrogen present in a plant or in a soil.

Calcitic lime Calcium carbonate, $CaCO_3$.

Canes Previous season's growth from the arms or trunk of a grape plant.

Capillarity The movement of water into soil pores caused by the attractive forces between the water molecules and the soil particles (adhesion) and the attractive forces between water molecules (cohesion).

Capillary water Water held in the capillaries of the soil at soil moisture tensions of one-third to 31 bars. Capillary water meets essentially the entire water needs of plants.

Capitulum A globose or disk inflorescence with a very short axis and sessile flowers.

Capsule Simple, dry, dehiscent fruit with two or more locules which split in various ways.

Carotenoids Pigments which range in color from yellow to red, found in chromoplasts as well as chloroplasts.

Caryopsis An indehiscent dry fruit with one seed which is completely fused to the inner surface of the pericarp.

Cation exchange capacity The number of negatively charged sites on a soil which can react with and hold cations. The cation exchange capacity is high for clays and humus, and low for sand.

Catkin A type of spike inflorescence that has unisexual flowers with a perianth.

Cell The smallest unit of living matter capable of continued independent life and growth; each is a self-contained and at least a partially self-sufficient unit bounded by a cell wall.

Cell differentiation The changes a cell undergoes during growth, which result in a cell with a specialized form and function.

Centromere That portion of the chromosome to which the spindle fiber is attached.

Certified seed The progeny of registered seed stock. It is the final stage in the expansion program and is certified with a metal seal and blue tag.

Chelate A complex organic molecule which can be combined with a cation such as Fe^{++} but will not ionize. Chelates are used to supply micronutrients where fixation by the soil will make the unchelated ions unavailable.

Chewing insects Insects, such as worms, grasshoppers, and Japanese beetles, with chewing mouth parts.

Chilling injury Damage to certain horticultural products, such as banana, papaya, cucumber and sweet potato, which results from exposure to cold but above-freezing temperatures.

Chilling requirement A cold period required by certain plants and plant parts in order to break physiological dormancy or rest. The chilling requirement is expressed in terms of the required number of hours at 7°C or less.

Chlorenchyma cells Parenchyma cells that contain chloroplasts and compose the palisade and spongy mesophyll cells of leaves. Principal site for photosynthesis.

Chlorophyll The green pigment in plants which absorbs the radiant energy that is ultimately fixed in the form of reduced carbon compounds.

Chloroplast The organelle of a cell in which photosynthesis occurs.

Chromatids The two paired parts of the chromosome at the time of its longitudinal replication.

Chromatin A threadlike network of genetic material in the nucleus composed of DNA and proteins.

Chromoplast A plastid containing only carotenoid pigments (yellow, red, and orange), as distinguished from a chloroplast, which contains green pigments as well.

Chromosomes One of the nuclear structures of definite number, consisting of chromatin and bearing hereditary units or genes.

Cladophyll A leaflike structure which may bear flowers, fruits, and temporary leaves.

Class In the scientific classification system, a group of plants made up of orders.

Clay Fine soil particles less than 0.002 mm in diameter.

Clean cultivation Periodic soil tillage to eliminate all vegetation other than the crop being grown.

Climate The long-term average weather conditions.

Clone Plant cultivar derived from a single individual and propagated entirely by vegetative means.

CO_2 compensation point The CO_2 concentration at which there is no net CO_2 flux, as photosynthesis balances respiration. In C_3 plants this is often about 50 ppm; in C_4 plants, it is close to zero.

Cold storage An insulated storage utilizing mechanical refrigeration to maintain a stable, cold temperature for long-term storage.

Collenchyma cells Modified parenchyma cells which are functional during primary growth in a plant.

Colloid An insoluble particle small enough to remain suspended in a liquid without agitation.

Commercial floriculture Area of horticulture which includes the commercial production and distribution of cut flowers, flowering pot plants, foliage plants, and bedding plants.

Common storage An unrefrigerated storage in which outside air is used to keep temperatures cool.

Companion cells Small, slender, living cells associated with the sieve cell in the phloem of angiosperms.

Complete fertilizer A fertilizer which contains nitrogen, phosphorus, and potassium.

Complete flower Flower composed of a short axis or receptacle from which arise four sets of floral parts—sepals, petals, stamens, and pistils.

Compound leaf   A leaf with the blade divided into several leaflets or sections.

Compound ovary   An ovary with two or more locules.

Compound tissue   Tissue composed of two or more cell types.

Compound umbel   Type of inflorescence where a series of simple umbels arise from the same point on the main axis.

Concentrate sprayer   A sprayer designed to deliver pesticides to a crop at normal amounts per hectare but in much lower volumes of water.

Contact herbicide   An herbicide that kills only the tissues with which it comes into contact.

Contact poison   An insecticide which is absorbed by the insect through the skin or body openings, rather than being ingested.

Controlled atmosphere storage   A cold storage in which the concentrations of atmospheric gases are adjusted to extend the storage life of fresh produce. Usually oxygen is lowered and carbon dioxide raised.

Cork cambium   A cylindrical layer of cells of the cortex or of the phloem which develop or reinitiate the capacity to divide.

Cork (phellem)   Plant part composed of cells with suberized walls formed by the cork cambium.

Corms   Short, fleshy, underground stems with few nodes and very short internodes.

Cortex   The primary parenchyma tissue lying between the epidermis and the vascular tissue of stems and roots.

Corymb   Type of raceme inflorescence where the pedicels of the lower flowers are longer than the pedicels of the upper flowers, resulting in a flattopped inflorescence.

Cotyledon   Embryonic leaves which may serve as food-storing organs or may develop into photosynthetic structures as the seed germinates.

Cover crop   A crop, such as wheat or rye, grown to reduce soil erosion, conserve nutrients, and provide organic matter. Cover crops are grown during the season when a cash crop is not being grown or between the rows of crops, such as peaches.

Critical level   A concentration of a nutrient element below which deficiency symptoms are likely, or a response to additions of the nutrient may be expected.

Crossing over   The exchange between nonsister chromatids which transfers genes from one chromatid to another.

Cross-pollination   The process in which pollen is transferred from an anther of one flower to the stigma of a second flower of a different cultivar.

Cultivar   A plant derived from a cultivated variety that has originated and persisted under cultivation, not necessarily referable to a botanical species, and of botanical or horticultural importance, requiring a name.

Cuticle   The waxy covering on leaves or fruit, which protects the tissue against excess moisture loss.

Cutin   A group of waxy substances which are deposited both within the epidermal cell walls and on the outer surface of the walls and minimizes water loss.

Cuttings   Detached vegetative plant parts which when placed under conditions favorable for regeneration will develop into a complete plant with characteristics identical to the parent plant.

Cyme   A broad, more-or-less flat-topped determinate inflorescence in which the central flowers bloom first.

Cytokinesis   Cell division, as distinguished from mitosis or nuclear division.

Cytoplasm   The living matter or physical substance of the cell.

Daily-flow irrigation   See *trickle irrigation.*

Dark reactions   The second series of photosynthetic reactions in which CO_2 is actually "fixed." The energy to drive the dark reactions comes from the end product of the light reactions.

Dark respiration   The complex series of reactions in which carbohydrates, especially glucose, are broken down to release energy, much of which is trapped in high-energy compounds like ATP. Dark respiration occurs in both the dark and light periods.

Day-neutral plant   A plant which will flower under any daylength.

Dehiscent   Type of dry fruit in which the carpel splits along definite seams at maturity.

Denitrification   The conversion of nitrate nitrogen to gaseous forms which leads to nitrogen losses from soils.

Deoxyribonucleic acid (DNA)   A nucleic acid consisting of the sugar deoxyribose, together with phosphate, adenine, guanine, cytosine, and thymine.

Dermal system   The outermost layer of cells, the epidermis, of floral parts, leaves, fruits, seeds, and of stems and roots until they begin secondary growth.

Determinate growth   Limited growth.

Dewpoint   The temperature at which a given mixture of air and water vapor will reach 100 percent relative humidity or at which condensation will start to occur.

Dichasium   A type of inflorescence with a peduncle with a terminal flower and a pair of lateral branches below each bearing lateral branch.

Dicotyledon   Class of plants with embryos which have two cotyledons.

Dictyosomes   Cell organelles referred to as *Golgi bodies.* They are groups of flat, disk-shaped sacs formed by cytoplasmic membranes which serve as a source of plasma membrane materials and collection centers for complex carbohydrates.

Dihybrid   Plant that differs by two pairs of genes.

Dioecious   Type of sex expression where plants produce staminate and pistillate flowers on separate plants.

Disbudding   The removal of vegetative or flower buds.

Diseased plant   A plant which is abnormal because of an infectious pathogen.

Division   Taxon or taxonomic group of the plant kingdom, made up of classes. The Latin names of divisions end in *-phyta.*

Dolomitic lime Calcium carbonate ($CaCO_3$) with sizable quantities of dolomite—$CaMg(CO_3)_2$—present. Dolomitic lime is recommended where magnesium tends to be deficient.

Dormancy A state of suspended growth or the lack of outwardly visible activity caused by environmental or internal factors.

Double-cross Combining of two single crosses in hybridization between two inbred lines.

Double fertilization Union of the two male gametes with the female gamete and the polar nuclei.

Double-working Type of grafting where the graft combination contains an interstock or intermediate stem piece which is grafted between the scion and stock.

Drip irrigation See *trickle irrigation*.

Drupe Type of fruit with a thin exocarp, a mesocarp that is thick and fleshy, and an endocarp that is hard and stony.

Dry fruits Classification of fruit in which the pericarp is often hard and brittle at maturity.

Duster A type of pesticide applicator in which the pesticide is applied as a dry powder, usually mixed in a diluent such as talc.

Electromagnetic spectrum The range of radiant energy ranging from wavelengths of < 0.001 nm to $> 100,000$ nm.

Embryo sac Cavity within the ovule.

Emitter The water delivery mechanism or outlet in a trickle irrigation system.

Emulsifiable concentrate A liquid formulation which mixes with water to form an emulsion but does not dissolve to form a solution.

Endocarp The inner area of the fruit pericarp.

Endodermis Single layer of cells between the cortex and pericycle in the roots.

Endoplasmic reticulum Structure extending throughout entire units of cytoplasm and functioning in the transport of cell products, as a surface for protein synthesis by the ribosomes, in the separation of enzymes and enzyme reactions, in support, and in the moving of cell membrane components into position, as in cell division.

Entomology The study of insects and their control.

Enzymes Proteinaceous molecules which act as catalysts for specific chemical reactions.

Epicotyl The part of the axis of an embryo above the region of attachment of the cotyledons.

Epidermis The outer layer of cells of all parts of a young plant and of some parts of older plants, such as leaves and fruits. These cells are usually covered with cutin.

Epigynous Classification of a flower in which the perianth and stamens are attached above the ovary.

Evapotranspirational potential The water loss per unit time of evaporation and transpiration combined.

Exocarp The outer skinlike region of the fruit pericarp.

External dormancy The inability to grow caused by unfavorable external conditions such as moisture supply, temperature, oxygen, or light.

Fallow A system in which land is left without a crop or weed growth for extended periods to accumulate moisture.

Family A group of closely related genera or, in a few cases, a single genus.

Fan-and-pad cooling system A cooling device used in greenhouses. Air is pulled through wet pads by means of fans. As water evaporates, large quantities of heat are absorbed (heat of vaporization).

Far-red light Radiant energy at a predominant wavelength of 730 nm.

Fermentation A process in which sugars are converted to ethyl alcohol by the action of specific yeasts. To prevent the conversion of ethanol to acetic acid by aerobic organisms, oxygen must be excluded.

Fiber cells Sclerenchyma cells which are slender and elongated with tapering ends which overlap and are often fused with one another.

Fibrous root System in which primary and lateral roots develop more or less equally and have a limited quantity of cortex.

Field capacity The amount of water which a soil can hold against gravity. At field capacity the soil moisture tension is one-third bar, and this is the upper limit of available water. Field capacity is expressed as a percentage of the dry weight of a soil.

Field crate A container, holding from 10 to 25 kg, used to haul fresh fruits or vegetables to the storage or packing house.

Filament The part of the stamen which holds the anther in a position favorable for pollen dispersal.

Fire blight A major disease of pear and apple caused by the bacterium *Erwinia amylovora.*

Fleshy fruits Classification of fruits which includes the berry, pepo, hesperidium, drupe, and pome. They have a pericarp that is soft and fleshy at maturity.

Fleshy root A root that accumulates and stores a rich supply of reserve food for the plant.

Floriculture The study of growing, marketing, and arranging flowers and foliage plants.

Florigen A term given to the so-called "flowering hormone," which has been hypothesized for decades but has not yet been isolated.

Flower A shoot of determinate growth with modified leaves that is supported by a short stem; the structure involved in the sexual reproductive processes of angiosperms.

Follicle Simple, dry, dehiscent fruit having one locule which splits along one suture.

Footcandle A unit of measure for visible light. (One footcandle = 10.7 lux; see also *lux.*) Full midday sunshine in summer approximates 10,000 fc.

Forcing The manipulation of environmental factors which makes it possible to produce a marketable pot plant or cut flower out of season.

Form A member of a population that differs from other members to a degree not great enough that it can be called a *cultivar.*

Foundation seed Seed stock handled to most nearly maintain specific genetic identity and purity under supervised or approved production methods certified by the agency.

Frost pocket A depression in the terrain into which cold air drains but from which it cannot escape, thus causing it to be an area very subject to freeze injury.

Fruit An expanded and ripened ovary with attached and subtending reproductive structures.

Fumigant Organic compounds with high vapor pressures which are gases at temperatures above 5°C. These are used for insect or disease control in confined areas.

Fundamental system All tissues of the plant encased by the epidermis, other than those of the vascular system.

Gametes Male and female sex cells.

Gametophyte phase In alternation of generations, the sexual phase producing gametes. Gametophytic nuclei are haploid (1N).

Generative nucleus The nucleus of pollen grains which by division forms the sperms.

Genes Basic units of hereditary material that dictate the characteristics of individuals.

Genotype The genetic makeup, determined by the assemblage of genes it possesses.

Geotropism The movement of plants in response to gravity.

Germination The intiation of active growth by the embryo, resulting in the rupture of seed coverings and the emergence of a new seedling plant capable of independent existence.

Grafting The joining of two separate structures, such as a root and stem or two stems, so that by tissue regeneration they form a union and grow as one plant.

Gravitational water Water in excess of capillary water and thus held at soil moisture tensions of less than one-third bar. Gravitational water percolates 24 to 48 h after a rain or irrigation.

Greenhouse effect The effect of the earth's atmosphere on incoming and outgoing radiation. Solar radiation is predominantly of short wavelengths and quite readily passes through the atmosphere. Terrestrial radiation is of much longer wavelengths and is trapped or reflected by the atmosphere. The atmosphere acts much like the glass in a greenhouse because of its selective transmission of radiant energy.

Growing season The period from the last spring freeze until the first freeze in the fall. In the United States this ranges from about 100 to 365 days.

Guard cells Cells which bound stomates and by their turgor pressure determine whether a stomate is open or closed.

Guttation The exudation of water through the leaves via structures called *hydathodes*. Guttation occurs largely at night because of root pressure when little or no transpiration is occurring.

Gymnosperm A plant that bears naked seeds without an ovary.

Gynomonoecious Type of sex expression where plants contain perfect as well as imperfect pistillate flowers on the same plant.

Hardening The result of a great many changes which occur in a plant as it develops resistance to adverse conditions, especially cold.

"Hardening off" The treatment of tender plants to enable them to survive a more adverse environment. Treatments involve withholding nutrients, lowering temperatures, allowing temporary wilting, and other methods to slow growth rate.

Hardpan An impervious layer in a soil which restricts root penetration as well as movement of air and water.

Hardwood cuttings Cuttings made from woody deciduous species and narrow-leaved evergreen species, such as grape and hemlocks.

Heading back Type of pruning cut where the terminal portion of the shoot is removed but the basal portion is not.

Heat of fusion The amount of heat required to change 1 g of a substance at its melting point from the solid to the liquid state, or vice versa. The heat of fusion of water is 80 cal.

Heat of vaporization The amount of heat required to change 1 g of a substance at its boiling point from the liquid to the vapor state, or vice versa. The heat of vaporization of water is 540 cal.

Herbaceous cuttings Cuttings made from succulent herbaceous plants, such as chrysanthemums, coleus, and geraniums.

Herbaceous perennials Plants with soft, succulent stems whose tops are killed back by frost in many temperate and colder climates, but whose roots and crowns remain alive and send out top growth when favorable growing conditions return.

Herbarium A collection of plant specimens that have been taxonomically classified, pressed, dried, and mounted on a sheet of herbarium paper.

Herbicide A material which will kill plants. Herbicides may kill essentially all plants or be quite selective in their activity.

Hesperidiums Fruits with a leathery rind.

Heterotrophic bacteria Bacteria which obtain energy by degradation of organic matter.

Heterozygous Having contrasting genes of a gene pair present in the same organism.

High-pressure sprayer A sprayer utilizing a high-pressure pump to force the spray through nozzles for both atomization and delivery to the plant.

Homozygosity Condition in which identical genes of a gene pair are present, such as a yellow-seeded pea plant with genes (YY) for yellow endosperm and (yy) for white-seeded plants with white endosperm.

Homozygous Having identical genes of a gene pair present in the same organism.

Horticulture The intensive cultivation of plants. (*hortus*, "garden"; *cultura*, "cultivation.")

Humus A dark-colored, amorphous organic material which is the product of organic matter degradation by both microorganisms and chemical reactions. Humus is resistant to further degradation.

Hybrid   In its simplest form, a first-generation cross between two genetically diverse parents.

Hydrated lime   Burned lime which has been reacted with water to form $Ca(OH)_2$ or $Ca(OH)_2 + Mg(OH)_2$.

Hydrocooling   A cooling system for fresh produce in which the product is flooded with large volumes of cold water to remove field heat.

Hydrohandling   System of unloading containers and conveying fruits and vegetables in water to minimize bruising.

Hygroscopic coefficient   The amount of water present in a soil when the soil moisture tension is 31 bar. Such water is unavailable to plants.

Hygroscopic materials   Substances, such as salt, which attract water.

Hygroscopic water   Soil moisture which exists as a very thin film around soil particles and is unavailable to plants. At the upper limit of hygroscopic water, soil moisture tension is 31 bar.

Hygrothermograph   A device which continuously records both temperature and relative humidity.

Hypobaric storage   A new cold-storage principle in which atmospheric pressure is reduced. The subatmospheric pressure lowers respiration by reducing both oxygen and ethylene concentration in plant tissues.

Hypocotyl   The axis of an embryo below the cotyledons.

Hypogynous   Classification of a flower in which the sepals, petals, and stamens are attached to the receptacle below the ovary.

Imperfect flower   Flower lacking either the stamen or the pistil.

Incomplete flower   Flower lacking one or more of the four sets of floral parts.

Indehiscent   Type of dry fruit in which the fruit wall does not split at any certain point or seam at maturity.

Indeterminate growth   Growth that is potentially limitless.

Inflorescence   The arrangement of the flowers on the floral axis—a flower cluster.

Insolation   The radiation received from the sun.

Integument   The wall of the mature ovule which surrounds the embryo sac.

Interstock   Intermediate stem piece which is grafted between the scion and stock.

Lamellae   Disk-shaped structure in the chloroplast.

Landscape design   The profession concerned with the planning and planting of outdoor space to secure the most desirable relationship between land forms, architecture, and plants to best meet human needs for function and beauty.

Lath house   A structure consisting of a frame supporting strips of wood which are spaced to provide about 50 percent shade. Snow fence is a common material used in constructing lath houses.

Laticifer cells   Parenchyma cells which are specialized to synthesize and store latex.

Layering A vegetative method of propagation which produces new individuals by producing adventitious roots before the new plant is severed from the parent plant.

Leaching The downward movement of nutrients or salts through the soil profile in soil water. Leaching accounts for nutrient losses but can also be beneficial in ridding a soil of excess salts.

Leaf arrangement The arrangement of leaves on the stem.

Leaf-bud cuttings Cuttings consisting of a leaf blade, petiole, and a short piece of the stem, with the attached axillary bud placed with the bud end in the medium and covered enough to support the leaf.

Leaf cuttings Entire leaves with or without the petioles.

Leaves Vegetative plant parts that are lateral outgrowths of stems which have developed special structural adaptations for photosynthesis.

Legume A simple, dry, dehiscent fruit with one locule which splits along two sutures.

Leucoplast A colorless plastid.

Light compensation point A level of irradiance at which there is no net flux of CO_2 from a leaf. At the light compensation point, photosynthesis equals respiration.

Light meter An instrument to measure visible light, usually in units of footcandles or lux.

Light reactions The first series of photosynthetic reactions in which light energy is converted to chemical energy.

Light saturation A level of irradiance above which there is no further increase in net photosynthesis.

Lignin A complex organic polymer which provides rigidity, toughness, and strength to cell walls.

Lime Ground limestone which is used to raise soil pH. Limestone consists of $CaCO_3$, with varying amounts of $CaMg(CO_3)_2$ present.

Line Plant cultivar of uniform appearance which has horticultural value and can be reproduced uniformly by seed.

Locules The cavities of the ovary of the pistil of a flower.

Long-day (short-night) plant A plant which requires a day longer than its critical daylength (or, more exactly, a night shorter than its critical dark period) in order to flower.

Lux A unit of measure for visible light. Full midday sunshine in summer approximates 108,000 lux or 108 klux (10.7 lux = 1 fc). See also *footcandle*.

Luxury consumption The uptake of a nutrient in quantities in excess of that required for optimum growth and productivity. Luxury consumption of potassium is common.

Lysosomes Single-membrane organelles in the cytoplasm which contain hydrolytic enzymes capable of digesting other cellular particles.

Macronutrient Essential nutrient which is needed in relatively large amounts, e.g., nitrogen, potassium.

Malling-Merton series A group of apple rootstocks which were the result of crossing the Malling series with 'Northern Spy' to incorporate resistance to wooly aphids. The Malling-Merton series is largely semivigorous to vigorous.

Malling series A group of apple rootstocks which originated at the East Malling Research Station in England. These vary in vigor from the very dwarfing M 27 up to others which are very vigorous.

Market gardening The growing of an assortment of vegetables for local or roadside markets.

Megagametophyte The female gametophyte.

Megaspore mother cell A diploid cell which by meiosis produces four megaspores.

Meiosis A type of cell division which produces the gametophytic phase in both the male and the female reproductive parts of the flower. Meiosis includes one reduction division of chromosome number followed by one duplication division.

Meristematic cells Isodiametric cells which contain dense protoplasm with small vacuoles and have thin walls consisting of only the primary layer. They function in cell division.

Mesocarp The center portion of the fruit pericarp.

Mesophyll A tissue consisting of palisade parenchyma cells and spongy parenchyma cells between the upper and lower epidermal layers of the leaf.

Microgametophyte The male gametophyte.

Micronutrient An essential nutrient which is needed in small amounts; also called a *trace* or *minor element*; e.g., boron, molybdenum.

Microsporangia The four chambers of the four lobes of the young anther.

Microspore mother cell A diploid cell which by meiosis produces four microspores.

Microtubules Cell organelles located in the cytoplasmic matrix of nondividing cells and in the spindle fibers of dividing cells which may be involved in the growth of the cell wall, cell plate development, and mitosis.

Middle lamella A viscous (jellylike) substance, consisting of calcium and magnesium pectates, binding the primary cell wall of one cell to that of the adjacent cell.

Minimum tillage A soil management system in which the winter cover crop or sod is killed by herbicide and the crop (often corn) is seeded directly into the sod without plowing or disking.

Mitochondria Organelles which serve as the cell's "powerhouse" and are the site of oxidation of organic molecules which release energy and the incorporation of the energy into molecules of adenosine triphosphate (ATP), the main chemical energy source for all cells.

Mitosis Nuclear division, involving replication of chromosomes and their separation into two groups of equal numbers to form two daughter nuclei.

Mixed bud Bud containing primordial tissue for both leaves and flowers.

Monochasium Type of inflorescence with a peduncle bearing a terminal flower with a lateral flower producing branches below it.

Monocotyledon   Class of plants with embryos which have one cotyledon.

Monoecious   Type of sex expression in plants by which they produce staminate and pistillate flowers on the same plant.

Mosaics   A group of diseases caused by virus infection. Symptoms include mottling of the leaves and flowers.

Mound or stool layering   A layering procedure in which the new shoots develop in the spring and soil is mounded around their bases, excluding light and enhancing root formation.

Mulch   A material applied to the surface of a soil for a variety of purposes such as conservation of moisture, stabilization of soil temperature, and suppression of weed growth.

Mutation   A spontaneous change in the genetic makeup of the cell.

Mycorrhiza   A fungus living in a mutualistic relationship with the roots of the vascular plant.

Natural System   Classification system which attempts to show relationships among plants through the use of selected morphological structures.

Nematode   Microscopic animals, mostly worm-shaped, which can be parasitic on plants as well as animals. Nematode damage to many crops can be severe.

Net photosynthesis   Gross photosynthesis minus respiration. Net photosynthesis is determined by measuring net CO_2 uptake.

Nitrification   The conversion of ammonium ions to nitrite and then to nitrate ions $NH_4 \rightarrow NO_2^- \rightarrow NO_3^-$.

Nitrogen fixation   A process carried out by the symbiotic relationship between higher plants (particularly legumes) and certain bacteria in which atmospheric N_2 is fixed in a form usable by plants. Certain free-living bacteria and blue-green algae can also fix nitrogen.

Nucleoli   One or more spherical bodies in the nucleus, composed of ribosomal RNA and proteins.

Nucleus   Protoplasmic structure which is the center of much of the physiological activity of the cell and functions in the transmission of hereditary characteristics.

Nursery grower   A person who produces or distributes ornamental plants.

Nut   An indehiscent dry fruit similar to the achene except that the pericarp is hard throughout.

Offsets   Bulblets grown to full size.

Offshoots   Short, horizontal stems which occur in whorls or near whorls at the crown of stems.

Olericulture   The study of vegetable production.

Opposite leaves   Leaves arising from opposite sides of the same node.

Order   A group of closely related families with some common traits but some marked differences. The Latin names of orders end in *-ales*.

Organic matter   Carbonaceous materials of either plant or animal origin, which exist in all stages of decomposition in soils.

Ovary   Part of the pistil that contains one or many small bodies known as ovules.

Ovule   The immature seed in the ovary.

Palisade cell   Elongated cells that generally form the upper part of the mesophyll in leaves.

Palmate leaf   A compound leaf with all the leaflets arising from one point at the end of the petiole.

Palmately veined leaves   Leaves with netted venation having several large veins radiating into the blade from the petiole at the level where petiole and blade join.

Panicle   Type of inflorescence with either a cluster of racemes or corymbs, distinctly branched.

Parallel venation   Leaves with large veins that are essentially parallel to one another and are not connected by lateral veins.

Parenchyma cells   Thin-walled cells having large central vacuoles which force the cytoplasm and nuclei against the inner surface of their walls. They function in food storage, photosynthesis, wound healing, and the forming of adventitious structures.

Parietal placentation   Placentation in which the ovules arise from the inner surface of the ovary wall.

Parthenocarpic   Term describing fruit that develops without fertilization.

Parthenogenesis   The production of offspring by females in the absence of males, as can occur with aphids.

Peduncle   The short stem of the flower cluster.

Pepos   Berries that have a hard rind around the fruit.

Perennials   Plants which do not die after flowering but live from year to year.

Perfect flower   A flower that has both pistil (or pistils) and stamens but may lack sepals, petals, or both.

Perianth   Structure consisting of the sepals and petals.

Pericarp   The fruit wall, consisting of three distinct layers: the exocarp, the mesocarp, and the endocarp.

Pericycle   Thin layer of parenchyma cells that separates the endodermis from the vascular components in roots.

Periderm   A tissue consisting of cork, cork cambium, and phelloderm.

Perigynous   Classification of a flower in which the receptacle is extended to form a cuplike structure around a portion of ovary.

Permanent sod   A soil management system in which a sod is periodically mowed but no tillage is carried out. Benefits include prevention of soil erosion, maintenance of good organic-matter levels, and excellent soil structure.

Permanent wilting point   See *wilting coefficient.*

Pest management   The control of a pest or group of pests by a broad spectrum of techniques ranging from biological means to pesticides. The goal is to keep damage below economic levels without completely eliminating the pest.

Petals   Structures collectively making the corolla, which protect the inner reproductive structures and often attract insects by either their color or their nectar and thus facilitate pollination.

Petiole The leaf stalk.

pH A measure of acidity or alkalinity, expressed as the negative log of the hydrogen ion concentration. A pH of 7 is neutral; less than 7 is acidic; more than 7 is basic.

Phenotype The external appearance of a plant, governed by its internal genotype.

Phloem A compound vascular system of plants composed of sieve tubes, companion cells, fibers, and parenchyma cells.

Phloem fibers Nonliving, thick-walled, elongated cells which are in groups within the phloem.

Photoperiodism The developmental responses of plants to the relative lengths of the light and dark periods.

Photorespiration (light respiration) Respiratory utilization of photosynthetic products which occurs only during the light period. C_3 plants have high rates of photorespiration, whereas C_4 plants have little or none.

Photosynthetically active radiation (PAR) Radiant energy between 400 and 700 nm to which the photosynthetic apparatus responds.

Phototropism The bending of plants in the direction of more intense light or toward the light if irradiated from only one side.

Phyllotaxy The arrangement of leaves on the stem.

Phylogenetic system System which classifies plants according to their evolutionary pedigree, reflecting genetic relationships between and among plants and establishing their progenitors.

Phytochrome A photoreceptive pigment which receives radiant stimuli leading to many photomorphogenic responses.

Pick your own A system of direct marketing in which the customer harvests the product. This marketing method is well adapted to strawberries, raspberries, some tree fruits, and many vegetables.

Pinching Breaking off the terminal growing point, thus allowing the axillary buds to start to grow.

Pinnate leaf A compound leaf with the leaflets arranged along both sides of the midrib.

Pinnately veined leaves Leaves with netted venation where secondary veins extend laterally from a single midrib.

Pistil The female reproductive organ, consisting of the stigma, style, and ovary.

Pistillate flowers Flowers in which only the pistils are present; there are no stamens.

Pith rays Areas of cells between adjacent vascular bundles.

Placenta The tissue of the ovary to which an ovule (or ovules) is attached.

Plant growth A permanent increase in volume, dry weight, or both.

Plant key An analytical device whereby a choice between two or sometimes more contradictory propositions has to be made in each step.

Plant pathogen A microorganism which causes a plant disease.

Plant propagation Increase in numbers or perpetuation of a species by reproduction.

Plant taxonomy The science that includes classification, nomenclature, and identification of plants.

Plasma membrane A differentially or selectively permeable, living, flexible membrane essentially composed of lipids and proteins.

Plasmodesmata Pores or pits in nonlignified cell walls in plants through which cytoplasmic strands extend.

Plastic mulch Thin polyethylene film, which may be clear or black, that is used as a mulch, especially for vegetables. Benefits include moisture retention, increased soil temperature, and—with black plastic—complete weed control.

Plastids Cell organelles involved in food synthesis; in storage of fats, starches, proteins, and various pigments; or in both.

Pollen tube The tube, formed by the pollen grain, which grows through stigma, style, ovary, and micropyle to the female gametophyte.

Pollination The transfer of pollen from the anther to the stigma either within the same flower or between different flowers.

Polyploidy Condition in which individuals have more than two sets of chromosomes in their cells.

Pome Type of fruit in which the portions produced by the pericarp are enclosed within fleshy parts that are derived from parts of the flower other than the ovary.

Pomology The study of fruit production.

Pore space The openings in a soil not occupied by solid particles. Pore spaces are occupied by varying proportions of water and air.

Pot-bound Having restricted root growth and a circular pattern of root growth caused by a too-small container.

Prepackaging Packing of a product in a consumer package by the packer or shipper rather than by the retailer. Examples are net bags, polyethylene bags, and trays with overwraps.

Primary tissue Permanent tissue developed directly from the apical meristem.

Protoplasm The living substance of the cell.

Protoplast The living unit inside the cell wall. It is composed of protoplasm and plasma membrane.

Pruning Removal of plant parts such as buds, developed shoots, and roots to maintain a desirable form by controlling the direction and amount of growth.

Psychrometer A device using wet-bulb and dry-bulb thermometers to determine relative humidity.

Puddled soil A soil in which the structure has been destroyed by poor soil management. The pore spaces are closed up and both aeration and water movement are poor.

Raceme Type of inflorescence in which stalked flowers are on pedicels approximately equal in length on a single floral axis.

Radiation One of the forms in which energy is transferred. Radiation is characterized by both wavelength and frequency and includes the range from short wavelength and high energy (e.g., gamma rays) to long wavelength and low energy (e.g., radio waves). Visible light is a very small part of the total spectrum of radiation.

Radially symmetrical (actinomorphic)   Term describing a flower which may be divided into similar parts by division along more than one longitudinal plane, and in which all the floral parts of each group are alike in size and shape.

Radiational freeze   A freeze associated with calm conditions, radiational cooling, and a temperature inversion.

Radicle   Lower portion of the hypocotyl.

Receptacle   The enlarged apex of the pedicel where the floral parts arise, sometimes called the *torus*.

Red light   Radiant energy at a predominant wavelength of 660 nm.

Relative humidity   The amount of water vapor present in the air, expressed as a percentage of the maximum water vapor that the air could hold at the same temperature and pressure.

Rest   A state of suspended growth or outwardly visible activity due to internal physiological factors. Rest is broken by exposure to temperatures of 7°C or less for an extended period of time (chilling requirement). Also referred to as *physiological dormancy*.

Rhizobium   A bacterium which can infect legumes in a symbiotic relationship and fix atmospheric nitrogen.

Rhizomes   Horizontal stems that grow partly or entirely underground. They are often thickened and serve as storage organs.

Ribonucleic acid (RNA)   A nucleic acid consisting of the sugar ribose, together with phosphate, adenine, guanine, cytosine, and uracil.

Ribosomes   Small, dense, globular particles floating in the cytoplasm and associated with the membranes of the endoplasmic reticulum.

Root   Vegetative plant part which anchors the plant, absorbs water and minerals in solution, and often stores food.

Root cuttings   Cuttings 5 to 15 cm long made from root sections in the fall or winter.

Root hair   An extension of the epidermal cells on young roots immediately behind the root tip. Root hairs are vitally important in both water and nutrient uptake.

Root pressure   A force generated in the roots and stem of plants which can partially account for the rise of water in plants.

Runner   A slender stolon with elongated internodes. These root at the nodes which touch the ground.

Saline-sodic soil   An alkali soil with more than 2000 ppm soluble salts and high sodium. Saline-sodic soils must be treated with gypsum or sulfur before leaching if they are to be reclaimed.

Saline soil   An alkali soil with more than 2000 ppm soluble salts, but relatively low sodium. Such soils can often be reclaimed by leaching.

Samara   An indehiscent dry fruit with either one or two seeds, in which pericarp bears a flattened winglike outgrowth.

Sand   Coarse soil particles ranging from 0.05 to 2 mm in diameter.

Scales   Modified leaves that protect structures.

Scarification   The chemical or physical treatment given to some seeds in order to break or weaken the seed coat sufficiently for germination to occur. The intact seed coat may inhibit penetration of water or oxygen or may prevent emergence of the embryo.

Scion (cion)   The upper part of the union of a graft.

Sclerenchyma cell   Strengthening cell with a thick, usually lignified, wall having no protoplast at maturity. The two definable types of sclerenchyma cells are fibers and sclereids.

Seed   Plant embryo with associated stored food encased in a protective seed coat.

Selective herbicide   An herbicide which kills only certain groups of plants. For example 2,4,-D kills broadleaf plants, but not grasses.

Self-fertility   The ability of a plant to set viable seed or fruit with pollen from the same cultivar.

Self-pollination   The process by which pollen is transferred from an anther to a stigma of the same flower or another flower of the same cultivar.

Self-sterility   The lack of the ability of a plant to set viable seed or fruit with pollen from the same cultivar.

Semihardwood cuttings   Cuttings made from woody, broad-leaved evergreen species such as ligustrum and holly.

Semipermeable membrane   A membrane which allows free movement of water but which restricts passage of solutes.

Sepals   Structures which usually form the outermost whorl of the flower; collectively, the calyx.

Separation   Use of bulbs and corms in propagation utilizing the naturally detachable parts.

Serrate leaf   Serrations or "teeth" along the margin of the leaf.

Sessile   Without a petiole (as in some leaves), or without a pedicel (as in some flowers and fruits).

Sexual reproduction   The reproduction of plants through a sexual process involving meiosis.

Shade cloth   A fabric woven from saran fibers to provide shade levels ranging from less than 20 percent to more than 90 percent.

Shoot   Stem, 1 year old or less, that possesses leaves.

Short-day (long-night) plant   A plant which requires a day shorter than its critical daylength (or, more exactly, a night longer than its critical dark period) in order to flower.

Sieve cells   The main component of the phloem; elongated, slender cells with thin, nonlignified walls.

Silique   Simple, dry, dehiscent fruit with two fused locules which separate at maturity, leaving a persistent partition between them.

Silt   Soil particles ranging from 0.002 to 0.05 mm in diameter.

Simple bud   Bud containing either leaf or flower primordia, but not both.

Simple layering   A method similar to tip layering, except that the stem behind the end of the branch is covered with soil and the tip remains above ground.

Simple leaves   Leaf blades consisting of one unit.

Simple ovary   Ovary having only one locule.

Simple tissue   Tissue composed entirely of one cell type.

Single cross   Hybridization between two inbred lines.

Single fruits   Classification of fleshy or dry fruits which form a single, ripened ovary.

Slow-release fertilizer   A fertilizer which is made by coating the particles with a wax or other insoluble or very slowly soluble material to provide a predictable, slow release of the encapsulated materials.

Sodic soil   An alkali soil with high sodium but low soluble salts. Sodic soils are called *black alkali* soils because of the organic matter deposit on the surface. Sodic soils are very difficult to reclaim.

Softwood cuttings   Cuttings taken from soft, succulent, new spring growth of deciduous or evergreen species of woody plants.

Soil   The outer, weathered layer of the earth's crust which has the potential to support plant life. Soil is made up of inorganic particles, organic matter, microorganisms, water, and air.

Soil horizon   Layers of soil which constitute the soil profile from the surface to the bedrock.

Soil management   The practices used in treating a soil, which may include various types of tillage and production systems.

Soil moisture tension   The force with which a soil holds on to the soil moisture present. As soil moisture declines, soil moisture tension increases.

Soil pasteurization   The heat or chemical treatment of soil to destroy all harmful organisms but not necessarily *all* soil organisms.

Soil structure   The arrangement of individual soil particles.

Soil texture   The makeup of soil in terms of sand, silt, and clay.

Spadix   A fleshy spike inflorescence with very small male flowers above and female flowers below, embedded in the spike axis.

Specific leaf weight   The dry weight per unit leaf area.

Spermatophyta   That division of the plant kingdom which includes plants reproduced by seeds.

Spike   Inflorescence of an indeterminate raceme of sessile flowers attached to the floral axis with the oldest flowers at the base.

Spikelet   Diminutive type of spike inflorescence found in grasses.

Spines   Sharp-pointed woody structures, usually modified from a leaf or part of a leaf.

Spores   Unicellular or few-celled structures of many types and forms, usually involved in asexual reproduction.

Sporophyte phase   In alternation of generations, the asexual phase. Nuclei of the sporophytic phase are diploid (2N).

Sprayer   A type of pesticide applicator in which the pesticide is mixed in water and distributed on the plant.

Sprinkler irrigation (aerial)   The application of water to a crop from overhead by use of a wide range of systems.

Spurs Stems with short internodes, usually from older wood that bears leaves, fruit, or both.

Spur-type tree A fruit tree with a compact growth habit caused by shorter internodes and more spurs than those contained in standard growing trees of the same cultivar.

Stamen Part of the flower consisting of the anther in which pollen is produced and a slender filament which holds the anther in a position favorable for pollen dispersal. Male reproductive organ.

Staminate flower Flower in which only the stamens are present; there are no pistils.

Stele The vascular tissues and associated fundamental tissues of the root and stem.

Stem Vegetative part of the axis of a plant which develops from the epicotyl of the embryo or from a bud of an already existing stem or root.

Stem cuttings Segments of shoots containing lateral or terminal buds.

Stigma The pollen-receiving site of the pistil.

Stipules Leaflike appendages often found on either side of the base of the petiole; they may subtend the leaf, but they are not actual parts of the leaf.

Stock (rootstock) The lower part of a graft.

Stolon An aboveground stem which reclines or becomes prostrate and may form roots at the nodes that may come into contact with the ground.

Stomach poison An insecticide which must be ingested by the target insect and is thus effective against chewing insects.

Stomate A microscopic opening in leaves through which gas exchange for photosynthesis, respiration, and transpiration occurs. Stomates are bounded by guard cells, the turgor pressure of which opens or closes the stomates.

Stratification The storing of seeds at low temperatures (2 to 4°C) under moist conditions in order to break physiological dormancy or rest.

Stroma The granular portion of the chloroplast where transformation of carbon dioxide to carbon-containing compounds occurs.

Style The slender part of a pistil between the stigma and the ovary, through which the pollen tube grows.

Suberin A thin, varnishlike layer which seals the moisture of the cutting inside the tissue and keeps rot-producing organisms out.

Subsurface irrigation Application of irrigation water to a crop by artificially elevating the water table. Water moves upward into the root zone by capillarity.

Subtropical fruit A fruit plant intermediate between tropical and temperate species. Some subtropicals are evergreen, others deciduous. Minimum temperatures which can be withstood are usually −4 to −5°C. Examples are citrus fruits.

Succulent leaves Water storage structures characteristic of plants in arid and semiarid regions.

Sucking insects Insects with sucking mouth parts which suck sap from plants. Examples are aphids, scale insects, and leaf hoppers.

Surface irrigation The application of water directly to the soil surface, e.g., flooding the entire area or flooding furrows.

Systemics Materials readily translocated throughout a plant. These may be soil-applied or sprayed on the aboveground parts.

Taproot Primary root that persists and maintains its dominance.

Taxon A category or taxonomic group (plural, *taxa*).

Temperate fruit A fruit plant which requires a cool period and is deciduous. Examples are apple, pear, and peach.

Temperature inversion The condition at night in the lower atmosphere which is the reverse of "normal" daytime conditions. During the day, the warmest air is usually at the surface of the ground and temperatures decline with increasing altitude. During a temperature inversion, surface layers cool, and thus there is warm air above the cooler surface air.

Tendrils Slender, twining leaf modifications used for support, as in the terminal leaflets of the grape.

Terminal buds Large and vigorous buds at the tips of stems, responsible for terminal growth.

Tetrad The four microspores collectively.

Thermistor An electrical temperature sensor based on the changing resistance of certain materials as their temperatures change.

Thermocouple An electrical temperature sensor.

Thinning Type of pruning cut in which entire shoots are removed; an extension or complement of heading back.

Thorn A short, sharp-pointed branch.

Thyrse A compact, condensed, determinate cyme or panicle inflorescence.

Tip layering Layering in which rooting takes place near the tip of the current season's shoot, which naturally falls to the ground.

Tissue An organized group of cells with similar origin and function.

Tissue culture The growing of masses of unorganized cells (callus) on agar or in liquid suspension. Useful for the rapid asexual multiplication of plants.

Tonoplast Membrane surrounding the vacuole.

Topworking Grafting procedure by which branches of trees are changed to a more desirable cultivar.

Totipotency The ability to generate or regenerate a whole organism from a part.

Tracheids Fundamental cell type in the xylem.

Translocated herbicide An herbicide which is absorbed and translocated throughout the plant.

Transpiration The giving off of water vapor by plants. Transpiration occurs through the cuticle, stomates, and lenticels, but the great majority is through the open stomates.

Transpirational pull A tension within a plant which is generated by transpiration and exerts a pulling force. Transpirational pull is a major factor in the rise of water in plants.

Transpiration ratio The ratio of units of water taken up per unit of dry matter produced by a plant. Also called the *water requirement*.

Trench layering A layering method in which a number of new plants may be obtained from a given stock plant of a certain species.

Tribe A subdivision of a subfamily.

Trickle irrigation The application of small quantities of water directly to the root zone through various types of delivery systems on a daily basis.

Tropical fruit A fruit plant which is evergreen and cannot withstand freezing temperatures; examples are banana, pineapple, and coffee.

Truck crop production Large-scale production of a limited number of vegetable crops for wholesale markets and shipping.

Tube nucleus One of the nuclei of a pollen grain, thought to influence the growth and development of the pollen tube.

Tubers Enlarged underground stems serving as storage organs of starch or related materials.

Twig Stem, 1 year old or less, without leaves.

Umbel Type of inflorescence in which the pedicels arise from a common point and are about equal in length.

Uniform group A first-generation hybrid (F_1) resulting from a cross of two other cultivars which can be reconstituted on each occasion by crossing two constant parents maintained either by inbreeding or as clones.

Vacuum cooling A cooling system for fresh leafy vegetables, such as lettuce. The product is put into a vacuum chamber, and the atmospheric pressure is lowered to about 4.6 mm Hg. As water evaporates, the heat of vaporization quickly removes heat from the product.

Vacuole Part of the cell filled with cell sap composed of a water solution of inorganic salts, various organic solutes, and crystals.

Vapor pressure (water) The pressure exerted by water vapor.

Vapor-pressure gradient (water) The difference between the actual water vapor pressure and the vapor pressure needed to saturate the air at the same temperature.

Vascular bundle A strand of conducting and strengthening tissue consisting of primary xylem, primary phloem, and often cambium.

Vascular cambium Cambium that gives rise to secondary phloem and xylem.

Vascular system System composed of the xylem and phloem which conducts water, mineral salts in solution, and foods in solution, in addition to providing some support and strength.

Vegetable The edible portion of an herbaceous garden plant.

Vernalization The requirement of certain plants for a cold period before they will flower.

Vessel Long continuous tube in the xylem through which water and minerals move.

Virus An infectious, submicroscopic particle containing both protein and ribonucleic acid (RNA).

Visible light   The part of the electromagnetic spectrum between wavelengths of approximately 380 and 760 nm to which the human eye responds.

Viscosity   The "thickness" of a liquid, which can also be expressed as its "resistance to flow."

Water table   The upper edge of free water in the soil. If a hole is dug, water will fill the hole to the level of the water table.

Weather   The short-term atmospheric conditions, including temperature, relative humidity, wind, sky conditions, precipitation, and atmospheric pressure.

Weed   A plant growing where it is not wanted.

"Wet feet"   A condition where plants are exposed to excess soil moisture caused by flooding or a high water table.

Wettable powder   A powder which will mix with water to form a suspension but does not dissolve, so that continuous agitation is required.

Whorled arrangement   Three or more leaves at a node.

Wilting coefficient   The amount of water present in a soil when a plant cannot obtain enough water to regain turgidity in a saturated atmosphere. Soil moisture tension is about 15 bar. The wilting coefficient is sometimes referred to as the *permanent wilting point*. This is the lower limit of available water and, like field capacity, is expressed as a percentage of the dry weight of the soil.

Wind machine   A large fan, usually permanently mounted on a tower, which mixes air within a temperature inversion to prevent freeze damage by raising the air temperature at the crop level.

Winter desiccation   Injury to plants, particularly evergreens, by moisture loss from the aboveground portions which cannot be replaced because of frozen or very cold soil. Symptoms are browning and death of leaves and small branches.

Woody perennials   Plants with woody fiber.

Xylem   A compound vascular tissue of plants composed of tracheids, vessels, fibers, and parenchyma cells.

Xylem fibers   Elongated, strengthening cells with thickened walls.

Yellows   A group of diseases caused by virus or viruslike infection. Symptoms include yellowing, curling, and stunting of plants.

Zygomorphic   See *bilaterally symmetrical.*

Zygote   A fertilized egg; a cell resulting from the fusion of gametes.

APPENDIX: CONVERSIONS

Temperature: Fahrenheit and Celsius To convert a given temperature from one scale to the other, locate it in the center columns of Appendix Table 1 ("C or F"), and locate the equivalent in the column to the left ("C") or the column to the right ("F"), as appropriate. For example, to convert 35°C to Fahrenheit, find 35 in the center column, and read 95 in the column to the right; to convert 14°F to Celsius, find 14 in the center column and read -10 in the column to the left. To convert from one scale to the other without a table, use one of the following equations, as appropriate (see the last section of Appendix Table 2):

$$°F = 1.80°C + 32$$
$$°C = 0.555(°F - 32)$$

Weight, volume, length, area, pressure, yield, and temperature Appendix Table 2 shows conversions for United States units and metric units for weight, volume, length, area, pressure, yield (or rate), and temperature. To convert from United States units to metric units: find the appropriate unit in column 2, and multiply the measurement by the figure in the column to the right of it. To convert from metric units to United States units, find the appropriate measure in column 1 and multiply by the figure in the column to the left of it.

APPENDIX TABLE 1
CONVERSIONS BETWEEN CELSIUS (C) AND FAHRENHEIT (F) TEMPERATURES

C	C OR F	F	C	C OR F	F	C	C OR F	F
−73.3	−100	−148.0	− 6.1	21	69.8	16.1	61	141.8
−70.6	− 95	−139.0	− 5.6	22	71.6	16.7	62	143.6
−67.8	− 90	−130.0	− 5.0	23	73.4	17.2	63	145.4
−65.0	− 85	−121.0	− 4.4	24	75.2	17.8	64	147.2
−62.2	− 80	−112.0	− 3.9	25	77.0	18.3	65	149.0
−59.5	− 75	−103.0	− 3.3	26	78.8	18.9	66	150.8
−56.7	− 70	− 94.0	− 2.8	27	80.6	19.4	67	152.6
−53.9	− 65	− 85.0	− 2.2	28	82.4	20.0	68	154.4
−51.1	− 60	− 76.0	− 1.7	29	84.2	20.6	69	156.2
−48.3	− 55	− 67.0	− 1.1	30	86.0	21.1	70	158.0
−45.6	− 50	− 58.0	− 0.6	31	87.8	21.7	71	159.8
−42.8	− 45	− 49.0	0	32	89.6	22.2	72	161.6
−40.0	− 40	− 40.0	0.6	33	91.4	22.8	73	163.4
−37.2	− 35	− 31.0	1.1	34	93.2	23.3	74	165.2
−34.4	− 30	− 22.0	1.7	35	95.0	23.9	75	167.0
−31.7	− 25	− 13.0	2.2	36	96.8	24.4	76	168.8
−28.9	− 20	− 4.0	2.8	37	98.6	25.0	77	170.6
−26.1	− 15	5.0	3.3	38	100.4	25.6	78	172.4
−23.3	− 10	14.0	3.9	39	102.2	26.1	79	174.2
−20.6	− 5	23.0	4.4	40	104.0	26.7	80	176.0
−17.8	0	32.0	5.0	41	105.8	27.2	81	177.8
−17.2	1	33.8	5.6	42	107.6	27.8	82	179.6
−16.7	2	35.6	6.1	43	109.4	28.3	83	181.4
−16.1	3	37.4	6.7	44	111.2	28.9	84	183.2
−15.6	4	39.2	7.2	45	113.0	29.4	85	185.0
−15.0	5	41.0	7.8	46	114.8	30.0	86	186.8
−14.4	6	42.8	8.3	47	116.6	30.6	87	188.6
−13.9	7	44.6	8.9	48	118.4	31.1	88	190.4
−13.3	8	46.4	9.4	49	120.2	31.7	89	192.2
−12.8	9	48.2	10.0	50	122.0	32.2	90	194.0
−12.2	10	50.0	10.6	51	123.8	32.8	91	195.8
−11.7	11	51.8	11.1	52	125.6	33.3	92	197.6
−11.1	12	53.6	11.7	53	127.4	33.9	93	199.4
−10.6	13	55.4	12.2	54	129.2	34.4	94	201.2
−10.0	14	57.2	12.8	55	131.0	35.0	95	203.0
− 9.4	15	59.0	13.3	56	132.8	35.6	96	204.8
− 8.9	16	60.8	13.9	57	134.6	36.1	97	206.6
− 8.3	17	62.6	14.4	58	136.4	36.7	98	208.4
− 7.8	18	64.4	15.0	59	138.2	37.2	99	210.2
− 7.2	19	66.2	15.6	60	140.0	37.8	100	212.0
− 6.7	20	68.0						

APPENDIX TABLE 2
CONVERSION BETWEEN UNITED STATES UNITS AND METRIC UNITS

TO CONVERT COLUMN 1 TO COLUMN 2, MULTIPLY BY:	COLUMN 1: METRIC UNITS	COLUMN 2: U.S. UNITS	TO CONVERT COLUMN 2 TO COLUMN 1, MULTIPLY BY:
WEIGHT			
1.102	ton (metric)	ton (U.S.)	0.907
2.205	kilogram, kg	pound, lb	0.454
0.035	gram, g	ounce (avdp), oz	28.349
VOLUME			
0.264	liter, L or l	gallon (U.S.), gal	3.785
1.057	liter, L or l	quart (fluid), qt	0.946
2.838	hectoliter, hl	bushel, bu	0.352
0.034	milliliter, ml	ounce (fluid), oz	29.573
LENGTH			
0.621	kilometer, km	mile, mi	1.609
1.094	meter, m	yard, yd	0.914
3.281	meter, m	feet, ft	0.305
0.394	centimeter, cm	inch, in	2.540
AREA			
2.471	hectare, ha	acre, A	0.405
247.1	kilometer^2, km^2	acre, A	0.004
0.386	kilometer^2, km^2	mile^2, mi^2	2.590
10.765	meters^2, m^2	feet^2, ft^2	0.093
0.155	centimeters^2, cm^2	inches^2, in^2	6.452
PRESSURE			
0.987	bar	atmosphere, atm	1.013
14.504	bar	lb/in^2, psi	0.069
14.223	kg/cm^2	lb/in^2, psi	0.070
14.70	atmosphere, atm	lb/in^2, psi	0.068
0.968	kg/cm^2	atmosphere, atm	1.033
YIELD OR RATE			
0.446	ton (metric/hectare)	ton (U.S.)/acre	2.242
0.892	kg/ha	lb/A	1.121
1.149	hl/ha	bu/A	0.870
0.107	l/ha	gal/A	9.346
8.347	kg/l	lb/gal	0.120
TEMPERATURE*			
1.80° C + 32	Celsius, C	Fahrenheit, °F	0.555 (F − 32)

*See also Appendix Table 1 for temperature conversions.

INDEX

Respiration:
in C_3 plants, 153
in C_4 plants, 153
cytochrome system, 156–157
dark, 151, 153–156, *154*
effect of temperature of, 151–152
glycolysis, *154*
Krebs cycle, *155*–157
measurement of, 153
as modified by controlled atmosphere storage, 449
as modified by hypobaric storage, 450
organelles involved, 152
photo-, 151, 153
rates of, 153
of roots, effects of O_2 and CO_2, 227–228
of seeds, 342
substrates for, 152–153
summary equation, 152
Rest, carrot, 541
Rest period:
of buds, 265–266
Easter lily, 586
Return stack heaters for freeze protection, *199*
Rhizobium, 307, 325
Rhizoctonia, 601, 603
Rhizomes, asparagus, 562
Rhizopus, rot of sweet potato, 519
Ribosomes, 58
Roadside markets, 454–455
Roadside marketing, 9
for sweet corn, 536
of vegetables, 510
Root(s):
characteristics of, 75–77
distribution of, 226
effects of calcium deficiency, 315
environmental effects, 83
expanse, 225
hairs, 80–82, *81*
extent of, 225
production of, 225, 228
kinds of, 77–78
lack of dormancy, 206–207
lateral, 82–83
penetration, factors limiting, 226–228
pressure, 229
structure, 78–*80*
vascular structure, 83
Root injury in winter, 210
Root pruning in field nursery, 612, 622
Rootstocks:
apple, 476–478
peach, 480
(*See also* specific crops)
Rosa species, 577–578
Rose bush, waxing to suppress water loss, 239
Roses, *577*–578, 596–597
pruning of, 396

Rosette, 322
Row covers, 196–197
Rubus species, 492–493
Rutabaga, 543

Saintpaulia ionantha, 592–593
Sansevieria, 597, 598
Sauerkraut, 544
Scarification of seeds, 208, 336, 339
sweet potato, 517
Seasons, effects on daylength, *182*
Seed:
dormancy, 136
hybrid, 344
light requirements for germination, 208, 269
scarification of, 208
stratification, 208, 340
structure, *135*–136
Seed certification program, 344
Seed germination, factors affecting: dormancy, 340
food reserves, 335
light, 340
requirement, 269
light-temperature interaction, 269
oxygen, 339
requirements for, 334
rest, 340
temperature, *339*
requirements, 191
water, 336, 339
Seed longevity, *342*, 343
Seed planting:
depth of, 345
methods of, 345
pelleting of seed, 345
seed tapes, 345
time of, 345
Seed production:
cleaning procedures, 341–342
environmental requirements, 340–341
harvest techniques, 341
Seed scarification, 208, 336, 339
Seed storage:
longevity of different species, *342*, 343
procedures used, 342, 343
Seed stratification, 208, 340
Seed treatment:
disinfectants, 343
protectants, 343
Seed viability, testing for, *338*, 344
Seedling development, 335–338
Senecio cruentus, 595–596
Sevin for insect control, 416
Shade:
cloth, 189, 599
effects: on fruit color, 271
on fruiting, 270–271
on leaf aging, 150–151

Water:
  vapor pressure gradient, 216–217, 219, 233–234, 236
  viscosity as affected by temperature, 229
Water loss of produce in storage, temperature effects, *236–237*
Water rights, 242
Water stress, effect on transpiration, 234
Water table:
  elevation by subsurface irrigation, 253
  fluctuating, 226–227
Water vapor, effects on radiation, 183
Watermelon:
  climatic requirements for, 554
  harvest, 554
  marketing of, 554–555
  origin of, 554
  plant characteristics of, 554
  pollination, 554
  types of, 554
  U.S. industry, 554
Wax, 160–161
  on fruit surfaces, 237
Waxing to suppress water loss, 238–239
Weather, 177
Weather station, 172–173, *176–177*
Weed control:
  biological, 429–430
  chemical, 430–431
  competition, 429
  mechanical, 429
Weed damage:
  aesthetics, 428
  contamination, 428
  host for insects and diseases, 428
  poisonous species, 428
  reduced crop yields, 427–428
Wet feet, tolerance of tree fruits, 227
Wilting coefficient, 284
Wind, effects on transpiration, 234

Wind machines for freeze protection, 200–*201*
Windbreaks:
  to reduce winter desiccation, 210
  to suppress transpiration, 235
Winter desiccation, 209–210
Winter injury:
  actual freezing, 210
  bark splitting, 211
  desiccation, 209–210
  effects of excess nitrogen, 304
  frost heaving, 210–211
  from ice, 211
  of nursery stock, 450–451
  to roots, 211
  southwest injury, 211–212
  types of, 209–212
Woolly aphid, resistance of Malling-Merton series, 477
World population trends, 16

Xanthophylls, structure, 140
Xenia in sweet corn, 535
Xylem:
  blockage, 234
  in stems, 92–95, *94*

Yams, 517

Zea mays, 533–537
Zinc:
  availability in soils, 322–323
  deficiency symptoms, 322
  fertilizers, 322–323
  forms in soils, 322
  functions in plants, 322
  metabolic effects of deficiency, 322